Smart Manufacturing Innovation and Transformation:

Interconnection and Intelligence

Zongwei Luo
The University of Hong Kong, China

A volume in the Advances in Logistics, Operations, and Management Science (ALOMS) Book Series

Managing Director:	Lindsay Johnston
Production Editor:	Jennifer Yoder
Development Editor:	Austin DeMarco
Acquisitions Editor:	Kayla Wolfe
Typesetter:	Michael Brehm
Cover Design:	Jason Mull

Published in the United States of America by
Business Science Reference (an imprint of IGI Global)
701 E. Chocolate Avenue
Hershey PA 17033
Tel: 717-533-8845
Fax: 717-533-8661
E-mail: cust@igi-global.com
Web site: http://www.igi-global.com

Copyright © 2014 by IGI Global. All rights reserved. No part of this publication may be reproduced, stored or distributed in any form or by any means, electronic or mechanical, including photocopying, without written permission from the publisher. Product or company names used in this set are for identification purposes only. Inclusion of the names of the products or companies does not indicate a claim of ownership by IGI Global of the trademark or registered trademark.

Library of Congress Cataloging-in-Publication Data

Smart manufacturing innovation and transformation : interconnection and intelligence / Zongwei Luo, editor.
    pages cm
  Includes bibliographical references and index.
  Summary: "This book covers both theoretical perspectives and practical approaches to smart manufacturing research and development triggered by ubiquitous interconnection and intelligence, discussing the transformation of manufacturing, the latest developments in smart manufacturing innovation, current and emerging technology opportunities, and market imperatives that enable manufacturing innovation and transformation"-- Provided by publisher.
  ISBN 978-1-4666-5836-3 (hardcover) -- ISBN 978-1-4666-5837-0 (ebook) -- ISBN 978-1-4666-5839-4 (print & perpetual access) 1. Manufacturing processes--Automation. 2. Flexible manufacturing systems. 3. Production engineering. 4. Manufacturing industries--Technological innovations. I. Luo, Zongwei, 1971-
 TS183.S56 20114
 670--dc23
                                                            2013050633

This book is published in the IGI Global book series Advances in Logistics, Operations, and Management Science (ALOMS) (ISSN: 2327-350X; eISSN: 2327-3518)

British Cataloguing in Publication Data
A Cataloguing in Publication record for this book is available from the British Library.

All work contributed to this book is new, previously-unpublished material. The views expressed in this book are those of the authors, but not necessarily of the publisher.

For electronic access to this publication, please contact: eresources@igi-global.com.

Advances in Logistics, Operations, and Management Science (ALOMS) Book Series

John Wang
Montclair State University, USA

ISSN: 2327-350X
EISSN: 2327-3518

Mission

Operations research and management science continue to influence business processes, administration, and management information systems, particularly in covering the application methods for decision-making processes. New case studies and applications on management science, operations management, social sciences, and other behavioral sciences have been incorporated into business and organizations real-world objectives. The **Advances in Logistics, Operations, and Management Science** (ALOMS) Book Series provides a collection of reference publications on the current trends, applications, theories, and practices in the management science field. Providing relevant and current research, this series and its individual publications would be useful for academics, researchers, scholars, and practitioners interested in improving decision making models and business functions.

Coverage

- Computing and Information Technologies
- Decision Analysis and Decision Support
- Finance
- Information Management
- Marketing Engineering
- Operations Management
- Organizational Behavior
- Political Science
- Production Management
- Services Management

IGI Global is currently accepting manuscripts for publication within this series. To submit a proposal for a volume in this series, please contact our Acquisition Editors at Acquisitions@igi-global.com or visit: http://www.igi-global.com/publish/.

The Advances in Logistics, Operations, and Management Science (ALOMS) Book Series (ISSN 2327-350X) is published by IGI Global, 701 E. Chocolate Avenue, Hershey, PA 17033-1240, USA, www.igi-global.com. This series is composed of titles available for purchase individually; each title is edited to be contextually exclusive from any other title within the series. For pricing and ordering information please visit http://www.igi-global.com/book-series/advances-logistics-operations-management-science/37170. Postmaster: Send all address changes to above address. Copyright © 2014 IGI Global. All rights, including translation in other languages reserved by the publisher. No part of this series may be reproduced or used in any form or by any means – graphics, electronic, or mechanical, including photocopying, recording, taping, or information and retrieval systems – without written permission from the publisher, except for non commercial, educational use, including classroom teaching purposes. The views expressed in this series are those of the authors, but not necessarily of IGI Global.

Titles in this Series
For a list of additional titles in this series, please visit: www.igi-global.com

Smart Manufacturing Innovation and Transformation Interconnection and Intelligence
Zongwei Luo (The University of Hong Kong, China)
Business Science Reference • copyright 2014 • 335pp • H/C (ISBN: 9781466658363) • US $225.00 (our price)

Handbook of Research on Design and Management of Lean Production Systems
Vladimír Modrák (Technical University of Košice, Slovakia) and Pavol Semančo (Technical University of Košice, Slovakia)
Business Science Reference • copyright 2014 • 487pp • H/C (ISBN: 9781466650398) • US $325.00 (our price)

Cases on Management and Organizational Behavior in an Arab Context
Grace C. Khoury (Birzeit University, Palestine) and Maria C. Khoury (Independent Researcher, Palestine)
Business Science Reference • copyright 2014 • 423pp • H/C (ISBN: 9781466650671) • US $175.00 (our price)

Management Science, Logistics, and Operations Research
John Wang (Montclair State University, USA)
Business Science Reference • copyright 2014 • 481pp • H/C (ISBN: 9781466645066) • US $225.00 (our price)

Strategic Performance Management and Measurement Using Data Envelopment Analysis
Ibrahim Osman (American University of Beirut, Lebanon) Abdel L. Anouze (American University of Beirut, Lebanon) and Ali Emrouznejad (Aston University, UK)
Business Science Reference • copyright 2014 • 359pp • H/C (ISBN: 9781466644748) • US $185.00 (our price)

Outsourcing Management for Supply Chain Operations and Logistics Service
Dimitris Folinas (Department of Logistics, ATEI-Thessaloniki, Greece)
Business Science Reference • copyright 2013 • 596pp • H/C (ISBN: 9781466620087) • US $185.00 (our price)

Operations Management Research and Cellular Manufacturing Systems Innovative Methods and Approaches
Vladimir Modrák (Technical University of Kosice, Slovakia) and R. Sudhakara Pandian (Kalasalingam University, India)
Business Science Reference • copyright 2012 • 368pp • H/C (ISBN: 9781613500477) • US $185.00 (our price)

Fashion Supply Chain Management Industry and Business Analysis
Tsan-Ming Choi (The Hong Kong Polytechnic University, Hong Kong)
Information Science Reference • copyright 2012 • 392pp • H/C (ISBN: 9781609607562) • US $195.00 (our price)

www.igi-global.com

701 E. Chocolate Ave., Hershey, PA 17033
Order online at www.igi-global.com or call 717-533-8845 x100
To place a standing order for titles released in this series, contact: cust@igi-global.com
Mon-Fri 8:00 am - 5:00 pm (est) or fax 24 hours a day 717-533-8661

Editorial Advisory Board

Evon Abu-Taieh, *Jordan University, Jordan*
Francis TK Au, *The University of Hong Kong, Hong Kong*
Jinjun Chen, *University of Technology, Sydney, Australia*
Rajit Gadh, *UCLA, USA*
Shuichi Ishida, *Cambridge University, UK*
Jason J. Jung, *Yeungnam University, South Korea*
Malgorzata Pankowska, *University of Economics in Katowice, Poland*
Michele Ruta, *World Trade Institute, Switzerland*
Venky N. Shankar, *Penn State University, USA*
Jing Shi, *North Dakota State University, USA*
Samuel Fosso Wamba, *University of Wollongong, Australia*
John Williams, *MIT, USA*
Edward C Wong, *Hong Kong University of Science and Technology, Hong Kong*
Yuchun Xu, *Cranfield University, UK*
Laurence T. Yang, *St. Francis Xavier University, Canada*
Winston Zhang, *Harbin Institute of Technology, China*
Yin Zhoupin, *Huazhong University of Science and Technology, China*

Table of Contents

Foreword xvi

Preface xix

Acknowledgment xxv

Section 1
Introduction to Smart Manufacturing

Chapter 1
Introduction to Smart Manufacturing: Value Chain Perspective for Innovation and Transformation.... 1
Zongwei Luo, The University of Hong Kong, China

Section 2
Smart Manufacturing Optimization

Chapter 2
Robust Optimization for Smart Manufacturing Planning and Supply Chain Design in Chemical
Industry 21
Tianxing Cai, Lamar University, USA

Chapter 3
Meta-Heuristic Structure for Multiobjective Optimization Case Study: Green Sand Mould System .. 38
T. Ganesan, Universiti Technologi PETRONAS, Malaysia
I. Elamvazuthi, Universiti Technologi PETRONAS, Malaysia
K. Z. KuShaari, Universiti Technologi PETRONAS, Malaysia
P. Vasant, Universiti Technologi PETRONAS, Malaysia

Chapter 4
Hybrid Evolutionary Optimization Algorithms: A Case Study in Manufacturing Industry 59
Pandian Vasant, Universiti Teknologi PETRONAS, Malaysia

Chapter 5
A Framework for the Modelling and Optimisation of a Lean Assembly System Design with Multiple Objectives 96

Atiya Al-Zuheri, University of South Australia, Australia & Ministry of Science and Technology, Iraq
Lee Luong, University of South Australia, Australia
Ke Xing, University of South Australia, Australia

Section 3
Smart Manufacturing Enabling Technologies

Chapter 6
Design of Anti-Metallic RFID for Applications in Smart Manufacturing 127

Bo Tao, Huazhong University of Science and Technology, China
Hu Sun, Huazhong University of Science and Technology, China
Jixuan Zhu, Huazhong University of Science and Technology, China
Zhouping Yin, Huazhong University of Science and Technology, China

Chapter 7
Towards Smart Manufacturing Techniques Using Incremental Sheet Forming 159

J.B. Sá de Farias, University of Aveiro, Portugal
S. Marabuto, University of Aveiro, Portugal
M.A.B.E. Martins, University of Aveiro, Portugal
J.A.F Ferreira, University of Aveiro, Portugal
A. Andrade Campos, University of Aveiro, Portugal
R.J. Alves de Sousa, University of Aveiro, Portugal

Chapter 8
Software Development Tools to Automate CAD/CAM Systems 190

N. A. Fountas, School of Pedagogical and Technological Education (ASPETE), Greece
A. A. Krimpenis, School of Pedagogical and Technological Education (ASPETE), Greece
N. M. Vaxevanidis, School of Pedagogical and Technological Education (ASPETE), Greece

Section 4
Smart Manufacturing Interconnection

Chapter 9
The Interaction between Design Research and Technological Research in Manufacturing Firm 226

Satoru Goto, Ritsumeikan University, Japan
Shuichi Ishida, Ritsumeikan University, Japan
Kiminori Gemba, Ritsumeikan University, Japan
Kazar Yaegashi, Ritsumeikan University, Japan

Chapter 10
The Role of Brand Loyalty on CRM Performance: An Innovative Framework for Smart Manufacturing 252
    Kijpokin Kasemsap, Suan Sunandha Rajabhat University, Thailand

Chapter 11
Smart, Innovative and Intelligent Technologies Used in Drug Designing 285
    S. Deshpande, Data Consulting, New Delhi, India
    S. K. Basu, University of Lethbridge, Canada
    X. Li, Industrial Crop Research Institute, Yunan Academy of Agricultural Sciences, China
    X. Chen, Institute of Food Crops, Yunan Academy of Agricultural Sciences, China

Section 5
Smart Manufacturing Sustainability

Chapter 12
Fair Share of Supply Chain Responsibility for Low Carbon Manufacturing 303
    Yu Mei Wong, The University of Hong Kong, Hong Kong

Chapter 13
Antecedents of Green Manufacturing Practices: A Journey towards Manufacturing Sustainability.. 333
    Rameshwar Dubey, Symbiosis Institute of Operations Management, India
    Surajit Bag, Tega Industries Limited, India

Compilation of References 355

About the Contributors 397

Index 405

Detailed Table of Contents

Foreword xvi

Preface xix

Acknowledgment xxv

Section 1
Introduction to Smart Manufacturing

Chapter 1
Introduction to Smart Manufacturing: Value Chain Perspective for Innovation and Transformation.... 1
    Zongwei Luo, The University of Hong Kong, China

Fast advances in information technology (RFID, sensor, Internet of things, and the Cloud) have led to a smarter world vision with ubiquitous interconnection and intelligence. Smart manufacturing refers to advanced manufacturing with wise adoption of information technologies throughout end to end product and service life-cycles, capturing manufacturing intelligence for wise production and services. It represents a field with intense competition in this century of national competitiveness. In this chapter, an introduction to smart manufacturing innovation and transformation is presented. An example is used to illustrate what is happening in China's manufacturing industry, with insights about China's strategy of advanced manufacturing research and development. The chapter emphasizes the value chain analysis for setting smart manufacturing strategies. A case study is conducted in detail to showcase a value chain analysis of RFID enabled SIM-smart card manufacturing for China's mobile payment industry.

Section 2
Smart Manufacturing Optimization

Chapter 2
Robust Optimization for Smart Manufacturing Planning and Supply Chain Design in Chemical
Industry 21
    Tianxing Cai, Lamar University, USA

The depletion of natural resource, the complexity of economic markets and the increased requirement for environment protection have increased the uncertainty of chemical supply and manufacturing. The consequence of short-time material shortage or emergent demand under extreme conditions, may cause local areas to suffer from delayed product deliveries and manufacturing disorder, which will both cause tremendous economic losses. In such urgent events, robust optimization for manufacturing planning and supply chain design in chemical industry, targeting the smart manufacturing, should be a top priority. In this chapter, a novel methodology is developed for robust optimization of manufacturing planning and supply chain design in chemical industry, which includes four stages of work. First, the network of the chemical supply chain needs to be characterized, where the capacity, quantity, and availability of various chemical sources is determined. Second, the initial situation under steady conditions needs to be identified. Then, the optimization is conducted based on a developed MILP (mixed-integer linear programming) model in the third stage. Finally, the sensitivity of the manufacturing and transportation planning with respect to uncertainty parameters is characterized by partitioning the entire space of uncertainty parameters into multiple subspaces. The efficacy of the developed methodology is demonstrated via a case study with in-depth discussions.

Chapter 3
Meta-Heuristic Structure for Multiobjective Optimization Case Study: Green Sand Mould System .. 38
 T. Ganesan, Universiti Technologi PETRONAS, Malaysia
 I. Elamvazuthi, Universiti Technologi PETRONAS, Malaysia
 K. Z. KuShaari, Universiti Technologi PETRONAS, Malaysia
 P. Vasant, Universiti Technologi PETRONAS, Malaysia

In engineering optimization, one often encounters scenarios that are multiobjective (MO) where each of the objectives covers different aspects of the problem. It is hence critical for the engineer to have multiple solution choices before selecting of the best solution. In this chapter, an approach that merges meta-heuristic algorithms with the weighted sum method is introduced. Analysis on the solution set produced by these algorithms is carried out using performance metrics. By these procedures, a novel chaos-based metaheuristic algorithm, the Chaotic Particle Swarm (Ch-PSO) is developed. This method is then used generate highly diverse and optimal solutions to the green sand mould system which is a real-world problem. Some comparative analyses are then carried out with the algorithms developed and employed in this work. Analysis on the performance as well as the quality of the solutions produced by the algorithms is presented in this chapter.

Chapter 4
Hybrid Evolutionary Optimization Algorithms: A Case Study in Manufacturing Industry 59
 Pandian Vasant, Universiti Teknologi PETRONAS, Malaysia

The novel industrial manufacturing sector inevitably faces problems of uncertainty in various aspects such as raw material availability, human resource availability, processing capability and constraints and limitations imposed by the marketing department. These problems have to be solved by a methodology which takes care of such unexpected information. As the analyst faces this man made chaotic and due to natural disaster problems, the decision maker and the implementer have to work collaboratively with the analyst for taking up a decision on an innovative strategy for implementation. Such complex problems of vagueness and uncertainty can be handled by the hybrid evolutionary intelligence algorithms. In this chapter, a new hybrid evolutionary optimization based methodology using a specific non-linear membership function, named as modified S-curve membership function, is proposed. The modified S-curve membership function is first formulated and its flexibility in taking up vagueness in parameters is established by an analytical approach. This membership function is applied for its useful performance through industrial production problems by employing hybrid evolutionary optimization algorithms. The novelty and the originality of this non-linear S-curve membership function are further established using a real life industrial production planning of an industrial manufacturing sector. The unit produces 8 products using 8 raw materials, mixed in various proportions by 9 different processes under 29 constraints. This complex problem has a cubic non-linear objective function. Comprehensive solutions to a non-linear real world objective function are achieved thus establishing the usefulness of the realistic membership function for decision making in industrial production planning.

Chapter 5
A Framework for the Modelling and Optimisation of a Lean Assembly System Design with Multiple Objectives 96
 Atiya Al-Zuheri, University of South Australia, Australia & Ministry of Science and Technology, Iraq
 Lee Luong, University of South Australia, Australia
 Ke Xing, University of South Australia, Australia

The newest assembly system is lean assembly, which is specifically designed to respond quickly and economically to the fluctuating nature of the market demands. Successful designs for these systems must be capable of satisfying the strategic objectives of a management in manufacturing company. An example of such systems is the so-called walking worker assembly line WWAL, in which each cross-trained worker travels along the line to carry out all tasks required to complete a job. Design approaches for this system have not been investigated in depth both of significant role in manual assembly process design; productivity and ergonomics. Therefore these approaches have had a limited success in actual applications. This chapter presents an innovative and integrated framework which offers significant potential improvement for productivity and ergonomics requirements in WWAL design. It establishes a systematic approach clearly demonstrating the implementation of a developed framework based on the simultaneous application of mathematical and meta- heuristic techniques.

Section 3
Smart Manufacturing Enabling Technologies

Chapter 6
Design of Anti-Metallic RFID for Applications in Smart Manufacturing 127
 Bo Tao, Huazhong University of Science and Technology, China
 Hu Sun, Huazhong University of Science and Technology, China
 Jixuan Zhu, Huazhong University of Science and Technology, China
 Zhouping Yin, Huazhong University of Science and Technology, China

Anti-metallic passive RFID tags play a key role in manufacturing automation systems adopting RFID techniques, such as manufacturing tool management, logistics and process control. A novel long range passive anti-metallic RFID tag fabrication method is proposed in this chapter, in which a multi-strip High Impendence Surface (HIS) with a feeding loop is designed as the antenna radiator. Firstly, the bandwidth enhancement methods for passive RFID tags based on micro strips are discussed. Then, a RFID tag design based on multi-strip antenna is proposed and its radiation efficiency is analyzed. After that, some key parameters of the RFID antenna proposed are optimized from the viewpoint of radiation efficiency and impedance match performance. Targeted for manufacturing plants with heavy metallic interfering, the proposed RFID tag can significantly enhance the radiation efficiency to improve the reading range as well as the bandwidth. Finally, some RFID tag prototypes are fabricated and tested to verify their performance and applicability against metallic environment, and the experimental results show that these fabricated RFID tags have outstanding reading performance and can be widely used in manufacturing plant full of heave metallic interfering.

Chapter 7
Towards Smart Manufacturing Techniques Using Incremental Sheet Forming 159
    J.B. Sá de Farias, University of Aveiro, Portugal
    S. Marabuto, University of Aveiro, Portugal
    M.A.B.E. Martins, University of Aveiro, Portugal
    J.A.F Ferreira, University of Aveiro, Portugal
    A. Andrade Campos, University of Aveiro, Portugal
    R.J. Alves de Sousa, University of Aveiro, Portugal

The current world's economical crisis raised the necessity from the industry to produce components cheaper and faster. In this sense, the importance of smart manufacturing techniques, proper articulation between CAD/CAM techniques and integrated design and assessment becomes critical. The Single Point Incremental Forming (SPIF) process represents a breakpoint with traditional forming processes, and possibly a new era in the small batches production or customized parts, being already used by automotive industry for light components. While classical stamping processes need a punch, a die, a holder and a press, in the SPIF process the final geometry is achieved incrementally through the action of a punch with a spherical head. Since the blank is clamped at the edges, there is no need to employ a die with the shape of the final part. However, this process must be further improved in terms of speed and dimensional accuracy. Because the process is cheap and easy to implement, it is currently the subject of intensive experimental and numerical research, but yet not deeply understood. This chapter gives an overview on the techniques currently being employed to optimize the process feasibility.

Chapter 8
Software Development Tools to Automate CAD/CAM Systems 190
    N. A. Fountas, School of Pedagogical and Technological Education (ASPETE), Greece
    A. A. Krimpenis, School of Pedagogical and Technological Education (ASPETE), Greece
    N. M. Vaxevanidis, School of Pedagogical and Technological Education (ASPETE), Greece

In today's modern manufacturing, software automation is crucial element for leveraging novel methodologies and integrate various engineering software environments such Computer aided design (CAD), Computer aided process planning (CAPP), or Computer aided manufacturing (CAM) with programming modules with a common and a comprehensive interface; thus creating solutions to cope with repetitive tasks or allow argument passing for data exchange. This chapter discusses several approaches concerning engineering software automation and customization by employing programming methods. The main focus is given to design, process planning and manufacturing since these phases are of paramount importance when it comes to product lifecycle management. For this reason, case studies concerning software automation and problem definition for the aforementioned platforms are presented mentioning the benefits of programming when guided by successful computational thinking and problem mapping.

Section 4
Smart Manufacturing Interconnection

Chapter 9
The Interaction between Design Research and Technological Research in Manufacturing Firm 226
Satoru Goto, Ritsumeikan University, Japan
Shuichi Ishida, Ritsumeikan University, Japan
Kiminori Gemba, Ritsumeikan University, Japan
Kazar Yaegashi, Ritsumeikan University, Japan

Design has significantly affected innovation and the discipline of design management focuses on meanings that it brings about a drastic change in life style of consumers. Although the relationship between design and technology is one of the important issues for the innovation of meanings, there were only a few studies which suggested the comprehensive model that includes design and technology. Verganti proposed the concept of design driven innovation, which regarded a design process of NPD as a research activity, and demonstrated a relationship between the technological research and the design research. In particular, he examined deeply the mechanism of the design research. In order to deepen his discussion, this chapter aims to suggest some propositions and a comprehensive model related to the interaction between the design research and the technological research. The authors utilize the concepts of exploration and exploitation for their framework. It shows that an augmenting of both researches may create effectively radical meanings or technologies and an integration of both researches may create radical meanings and technologies concurrently. In the case study of FPD industry, this chapter examines how some companies create competitive advantages by both researches and the commoditization of technology may cause the transition to the design research from the technological research as source of the competitive advantages. Additionally, the chapter suggests the strategic and organizational issues for conducting the design research and the technological research interactively in the discussion section.

Chapter 10
The Role of Brand Loyalty on CRM Performance: An Innovative Framework for Smart Manufacturing 252
Kijpokin Kasemsap, Suan Sunandha Rajabhat University, Thailand

This chapter introduces the framework and causal model of customer value, customer satisfaction, brand loyalty, and customer relationship management performance in terms of the innovative manufacturing and marketing solutions. It argues that dimensions of customer value, customer satisfaction, and brand loyalty have mediated positive effect on customer relationship management performance. Furthermore, brand loyalty positively mediates the relationships between customer value and customer relationship management performance and between customer satisfaction and customer relationship management performance. Customer value is positively correlated with customer satisfaction. Understanding the theoretical learning is beneficial for organizations aiming to increase customer relationship management performance and achieve business goals.

Chapter 11
Smart, Innovative and Intelligent Technologies Used in Drug Designing 285
    S. Deshpande, Data Consulting, New Delhi, India
    S. K. Basu, University of Lethbridge, Canada
    X. Li, Industrial Crop Research Institute, Yunan Academy of Agricultural Sciences, China
    X. Chen, Institute of Food Crops, Yunan Academy of Agricultural Sciences, China

Smart and intelligent computational methods are essential nowadays for designing, manufacturing and optimizing new drugs. New and innovative computational tools and algorithms are consistently developed and applied for the development of novel therapeutic compounds in many research projects. Rapid developments in the architecture of computers have also provided complex calculations to be performed in a smart, intelligent and timely manner for desired quality outputs. Research groups worldwide are developing drug discovery platforms and innovative tools following smart manufacturing ideas using highly advanced biophysical, statistical and mathematical methods for accelerated discovery and analysis of smaller molecules. This chapter discusses novel innovative applications in drug discovery involving use of structure-based drug design which utilizes geometrical knowledge of the three-dimensional protein structures. It discusses statistical and physics based methods such as quantum mechanics and classical molecular dynamics which can also play a major role in improving the performance and in prediction of computational drug discovery. Lastly, the authors provide insights on recent developments in cloud computing with significant increase in smart and intelligent computational power thus allowing larger data sets to be analyzed simultaneously on multi processor cloud systems. Future directions for the research are outlined.

Section 5
Smart Manufacturing Sustainability

Chapter 12
Fair Share of Supply Chain Responsibility for Low Carbon Manufacturing 303
    Yu Mei Wong, The University of Hong Kong, Hong Kong

Large amounts of carbon emissions and pollution are generated during the manufacturing process for consumer goods. Low carbon manufacturing has been increasingly enquired or requested by stakeholders. However, international trade blurs the responsibility for carbon emissions reduction and raises the questions of responsibility allocation among producers and consumers. Scholars have been examining the nexus of producer versus consumer responsibility among supply chains. Recently, there have been discussions on the share of producer and consumer responsibility. Both producer and consumer responsibility approaches have intrinsic shortcomings and are ineffective in curbing the rise of carbon emissions in supply chains. Shared responsibility based on the equity principle attempts to address these issues. This chapter relates a case study of carbon impact on China's export and economy with scenarios which show that the benefits of carbon reduction by producers can trickle down along the supply chain and motivate the sharing responsibility under certain circumstances. The share of producer and consumer responsibility for low carbon manufacturing can be enabled when embodied carbon emissions in goods and services are priced and such accurate information is available. A mechanism engaging the global participation is recommended. The author calls for further research on the system pricing embodied carbon emission, the universal standard to calculate the embodied carbon emissions and to disclose the information, and the way to secure global cooperation and participation.

Chapter 13
Antecedents of Green Manufacturing Practices: A Journey towards Manufacturing Sustainability.. 333
Rameshwar Dubey, Symbiosis Institute of Operations Management, India
Surajit Bag, Tega Industries Limited, India

The purpose of this chapter is to identify green supply chain practices and study their impact on firm performance. In this study, the authors have adopted a two-pronged strategy. First, they reviewed extant literature published in academic journals and reports published by reputed agencies. They identified key variables through literature review and developed an instrument to measure the impact of GSCM practices on firm performance. The authors pretested this instrument using five experts drawn from industry having expertise in GSCM implementation and two academicians who have published their articles in reputed journals in the field of GSCM and sustainable manufacturing practice. After finalizing the instrument, the study then randomly targeted 175 companies from CII Institute of Manufacturing database and obtained response from 54 which represent 30.85% response rate. The authors also performed non-response bias test to ensure that non-response bias is not a major issue. They further performed PLSR analysis to test our hypotheses. The results of the study are very encouraging and provide further motivation to explore other constructs which are important for successful implementation of GSCM practices.

Compilation of References 355

About the Contributors 397

Index 405

Foreword

A supply chain is a complex system composed by organizations and people with their activities involved in transferring a product or service toward a final customer. The key to make a successful supply chain relies on an extended collaboration, implying the integration among actors involved in the productive and logistic network. An integrated and flexible management of production and logistics (physical and information flows) has to be set-up both inside and outside factory boundaries. Specialized production and distribution processes suffer from the limited interaction allowed by rigid networks. As a result, nowadays a relevant component of competition in the market occurs among information chains. The supply chain can no longer be represented as static or linear, but it needs to be evaluated dynamically, as a complex system made of interactions and connections among actors operating along the chain itself.

Many empirical investigations have demonstrated there is a positive correlation between enterprise performances and its propensity and attitude to be integrated into larger systems. This is the reason why enterprises are more and more attentive to the opportunity offered by both coordination and cooperation among their internal functions and the other external actors contributing in different ways to the business. Hence, information has been an increasingly strategic asset in the last few years. It covers a determinant position in manufacturing, logistics, and marketing. The physical flow of raw materials, products, and their related information is considered a strategic element for quality standards of products/services, for business analysis and evaluation, and finally to allow corrective actions. In particular, current trends in the consumer products market assign a growing significance to smart production chains. The even more increasing dimension of manufacturing groups –mostly due to the reached degree of decentralization of production- makes them privileged centers of value accumulation, acting as sources for the information flow for the whole chain. As a result, the main manufacturers are investing in new technology in order to boost the information exchange and are mandating the adoption of interoperable solutions to commercial partners.

In general, the Ubiquitous Computing (UbiComp) paradigm favors the pervasiveness of information in a given context. As originally introduced by Mark Weiser, UbiComp requires both information and computational capabilities, which are deeply integrated into common objects and/or actions and the user will interact with many computational devices simultaneously, exploiting data automatically extracted from "smart objects" permeating the environment during her ordinary activities. In addition, AutoID (Automatic IDentification) technologies play a relevant role in creating a virtual counterpart to physical objects. Because of their features, they provide better levels of automation in the product chains and help prevent human errors.

However, a more advanced exploitation of the very large amount of data now put at one's disposal by the permeation of identification technologies in the production processes is strongly required. The trivial one-to-one association between the physical and digital worlds restrains a powerful adoption of such data wealth. Massive data extraction and analysis is quite difficult without the support of proper management and aggregation schemes. Concrete improvements in information-enhanced production and supply chains rely on global trend inspections over the chains themselves, which require multidimensional analyses of huge amounts of data generated and stored in central DBMSs. Serious data management issues are then inevitably inherited and must be faced.

Hence, innovative models are required not only to let information permeate a manufacturing system, but also to leverage the derived informative asset in a fruitful way. By exploiting a distributed architecture, a unified framework should enable both quick run-time analyses (with respect to a local fragment of the overall information thesaurus) and stand-alone massive business logic elaborations (with respect to a centralized DBMS) following needs and requirements of the supply chain actors.

Noteworthy, it should chase the possibility to add semantically rich and unambiguous information able to follow a product in each step of its life cycle. Such a manufacturing model could allow manufacturers to trace and discover the information flow—associated to products thanks to their embedded informative counterparts—along the supply chain. Different analyses could be so formalized (*e.g.,* product-centric, node-centric, path-oriented, time-oriented). Exploiting proper queries, product and process information could be read, updated, and integrated during manufacturing, packaging, and distribution, thus allowing full traceability up to sales, as well as intelligent and de-localized interrogation of product data. Several tangible (economic) and intangible benefits are expected. During product manufacturing and distribution, a wide-area support network interconnecting commercial partners is not strictly needed: this is a significant innovation with respect to common management solutions. A structured and detailed description of product features, allow goods to auto-expose their description to any computing environment they reach. This favors decentralized approaches and enables context-aware applications, based on less expensive and more manageable mobile ad-hoc networks. In addition to improved traceability, a smart manufacturing and distribution system hence provides unique value-added capabilities.

This book addresses modern problems affecting supply chains and manufacturing systems from the perspective of innovation and transformation. Nowadays, specialized production and distribution processes suffer from the limited interaction allowed by old rigid networks. As a result, a relevant component of competition in the market occurs among information chains. Smart solutions, novel approaches, and technical and technological enhancements are seen as a means to overcome most common drawbacks in systems and infrastructures—such as the manufacturing ones—having a long and entrenched history and classically being resilient to modernization and evolution. The book envisions interconnection and intelligence as two fundamental elements to be pursued to boost the innovation and transformation of manufacturing systems. The key to make a successful production and distribution chain must be researched in an extended collaboration, implying the integration among actors involved in the productive and logistic network. A flexible management of production and logistics (physical and information flows) has to be set-up both inside and outside factory boundaries.

The book explains that a relevant aspect of a smart production chain is the information sharing, which allows the optimization of actions and the improvement of performance both in terms of provided features and by enabling innovative services available for all involved actors. The envisioned production schemes can support a range of use cases, for different stakeholders along a product life cycle. The coherent development of smart manufacturing models allows a strengthening of the information to be shared between the actors involved in production chains, reducing the costs of adoption of technology in business. Furthermore, an increase in transparency and trust is achieved not only between supply chain partners, but also between retailers and customers.

The book is organized twelve chapters; they were carefully selected to provide a wide scope to the general theme of the book with minimal overlap to reduce duplications. Each contributor was asked that his/her chapter should include a state-of-the-art survey as well as ongoing developments. Most promising approaches in the field of innovation of manufacturing system were so selected and brought together to evidence a fundamental aim that such allowed interconnection may be a direct competitive advantage for companies that adopt the technology.

Michele Ruta
Technical University of Bari, Italy

Michele Ruta *received the laurea degree in Electronics Engineering from the Technical University of Bari in 2002 and a Ph.D. in Computer Science from Technical University of Bari in 2007. He is currently assistant professor at Technical University of Bari. His research interests include the Semantic Web of Things and model checking. On these topics, he has co-authored papers in international journals and conferences; he is member of editorial boards of several international journals and books. He is involved in various national and international research projects related to his research interests. He co-authored works that received the best paper award at the ICEC2007 and SEMAPRO2010 conferences and has been program committee member of several international conferences and workshops in areas related to his research interests.*

Preface

INTRODUCTION

Fast advances in information technology (RFID, sensor, Internet of things, and the Cloud) have led to a smarter world vision with ubiquitous interconnection and intelligence. Smart manufacturing refers to advanced manufacturing with wise adoption of information technologies throughout end-to-end product and service life cycles, capturing manufacturing intelligence for wise production and services. Smart manufacturing represents a field with intense competition in this century of national competitiveness.

Thus, it is the right time to have a book with innovative findings in advanced manufacturing research and development. The main intent is to promote an international knowledge exchange community involving multidisciplinary participation from researchers, practitioners, and academicians with insight addressing issues in real life problems towards smarter manufacturing, triggered, and facilitated by ubiquitous interconnection and intelligence, incubating manufacturing innovation, and transformation.

INNOVATION AND TRANSFORMATION

Manufacturing in recent decades has made amazing progress. At present, manufacturing has adopted and leveraged the latest achievements in materials, mechanics, physics, chemistry, and computer simulation technology, network technology, control technology, nanotechnology, biotechnology, and sensor technology. New manufacturing mechanisms, manufacturing tools, processes, and equipment continue to emerge. Manufacturing as a technology has developed into a new engineering science subject – Manufacturing Science. Cross-regional distribution of manufacturing resources in the era of economic globalization has forced manufacturing collaboration to become a daily necessary means. The networked collaborative tools and systems are also increasingly rich. Intelligent Manufacturing has been recognized as the direction of manufacturing technology innovation, the maturity of the related theories, and technologies will be one of the signs of the advent of knowledge economy.

Advanced manufacturing's role as the backbone of a country has been re-recognized and has aroused wide attention of major developed and developing countries. These countries have already started a series of advanced manufacturing technology research programs. Europe and the United States proposed "re-industrialization" in recent years to seize the commanding control of global industrial technology, and to further capitalize on the high-end manufacturing. In the U.S., President Barack Obama has announced infrastructure and technology policy and steps to restore the center of manufacturing in the U.S. economy. In the face of fierce international competition in the 21st century, Chinese government has planned accordingly and launched a series of major and key projects, carrying out special studies in the frontier of advanced manufacturing technology and equipment.

INTERCONNECTION AND INTELLIGENCE

Development of the manufacturing industry so far has involved multi-disciplinary applications, adopting the latest achievement in materials, mechanics, physics, chemistry, and computer simulation technology, network technology, control technology, nanotechnology, biotechnology, and sensor technology. Fast advances in Internet/Internet of Things, and next generation of information technology will lead to industry revolution, promoting the manufacturing industry to shift from the traditional manufacturing towards industrial chain based manufacturing. The trend and requirements of digital manufacturing, service manufacturing, and intelligent manufacturing will become more apparent and prominent.

Pursuit of the consistency of business and customer goals is the eternal theme of service manufacturing. High quality, high efficient, and personalized manufacturing represents the most significant features of service manufacturing. Currently, the volatility of consumer demand is widening, so is financial volatility. Expectation on customer service and product quality is continuously higher and higher. Business models are also diversified. All of these present great challenges to manufacturing. Improving services, responding to consumers' changing expectations, and demand patterns needs more and better resources and information. The development and application of Internet of Things and cloud computing in manufacturing provides powerful tools to achieve those. With Internet of Things application in the manufacturing sector deepening, it will give rise to new data, new technologies, new products, new applications, and new business models. Consequently, it will generate huge support and management needs. Cloud computing as a new computing paradigm is developing rapidly.

Intelligent manufacturing provides intelligent means for the optimal use of manufacturing resources, leveraging recycling economy theory and re-manufacturing concept for traditional industries to provide resource-saving, energy saving, environment-friendly technologies. It will also provide advanced technology and equipment for new energy, new materials, biomedicine, a new generation of information networks, smart grid, green means of delivery, eco-friendly, ocean and aviation, public security, and other strategic developments of emerging industries.

ABOUT SMART MANUFACTURING

This book on smart manufacturing focuses on innovation and transformation from the perspective of interconnection and intelligence leading to manufacturing sustainability. The whole book consists of five sections: 1) Introduction to Smart Manufacturing, 2) Smart Manufacturing Optimization Techniques, 3) Smart Manufacturing Enabling Technologies, 4) Smart Manufacturing Interconnection, and 5) Smart Manufacturing Sustainability.

Section 1, "Introduction to Smart Manufacturing," includes one chapter:

- **Chapter 1:** "Introduction to Smart Manufacturing: Value Chain Perspective for Innovation and Transformation" by Zongwei Luo

In chapter 1, an introduction to smart manufacturing innovation and transformation is presented. An example is used to illustrate what is happening in China's manufacturing industry, with insights about China's strategy of advanced manufacturing research and development. Emphasis is laid upon the value chain analysis for setting smart manufacturing strategies. A case study is conducted in detail to showcase a value chain analysis of RFID-enabled SIM-smart card manufacturing for China's mobile payment industry.

Section 2, "Smart Manufacturing Optimization," includes the following four chapters:

- **Chapter 2:** "Robust Optimization for Smart Manufacturing Planning and Supply Chain Design in Chemical Industry" by Tianxin Cai
- **Chapter 3:** "Meta-Heuristic Structure for Multiobjective Optimization Case Study: Green Sand Mould System" by T. Ganesan, I. Elamvazuthi, K. Z. KuShaari, and P. Vasant
- **Chapter 4:** "Hybrid Evolutionary Optimization Algorithms: A Case Study in Manufacturing Industry" by Pandian Vasant
- **Chapter 5:** "A Framework for the Modeling and Optimisation of a Lean Assembly System Design with Multiple Objectives" by Atiya Al-Zuheri, Lee Luong, and Ke Xing

In chapter 2, a novel methodology is developed for robust optimization for manufacturing planning and supply chain design in the chemical industry, which includes four stages of work. First, the network of chemical supply chain needs to be characterized, where the capacity, quantity, and availability of various chemical sources are determined. Second, the initial situation under steady conditions needs to be identified. Then, the optimization is conducted based on a developed MILP (Mixed-Integer Linear Programming) model in the third stage. Finally, the sensitivity of the manufacturing and transportation planning with respect to uncertainty parameters is characterized by partitioning the entire space of uncertainty parameters into multiple subspaces. The efficacy of the developed methodology is demonstrated via a case study with in-depth discussions.

Chapter 3 focuses on "Meta-Heuristic Structure for Multiobjective Optimization Case Study: Green Sand Mould System." In engineering optimization, it is often that one encounters scenarios that are Multi-Objective (MO), where each of the objectives covers different aspects of the problem. It is, hence, critical for the engineer to have multiple solution choices before selecting of the best solution. In this chapter, an approach that merges meta-heuristic algorithms with the weighted sum method is introduced. Analysis on the solution set produced by these algorithms was carried out using performance metrics.

Chapter 4 focuses on "Hybrid Evolutionary Optimization Algorithms: A Case Study in Manufacturing Industry." The novel industrial manufacturing sector inevitably faces problems of uncertainty in various aspects such as raw material availability, human resource availability, processing capability and constraints, and limitations imposed by marketing department. This problem has to be solved by a methodology, which takes care of such unexpected information. As the analyst faces manmade chaotic and due to natural disaster problems, the decision maker and the implementer have to work collaboratively with the analyst for taking up a decision on an innovative strategy for implementation. Such complex and hard problems of vagueness and uncertainty can be handled by the hybrid evolutionary intelligence algorithms.

Chapter 5 focuses on a framework for the modeling and optimization of a lean assembly system design with multiple objectives. The newest assembly system is lean assembly, which is specifically designed to respond quickly and economically to the fluctuating nature of the market demands. Successful designs for these systems must be capable of satisfying the strategic objectives of a management in manufacturing company. An example of such systems is the so-called Walking Worker Assembly Line (WWAL), in which each cross-trained worker travels along the line to carry out all tasks required to complete a job. Design approaches for this system have not been investigated in depth, both of significant role in manual assembly process design, productivity, and ergonomics. Therefore, these approaches have had limited success in actual applications. This chapter presents an innovative and integrated framework, which offers significant potential improvement for productivity and ergonomics requirements in WWAL design. It establishes a systematic approach clearly demonstrating the implementation of a developed framework based on the simultaneous application of mathematical and meta-heuristic techniques.

Section 3, "Smart Manufacturing Enabling Technologies," includes the following three chapters:

- **Chapter 6:** "Design of Anti-Metallic RFID for Applications in Smart Manufacturing" by Bo Tao, Hu Sun, Jixuan Zhu, and Zhouping Yin
- **Chapter 7:** "Towards Smart Manufacturing Techniques using Incremental Sheet Forming" by J.B. Sá de Farias, S. Marabuto, M.A.B.E. Martins, J.A.F Ferreira, A. Andrade Campos, and R.J. Alves de Sousa
- **Chapter 8:** "Software Development Tools to Automate CAD/CAM Systems" by N.A. Fountas, A.A. Krimpenis, and N.M. Vaxevanidis

Chapter 6 is focused on "Design of Anti-Metallic RFID for Applications in Smart Manufacturing." A novel long range passive anti-metallic RFID tag fabrication method is proposed in this chapter, in which a multi-strip High Impendence Surface (HIS) with a feeding loop is designed as the antenna radiator. Firstly, the bandwidth enhancement methods for passive RFID tags based on micro strips are discussed. Then, a RFID tag design based on multi-strip antenna is proposed and its radiation efficiency is analyzed. After that, some key parameters of the RFID antenna proposed are optimized from the viewpoint of radiation efficiency and impedance match performance. Targeted for manufacturing plants with heavy metallic interfering, the proposed RFID tag can significantly enhance the radiation efficiency to improve the reading range as well as the bandwidth.

Chapter 7 is focused on smart manufacturing techniques to produce components cheaper and faster. In this sense, the importance of smart manufacturing techniques, proper articulation between CAD/CAM techniques, and integrated design and assessment becomes critical. Concerning components obtained after sheet metal forming operations, like in the automotive industry, it is mandatory to shorten even more products' lifetime cycle, especially for small batches or rapid prototyping. Considering the technological viewpoint but also economical competitiveness, the Single Point Incremental Forming (SPIF) process represents a breakpoint with traditional forming processes, and possibly a new era in the small batches production or customized parts, being already used by automotive industry for light components. In this chapter, an overview is given on the techniques currently being employed to optimize the process feasibility. Conclusions are made about the future trends in process development, integrating CAD, CAM, and optimization techniques aiming to improve geometrical accuracy and mechanical reliability.

Chapter 8 focuses on automation and customization of engineering software using programming. In today's modern manufacturing, software automation is a crucial element for leveraging novel methodologies and integrating various engineering software environments such as Computer-Aided Design (CAD), Computer-Aided Process Planning (CAPP), or Computer-Aided Manufacturing (CAM) with programming modules with a common and a comprehensive interface, creating solutions to cope with repetitive tasks or allow argument passing for data exchange. This chapter discusses several approaches concerning engineering software automation and customization by employing programming methods. Main interest is given to design, process planning, and manufacturing, since these phases are of paramount importance when it comes to product lifecycle management. Thereby, case studies concerning software automation and problem definition for the aforementioned platforms are presented, mentioning the benefits of programming when guided by successful computational thinking and problem mapping.

Section 4, "Smart Manufacturing Interconnection," includes the following three chapters:

- **Chapter 9:** "The Interaction between Design Research and Technological Research in Manufacturing Firm" by Satoru Goto, Shuichi Ishida, Kiminori Gemba, and Kazar Yaegashi
- **Chapter 10:** "The Role of Brand Loyalty on CRM Performance: An Innovative Framework for Smart Manufacturing" by Kijpokin Kasemsap
- **Chapter 11:** "Smart, Innovate, and Intelligent Technologies Used in Drug Design" by S. Deshpande, S.K. Basu, X.P. Li, and X. Chen

Chapter 9 is focused on the interaction between design research and technological research in manufacturing firms. Design has significantly affected innovation, and the discipline of design management focuses on meanings that it brings about a drastic change in the lifestyle of consumers. Although the relationship between design and technology is one of the important issues for the innovation of meanings, there were only a few studies suggesting the comprehensive model that includes design and technology. This chapter shows that an augmenting of both researches may create effectively radical meanings or technologies, and an integration of both researches may create radical meanings and technologies concurrently. The case study of FPD industry examines how some companies create competitive advantages by both researches and the commoditization of technology may cause the transition to the design research from the technological research as source of the competitive advantages. Additionally, the strategic and organizational issues are suggested for conducting the design research and the technological research interactively in the discussion section.

Chapter 10 focuses on "The Role of Brand Loyalty on CRM Performance: An Innovative Framework for Smart Manufacturing." This chapter introduces the framework and causal model of customer value, customer satisfaction, brand loyalty, and customer relationship management performance in terms of the innovative manufacturing and marketing solutions. It argues that dimensions of customer value, customer satisfaction, and brand loyalty have mediated positive effect on customer relationship management performance. Furthermore, brand loyalty positively mediates the relationships between customer value and customer relationship management performance and between customer satisfaction and customer relationship management performance.

Chapter 11 focuses on "Smart, Innovative, and Intelligent Technologies used in Drug Designing." Smart and intelligent computational methods are essential nowadays for designing, manufacturing, and optimizing new drugs. New and innovative computational tools and algorithms are consistently developed and applied for the development of novel therapeutic compounds in many research projects. Rapid developments in the architecture of computers have also provided complex calculations to be performed in a smart, intelligent, and timely manner for desired quality outputs. Research groups worldwide are developing drug discovery platforms and innovative tools following smart manufacturing ideas using highly advanced biophysical, statistical, and mathematical methods for accelerated discovery and analysis of smaller molecules.

Section 5, "Smart Manufacturing Sustainability," includes the following two chapters:

- **Chapter 12:** "Fair Share of Supply Chain Responsibility for Low Carbon Manufacturing" by Yumei Wong

- **Chapter 13:** "Antecedents of Green Manufacturing Practices: A Journey towards Manufacturing Sustainability" by Rameshwar Dubey and Surajit Bag

Chapter 12 is focused on "Fair Share of Supply Chain Responsibility for Low Carbon Manufacturing." A large amount of carbon emissions is emitted, and pollution is generated during the manufacturing process for consumer goods. Low carbon manufacturing has been increasingly asked about or required by stakeholders. However, international trade blurs the responsibility for carbon emissions reduction and raises the questions of responsibility allocation among producers and consumers. Scholars have been examining the nexus of producer versus consumer responsibility among supply chain, and recently, there are discussions on the share of producer and consumer responsibility. Producer or consumer responsibility approach has intrinsic shortcomings and is ineffective in curbing the rise of carbon emissions in supply chains. Shared responsibility based on the equity principle attempts to address these issues. A case study of carbon impact on China's export and economy with scenarios shows that the benefits of carbon reduction taken by the producers can be trickled down along the supply chain and the motivation of the sharing responsibility can be created under certain circumstances. The share of producer and consumer responsibility for low carbon manufacturing can be enabled when embodied carbon emissions in goods and services are priced and such accurate information is available. A mechanism engaging global participation is recommended. The author calls for further research on the system pricing embodied carbon emission, the universal standard to calculate the embodied carbon emissions and to disclose the information, and the way to secure global cooperation and participation.

Chapter 13 focuses on "Antecedents of Green Manufacturing Practices: A Journey towards Manufacturing Sustainability." The purpose of this chapter is to explore manufacturing practices that help firms to achieve better environmental and business performance. In short, it can be termed as antecedents "green manufacturing" practices that enhance firm performance. In this chapter, the authors have adopted a secondary literature survey approach to identify variables and identify research gaps. Based on the constructs and items identified through the literature survey, the researchers have developed a structured questionnaire, which was pretested before use as the final survey.

Zongwei Luo
The University of Hong Kong, China

Acknowledgment

I would like to express my gratitude to a number of people who have contributed to the completion of this book in various ways and to thank them all for their assistance and encouragement.

First, I wish to thank all of the authors for their excellent contributions to this book. All of you also served as reviewers for manuscripts written by other authors. Thank you all for your contributions and your constructive reviews.

Second, I wish to thank all Editorial Advisory Board (EAB) members. Many of you have delivered responsive and valuable reviews.

Finally, I wish to thank the staff of IGI Global for their help and guidance. Special thanks goes to Austin!

Zongwei Luo
The University of Hong Kong, China

Section 1

Introduction to Smart Manufacturing

Chapter 1
Introduction to Smart Manufacturing:
Value Chain Perspective for Innovation and Transformation

Zongwei Luo
The University of Hong Kong, China

ABSTRACT

Fast advances in information technology (RFID, sensor, Internet of things, and the Cloud) have led to a smarter world vision with ubiquitous interconnection and intelligence. Smart manufacturing refers to advanced manufacturing with wise adoption of information technologies throughout end to end product and service life-cycles, capturing manufacturing intelligence for wise production and services. It represents a field with intense competition in this century of national competitiveness. In this chapter, an introduction to smart manufacturing innovation and transformation is presented. An example is used to illustrate what is happening in China's manufacturing industry, with insights about China's strategy of advanced manufacturing research and development. The chapter emphasizes the value chain analysis for setting smart manufacturing strategies. A case study is conducted in detail to showcase a value chain analysis of RFID enabled SIM-smart card manufacturing for China's mobile payment industry.

1. INTRODUCTION

Manufacturing in recent decades has made amazing progress. At present, manufacturing has adopted and leveraged more and more the latest achievements in materials, mechanics, physics, chemistry, and computer simulation technology, network technology, control technology, nanotechnology, biotechnology, and sensor technology. New manufacturing mechanism, manufacturing tools, processes and equipment continue to emerge. Manufacturing as a technology has developed into a new engineering science subject- Manufacturing Science. Cross-regional distribution of manufacturing resources in the era

DOI: 10.4018/978-1-4666-5836-3.ch001

of economic globalization, has forced manufacturing collaboration a daily necessary means. The networked collaborative tools and systems are also increasingly rich. Intelligent Manufacturing has been recognized as the direction of manufacturing technology innovation, the maturity of the related theories and technologies will be one of the signs of the advent of knowledge economy (Luo, 2013).

Advanced manufacturing's role as the backbone of a country has been re-recognized and has aroused wide attention of major developed and developing countries. These countries have already started a series of advanced manufacturing technology research programs. Europe and the United States proposed "re-industrialization" in recent years to seize the commanding control of global industrial technology, and to further capitalize on the high-end manufacturing. In the U.S., President Barack Obama has announced infrastructure and technology policy and steps to restore the center of manufacturing in the U.S. Economy (Curtis, 2013). In the face of fierce international competition in the 21st century, Chinese government has planned accordingly and launched a series of major and key projects, carrying out special studies in the frontier of advanced manufacturing technology and equipment.

In this chapter, an introduction to smart manufacturing innovation and transformation is presented. An example is used to illustrate what is happening in China's manufacturing industry, with analysis of China's strategy of advanced manufacturing research and development. Emphasis is put upon value chain analysis for setting smart manufacturing strategies. In this context a case study is conducted in detail to show case a value chain analysis of smart card manufacturing for China's mobile payment industry.

Organization of this chapter is as follows. Section 2 is devoted to an introduction to manufacturing for China, especially on manufacturing restructuring and opportunities. Section 3 is presented to introduce smart manufacturing research. Research trend of digital manufacturing, service manufacturing and intelligent manufacturing is discussed in detail. Section 4 is focused on smart manufacturing research agenda and fields. Section 5 introduces a case study of smart card manufacturing in China. Section 6 presents smart card value chain analysis for China's mobile payment industry. Section 7 concludes the chapter with manufacturing strategies discussions.

2. MANUFACTURING FOR CHINA

2.1 Cornerstone of China's National Economy

The manufacturing industry is one of the pillar industries of China's national economy, the carrier for science and technology development and a bridge to productivity conversion. In 2010, China's share of global manufacturing output rose to 19.8 percent. Comparing with 19.4 percent of the U.S., China has become the world's manufacturing superpower. With over 200 kinds of industrial products output as well as export ranking top one in the world, China has become a truly big manufacturing center of the world. However, there exists a considerable gap in China's manufacturing technology and equipment manufacturing capacity, compared with other manufacturing powers in the world. Many of research and development activities are visible in the area of advanced manufacturing theory and technology. Research institutes and universities are sought for providing the source of innovation for the production of China's manufacturing equipment to break through the blockade of high end manufacturing technologies by developed countries. This is of great significance in helping ensure national security and sustain healthy and stable development for China. Furthermore efforts are needed to carry out cutting edging research and development in smart manufacturing with aims to further expand and deepen the manufacturing technology innovation, help China develop into a high end manufacturing power, accelerate the transform from "Made in China" towards "Created in China."

2.2 Manufacturing Industry Restructuring and Opportunities

Today the world is at a point facing worldwide economic structural adjustment, change and reorganization, presenting historical opportunities to China. China's national export-oriented economy urgently needs transformation, with urgent need to adjust industrial structures and to enhance domestic demand to sustain China's economic development. Advanced manufacturing, strategic emerging industries and modern services represent the direction of the scientific, technological and industrial change, which the state of China will vigorously pursue and develop in order to expand and create market demand. Deep integration and synergy among these industries will support China's technological and industrial change, helping China to occupy a favorable position in this competitive world with ongoing global restructuring and international industrial transfer and the international industry adjustment.

In China, manufacturing is the cornerstone of national economy. However, it seats in the low end of industrial chain. Its potential pulling power for alleviating domestic demand on the national economy is not adequately reflected. And as to the nation's equipment manufacturing sector, the capacity for independent innovation is weak. Big but not strong characterizes current China's manufacturing, with urgent need to improve its competitiveness in the world. In the Pearl River Delta (PRD), especially in Guangdong, there exist many Hong Kong invested enterprises, with close ties with Hong Kong. Current market situation still demands the implementation for upgrading, transformation and relocation policies. Facing the risk of economic slowdown, Guangdong is actively implementing the policy of "double relocation," with notable statement like "relocation would yield hope, staying at current would face the risk of death, and upgrade is the only direction to survive." In the process of this industrial restructuring, based on its well developed and strong services industry, Hong Kong is facing a considerable responsibility and great opportunity to assist those enterprises based in the Pearl River Delta by vigorously developing needed manufacturing services to help them restructure and upgrade. Currently Hong Kong is also vigorously pursuing to advance new generations of information technology, including cloud computing. It is another responsibility and opportunity for Hong Kong to leverage this scientific and technological progress to interface and meet the business needs of the enterprises in PRD, and to further expand to other regions of the Mainland.

2.3 Significance of Advanced Manufacturing Innovation

Smart manufacturing research starts from a perspective of industrial chains, fostering a deep synergy among Internet of things, intelligent manufacturing and services science. Through mechanism innovation, Smart manufacturing research often emphasizes collaborative research and innovation, leveraging resources worldwide, to advance manufacturing technology. It is a natural strategy for China to seek and incubate domestic and international R & D cooperation, make good use of available resources to establish smart manufacturing innovation platform, and aim to form a good research science and technology innovation system, improve scientific and technological innovation chains.

Evidently China is establishing national innovation platforms in smart manufacturing, an integrated area of the Internet of Things, intelligent manufacturing and service science for the development of theory and common technology research and application. Objective is to provide a solid theoretical basis and common technologies and innovative applications for the domestic manufacturing industries to upgrade, with original results achieved to help China's manufacturing sector to move to the high end in the industrial chain with high value add.

China also needs to foster a talent based and academic exchange center with world acclaim in the field of smart manufacturing. Through attracting outstanding academic leaders and innovative teams, China provide facilities for undertaking major national basic and applied research projects, and evolve into an international well-known academic exchange center.

China is also developing nation-wide strategic leading-edge technology incubation bases for smart manufacturing. Through deep integration and synergy among Internet of Things, intelligent manufacture and service science, smart manufacturing research could achieve major breakthroughs in key areas for the integration of advanced manufacturing, strategic emerging industries and modern service industry development with strategic cutting-edge technology incubation.

3. SMART MANUFACTURING RESEARCH

In the face of fierce worldwide competition in the 21st century, Chinese government has planned accordingly and launched a series of major and key projects, carrying out special studies in the frontier of advanced manufacturing technology and equipment. In April of this year, the Ministry of Science and Technology, in combination with other ministries, has issued a "12/5" Manufacturing Information Technology Project Planning, outlining manufacturing information technology projects over the next five years for the development of manufacturing services, intelligent manufacturing. In recent years, the National Hi-Tech Research and Development Program (863 Program) in 2008 and 2006, has called for "advanced manufacturing technology major projects in the field of radio frequency identification (RFID) technology and applications," by the development and application of RFID technology to promote transformation and upgrading of China's advanced manufacturing (China 863, 2013). 863 Program in advanced manufacturing technology in 2010 set research themes of "WIA-based wireless monitoring and control technologies, devices and systems R & D" and "Cloud manufacturing service platform key technologies" and so on. 863 Program in advanced manufacturing technology in 2009 set a "service robots" key project theme. In 2007, 863 Program in advanced manufacturing technology had four topics including intelligent robot technology, extreme manufacturing technology and major products and facilities life forecasting techniques. China National focused basic research and development plan (973 Program) projects, such as 2005 digital manufacturing basic research and 2003 high-performance electronic product design manufacturing microscopic technology, digital new principle and new method (China 973, 2013). National Natural Science Foundation major focused project, in 2003, had called for the key technologies in advanced electronics manufacturing and important scientific and technical issues in digital manufacturing theory in the network environment in 2003 (China NSFC, 2013).

Development of the manufacturing industry so far has involved with multi-disciplinary applications, manufacturing technologies have more and more adopted the latest achievement in materials, mechanics, physics, chemistry, and computer simulation technology, network technology, control technology, nanotechnology, biotechnology, sensor technology. Currently, with booming economy, China has respected more a science and technology led economic development. Internet/Internet of Things, and next generation of information technology will lead to industry change, promoting the manufacturing industry to shift from the traditional manufacturing towards industrial chain based manufacturing. The trend and requirements of digital manufacturing, service manufacturing, and intelligent manufacturing will become more apparent and prominent. Digital manufacturing will deepen the manufacture of hard power, and service manufacturing is to enhance the manufacturing soft power. Their integration will be the direction of intelligent manufacturing development, in order to achieve the optimal balance of hard and soft power. It provides a good tool for China's manufacturing industry to achieve global competitive advantages.

3.1 Digital Manufacturing

Digital manufacturing is a strategic choice for manufacturing innovation, including digital design, digital management, digital production, digital manufacturing equipment, and digital enterprise. Further integration of information technology into digital manufacturing will promote digital manufacturing from traditional manufacturing to both ends of the industry chain, evolving towards industrial chain based digital manufacturing. This industrial chain based manufacturing would help effective coordination of worldwide design and manufacturing resources, improve the capacity of complex product development and product life-cycle optimization, shorten the development cycle and reduce development costs, which can greatly enhance the capability of independent innovation and market competitiveness of China's manufacturing industry.

High-precision, high efficiency, high reliability are the eternal theme of manufacturing technology development. High-precision, high efficiency, high reliability are the most remarkable features of modern digital manufacturing equipment. Numeric control technology is the core technology of digital manufacturing equipment. It has created a precedent for the manufacturing equipment to the digital development direction. Digital prototyping with the help of advanced calculation models, calculation methods and digital preview means could reveal the essence of the manufacturing process, including prediction processing, assembly and even product life cycle process, providing necessary support for manufacturing process optimization. Digital prototyping technology reflects cross and inter-disciplinary integration and synergy among manufacturing science, materials science and computing science, is an internationally recognized cutting-edge technology, and has been playing an increasingly important role in equipment manufacturing in energy and transportation industries. Processing technology development is toward the direction of a new generation of manufacturing equipment with ultra-sophistication and ultra-high speed. At present, super precision processing has achieved a stable sub-micron processing, and is developing toward nanofabrication expanding processing materials from metal to non-metallic. New processing and manufacturing processes continue to emerge. The rapid prototyping manufacturing processes gradually mature. More and more other new manufacturing processes emerge, including 3D engraving and processing, three-beam processing and modification, new bionic manufacturing, bio-manufacturing, micro-nano manufacturing. These provide a steady stream of powerful advanced manufacturing equipment and technology innovation, and are the hot spots of current international advanced manufacturing technology research.

3.2 Service Manufacturing

Development of service manufacturing would be in the direction toward industry chain based manufacturing providing capacity of customer-oriented, on-demand products and services, improving the consistency between the enterprise and customer goals and achieving a win-win situation. Service manufacturing would stay close to the market and demand, enhance the understanding of the needs and ability to create harmonious win-win development capabilities on the basis of collaborative innovation with customer, and form capability to grasp control of the industrial chain and achieve competitive advantage.

Pursuit of the consistency of business and customer goals is the eternal theme of service manufacturing. High-quality, high efficient and personalized manufacturing represents the most significant features of service manufacturing. Currently, the volatility of consumer demand is widening, so is financial volatility. Expectation on customer service and product quality is continuously higher and higher. Business models are also diversified. All of these present great challenges to manufacturing. Improving services, responding to consumers changing expectations and demand patterns needs more and better resources and information. Having capability of accessing such resources and information will be cornerstone of service manufacturing. Resources and information management and analysis, effective demand identification and management, intelligent manufacturing decision making are the core of service manufacturing services. The development and application of Internet of Things and cloud computing in manufacturing provides powerful tools to achieve those. Internet of Things, including Radio Frequency Identification (RFID) is a major focus of China's next generation IT innovation direction, providing a huge room for innovation in the field of chips, sensors, near field transmission, mass data processing, and comprehensive integration and application. With Internet of Things applications in the manufacturing sector deepening, it will give rise to a large number of new data, new technologies, new products, new applications, and new business models. Consequently it will lead to huge support and management needs. Cloud computing as a new computing paradigm is developing rapidly, but still in its infancy with a few problems to solve, which mainly include the virtualization technology and massively scalable programming techniques. With the development of Internet of things, corresponding theory and technology of cloud computing for the Internet of Things are necessary and a converged Internet of things with cloud computing is the direction of applying cloud computing and the Internet of Things in manufacturing.

3.3 Intelligent Manufacturing

Intelligent manufacturing evolved from the traditional artificial intelligence. Application of artificial intelligence simulates a human expert's intelligence activities, including analysis, judgment, reasoning, deliberation and decision-making. It is used to replace or extend the human in the manufacturing environment, at the same time, providing collection, storage, improve, share, inheritance and development of human experts manufacturing intelligence. However, due to the complexity of the human brain thinking activities, there still demands a great deal of improvement in artificial intelligence and simulation. It results that the intelligent manufacturing is usually demonstrated as intelligent machines to show the performance of flexible and intelligent manufacturing systems in the manufacturing unit and flexible and intelligent network-based integration.

Intelligent Manufacturing is a deep integration of advanced manufacturing technology, biotechnology, nanotechnology and IT, and application of modern sensor technology, network technology, automation technology, intelligent technology and other advanced technology integrating with traditional manufacturing technology to enhance manufacturing to provide manufacturing intelligence to adapt to the complex environment for safe and reliable, cost-effective and sustainable manufacturing. Intelligent manufacturing is well suited for product life cycle wide ubiquitous sensing based manufacturing, with features like information-driven, capacity for man-machine cooperation, self-organized and flexible, self-learning ability, self-maintenance capabilities and others.

Volatility of consumer demand and uncertainty of economic activity continue to grow. Diversified varieties, small batch manufacturing demand will become a normal pattern. Flexible and intelligent manufacturing to realize personalized service manufacturing is a reasonable choice, the development trend of future manufacturing.

Human-machine integration is a key to intelligent manufacturing, still to be improved. Development direction of intelligent manufacturing towards a high degree of man-machine integration into human-machine integrated manufacturing. The sign of maturity for intelligent manufacturing systems is reflected as real time and seamless capacity to introduce human's thinking in manufacturing involvement and intervention.

Intelligent manufacturing provides intelligent means for the optimal use of manufacturing resources, leveraging recycling economy theory and re-manufacturing concept for traditional industries to provide resource-saving, energy saving, environment-friendly technologies. It will also provide advanced technology and equipment for new energy, new materials, biomedicine, a new generation of information networks, smart grid, green means of delivery, eco-friendly, ocean and aviation, public security and other strategic development of emerging industries.

In October 2010, the State Council of China has decided to speed up the cultivation and development of strategic emerging industries, including the development of intelligent manufacturing equipment. In March 2011, the 12/5 National Economic and Social Development Plan suggested to further the development of intelligent manufacturing equipment. In July 2011, the national 12/5 Science and Technology Development Plan set intelligent manufacturing as one of the main support directions.

4. SMART MANUFACTURING RESEARCH AGENDA

In recent years, China maintains a booming out looking of rapid economic development. In 2010, its national manufacturing sector has surpassed the U.S. as the top 1 manufacturing country in the world. At the same time, China is eager to move towards high-end manufacturing power. Based on clear understanding of the trends of manufacturing technology development (digital, service and intelligent manufacturing), we envision research themes of smart manufacturing would be built around synergy among Internet of Things, intelligent manufacturing, and services science, with following three research areas along the direction of smart manufacturing:

- Internet of Things and digital manufacturing theory and technology
- Intelligent manufacturing theory and technology
- Advanced manufacturing theory and technology.

4.1 Internet of Things and Digital Manufacturing Theory and Technology

Embedded and Micro-Electromechanical Systems and Technology

Along the line for the development of high end manufacturing equipment, it is necessary to carry out research of integration and application of micro-electromechanical systems and embedded technologies for product innovation. Research is required for systems and technologies for networked manufacturing oriented highly reliable real-time control, remote sensing, and intelligent diagnosis. Through system miniaturization of the system, and system and component integration, new principles and new methods are to be explored. Other relevant areas include conducting research of monolithic chip integration, unified system of MEMS devices manufacturing and packaging, and related processes.

Internet of Things and Cloud Computing

Internet of Things, including Radio Frequency Identification (RFID) technology will advance digital manufacturing to a deeper and wider digitization for digital design, digital manufacturing, digital testing, digital testing, digital diagnostics and digital management, on the basis of the ubiquitous sensing, objects interconnection and information fusion. Through the integration of Internet of Things (IoT) and cloud computing, research is to address the challenges of massive data management, interconnection and massive information fusion. For the complex manufacturing and expanded the industrial chain environment, R&D is necessary to study the theory and technology of digital identification, digital identification and sensing technology integration.

Computer-Aided Design, Engineering and Process Design

IoT enhanced CAD/CAE/CAM/CAPP systems are to be developed. Research is called on virtual reality, 3D modeling, kinematics simulation, dynamic simulation, finite element analysis, simulation control, human-computer interaction technology, and virtual reality technology. Efforts are necessary to study advanced design methods integrating modeling, simulation and optimization technology in to a unified design system.

4.2 Intelligent Manufacturing Theory and Technology

Intelligent Manufacturing Equipment Software Environment and Operating System

Research and development of intelligent manufacturing operating system will provide software platform with development supporting environment and tools to support manufacturing equipment and process design, control and execution systems for intelligent manufacturing equipment. Studying intelligent programming and programming optimization techniques, optimization techniques of manufacturing processes, remote health monitoring and diagnosis of manufacturing equipment, and manufacturing reliability data collection and analysis can provide a common operating, control and software development support environment.

Green Manufacturing and Sustainable Development

Establishment of green supply chain management theory for advanced manufacturing is urgent. Research is called for Internet of Things enhanced product life cycle management, control and dynamic optimization and full life cycle product data management and enterprise resource management and optimization. With circular economy and re-manufacturing concept, it is necessary to study and development of theory and technology for providing advanced manufacturing equipment for traditional industries as well as new energy, new materials, biomedicine, next generation of information networks, smart grid, green transportation, public safety and other emerging industries with resource conservation, energy saving and environment-friendly technology.

4.3 Advanced Manufacturing Theory and Technology

Manufacturing Information and Services

Research efforts are necessary to study Internet of Things enhanced service manufacturing, research theory and methods to influence and control of manufacturing industry chains. Leveraging integration and information services, manufacturing services system based services value chain would be better understood and analyzed. Studying the value of information via researching information space-time characteristics could reveal its relationship with the value of information. Information pricing via analyzing information flow, capital flow, logistics, business flow and value-added flow interactions and time impact enables development of theory and technology for high value-added services and applications. Development and application of risk transfer and analysis methods via studying the interaction law among information flow, capital flow, logistics, business flow and value-added flow could be one of feasible means.

E-Commerce-Oriented Service Manufacturing

E-commerce oriented service manufacturing demands customer-oriented manufacturing theory and technology system to improve the consistency of corporate and customer goals, effective close to the market and demand, improve the understanding of the needs and creativity in providing collaborative innovation with customers and provide ways and means of grasp control of the industrial chain and competitive advantage on the basis of a harmonious win-win development ability. Thus, it is necessary to study e-commerce and supply chain integration theory and e-commerce oriented manufacturing system.

5. CASE STUDY ON RFID ENABLED SIM-CARD

With the coming of 3G time, the mobile phone has already not merely provided the communication services; the high band width wireless data channel impels the fast development of other mobile services. Instead of the cash, check and credit card, the application of mobile payment through handset will change the current spending habit of customers in order to bring about the revolution in the payment field in the e-commerce era. As a new commercial payment service, mobile payment has a huge market potential and has drawn the attention of mobile operators, banking institutions and IT service providers. Due to the gradual saturation of remote mobile payment as well as the large requirement of modern people, the near field mobile payment is gradually walking into the mobile payment market. To occupy the advantageous position in the near field mobile payment market, competition has been introduced by mobile operators, financial institutions, the 3rd party payment service providers and some other participants. A proper business pattern and technical solution undoubtedly will promote the development and application of mobile payment. In this study, the relevant mobile payment background including the current development situation and popular contactless technical solutions are introduced. Moreover, the details about mobile payment value chain are described. And three major business models of mobile payment via corresponding smart card choice are analyzed deeply through using SWOT analysis in order to get a proper business pattern. This case study applies AHP method to evaluate two main-stream SIM based mobile payment solutions (SIMpass and RF-SIM) in accordance of SWOT criteria. The result of the evaluation shows the existing issues in the process of promotion and the future tendency. With the development of Near Field Contactless technology, people realized that they can implant a RFID module into SIM card to implement mobile payment using their cell phones. Therefore mobile operators pay widespread attention to the NFC technology and regard SIM-card based payment research as one of their main directions for value-added service of mobile phone.

Developing SIM-card based payment application is a win-win choice. After decades of development, the voice service for mobile phone user is so improved and has less space to gain more profit. And with the widely application of NFC payment, especially the rapid progress of application in public bus and subway, RFID technology becomes more and more familiar and acceptable by users and the supply chain is getting mature. Therefore, more and more telecom operators take SIM-card based mobile payment application as a strategic product for the future mobile lifestyle. Secondly, RFID card and mobile phone both emphasize private proprietary features. Some typical uses of NFC card, such as Identity Card, Bank Credit Card, and Access Card and so on, are all based on personal identification application; meanwhile, mobile phone as a personal communications tool, while to some extent has become a symbol of personal identity. Third, SIM card is a natural module in the phone that can be used to store identification information and payment information. With the improvement of technology, several NFC applications can be implanted into one SIM card and will not interfere with each other.

Therefore, mobile operators can use SIM-card based mobile payment products to enrich their value-added services and increase user stickiness; and at the same time NFC payment can rely on the strong network and the driving force of mobile operators to promote the development of the industry chain. Given the complexity of the industrial value chain, SIM-card based mobile payment develops relatively later than the entire mobile payment industry.

Several SIM-card based mobile payment solutions are developed in different areas of China theses years. However, the incompatibility of each solution in different areas actually decreases the convenience of mobile payment and may block the sustainable development of mobile payment industry. So, how to find out an optimal solution which mostly fits the Chinese market and help other solutions become collaboration is becoming more important than ever. And our study is focus on this.

We first point out the three most popular mobile payment solutions and list the advantages and disadvantages of them. Then we apply the value chain theory into the SIM-based mobile payment industry and find out the crucial participant(s) by value chain analysis. After that, three major business models of mobile payment are analyzed deeply through using SWOT analysis in order to get a proper business pattern.

6. SMART CARD MANUFACTURING VALUE CHAIN FOR MOBILE PAYMENT INDUSTRY

Since the adoption of RFID technology in the mobile payment field, a lot of investors have paid widespread attention to the near-field payment business especially focused on the SIMpass and RF-SIM solutions. Review the most references including industry magazines, on-line surveys and market analysis reports from a variety of institutions, they are majorly put their concentration on the comprehensive analysis of the recent market environment as well as the technology comparison among different kind of NFC realization. Since now, a lot of statistics have been collected to reflect present status of industry development such as the statistical charts to show the different acceptance for various SIM based mobile payment application in different regions. Even many materials detailed analyze the major factors that will affect the industry development and popularization in different aspects like policy support, risk evaluation, social recognition and so on, few references can achieve the quantitative analysis for these factors objectively and then provide the comprehensive framework to get the clear strategic planning for the main-stream solutions, SIMpass and RF-SIM.

To fulfill this gap, we apply SWOT-AHP methodology (Kurttila, Pesonen, & Kangas, 2000) in the SIM-card based mobile payment industry in China, and set different intensity level to each factor for different solutions respectively. Therefore, we can evaluate and compare the solutions using calculated results from various criteria which were quantified. Using the quantified data will help people better to understand the SWOT aspects of each mobile payment solutions and become more convincing. Furthermore, we can develop future strategies for each solution based on the quadrangle derived from the calculated SWOT analysis.

6.1 Technology Analysis

Although the technology of SIM-card based Mobile Payment becomes more and more mature, the technologies for developing application with the combination of mobile phone still stays in the exploration stage. Currently, a variety of near field mobile payment technology solutions are being employed, among which three categories of near field contactless mobile payment implementations have been introduced. The most popular SIM based RFID Mobile Payment Solutions are: eNFC, SIMpass and RF-SIM.

e-NFC

e-NFC (NFC Forum 2013)is a compromising solution designed by some card vendors, chip vendors together with mobile network operators after going through the frustration of the traditional NFC technologies. It uses very similar approach for implementation. The chip and antenna which support mobile payment are put in the terminal, the security is controlled by (U)SIM, and the communication between the NFC chip and the SIM card compiles with SWP(Single Wire Protocol).

e-NFC, as the International Standard defined by 3GPP, has gradually come to maturity, but the complete commercial solution is still in the pipeline. The mobile phone and SIM card have to be modified which will inevitably increase the cost. Furthermore, currently the lack of maturity of the Industry Chain slows down the development of e-NFC solution. Additional, the domestic application may encounter the patent embarrassment from foreign manufacturers.

SIMpass

SIMpass (SIMpass, 2013) is a contactless technology solution based on RFID-SIM card. It provides a complete management platform which implements the management of multi-application and secure data over the air (OTA). SIMpass supports two interfaces: contact interface and contactless interface.

Once the Dual-interface SIM card is put into the SIM card slot in the mobile phone, the contact interface would implement the SIM application and provide proper functions of the card. Meanwhile, the contactless interface could support non-telecom applications such as e-wallet. In addition, SIMpass card can deal with non-scene non-telecom transaction such as OTA contact backup via SMS through STK user interface.

Since the contactless interface complies with ISO14443 standard, additional RF antenna coordinated with the card is required to provide power and the clock and data signal. There are two choices: SIMpass with antenna or Custom-made Mobile phones.

For the former one, no modifications to the mobile phones are needed. The antenna is connected to the SIMpass card and to be attached between the battery and back cover of the mobile phone. It is cost effective and can be easily adopted by end customers, which make it in favorable position. In this scheme, the reliability of antenna and the convenience for the customers to install the card and the antenna should be taken into account. And also, this scheme occupies the C4 and C8 contacts which cause the confliction with high-capacity memory card in the International Standard Application. Because of the low perceiving level, metallic material cannot be used for cell phone shell.

For the custom-made mobile phone, the SIMpass antenna is either integrated into the phone battery or the main board, which can be modified by mobile phone manufacturers. A connection route between the antenna and the antenna contact on the card is also designed in the phone. Mobile phone combined with the contactless applications makes it more reliable and easy to use. Since a custom-made mobile phone is needed, a customer has to replace his mobile phone if he wants to use the mobile payment, which will increase the cost of promotion and even discourage the promotion of the mobile payment.

Once the phone is in service, Contactless applications and telecom applications in the SIMpass will work simultaneously. In another words, contactless transactions could be done at the time of making a call or sending/receiving messages. When the phone is powered off, SIMpass can still work well just like a normal contactless card.

The high-level security chip produced by Infineon Technologies is employed in SIMpass solution which guarantees the security feature. And also, 13.56MHz communication frequency means the mature marketing application environment and the compatibility. The existing contactless mobile payment environment, such as subway, public transit, supermarket and bank, almost supports this frequency band.

RF-SIM

RF-SIM card (RF-SIM, 2013) can realize the short distant communication which is embodied with new RF technology that the users only need a smart card and make the handset they are using into an NFC-based handset with normal SIM card function. RF-SIM use miniature RF modules and built-in antenna to connect the external device communication. Some SIM cards is designed for mobile phones to normal communication, authentication, and only for the physical connection; Built-in software for managing is high safety of RFID and other logic-based VIP membership.

Its main communication features are as follows:

- Using of 2.4G frequency band, automatic frequency selection, high reliability of connection and communication.
- Two communication methods: support auto-sensing and active to connect.
- Model of two-way communication from 10cm~ 500cm, can be adjusted depending on the application.
- One-way data broadcasting (radius 100m).

- Air transport and auto TDES data encryption, anti-eavesdropping data, the mutual authentication conducts when card accessing.

One key advantage of RF-SIM is that it can be easily retrofitted to existing mobile phones. However, since it operates at 2.4GHz rather than NFC's 13.56MHz, RF-SIM terminals are incompatible with NFC terminals. POS system needs to be reconstructed in order to realize the compatibility with common public facilities. At this stage, this 2.4GHz frequency increases the difficulty of cooperation. In addition, the wide range of communication seems likely to cause safety problems.

6.2 Value Change Analysis

A value chain can be defined as the full range of activities which are required to bring a product or service from conception, through the different phases of production (involving a combination of physical transformation and the input of various producer services), delivery to final customers, and final disposal after use. In China, the industry value chain of SIM-card based mobile payment is mainly constructed by hardware device provider, mobile operators, third-party payment platforms, financial institutes, merchants and user.

Device Providers

Device providers play a role of hardware basic support in mobile payment industry value chain. As a communication value-added services, mobile payment naturally cannot be separated from the operating carriers and tools. So, it becomes very important that to provide high-quality mobile devices in order to ensure that the hardware used in mobile payments can be successfully achieved using standard. In the past, because of networks and technologies, device providers did not take developing payment technology as an important module for mobile devices. However, along with mobile communications evolution from 2G to 3G and mobile data services continues to rise, mobile device manufacturers began to provide operators mobile equipment as well as data service platforms and business solutions that including mobile payment services, and this laid the foundation for operators to provide mobile payment services.

Mobile Operators

The main task of Mobile operators is to build a mobile payment platform which mobile payments provide secure communications channels. Mobile operators can be regard as an important bridge which connects users, mobile operators, financial institutions and service providers, and play a critical role in promoting the development of mobile payment services.

In the mobile payment business, mobile operators gain their benefit mainly from the following four aspects: first, from the service provider's commission which is generally from 3% to 20%. Second, mobile payment services based on voice, SMS, and WAP can bring operators benefit from charging for data traffic. Third, the mobile payment services can stimulate the user to generate more demand for data services, thereby contributing to the development of other mobile Internet services. Fourth is to help mobile operators retain existing customers and attract new customers, and increase the competitiveness of enterprises.

Financial Institutes

To allow bank accounts associate with users' mobile phone numbers, banks need to establish a complete, flexible security system for mobile payment platform to ensure the safety and smoothness of the user payment process. Obviously, compared with the mobile operators, banks not only have cash, credit card and check-based payment system, but also have personal users and business resources.

Financial institutions gain their benefit from five aspects in the mobile payment industry. First, from the pre-paid amount stored in the mobile phone bank account which increase the savings amount then in no doubt help banks gain benefit. Second, gain from the profit-sharing of each mobile payment transactions. Third, more bank cards can be activated for the application of mobile payment business. Fourth, mobile payment business is effective in reducing the bank branches, thereby reducing operating costs. Fifth, it can help consolidate and expand the user base, and improve the market competitiveness of banks.

Based on the advantages of customer group, financial institutions can realize their value in the mobile payment market. Mobile payment is first applied among users who use mobile phone and also have consumer demand for products from contracted merchants: A1 is a mobile payment account binding with a bank account; A2 is paid directly by the mobile account without having to have a bank account. The merchants' potential consumers with mobile phone can also become the potential consumers of mobile payment, and they form the second source of mobile payment users. The third source is the consumers with a bank account and willing to buy mobile devices for mobile payment.

Third-Party Payment Service Providers

Third-party payment service providers play an important role as the link between banks and mobile operators in the develop progress of mobile payment business. Independent third-party mobile payment service providers is able to integrate the resources from mobile operators and banks in different areas and to offer mobile phone users with rich mobile payment services in order to attract users to pay for the fees of applications.

From the experience of European countries, the first to promote and provide mobile payment services is not the mainstream mobile operators, but third parties such as Sweden Paybox portal. Regardless of which mobile operator the user using and the user's personal financial account belonging to which bank, as long as the registered in Paybox, the user can enjoy the rich services on the company's platform.

In China, there are a number of third-party payment service providers in Shanghai, Beijing, and etc., which provide user mobile payment services. However, based on the current situation, the mobile operators are more willing to become the service provider of mobile payment business.

Mobile payment service provider has two revenue sources: first, charge equipment and technology licensing fees from mobile operators, banks and merchants; second is extracting commission from the mobile operators for the usage of mobile payment services.

Merchants

On a first look, it seems that merchants have little to do with mobile payment and gain little profit from this market, and is only a participant in the industry value chain. However, it is not true. For the merchants, deploying mobile payment systems in the shopping malls and retail stores, to a certain extent, can reduce the time and manpower resources for constructing middle part of payment, reduce the costs of operating, service and management, and also improve the payment efficiency, thereby increasing profitability. Not only that, the payment simplification, in fact, is saving the consumer time and costs, and will achieve higher user satisfaction and contribute to the improvement of brand value for the merchants. It can be said that in the specific mobile payment process, merchants have the most potential interest, as well as the highest input-output ratio.

6.3 AHP Analysis

Analytic Hierarchy Process (AHP) is a mathematical method applied on the decision-making problems with multiple criteria. This theory was introduced originally around 1980 (Saaty, 1980). It can handle both qualitative attributes and quantitative ones. The aim of the AHP's development is to systematize the complicated issues by decomposing the problem into constituent parts, comprehensively assessing the quantitative information, and obtaining the most appropriate strategic decision among alternatives. Currently, many derived AHP methods have appeared (Triantaphyllou & Mann, 1995).

By collecting statistical data, research and other methods, identified a number of strategic key factors that attributed to SWOT, strengths(S) and weaknesses(W) from internal and opportunities(O) and threats(T) from external environment, respectively.

Strengths Analysis

Near-field mobile payment takes absolutely advantages with operational convenience compared against traditional payment solution. Even the remote payment which is generally popular among people.

The relevant near-field payment technology has been improving for several years. Either the signal communication technology or terminal device construction develops rapidly and maturely. To speed up the adoption of near-field mobile payment service, relevant industry standards and regulations are being perfect gradually.

"All-in-One card" is the core purpose of SIM card based mobile payment business as well as the final demand for modern customers. Near-field technology helps people to accomplish different kind of payment in various application environments. It can be utilized as credit card for shopping, Entrance guard card for ID authentication, the public transport IC card for public traffic and other functional cards at the same time.

Weakness Analysis

NFC solution and SIMpass custom-made solution need to replace the original handset which forces users o purchase those mobile phones in order to support near-field mobile payment service. Even the SIMpass with antenna solution cannot reach the complete separation between SIM card and handset. This disadvantage will discourage its acceptance by potential customers.

Security issue of near-field payment becomes more complex than traditional cash or credit trading. Currently, safety payment becomes the topic people generally care about. Short distance contactless solution expresses the relatively better security and reliability. As the communication range increases, signal transmission stability and information security need to be considered a lot.

Compared to the traditional payment business, near-field mobile payment requires supports from different sectors. Either to modify the SIM card or to replace the handset needs to increase fund input for devices (re)construction. The manufacture of terminal facilities like POS machines and RFID reader needs an injection of new funds. Also the cost of equipment's deployment is still in need.

Opportunities Analysis

Recently, mobile phones holders are more than a thousand millions meanwhile the bank card customers have covered all over China. Those are all the potential users for the near-field mobile payment business. If each mobile communication customer decides to apply for the mobile payment service, the market will have handsome business performance. To turn this potential user group into real user group, the promotion and development strength have to be enhanced.

To popularize the mobile payment business and cultivate the spending habit of users, multi-side collaboration is absolutely necessary. For now, interested parties, mainly represented by mobile operators and bank institutions have seek out the opportunities for cooperation with each other. Three mobile operator giants have already separately negotiated with China Union Pay institution for the further development of SIMpass solution. Mobile operator and bank institution establish the solid alliance, so, not to be outdone, other participants such as the 3rd party payment service provider starts to find the cooperation in other ways. The cooperation relationship plays as a catalyst for the expansion of mobile payment market.

The promotion of mobile payment application cannot be achieved without the support of relevant government policies. Not long ago, the government launched the license of 3rd payment. This measure, to a large extent, boosts the mobile operators and banks to form a same purpose and will inspire user's enthusiasm for mobile payment.

Threat Analysis

Currently in China mobile payment market, a complete industry chain has not been formed yet. Multi-side in the industry chain all have strong willing to take relatively advantageous position in this field for the maximum profit. Therefore, various business patterns led by different parties exist at the same time. The lack of unified business model hinders the popularization of mobile payment service.

Although the unified standard (13.56MHz) eases the tension between mobile payment and bank institution more or less and also promotes their collaboration to develop SIMpass solution. The competition among various mobile operators is growing in intensity. The missing connection among each mobile operator brings about unfavorable factors to the promotion. Suppose that China Telecom customers can only accomplish the mobile payment in those stores configured with special POS machine which is captive to China Telecom. This situation will crucially deduce the user acceptance. To open up the cooperation among different mobile operators can be a huge challenge in the future.

The competition comes from foreign competitors should not be neglected. The introduction of Near-field payment surges up the mobile payment application fads. Foreign countries including Japan, Korean, Germany and so on, all devote major efforts to develop and promote the mobile payment business. The most famous cell phone company in Japan, NTT DoCoMo, has focused on the exploration of NFC handset for quite a long time and has achieved very good results. Gradually, NTT DoCoMo puts its long-term goal on the international mobile payment market especially China. The mature technology, complete formation of industry chain and the wealthy experience undoubtedly will attack our native market and challenge our own near-field payment solutions.

7. RESULTS AND CONCLUSION

7.1 Manufacturing Strategies

Strategy in Favor of SIMpass

Apparently the strengths and opportunities occupy the relatively important position in the strategic planning. With the support of internal strengths, the development can be accelerated through gasping the opportunities. Due to the significance of convenience feature, fully utilizing its seemingly advantages should be one of the strategic measures.

The self-advantage of convenience should be employed to avoid or lighten the threats from outside factors. SIMpass will be mostly confronted with the lack of complete industry chain and mature business pattern. The main reason to cause this treat is the lack of harmony among each sector of industry chain. The stiff competition among each participant of industry chain forms various business models. How to balance each party's benefit becomes the tough issue.

The external opportunities can be used to improve the internal weaknesses in order to boost the development of these two solutions. For SIMpass, although government favored this project with a large support, the need to modify handsets will be an unfavorable factor which may greatly affect the extensive use of mobile payment. Therefore, this solution must seek a route to improve the current status so that SIMpass can be succeeded for common mobile phones. Another problem which is capable of drawing attention is the cost issue. How to make merchants increase their acceptance of mobile payment so that they are willing to deploy the hardware devices (POS machines) will be an important task.

Strategy in Favor of RF-SIM

Compared to weaknesses and opportunities, the strengths and threats of RF-SIM solution seems to be superior in total intensity. Convenience is the biggest advantage of RF-SIM method which expresses in the aspect of requiring no modification of current mobile phones. This strength should be maintained and enhanced by expanding various applications in one single SIM card. To realize the real "all in one card" will be the final goal of this solution.

Due to the working frequency (2.4GHz) of RF-SIM cannot be compatible with common POS machines, seeking for collaboration becomes a hard task. The relevant industry standard and regulation should be confirmed rapidly in order to accelerate multi-side cooperation. Mobile operators owns the intellectual property rights of RF-SIM core technology which helps to increase the self-competitive superiorities when confronts to threats of foreign brand.

RF-SIM currently is being mainly promoted by China Mobile which possesses the largest handset user base in China. Large potential user group is the most significant opportunity that should be seized. The number of cell phone users must be more than the amount of POS machines that means even though the fund input increases lot because of hardware reconstruction, the large user group helps to make the quite respectable profit space.

7.2 Comparison Analysis

It is obviously that the strength of RF-SIM exceeds that of SIMpass which is mainly expressed in the convenience feature. And also, according to the relevant intensity data, SIMpass needs to improve their signal transmission technology through which only plastic shell can be penetrated.

On the other side, SIMpass solution recently owns more opportunities than RF-SIM. This can be manifested in government support and multi-side cooperation. Undoubtedly, in the current stage, SIMpass develops faster than RF-SIM due to its compliance of industry standard (13.56MHz) and the relatively mature collaboration between mobile operators and China Union Pay. In the future, the large potential user group will be one of the major energy for RF-SIM's promotion. Meanwhile RF-SIM solution should be improved to take proper measures to quickly form the business cooperation relationship. Government need to give it more attentions.

The weaknesses of SIMpass need to be paid a few more attention to than RF-SIM. The major disadvantage is that SIMpass cannot completely achieve the separation between handset and SIM card. The modification of cell phone and installation of antenna more or less affect the convenience. In the aspect of cost increment, even though the modification of handsets to some extent raise the cost of termination, deployment of RF-SIM needs to reconstruct many hardware facilities like POS machine and RFID reader which requires much more initial capital.

Minor total intensity of differences in the aspect of threats exists between two solutions. The most serious challenges of both two solutions are the incomplete industry chain as well as the lack of main-stream business pattern. Nowadays, SIMpass solution operators are being positive to cooperate with each other for maximum profit. The business model led by the combination of bank and mobile operator is being formed gradually. Compared to SIMpass, RF-SIM needs to redouble the efforts in this part by strongly promotion. Also SIMpass requires keeping its vigilance on the competition from foreign brand. RF-SIM has some advantages when facing foreign industry competition but once suffers the internal competition, without the perfect collaboration relationship, its development will be limited. Considering that RF-SIM is the solution mainly dominating by mobile operators, bank institution will not give up occupying the leading post based on its current advantage (unified standard).

7.3 Conclusion

In this chapter, an introduction to smart manufacturing innovation and transformation is presented. An example is used to illustrate what is happening in China's manufacturing industry, with insights about China's strategy of advanced manufacturing research and development. Emphasis is laid upon the value chain analysis for setting smart manufacturing strategies. A case study is conducted in detail to show case a value chain analysis of smart card manufacturing for China's mobile payment industry.

After years of development, the basic environment has been formed for national mobile payment business. Till 2010, the total number of mobile phone users has been over a billion and bank card holders have covered the total population. Because of the rapid increment of user base, the scale of national mobile payment market will be expanded in the near future.

However, the mobile payment business is still at the initial stage for cultivating the user habit. There are two kinds of mobile payment method: remote and near-field. The remote payment has being matured but gradually saturated. Thus, near-field payment business becomes the most prevalent business in the mobile payment field.

Generally, two kinds of mobile near-field payment technical solutions exist in the current market: one is phone-based; the other is SIM card-based. As the representative to the former solution, NFC handset has been widely employed in Japan. In accordance with the China national situation and market mechanism, SIM card-based mobile payment will be the main stream in the near future. Both of two SIM card-based methods, SIMpass and RF-SIM, have their own strong points and in-negligible limitations. Nevertheless, enchanted by the large commercial profit, mobile operators, bank institutions, the 3rd party payment service providers and other groups are all demanding a large slice of the cake. Because of the multi-side participants, different commercial business patterns are produced responding to market needs. To find a proper pattern is being the top priority in the current China market.

To analyze precisely the current situation, we present a mobile payment industry value chain analysis framework, taking into consideration the different participants in the industry chain. The proposed value chain analysis framework accomplishes the character analysis through which the relationship among multi-side institutions is detailed illustrated. By analysis, mobile operators, bank institutions and the 3rd party payment service providers play the principal roles in the mobile payment market. This leads to a situation that several business models begin to take shape in order to cater to market needs. Combining with China's situation, three basic business patterns now are employed side by side. In the chapter, we apply the SWOT–AHP analysis tool to demonstrate the strengths, weakness, opportunities and threats of each business model in view of existing near-field payment market. As a result, the proper approach is brought to the upshot; that is only by mutual cooperation they can gain the maximum benefits. To further analyze the near-field payment solutions, SIMpass and RF-SIM, the relevant factors in each group of SWOT analysis result are all taken into SWOT-AHP framework to be quantified. According to the evaluation, we can get the proper strategies to further develop SIMpass & RF-SIM solutions. Based on the comparison analysis of two solutions, we can make best use of the advantages and bypass the disadvantages.

Manufacturing in recent decades has made amazing progress. The role of advanced manufacturing role as the backbone of a country has been re-recognized and has aroused wide attention of major developed and developing countries. These countries have already started a series of advanced manufacturing technology research programs. Europe and the United States proposed "re-industrialization" in recent years to seize the commanding control of global industrial technology, and to further capitalize on the high-end manufacturing. So is China government, who has planned accordingly and launched a series of major and key projects, carrying out special studies in the frontier of advanced manufacturing technology and equipment.

At present, manufacturing has adopted and leveraged more and more the latest achievements in materials, mechanics, physics, chemistry, and computer simulation technology, network technology, control technology, nanotechnology, biotechnology, and sensor technology. New manufacturing mechanism, manufacturing tools, processes and equipment continue to emerge. Manufacturing as a technology has developed into a new engineering science subject- Manufacturing Science. Cross-regional distribution of manufacturing resources in the era of economic globalization, has forced manufacturing collaboration a daily necessary means. The networked collaborative tools and systems are also increasingly rich. Intelligent Manufacturing has been recognized as the direction of manufacturing technology innovation, the maturity of the related theories and technologies will be one of the signs of the advent of knowledge economy (Luo, 2013).

Currently, with booming economy, China has respected more a science and technology led economic development. Internet/Internet of Things, and next generation of information technology will lead to industry change, promoting the manufacturing industry to shift from the traditional manufacturing towards industrial chain based manufacturing. The trend and requirements of digital manufacturing, service manufacturing, and intelligent manufacturing will become more apparent and prominent. Digital manufacturing will deepen the manufacture of hard power, and service manufacturing is to enhance the manufacturing soft power. Their integration will be the direction of intelligent manufacturing development, in order to achieve the optimal balance of hard and soft power.

REFERENCES

China 863. (2013). Retrieved October 31, 2013, from http://www.863.gov.cn/

China 973. (2013). Retrieved October 31, 2013, from http://www.973.gov.cn

China NSFC. (2013). Retrieved October 31, 2013, from http://www.nsfc.gov.cn

Curtis, C. (2013). *A plan to revitalize American manufacturing.* The White House Blog. Retrieved Oct 31, 2013, from http://www.whitehouse.gov/blog/2013/02/13/plan-revitalize-american-manufacturing

Kurttila, M., Pesonen, M., & Kangas, J. (2000). Utilizing the analytic hierarchy process (AHP) in SWOT analysis – A hybrid method and its application to a forest-certification case. *Forest Policy and Economics, 1*, 41–52. doi:10.1016/S1389-9341(99)00004-0

Luo, Z. (2013). Introduction to mechanism design for sustainability. In Z. Luo (Ed.), *Mechanism design for sustainability: Techniques and cases*. Dordrecht, The Netherlands: Springer. doi:10.1007/978-94-007-5995-4_1

NFC Forum. (2013). Retrieved October 31, 2013, from http://www.nfc-forum.org

RF-SIM. (2013). Retrieved October 31, 2013, from http://www.directel.hk/rfsim.php

Saaty, T. (1980). *The analytic hierarchy process: Planning, priority setting, resource allocation.* New York: McGraw-Hill.

SIMpass. (2013). Retrieved October 31, 2013, from http://www.watchdata.com/telecom/10022.html

Triantaphyllou, E., & Mann, S. H. (1995). Using the analytic hierarchy process for decision making in engineering applications: Some challenges. *International Journal of Industrial Engineering: Applications and Practice, 2*(1), 35–44.

KEY TERMS AND DEFINITIONS

Cloud Computing: A networked computing paradigm where computing resource provisioning is accomplished in a virutalized and transparent way.

Innovation: The application of better solutions that meet new requirements, inarticulate needs, or existing market needs.

Intelligence: Actionable knowledge.

Interconnection: A networked relationship.

Internet of Things: connected uniquely identifiable objects.

RFID: Radio Frequency Identification.

Smart Manufacturing: Advanced manufacturing with wise adoption of information technologies throughout end-to-end product and service life-cycles, capturing manufacturing intelligence for wise production and services.

Strategy Management: An ongoing process that evaluates and controls the businesses and industries in which a company is involved.

Transformation: Making fundamental changes in how industry is structured in order to help cope with a shift in the economic environment.

Section 2
Smart Manufacturing Optimization

Chapter 2
Robust Optimization for Smart Manufacturing Planning and Supply Chain Design in Chemical Industry

Tianxing Cai
Lamar University, USA

ABSTRACT

The depletion of natural resource, the complexity of economic markets and the increased requirement for environment protection have increased the uncertainty of chemical supply and manufacturing. The consequence of short-time material shortage or emergent demand under extreme conditions, may cause local areas to suffer from delayed product deliveries and manufacturing disorder, which will both cause tremendous economic losses. In such urgent events, robust optimization for manufacturing planning and supply chain design in chemical industry, targeting the smart manufacturing, should be a top priority. In this chapter, a novel methodology is developed for robust optimization of manufacturing planning and supply chain design in chemical industry, which includes four stages of work. First, the network of the chemical supply chain needs to be characterized, where the capacity, quantity, and availability of various chemical sources is determined. Second, the initial situation under steady conditions needs to be identified. Then, the optimization is conducted based on a developed MILP (mixed-integer linear programming) model in the third stage. Finally, the sensitivity of the manufacturing and transportation planning with respect to uncertainty parameters is characterized by partitioning the entire space of uncertainty parameters into multiple subspaces. The efficacy of the developed methodology is demonstrated via a case study with in-depth discussions.

DOI: 10.4018/978-1-4666-5836-3.ch002

1. INTRODUCTION

In recent years, the depletion of natural resources, the complexity of economic markets and the increased requirement for environment protection have increased the uncertainty of chemical supply and manufacturing: the natural inventory quantity of mineral material and crude oil will be continuously decreased, which pushes the refinery industry to innovate their current process performance in order to increase the operation capability to handle the raw material with the low quality; the speedy progress of advanced technology development has resulted in the fast revolution of supply-demand relationship in the chemical market, which directly impact the distribution and business operation of chemical supply chain; the global attention for environment protection and sustainable development has definitely formed the pattern of survival of the fittest, which means the process or product will be eliminated if it is harmful to the surrounding environment. The manufacturing planning and supply chain operation are tremendously impacted by these factors in terms of resource limit, economy and environmental sustainability, transportation, as well as their spontaneous characterizations of uncertainty. To restore the functionality and capability of these supply chain and manufacturing systems under the unexpected situation, the recovery of raw material delivery is one of the most important because all the other operations have to be supported by enough available raw materials. On the other hand, if the local energy shortage caused by an uncertain event cannot be effectively restored, the local areas will be at risks of delivery delay, economic losses, and even public issues. Therefore, the recovery and delivery time minimization of a suffered chemical supply chain should be a top priority of a smart manufacturing planning in the chemical industry.

Nowadays chemical supply chain is characterized by its diversity of chemical products and processes. It also involves a complex network system composed of chemical generation, chemical transformation, chemical transportation, and chemical consumption. The network does provide the great flexibility for chemical transformation and transportation; meanwhile, it presents a complex task for conducting agile dispatching when abnormal events have caused local material shortages that need to be restored timely. Conceivably, any type of dispatched chemical material under certain emergency condition has its own characteristics in terms of availability, quantity, transportation speed, and conversion rate and efficiency to other types of chemical manufacturing process. Thus, different types of chemical should be dispatched through a superior plan. For instance, raw material sources such as petroleum or coal can be directly transported to a suffered area; meanwhile, they can also be converted to a typical intermediate in a source region and then sent to the suffered area through an available supply chain. Sometimes, even the transportation of the same type of chemical may have different alternative routes for selection, which needs to be optimally determined from the view point of the entire energy system of chemical supply chain design and manufacturing planning.

Facing the challenges of emergency response to material shortage of chemical supply, decision makers often encounters various uncertainties that inevitably influence the performance of a being designated chemical dispatch plan. The uncertainties can upset the optimality and even the feasibility of the designed plan. Thus, quantitative analysis on the impact of uncertainties is of great significance for the study of robust optimization for smart manufacturing planning and supply chain design in the chemical industry. Technically, a viable approach is to conduct a full evaluation of the effects of uncertainties based on all their possibilities. This will provide decision makers a complete roadmap of the space of uncertainty parameters. Through this way, the objective function and the optimization parameters are represented as functions of uncertainty parameters; meanwhile, the regions in the space of the uncertainties characterized by these functions can be obtained.

Product and process innovations will arise from the creative use of smart manufacturing intelligence gathered from every point of the supply chain, from consumer preferences through production and delivery mechanism. In March 2012, President Obama spoke at a Rolls-Royce jet engine manufacturing plant about the importance of a strong American manufacturing industry that creates good jobs for workers making products that can be sold all over the world. In that speech, he announced new efforts to support manufacturing innovation and the Administration will take immediate steps to launch a pilot institute for manufacturing innovation as part of its We Can't Wait efforts. One of the purposes is aimed to create a smart manufacturing infrastructure and approaches that let operators make real-time use of "big data" flows from fully instrumented plants in order to improve productivity, optimize supply chains and improve energy, water and material use. The new National Network for Manufacturing Innovation will work to leverage new investment from industry, state and local government, and the research community. This initiative will be a collaboration between Commerce's National Institute of Standards and Technology, the National Science Foundation, the Department of Defense, and the Department of Energy (the White House, 2012).

2. APPROACH COMPARISON

From the literature survey, the reported work on smart manufacturing has included the methods of forecasting intermittent demand (Willemain et al., 1994), artificial intelligence (Krakauer, 1987), cloud computing (Xu, 2012), intelligent manufacturing with labels and smart tags (Markham et al., 2011). The contribution includes the application model predictive control strategy for supply chain optimization (Perea-Lopez, Ydstie, & Grossmann, 2003), dynamic model for requirements planning with application to supply chain optimization (Graves, Kletter, & Hetzel, 1998), supply chain optimization of continuous process industries with sustainable considerations (Zhou, Cheng & Hua, 2000), supply chain design and analysis (Beamon, 2009), strategic supply chain optimization for the pharmaceutical industries (Papageorgiou, Rotstein, & Shah, 2001), supply chain optimization in continuous flexible process networks (Bok, Grossmann, & Park, 2000), supply chain optimization in the pulp mill industry (Bredström et al., 2004), multi-objective optimization of multi-echelon supply chain networks with uncertain product demands and prices (Chen & Lee, 2004) and the integrated system solution for supply chain optimization in the chemical process industry (Berning et al., 2002). However, the studied optimization problems target to achieve the objective of maximal profit or handle the modeling under the uncertainty of demand, there is still lack of study on decision marking support for chemical dispatch optimization under emergency situations and especially for the emergency response planning for the uncertainties of material loss. Conceivably, this is not a trivial task, because it involves multiple types of chemicals, different shortage modes (e.g., short of source materials or incapable of transformation or transportation), and various chemical delivery scenarios (e.g., material transportation from one place to another place, 1→1 for short; or possibly 1→n, n→1, and n→n) based on the available infrastructure of chemical supply chain.

In this chapter, a novel methodology is developed for robust optimization for manufacturing planning and supply chain design in chemical industry, which includes four stages of work. First, the network of chemical supply chain needs to be characterized, where the capacity, quantity, and availability of various chemical sources are determined. Second, the initial situation under steady conditions needs to be identified. Then, the optimization is conducted based on a developed MILP (mixed-integer linear programming) model in the third stage. Finally, the sensitivity of the manufacturing and transportation planning with respect to uncertainty parameters is characterized by partitioning the entire space of uncertainty parameters into multiple subspaces. The efficacy of the developed methodology is demonstrated via a case study with in-depth discussions. The developed methodology is the first prototype in the field of robust optimization to handle the abnormal and uncertain situation for smart manufacturing planning and supply chain design in the chemical industry. Although it is based on some simplifications and assumptions, it has built a solid foundation for future in-depth study and real applications.

3. PROBLEM STATEMENT

The study considers a region containing multiple chemical plants involving various raw materials and chemical processes. To support the manufacturing operation in these areas, chemical production and consumption must exist in this region. This can be considered as a chemical distribution and supply chain network, where each chemical plant is considered as a node having its own chemical generation, transformation, consumption, and storage system; meanwhile, multiple chemical transportation infrastructures are available, such as highway, railway and pipelines, which enable the distribution of multiple chemical materials among those nodes. The feasibility of chemical transportation between two nodes is also called connections. Note that in such a supply chain, each node has limited capacities of storage for different types of chemical raw material, chemical generation, and chemical transformation; meanwhile, each connection also has a restrained capability.

Under uncertain situations, both nodes and connections of a given supply chain may suffer from potential abnormal situations. They may be reflected as the breakdown of transportation pipeline, losses of source materials, or failure of chemical transformation equipment. In such a damaged network of supply chain and manufacturing, some nodes may suddenly be in the trouble. To recover the normal operation as soon as possible, the entire region should be motivated to support the suffered areas, i.e., through available infrastructure, dispatching various raw material or chemical product from other areas to the in-need areas timely. Note that different chemical materials have different transportation ways and speeds based on an available infrastructure. To minimize the delivery time in a supply chain, optimal decisions will be made to determine how to select appropriate and smart manufacturing sources and conduct robust chemical supply chain operation, which involves an optimization problem. For clarity, the optimal manufacturing planning and supply chain operation problem can be summarized as below.

Assumptions:
1. At the typical situation, the demand and current inventory for any type of chemical material in each node are predictable.
2. The chemical type does not change during the transportation; chemical can only be transformed to another type at a node;
3. The capacity for different chemical transformation at each node and capability for chemical transportation among different nodes based on available infrastructure are predicable.
4. For any type of chemical, the transportation speed from or to a node is considered as constant.

Given Information:
1. Chemical or raw material demand and inventory for any type of chemical type at each node of supply chain;

2. The capability and efficiency of chemical transformation of manufacturing processing in each node;
3. The connection availability of the supply chain network, which is used to determine the transportation route and speed.

Information to be Determined:
1. The minimum recovery time for network restoration;
2. The supply chain's response strategy, including transformation plan in each node and transportation plan among different nodes;
3. Partition of optimal response strategies regarding the uncertainty of chemical material shortage conditions.

4. METHODOLOGY FRAMEWORK

This methodology has included four stages as shown in Figure 1. In the first stage, the supply chain network for responsible area is characterized. It has including several manufacturing plants which have the capability of manufacturing. The capacity, quantity, availability, and the convertibility of various sources in each node will be determined. The considered energy may include raw materials, intermediate chemicals and final chemical products. In the second stage, under emergency of local chemical material shortage, the initial condition for each node of the studied supply chain will be evaluated and estimated. The initial condition is generally an update of scenarios from the first stage, affected by the emergency events case by case. The shortage information should be collected in this stage.

Based on the estimated initial conditions, in the third stage optimization based on a developed MILP model will be conducted, which will integrate information of input and output of chemical material resource, demand, inventory, and transformation and transportation methods to minimize

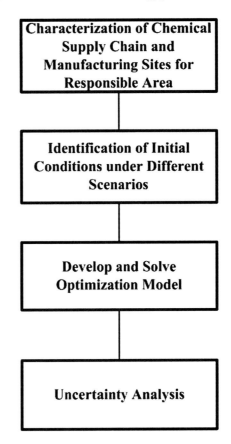

Figure 1. General methodology framework

the recovery time for local restoration. It is a general model, which can cover all the scenarios (e.g., 1→1, 1→n, n→1, or n→n). Available commercial solver, such as CPLEX, can be employed to solve the MILP model and obtain the global optimal solution. Since the optimal solution is obtained based on the shortage evaluations of the supply chain. Uncertainties, such as shortage percentage and supply to the system, have to be addressed during the decision making. They can affect the optimality and even feasibility of the obtained solution. Thus, the sensitivity of minimum restoration time to uncertainty parameters is characterized by partitioning the entire space of uncertainty parameters into multiple subspaces in the fourth stage. For each of the partitioned subspaces, the minimum restoration time with respect to the uncertainty parameters within the subspace has the same representation function. An algorithm has been developed for this purpose. It actually discloses a roadmap of uncertainty impacts to optimal solutions, based on which in-depth analysis can be conducted to support decision making.

5. OPTIMIZATION MODEL

The optimization problem is modeled as an MILP problem, which can be solved by commercial solvers like CPLEX (CPLEX, 2009; GAMS, 2009) to obtain the global optimal solution. All the sets, variables, and parameters employed in the MILP model are explained in both the context and nomenclature parts.

5.1 Objective Function

The objective function of this model is to minimize the restoration time used to satisfy the requirement of every type of chemical material at each node. The discrete time method is used, which divides the concerned time horizon into T small time periods. A binary variable, yt, is employed to indicate if the chemical delivery meets the requirement or not at the time period t. yt is 1 if all the requirement at each node in time period t has been satisfied; otherwise, yt is 0. Note that once yt becomes 1 in time period t, all $y_{t'}$ ($T \geq t' > t$) in the following time periods should also be 1. Therefore, the minimum recovery time can be modeled by Equation (1).

$$\min J = T - \sum_{t \in ST} y_t \qquad (1)$$

where time period set $ST = \{1, 2, ..., T\}$.

5.2 Material Balance at Each Node

Material balance should be met at each node, which means the chemical accumulation is equal to direct chemical input from other chemical nodes, plus converted chemical from other types of chemical at this node, and chemical supply from outside of the studied supply chain system; minus direct chemical output from this node, consumed chemical for converting to other types of chemical, and the chemical consumption demand at this node.

$$CV_{i,k,t} = CV_{i,k,t-1} + \sum_{j \in N, j \neq i} \left(CI_{j,i,k,t} - CO_{i,j,k,t} \right)$$
$$+ \sum_{l \in M, l \neq k} \left(CG_{i,l,k,t} - CC_{i,k,l,t} \right) + CS_{i,k,t} - CD_{i,k,t},$$
$$\forall i, k, t \qquad (2)$$

where $CV_{i,k,t}$ and $CV_{i,k,t-1}$ are the inventory quantity of type k chemical at node i at time t and t-1, respectively. Obviously, the difference ($CV_{i,k,t} - CV_{i,k,t-1}$) means the accumulation of type k chemical at node i; $CI_{j,i,k,t}$ represents the input quantity of type k chemical from node j to node i at time t; $CO_{i,j,k,t}$ represents the output quantity of type k chemical from node i to node j at time t; $CG_{i,l,k,t}$ represents the generated amount of type k chemical from the other chemical types at node i at time t; $CC_{i,k,l,t}$ represents the consumed amount of type k chemical for converting to other chemical types at node i at time t; $CS_{i,k,t}$ represents the chemical supply of type k chemical from outside of the studied supply chain to node i at time t; $CD_{i,k,t}$ represents the demand consumption of type k chemical at node i at time t.

5.3 Chemical Conversion and Transportation Constraints

The chemical conversion and transportation items in Equation (2) have special constraints that should be followed. Equation (3) shows that the generated chemical quantity is the product of available chemical quantity and conversion efficiency at node i at time t. Certainly, available chemical quantity $XG_{i,l,k,t}$ should not be over the inventory of $CV_{i,l,t-1}$ as shown in Equation (4); meanwhile, the chemical generation cannot be above its capacity limit $CG^u_{i,l,k}$, which is expressed by Equation (5).

$$CG_{i,l,k,t} = XG_{i,l,k,t}\, \eta_{i,l,k}, \quad \forall\, i, l, k, t \qquad (3)$$

$$XG_{i,l,k,t} \leq CV_{i,l,t-1}, \quad \forall\, i, l, k, t \qquad (4)$$

$$CG_{i,l,k,t} \leq CG^u_{i,l,k}, \quad \forall\, i, l, k, t \qquad (5)$$

where $XG_{i,l,k,t}$ represents the available energy from $CV_{i,l,t-1}$ that can be converted to generate type k chemical at node i at time t; $\eta_{i,l,k}$ is the associated chemical conversion efficiency.

Equation (6) shows another feature of chemical transportation: i.e., for a transportation from node i to node j, the input chemical quantity to node j should be equal to the output chemical quantity from node i, and there is a time delay between the departure and the arrival of such a transportation, denoted by $\tau_{i,j,k}$, which is mainly determined by the distance between nodes i to j, and the traffic conditions at that time. It can be roughly estimated and fixed during the modeling of this paper.

$$CI_{i,j,k,t} = CO_{i,j,k,t-\tau_{i,j,k}}, \quad \forall\, i, j, k, t \qquad (6)$$

5.4 Chemical Inventory, Consumption, Input, and Output Constraints

Chemical inventory at each node should not be beyond its limit, which is given by Equation (7).

$$CV_{i,k,t} \leq CV^u_{i,k}, \quad \forall\, i, k, t \qquad (7)$$

where $CV^u_{i,k}$ represents the upper limit for the chemical inventory quantity of chemical type k at node i. Based on the chemical inventory, the chemical consumption at each node should be restricted as:

$$CD_{i,k,t} = \alpha\, CV_{i,k,t-1}, \quad \forall\, i, k, t \qquad (8)$$

where α is a given chemical consumption factor. Equation (8) shows that the amount of demand chemical for consumption is a proportion of the chemical inventory quantity from the last time slot.

Let $CI^u_{j,i,k}$ and $TCI^u_{i,k}$ respectively represent the upper limit of $CI_{j,i,k,t}$ and its summation with respect to j. Then, Equations (9) and (10) suggest that each type of chemical fed to i should satisfy not only the limit of each individual $CI^u_{j,i,k}$, but also the total chemical input limit from all the other nodes.

$$CI_{j,i,k,t} \leq CI^u_{j,i,k}, \quad \forall\, i, j, k, t \qquad (9)$$

$$\sum_{j \in N,\, j \neq i} CI_{j,i,k,t} \leq TCI^u_{i,k}, \quad \forall\, i, k, t \qquad (10)$$

Similarly, Let $CO^u_{i,j,k}$ and $TCO^u_{i,k}$ respectively represent the upper limit of $CO_{i,j,k,t}$ and its summation with respect to j. Equations (11) and (12) suggest that each type of chemical output from i should satisfy not only the limit of each individual $CO^u_{i,j,k}$, but also the total chemical output limit directed to all the other nodes.

$$CO_{i,j,k,t} \leq CO^u_{i,j,k}, \quad \forall i, j, k, t \qquad (11)$$

$$\sum_{j \in N, j \neq i} CO_{i,j,k,t} \leq TCO^u_{i,k}, \quad \forall i, k, t \qquad (12)$$

5.5 Logic Constraints

Let $x_{i,k,t}$ be a binary variable to represent whether the supply of chemical type k at node i at time t meets the requirement or not: i.e., $x_{i,k,t}$ is 1 if it is satisfied; otherwise, $x_{i,k,t}$ is 0. Thus, Equation (13) shows the logic relation, where $RC_{i,k}$ is the required amount of chemical type k at node i; U is a sufficient big number.

$$-x_{i,k,t} U \leq RC_{i,k} - CV_{i,k,t} < (1 - x_{i,k,t}) U ,$$
$$\forall i, k, t \qquad (13)$$

As aforementioned, yt is the binary variable indicating if the overall chemical dispatch is accomplished at time t. Once yt becomes 1 in time period t, all $y_{t'}$ ($T \geq t' > t$) in the following time periods should also be 1. Therefore, Equations (14) and (15) must be fulfilled.

$$y_t |N||M| \leq \sum_{i \in N} \sum_{k \in M} x_{i,k,t} < (y_t + 1)|N||M|, \forall t \qquad (14)$$

$$y_t \geq y_{t-1}, \quad \forall t \geq 1 \qquad (15)$$

where $|N|$ and $|M|$ are the length of sets N and M (N is the set of all nodes and M is the set of all chemical types), respectively. Equation (14) shows that if yt is 1, $\sum_{i \in N} \sum_{k \in M} x_{i,k,t}$ could only be equal to the product of $|N|$ and $|M|$; otherwise, if yt is 0, $\sum_{i \in N} \sum_{k \in M} x_{i,k,t}$ will be rigorously less than $|N||M|$. Equation (15) indicates the logic that yt is a non-decreasing variable with respect to t.

6. CASE STUDY

To demonstrate the efficacy of the developed methodology, a case study has been conducted and discussed. In an chemical supply chain composed of three chemical plants (P, Q, and R), the infrastructure of chemical plant P got partially damaged and is currently experiencing chemical material shortage. The other two chemical plants (Q and R) are not affected by the event and thus are able to provide chemical materials to support chemical plant P. By following the first two stages of the proposed methodology, the initial conditions of this case study are summarized in Table 1, which gives all the parameters used in the optimization model.

During the regional shortage situation, assume all types of chemical suffer 10% loss in chemical plant P, meanwhile there is no chemical loss damage to the other cities. Note that the optimization is performed to deal with an MILP problem. Because the number of variables and constraints involved is not large, the global solution optimality can be guaranteed by employing the commercial solvers such as CPLEX used in this study. Based on the developed model with GAMS [15-16], the optimization result shows that the minimum recovery time for this chemical restoration of chemical plant P is 5 hours. The dynamic profiles for the restoration within the network are presented in Figures 2 through 10.

Figures 2 through 5 show the dynamic inventory (represented by bar chart) and consumption (represented by trend curves) of chemical type A (see Figure 2), chemical type B (Figure 3), chemical type C (Figure 4), and chemical type D

Table 1. Given data of the case study

Description	Parameter	Chemical Source	Unit	i/j P/Q or R	i/j Q/R or P	i/j R/P or Q
Chemical Conversion Yield from Other Chemical Sources (k: D)	$\eta_{i,l,k}$	l: A	kg/kg	0.45	0.36	0.54
		l: B	kg/kg	0.32	0.25	0.20
		l: C	kg/kg	0.80	0.70	0.60
Hourly Chemical A Generation Limit from Source Materials (k: D)	$CG^u_{i,l,k}$	l: A	kg/hr	250	230	280
		l: B	kg/hr	900	600	800
		l: C	kg/hr	80	90	100
Hourly Limit of Transportation Input from Plant j to Plant i	$CI^u_{j,i,k}$	k: A	kg/hr	20	20	20
		k: B	kg/hr	8	8	8
		k: C	kg/hr	40	40	40
		k: D	kg/hr	30	30	30
Hourly Limit of Transportation Output from Plant i to Plant j	$CO^u_{i,j,k}$	k: A	kg/hr	20	20	20
		k: B	kg/hr	8	8	8
		k: C	kg/hr	40	40	40
		k: D	kg/hr	30	30	30
Required Chemical Inventory Amount for Plant i	$RC_{i,k}$	k: A	kg	300	220	300
		k: B	kg	400	600	500
		k: C	kg	400	240	300
		k: D	kg	800	600	700
Hourly Chemical Supply for Plant i	$CS_{i,k,t}$	k: A	kg/hr	55	45	35
		k: B	kg/hr	250	350	450
		k: C	kg/hr	45	35	45
		k: D	kg/hr	0	0	0

(Figure 5) in different chemical plants during the period. Figures 2 through 4 indicate that the chemical sources of A, B, and C in chemical plant P have reached the required quantity within 5 hours.

From Figures 2 through 5, chemical quantity profiles in chemical plant Q and R only experience small upsets during the restoration period; while those of chemical plant P change a lot because it suffers from shortage. Especially from Figures 3 through 5, the source consumption of chemical A, B and C for chemical D generation in chemical plant P changes dramatically. This is because plant P needs to use the most efficient and available sources to generate product, so as to accomplish restoration as soon as possible.

To disclose more details of chemical transportation between each pair of plants in the supply chain, Figures 6 through 9 show accumulative transportation amounts of chemical A, chemical

Figure 2. Dynamic profiles of chemical A inventory and consumption in different plants

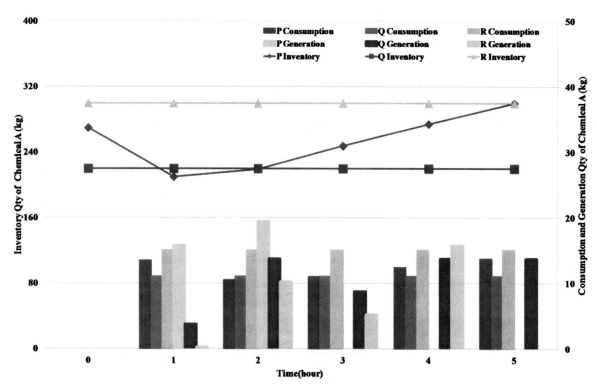

Figure 3. Dynamic profiles of chemical B inventory and consumption in different plants

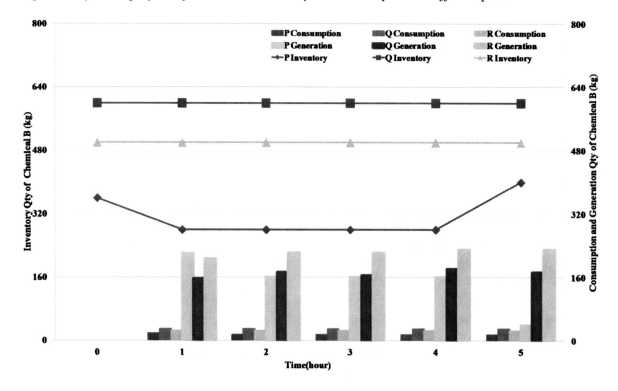

Figure 4. Dynamic profiles of chemical C inventory and consumption in different plants

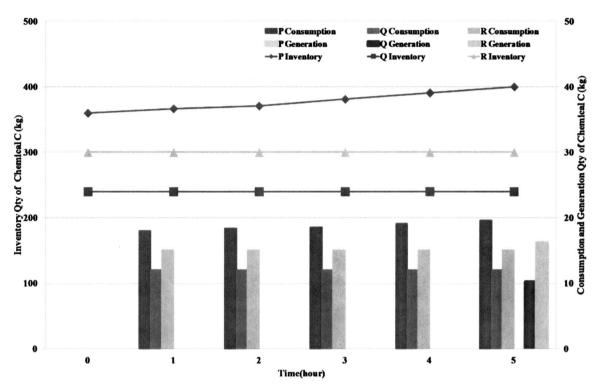

Figure 5. Dynamic profiles of chemical D inventory and consumption in different plants

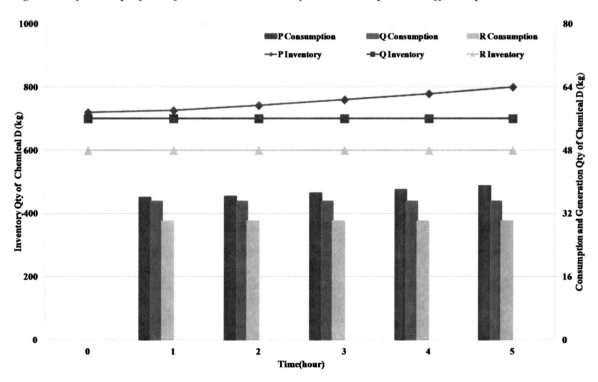

Figure 6. Accumulative transportation amount of chemical A between plants

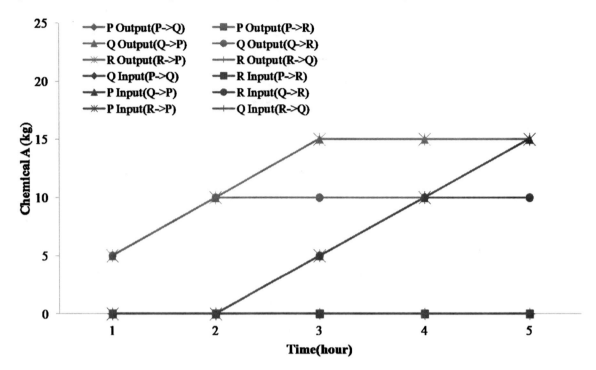

Figure 7. Accumulative transportation amount of chemical B between plants

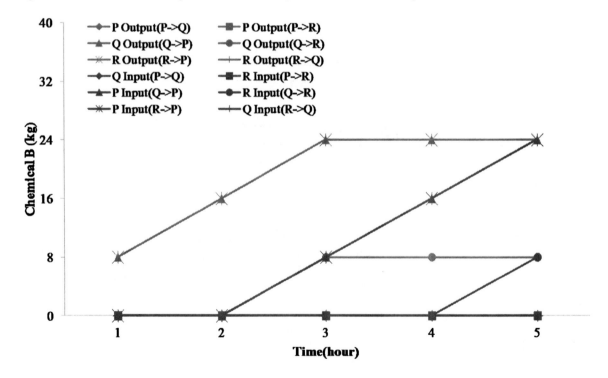

Figure 8. Accumulative transportation amount of chemical C between plants

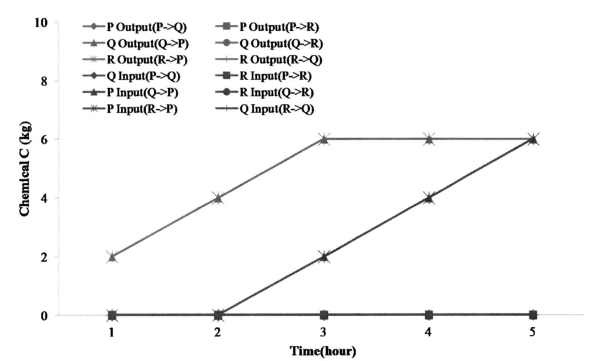

Figure 9. Accumulative transportation amount of chemical D between plants

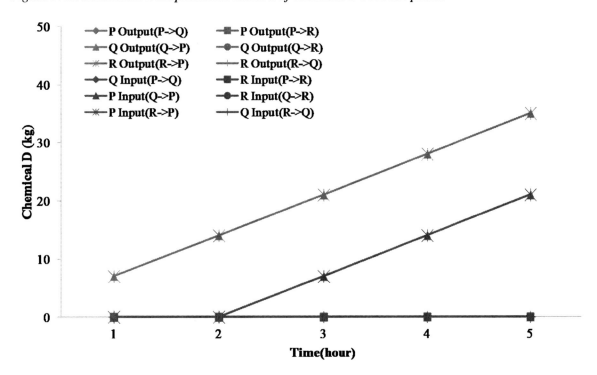

Figure 10. Minimum restoration time with respect to uncertainties

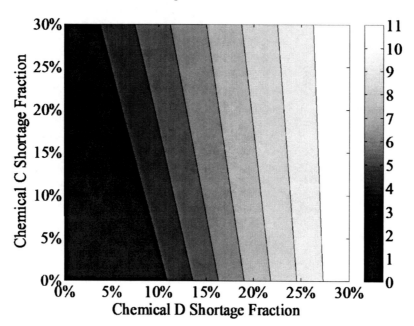

B, chemical C and chemical D, respectively. As time elapses, the accumulative transportation amount is always non-decreasing. It shows that the accumulative transportation amounts of various chemical resources from plant P to plant Q and from plant P to plant R keep zero until the inventory quantity in plant P has been satisfied; while the transportation amounts from plant Q and R to P are most significant, although the mutual transportations between Q and R are also nontrivial.

Because the source materials dispatched from one plant to another will take two hours to reach the destination, the curves in Figures 6 through 9 (respectively addressing chemical A, chemical B, chemical C and chemical D) clearly show such a transportation time delay. They also show that there are no chemical material output from plant P because of its status of shortage. Plant Q and R are dispatching the chemical material with the highest loading to plant P, which is consistent with the situation that the hourly transportation amounts from both plants Q and R to P are in their full capacities until the near end of the time period. Therefore, both accumulative transportation increments are changing linearly in a wide range in the plots.. After the 5th hour when plant P has restored the chemical material quantity, the chemical transportation activity from the other two cities reduces.

In this case study, two parameters, the material loss for chemical C and chemical D in plant P, are estimated before the running of optimization model. They are actually typical uncertainties that should be investigated to disclose their impact to the final optimization results. Assume they can vary within the range from 0 (no loss) to 0.3 (30% loss). Then a two-dimensional uncertainty space has been constructed, where the first uncertainty represents the loss fraction of chemical C and the second uncertainty represents the loss fraction of chemical D.

The 2D projection on the uncertainty space has been shown in Figure 10 for any emergency case specified by chemical C and chemical D shortage for plant P corresponds to a point in the plane. It helps to characterize the relation between the optimal restoration time and the studied two uncertain parameters. They provide decision makers a reference manual for the easy lookup of the possible impact from uncertainties and thus help conduct right decisions. For example, the 2D figure generally shows the less chemical C and chemical D shortages will result in less recovery time. The transportation behaviors among the supply chain network are actually moving to finally reach the left-bottom corner point. It should be known that any movement in a short time period can be decomposed into two directions: one is toward the iso-time direction that does not reduce recovery time at all; the other is gradient direction perpendicular to the iso-time direction, which represents the shortest recovery time and smart move direction. Based on such information, the dynamic impact of different smart manufacturing planning supply chain operation strategies can be evaluated, which will help decision makings under uncertainties.

7. CONCLUSION

In this chapter, a new methodology is developed robust optimization for smart manufacturing planning and supply chain operation in chemical industry. To minimize the chemical restoration time for the entire chemical manufacturing network, an MILP based model has been developed. The sensitivity of minimum restoration time to uncertainty parameters is also studied by partitioning the entire space of uncertainty parameters into multiple subspaces. The developed methodology is the first prototype in the area of risk assessment based chemical transportation optimization in order to achieve smart manufacturing planning and supply chain operation, although it is based on some simplifications and assumptions. The efficacy of the development is demonstrated through a virtual case study with in-depth discussions. It will help to handle the uncertainty of chemical supply and manufacturing due to the depletion of natural resource, the complexity of economic markets and the increased requirement for environment protection have enhanced, especially for the consequence of short-time material shortage under extreme conditions.

REFERENCES

Beamon, B. M. (1998). Supply chain design and analysis: Models and methods. *International Journal of Production Economics*, 55(3), 281–294. doi:10.1016/S0925-5273(98)00079-6

Berning, G., Brandenburg, M., Gürsoy, K., Mehta, V., & Tölle, F. J. (2002). An integrated system solution for supply chain optimization in the chemical process industry. *OR-Spektrum*, 24(4), 371–401. doi:10.1007/s00291-002-0104-4

Bok, J. K., Grossmann, I. E., & Park, S. (2000). Supply chain optimization in continuous flexible process networks. *Industrial & Engineering Chemistry Research*, 39(5), 1279–1290. doi:10.1021/ie990526w

Bredström, D., Lundgren, J. T., Rönnqvist, M., Carlsson, D., & Mason, A. (2004). Supply chain optimization in the pulp mill industry—IP models, column generation and novel constraint branches. *European Journal of Operational Research*, 156(1), 2–22. doi:10.1016/j.ejor.2003.08.001

Chen, C. L., & Lee, W. C. (2004). Multi-objective optimization of multi-echelon supply chain networks with uncertain product demands and prices. *Computers & Chemical Engineering*, 28(6), 1131–1144. doi:10.1016/j.compchemeng.2003.09.014

CPLEX. (2009). *Using the CPLEX callable library*. Incline Village, NV: CPLEX Optimization, Inc.

GAMS. (2009). *GAMS- A user's guide*. Washington, DC: GAMS Development Corporation.

Graves, S. C., Kletter, D. B., & Hetzel, W. B. (1998). A dynamic model for requirements planning with application to supply chain optimization. *Operations Research, 46*(3-Supplement-3), S35-S49.

Krakauer, J. (1987). *Smart manufacturing with AI*. Dearborn, MI: Society of Manufacturing Engineers.

Markham, C. E., Barber, D. G. B., Hise, J. H., Ihde, S. A., Lindsay, J. D., Nygaard, K. S., & Yosten, R. D. (2011). *U.S. patent no. 7,882,438*. Washington, DC: U.S. Patent and Trademark Office.

Papageorgiou, L. G., Rotstein, G. E., & Shah, N. (2001). Strategic supply chain optimization for the pharmaceutical industries. *Industrial & Engineering Chemistry Research, 40*(1), 275–286. doi:10.1021/ie990870t

Perea-Lopez, E., Ydstie, B. E., & Grossmann, I. E. (2003). A model predictive control strategy for supply chain optimization. *Computers & Chemical Engineering, 27*(8), 1201–1218. doi:10.1016/S0098-1354(03)00047-4

The White House. (2012). *President Obama speaks on manufacturing*. Retrieved from http://www.whitehouse.gov/photos-and-video/video/2012/03/09/president-obama-speaks-manufacturing#transcript

Willemain, T. R., Smart, C. N., Shockor, J. H., & DeSautels, P. A. (1994). Forecasting intermittent demand in manufacturing: A comparative evaluation of Croston's method. *International Journal of Forecasting, 10*(4), 529–538. doi:10.1016/0169-2070(94)90021-3

Xu, X. (2012). From cloud computing to cloud manufacturing. *Robotics and Computer-integrated Manufacturing, 28*(1), 75–86. doi:10.1016/j.rcim.2011.07.002

Zhou, Z., Cheng, S., & Hua, B. (2000). Supply chain optimization of continuous process industries with sustainability considerations. *Computers & Chemical Engineering, 24*(2), 1151–1158. doi:10.1016/S0098-1354(00)00496-8

KEY TERMS AND DEFINITIONS

Index, Set, and Space:

$i : j$ chemical plant node index

l, k chemical type indexes.

t time period index.

M set of chemical types.

N set of chemical plant nodes in the chemical manufacturing network.

$ST = \{t | t = 1, ..., T\}$ set of time periods with the last time period of T.

$CR(\theta)$ uncertainty space.

$CR(\theta^C)$ uncertainty subspace characterized by corner points of θ^C.

Parameters:

α a given chemical consumption factor.

$CI^u_{j,i,k}$ upper limit of $CI_{j,i,k,t}$.

$CO^u_{i,j,k}$ upper limit of $CO_{i,j,k,t}$.

$CV^u_{i,k}$ upper limit of $CV_{i,k,t}$.

$RC_{i,k}$ required chemical inventory amount for chemical k at node i.

SC threshold of the relative error for model regression.

T last time period for the considered time horizon.

$TCI^u_{i,k}$ total chemical k input limit from all the other nodes to node i.

$TCO^u_{i,k}$ total chemical k output limit from node i to all the other nodes.

U a sufficient big number for big M method in logic constraint equations.

Z^C regression model prediction results based on corner points of θ^C.

$\tau_{i,j,k}$ chemical k transportation delay time from nodes i to j.
θ uncertainties.
θ^L, θ^U lower and upper bounds of θ
$\eta_{i,l,k}$ chemical conversion efficiency from chemical l to chemical k at node i.

Variables:

$CD_{i,k,t}$ demanded chemical k in node i at time t.

$CC_{i,k,l,t}$ consumed chemical k for converting to other types of chemical in node i at time t.

$CG_{i,l,k,t}$ generated chemical k from chemical l at node i at time t.

$CI_{j,i,k,t}$ input chemical k from node j to node i at time t.

$CO_{i,j,k,t}$ output chemical k from node i to node j at time t.

$CV_{i,k,t}$ chemical k inventory at node i at time t.

$CS_{i,k,t}$ chemical k supply quantity for node i at time t.

$XG_{i,l,k,t}$ quantity of chemical l used to generate chemical k at node i at time t.

$x_{i,k,t}$ **binary variable:** if the supply of chemical k in node i at time t satisfies the requirement, then $x_{i,k,t}$ is 1; otherwise, $x_{i,k,t}$ is 0.

y_t **binary variable:** if all types of chemical supplies of all the chemical nodes are satisfied starting from time period t, y_t is 1; otherwise, y_t is 0.

Chapter 3
Meta-Heuristic Structure for Multiobjective Optimization Case Study:
Green Sand Mould System

T. Ganesan
Universiti Technologi PETRONAS, Malaysia

I. Elamvazuthi
Universiti Technologi PETRONAS, Malaysia

K. Z. KuShaari
Universiti Technologi PETRONAS, Malaysia

P. Vasant
Universiti Technologi PETRONAS, Malaysia

ABSTRACT

In engineering optimization, one often encounters scenarios that are multiobjective (MO) where each of the objectives covers different aspects of the problem. It is hence critical for the engineer to have multiple solution choices before selecting of the best solution. In this chapter, an approach that merges meta-heuristic algorithms with the weighted sum method is introduced. Analysis on the solution set produced by these algorithms is carried out using performance metrics. By these procedures, a novel chaos-based metaheuristic algorithm, the Chaotic Particle Swarm (Ch-PSO) is developed. This method is then used generate highly diverse and optimal solutions to the green sand mould system which is a real-world problem. Some comparative analyses are then carried out with the algorithms developed and employed in this work. Analysis on the performance as well as the quality of the solutions produced by the algorithms is presented in this chapter.

DOI: 10.4018/978-1-4666-5836-3.ch003

INTRODUCTION

Currently, a series of issues have been raised when dealing with systems involving multiobjective optimization (MO) in engineering (Eschenauer et al., 1990; Statnikov & Matusov, 1995). In industrial engineering, scenarios which are multiobjective in nature are often encountered (where each of the objectives portrays different aspects of the problem). It is essential for the engineer to have access to numerous solution choices before selecting of the best solution. Some known strategies in MO optimization are Strength Pareto Evolutionary Algorithm (SPEA) (Zitzler & Thiele, 1998), Non-dominated Sorting Genetic Algorithm II (NSGA-II) by Deb et al. (2002), Weighted Sum method (Fishburn, 1967; Triantaphyllou, 2000), Goal Programming (Luyben & Floudas, 1994) and Normal-Boundary Intersection method (NBI) (Das & Dennis, 1998). These methods use concepts of Pareto-optimality to trace the non-dominated solutions at the Pareto curve and objective functions aggregation (scalarization) to solve these MO problems.

In green sand mould systems, the quality of the product obtained from the moulding process is very dependent on the physical properties of the moulding sand (such as; hardness, permeability, green compression strength and bulk density). Faulty extent of the mentioned properties may result in casting defects such as; poor surface finish, blowholes, scabs, pinhole porosity, etc. Controllable variables such as; percentage of water, percentage of clay, grain fineness number and number of strokes heavily influence the physical properties of the moulded sand. Therefore, by classifying these parameters as the decision variables and the mould sand properties as the objective function, the MO optimization problem was formulated in Surekha et al. (2011). The purpose of this formulation is for the determination of the best controllable parameters for optimal final-product of the moulding process. A more rigorous study on the optimization and model development of mould systems can be seen in Sushil et al. (2010) and Rosenberg (1967).

This chapter aims to discuss the implementation of meta-heuristic techniques for the production of a solution set that dominantly approximate the Pareto frontier in the objective space. These techniques applied in conjunction with the weighted-sum method improve the dominance of the solution set during successive iterations. In addition, the characteristics (convergence, diversity) of the solutions with regard to their level of dominance were also measured using performance metrics. A suitable real-world case study for this framework is the green sand mould systems.

This chapter is organized as follows; Section 2 presents the metaheuristic algorithms employed while Section 3 provides details on the Chaotic PSO algorithm developed in this work. Section 4 discusses the measurement metrics followed by Section 5 which presents the problem formulation. The computational results are analyzed in Section 6 and the chapter ends with the Concluding Remarks in Section 7.

METAHEURISTIC TECHNIQUES

1. Genetic Algorithm (GA)

A genetic algorithm (GA) was hybridized with the NBI approach for the multiobjective optimization of green sand mould system. GAs is a type of population-based evolutionary search and optimization algorithm (Parappagoudar et al., 2007). An N-point crossover type was used in the generation of new offspring. To avoid the stagnation of the algorithm at the local minima, an N-bit flip mutation operator was employed. The GA scheme used in this work is as the following and the parameter settings of the GA used in this work are as in Algorithm 1 and Table 1 respectively.

Table 1. Genetic algorithm (GA) settings

Parameters	Values
Length of individual string	6 bit
No. of individuals in the population	6
Probability of mutation	0.3333
Probability of recombination	0.5
Initial string of individuals	Random
Bit type of individual's string	Real-coded
Cross-over type	N-point
Mutation type	N-bit flip
Selection type	Tournament

Algorithm 1. Genetic Algorithm

Step 1: Randomly initialize a population of n individuals.
Step 2: Fitness criterions are assigned to each of the individuals.
Step 3: Generate offspring by recombination from the current population.
Step 4: Mutate offspring.
Step 5: Perform parent selection (tournament selection).
Step 6: A new population of n individuals is selected.
Step 7: Set new population = current population.
Step 8: Evaluate fitness of offspring.
Step 9: If the termination conditions are satisfied halt and print solutions, else go to step 3.

2. Differential Evolution (DE)

DE is a class of evolutionary meta-heuristic algorithms first introduced by Storn & Price (1995). This core idea of this technique is the incorporation of perturbative methods into evolutionary algorithms. DE starts by the initialization of a population of at least four individuals denoted as P. These individuals are real-coded vectors with some size N. The initial population of individual vectors (the first generation denoted $gen = 1$) are randomly generated in appropriate search ranges. One principal parent denoted x^p_i and three auxiliary parents denoted x^a_i is randomly selected from the population, P. In DE, every individual in the population, P I would become a principle parent, x^p_i at one generation or the other and thus have a chance in mating with the auxiliary parents, x^a_i. The three auxiliary parents then engage in 'differential mutation' to generate a mutated vector, V_i.

$$V_i = x^a_1 + F(x^a_2 - x^a_3) \quad (1)$$

where F is the real-valued mutation amplification factor which is usually between 0 and 1. Next V_i is then recombined (or exponentially crossed-over) with x^p_i to generate child trial vector, x^{child}_i. The probability of the cross-over, CR is an input parameter set by the user. In DE, the survival selection mechanism into the next generation is called 'knock-out competition'. This is defined as the direct competition between the principle parent, x^p_i and the child trial vector, x^{child}_i to select the survivor of the next generation as follows:

$$x_i(gen+1) = \begin{cases} x^{child}_i(gen) \leftrightarrow f(x^{child}_i) \text{ better than } f(x^p_i) \\ x^p_i(gen) \leftrightarrow \text{ otherwise} \end{cases} \quad (2)$$

Therefore, the knock-out competition mechanism also serves as the fitness evaluation scheme for the DE algorithm. The parameter setting for the DE algorithm is given in Table 2. The algorithm of the DE method is shown in Algorithm 2.

Table 2. Differential Evolution (DE) Parameter Settings

Parameters	Values
Individual Size, N	6
Population Size, P	7
Mutation amplification factor, F	0.3
Cross-over Probability, CR	0.667

Algorithm 2: Differential Evolution (DE)
Step 1: Initialize individual size N, P, CR and F.
Step 2: Randomly initialize the population vectors, $x^G_{i.}$
Step 3: Randomly select one principal parents, $x^p_{i.}$
Step 4: Randomly select three auxiliary parents, $x^a_{i.}$
Step 5: Perform differential mutation & generate mutated vector, $V_{i.}$
Step 6: Recombine V_i with x^p_i to generate child trial vector, $x^{child}_{i.}$
Step 7: Perform 'knock-out' competition for next generation survival selection.
Step 8: If the fitness criterion is satisfied and $t= T_{max}$, halt and print solutions, else proceed to step 3.

3. Gravitational Search Algorithm (GSA)

The GSA algorithm is a meta-heuristic algorithm first developed in 2009 by Rashedi et al. (2009). This technique was inspired by the law of gravity and the idea of interaction of masses. This algorithm uses the Newtonian gravitational laws where the search agents are the associated masses. Thus, the gravitational forces influence the motion of these masses, where lighter masses gravitate towards the heavier masses (which signify good solutions) during these interactions. The gravitational force hence acts as the communication mechanism for the masses (analogous to 'pheromone deposition' for ant agents in ACO (Colorni et al., 1991) and the 'social component' for the particle agents in PSO (Kennedy & Eberhart, 1995). The position of the masses correlates to the solution space in the search domain while the masses characterize the fitness space. As the iterations increase, and gravitational interactions occur, it is expected that the masses would conglomerate at its fittest position and provide an optimal solution to the problem. Initially the GSA algorithm randomly generates a distribution of masses, $m_i(t)$ (search agents) and also sets an initial position for these masses, x_i^d. For a minimization problem, the least fit mass, $m_i^{worst}(t)$ and the fittest mass, $m_i^{best}(t)$ at time t are calculated as follows:

$$m^{best}(t) = \min_{j \in [1,N]} m_j(t) \quad (3)$$

$$m^{worst}(t) = \max_{j \in [1,N]} m_j(t) \quad (4)$$

For a maximization problem, it's simply vice versa. The inertial mass, $m'_i(t)$ and gravitational masses, $M_i(t)$ are then computed based on the fitness map developed previously.

$$m'_i(t) = \frac{m_i(t) - m^{worst}(t)}{m^{best}(t) - m^{worst}(t)} \quad (5)$$

$$M_i(t) = \frac{m_i(t)}{\sum_{j=1}^{N} m_j(t)} \quad (6)$$

such that,

$$M_{ai} = M_{pi} = M_{ii} = M_i : i \in [1, N] \quad (7)$$

Then the gravitational constant, $G(t+1)$ and the Euclidean distance $R_{ij}(t)$ is computed as the following:

$$G(t+1) = G(t) \exp\left(\frac{-\alpha t}{T_{max}}\right) \quad (8)$$

$$R_{ij}(t) = \sqrt{(x_i(t))^2 - (x_j(t))^2} \quad (9)$$

where α is some arbitrary constant and T_{max} is the maximum number of iterations, $x_i(t)$ and $x_j(t)$

are the positions of particle i and j at time t. The interaction forces at time t, $F_{ij}^d(t)$ for each of the masses are then computed:

$$F_{ij}^d(t) = G(t)\left(\frac{M_{pi}(t) \times M_{aj}(t)}{R_{ij}(t) + \varepsilon}\right) \times \left(x_j^d(t) - x_i^d(t)\right) \quad (10)$$

where ε is some small parameter. The total force acting on each mass i is given in a stochastic form as the following:

$$F_i^d(t) = \sum_{\substack{j=1 \\ i \neq j}}^{N} rand(w_j) F_{ij}^d(t) : rand(w_j) \in [0,1] \quad (11)$$

where $rand(w_j)$ is a randomly assigned weight. Consequently, the acceleration of each of the masses, $a_i^d(t)$ is then as follows:

$$a_i^d(t) = \left(\frac{F_i^d(t)}{M_{ii}(t)}\right) \quad (12)$$

After the computation of the particle acceleration, the particle position and velocity is then calculated:

$$v_i^d(t+1) = rand(w_j) + v_i^d(t) + a_i^d(t) \quad (13)$$

$$x_i^d(t+1) = x_i^d(t(t)) + v_i^d(t(t+1)) \quad (14)$$

where $rand(w_j)$ is a randomly assigned weight. The iterations are then continued until the all mass agents are at their fittest positions in the fitness landscape and some stopping criterion which is set by the user is met. The GSA algorithm is presented in Algorithm 3 and the parameter settings are given in Table 3.

Table 3. Gravitational search algorithm (GSA) settings

Parameters	Values
Initial parameter (G_o)	100
Number of mass agents, n	6
Constant parameter, α	20
Constant parameter, ε	0.01

Algorithm 3: Gravitational Search Algorithm (GSA)

Step 1: Initialize no of particles, m_i and initial positions, $x_i(0)$.
Step 2: Initialize algorithm parameters $G(0)$, α.
Step 3: Compute gravitational & inertial masses based on the fitness map.
Step 4: Compute the gravitational constant, $G(t)$.
Step 5: Compute distance between agents, $R_{ij}(t)$.
Step 6: Compute total force, $F_i^d(t)$ and the acceleration $a_i^d(t)$ of each agent.
Step 7: Compute new velocity $v_i(t)$ and position $x_i(t)$ for each agent.
Step 8: If the fitness criterion is satisfied and $t = T_{max}$, halt and print solutions else proceed to step 3.

4. Particle Swarm Optimization (PSO)

The PSO algorithm was initially developed in 1995 by Kennedy and Eberhart (1995). This technique originates from two different ideas. The first idea was based on the observation of swarming or flocking habits of certain types of animals (for instance; birds, bees and ants). The second concept was mainly related to the study of evolutionary computation. The PSO algorithm works by searching the search space for candidate solutions and evaluating them to some fitness function with respect to the associated criterion. The candidate solutions are analogous to particles in motion (swarming) through the fitness landscape in search for the optimal solution. Initially the PSO algorithm chooses some candidate solutions (candidate solutions can be randomly chosen or be set with some *a priori* knowledge). Then each particle's position and velocity (candidate solutions) are evaluated against the fitness function. Then if the fitness function is not satisfied, then update the individual and social component with some update rule. Next update the velocity and the position of the particles. This procedure is repeated iteratively until the all candidate solutions satisfy the fitness function and thus converges into a fix position. It is important to note that the velocity and position updating rule is crucial in terms of the optimization capabilities of the PSO algorithm. The velocity of each particle in motion (swarming) is updated using the following equation.

$$v_i(t+1) = wv_i(t) + c_1 r_1 [\hat{x}_i(t) - x_i(t)] + c_2 r_2 [g(t) - x_i(t)] \quad (15)$$

where each particle is identified by the index i, $v_i(t)$ is the particle velocity and $x_i(t)$ is the particle position with respect to iteration (t). The parameters w, c_1, $c2$, r_1 and r_2 are usually defined by the user. These parameters are typically constrained by the following inequalities:

$$w \in [0,1.2], c_1 \in [0,2], c_2 \in [0,2], r_1 \in [0,1], r_2 \in [0,1]. \quad (16)$$

The term $wv_i(t)$ in equation 15 is the inertial term which keeps the particle moving in the same direction as its original direction. The inertial coefficient w serves as a dampener or an accelerator during the particles motion. The term $c_1 r_1 [\hat{x}_i(t) - x_i(t)]$ also known as the cognitive component functions as memory. Hence, the particle tends return to the location in the search space where the particle had a very high fitness value. The term $c_2 r_2 [g(t) - x_i(t)]$ known as the social component serves to move the particle to the locations where the swarm has moved in the previous iterations. After the computation of the particle velocity, the particle position is then calculated as follows:

$$x_i(t+1) = x_i(t) + v_i(t+1) \quad (17)$$

The iterations are then continued until the all candidate solutions are at their fittest positions in the fitness landscape and some stopping criterion which is set by the user is met. The working algorithm of the PSO in this work is as the following algorithm.

Algorithm 4: Particle Swarm Optimization (PSO)
Step 1: Initialize no of particles, i and the algorithm parameters w, c_1, c_2, r_1, r_2, n_o
Step 2: Randomly set initial position $x_i(n)$ and velocity, $v_i(n)$
Step 3: Compute individual and social influence
Step 4: Compute position $x_i(n+1)$ and velocity $v_i(n+1)$ at next iteration
Step 5: If the swarm evolution time, $n > n_o + T$, update position x_i and velocity v_i and go to Step 3, else proceed to Step 6
Step 7: Evaluate fitness swarm.
Step 8: If fitness criterion satisfied, halt and print solutions, else go to step 3.

where n_o is some constant, n is the swarm iteration and T is the overall program iteration. If during the iteration process, the position of all the particles converges to some constant value, no further optimization occurs in the objective function, no constraints are broken and all the decision variables are non-negative then it can be considered that the fitness criterion are met. Then the solutions are at its fittest and thus the program comes to a halt and it prints the solutions. The parameter setting for the PSO algorithm is given in Table 4.

Table 4. Particle swarm optimization (PSO) settings

Parameters	Values
Initial parameter (c_1, c_2, r_1, r_2, w)	(1, 1.2, 0.5, 0.5, 0.8)
Number of particles	6
initial social influence $(s_1, s_2, s_3, s_4, s_5, s_6)$	(1.1, 1.05, 1.033, 1.025, 1.02, 1.017)
initial personal influence $(p_1, p_2, p_3, p_4, p_5, p_6)$	(3, 4, 5, 6, 7, 8)

CHAOTIC PARTICLE SWARM OPTIMIZATION (CH-PSO)

The algorithmic enhancement in this work involves the concept of the application of chaotic logistic maps (Lorenz, 1963; Flake, 1998; Jakobson, 1981) to the PSO algorithm to improve its diversification capabilities. Commonly in most PSO algorithms, a random number generator is used to establish the initial position and velocity distributions of the particles. This would give the initial diversity in the population of solutions represented by each of the particles. The diversification capabilities of the PSO are enhanced in this work by using plugging in the chaotic logistic map after the random number generator segment in the algorithm. This way, the random number generator produces the initial conditions, $N_i(0)$ for the chaotic logistic map. The logistic mapping is given as the following:

$$R_i(t) = \lambda N_i(t) \qquad (18)$$

$$N_i(t+1) = R_i(t)N_i(t)\big[1 - N_i(t)\big] \qquad (19)$$

$$R_i(t+1) = R_i(t) + \lambda' \qquad (20)$$

where $N(t)$ and $R(t)$ are variables in the logistic chaotic map, λ' and λ are relaxation constants specified by the user. The chaotic logistic function is then iterated for 400 times (N_{max}). The solutions

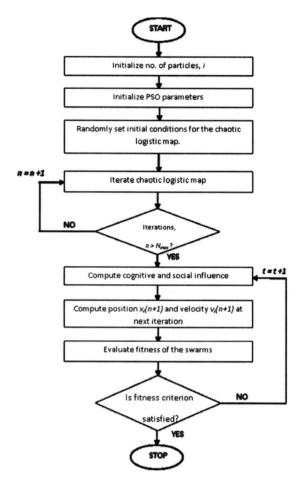

Figure 1. Flowchart for the Ch-PSO algorithm

obtained at the end of these iterations are then used as the initial distributions for the particle position and velocity ($x_i(0)$ and $v_i(0)$). Hence, another variant for the PSO algorithm (Chaotic PSO or Ch-PSO) with magnified diversification capabilities is developed in this work. The algorithm and the flowchart for this Ch-PSO method is given in Algorithm 5 and Figure 1 respectively:

Algorithm 5: Chaotic Particle Swarm Optimization (Ch-PSO)

Step 1: Set no of particles, i and the initialize parameter settings $w, c_1, c_2, r_1, r_2, n_o$

Step 2: Randomly initialize particles' position $x_i(t)$ and velocity $v_i(t)$

Step 3: Randomly set initial conditions for the chaotic logistic map.
Step 4: Iterate chaotic logistic map
Step 5: If the iterations, $n > N_{max}$, proceed to Step 6 else go to Step 4.
Step 6: Calculate cognitive and social components of the particles
Step 7: Compute position $x_i(t+1)$ and velocity $v_i(t+1)$ of the particles at next iteration
Step 8: Proceed with the evaluation of the fitness of each particle in the swarm.
Step 9: If the fitness conditions are satisfied and $t < T_{max}$, stop program and print solutions, else go to Step 6.

The parameter settings specified in the Ch-PSO algorithm is as in Table 5.

MEASUREMENT METRIC

1. Sigma Diversity Metric (SDM)

The diversity metric used in this work, is the sigma diversity metric (Mostaghim & Teich, 2003). The Sigma Diversity Metric (SDM) evaluates the locations of the solution vectors in the objective space relative to the sigma vectors. For lower dimensional objective spaces ($n < 3$), metrics that are based on spherical and polar coordinates can be used. However, as the dimensions increase beyond three ($n \geq 3$), the mentioned coordinate systems do not define the distribution of the solution vectors well (Mostaghim & Teich, 2005). In such scenarios, the SDM is highly recommended for effective computing of the solution distribution. To begin the computation of the SDM, two types of sigma lines would have to be constructed. First the sigma lines that represent the solution vectors, σ' and the sigma lines that represent the reference lines, σ. The sigma lines that represent the solution vectors can be computed as the following:

$$\sigma'_k(ij) = \frac{f_i^2 - f_j^2}{\sum_{l=1}^{n} f_l^2} \text{ such that } \forall i \neq j \quad (21)$$

where k denotes the index that represents the number of solution vectors, i, j and l denotes the index that represents the number of objectives and n denotes the total number of objectives. Then the magnitude sigma $|\sigma'_k|$ is computed as follows:

$$|\sigma'_k| = \sqrt{\sum_{i=1}^{m}\sum_{j=1}^{m} \sigma'_k(ij)} \quad (22)$$

Thus, for each line in the objective space (solution vector or reference line), there exists a unique

Table 5. Parameter settings for the Ch-PSO algorithm

Parameters	Values
Initial parameter (c_1, c_2, r_1, r_2, w)	(1, 1.2, 0.5, 0.5, 0.8)
Number of particles	6
initial social influence ($s_1, s_2, s_3, s_4, s_5, s_6$)	(1.1, 1.05, 1.033, 1.025, 1.02, 1.017)
initial personal influence ($p_1, p_2, p_3, p_4, p_5, p_6$)	(3, 4, 5, 6, 7, 8)
Constant, λ	5
Constant, λ'	0.01
N_{max}	400

sigma value. The central working principle is that the inverse mean distance of the solution vectors from the reference sigma vectors are computed. Since the reference sigma vectors are distributed evenly along the objective space, the mentioned inverse mean distance depicts the diversity of the solution spread. High values of the sigma diversity metric, indicates high uniformity and diversity in terms of the distributions of the solution vectors in the objective space.

2. Convergence Metric

The convergence metric used in this work was developed in Deb and Jain, (2002) and argues the convergence property of a solution set with respect to a reference set. In this work, since the Pareto optimal frontier was not known, a target vector P^* which was the most dominant vector was employed as the reference set. For a set of solutions for a single run of the program the formulation to compute the convergence metric is as the following:

$$d_i = \min_{j=1}^{|P^*|} \sqrt{\sum_{k=1}^{M}\left(\frac{f_k^i - f_k^j}{f_k^{\max} - f_k^{\min}}\right)} \quad (23)$$

where i and j denotes the subsequent objective function values, k is the index which denotes the objective function, f_k^{max} is the maximum objective function value, f_k^{min} is the minimum objective function value and M denotes the overall number of objectives. For this convergence metric, low the metric values indicate high convergence characteristics among the solution vectors.

3. Hypervolume Indicator (HVI)

The Hypervolume Indicator (HVI) is the only strictly Pareto-compliant indicator that can be used to measure the quality of solution sets in MO optimization problems (Zitzler & Thiele, 1998; Beume et al., 2007). Strictly Pareto-compliant can be defined such that if there exists two solution sets to a particular MO problem, then the solution set that dominates the other would a higher indicator value. The HVI measures the volume of the dominated section of the objective space and can be applied for multi-dimensional scenarios. When using the HVI, a reference point needs to be defined. Relative to this point, the volume of the space of all dominated solutions can be measured. The HVI of a solution set $x_d \in X$ can be defined as follows:

$$HVI(X) = vol\left(\bigcup_{(x_1,...,x_d) \in X} [r_1, x_1] \times ... \times [r_d, x_d]\right) \quad (24)$$

where $r_1,..., r_d$ is the reference point and $vol(.)$ being the usual Lebesgue measure. In this work the HVI is used to measure the quality of the approximation of the pareto front by the GSA and the DE algorithms when used in conjunction with the weighted sum approach.

GREEN MOULD APPLICATION PROBLEM

In the green sand mould system, the response parameters of the mould heavily influence the quality of the final product. IN Surekha et al., (2011), these parameters are selected as the objective functions. The responses parameters are; green compression strength (f_1), permeability (f_2), hardness (f_3) and bulk density (f_4). These objectives on the other hand are influenced by on the process variables which are; the grain fineness number (A), percentage of clay content (B), percentage of water content (C) and number of strokes (D). The objective functions and the range of the decision variables are shown as follows:

$$f_1 = 17.2527 - 1.7384A - 2.7463B$$
$$+32.3203C + 6.575D + 0.014A^2$$
$$+0.0945B^2 - 7.7857C^2 - 1.2079D^2$$
$$+0.0468AB - 0.1215AC - 0.0451AD$$
$$+0.5516BC + 0.6378BD + 2.689CD) \quad (25)$$

$$f_2 = 1192.51 - 15.98A - 35.66B$$
$$+9.51C - 105.66D + 0.07A^2 + 0.45B^2$$
$$-4.13C^2 + 4.22D^2 + 0.11AB + 0.2AC$$
$$+0.52AD + 1.19BC + 1.99BD - 3.1CD \quad (26)$$

$$f_3 = 38.2843 - 0.0494A + 2.4746B$$
$$+7.8434C + 7.774D + 0.001A^2$$
$$-0.00389B^2 - 1.6988C^2 - 0.6556D^2$$
$$-0.0015AB - 0.0151AC - 0.0006AD$$
$$-0.075BC - 0.1938BD + 0.65CD \quad (27)$$

$$f_4 = 1.02616 + 0.01316A - 0.00052B$$
$$-0.06845C + 0.0083D - 0.00008A^2$$
$$+0.0009B^2 + 0.0239C^2 - 0.00107D^2$$
$$-0.00004AB - 0.00018AC + 0.00029AD$$
$$-0.00302BC - 0.00019BD - 0.00186CD \quad (28)$$

$$52 \leq A \leq 94$$
$$8 \leq B \leq 12$$
$$1.5 \leq C \leq 3 \quad (29)$$
$$3 \leq D \leq 5$$

To obtain the size distributions of the silica sand and the grain fineness number, sieve analysis tests were carried out in Parappagoudar *et al.* (2007). Similarly, the authors also conducted gelling index tests for the determination the strength of clay. Next, experiments were conducted by varying the combination of the parameters using the central composite design. The mathematical model of the green mould system was developed where; the objective functions as given in equations (25)-(28) and the constraints as given in equation (29). The MO optimization problem statement for the green mould system problem is shown as follows:

$$\text{Max} \quad (f_1, f_2, f_3, f_4) \quad \text{subject to}$$

$$52 \leq A \leq 94$$
$$8 \leq B \leq 12$$
$$1.5 \leq C \leq 3 \quad (30)$$
$$3 \leq D \leq 5$$

The algorithms used in this work were programmed using the C++ programming language on a personal computer (PC) with an Intel dual core processor running at 2 GHz.

COMPUTATIONAL RESULTS AND ANALYSIS

In this work, the solution sets which are the approximations of the Pareto frontier were obtained using the GA, DE, PSO, GSA and Ch-PSO methods. The quality and characteristics of the spread of these solutions was determined using the measurement metrics. The nadir point (or the reference point) used in the HVI is a specific point where all the solutions sets produced by the algorithms dominates this point. The nadir point selected in this work is $(r_1, r_2, r_3, r_4) = (15, 40, 40, 1)$. The individual solutions (for specific weights) of GA, DE, PSO, GSA and Ch-PSO algorithms were gauged with the HVI and the best, median and worst solutions were determined.

The individual solutions for the GA algorithm and their individual HVI values are as in Table 6. The associated weights (w_1, w_2, w_3, w_4) for the best, median and worst solution are (0.1, 0.1, 0.4, 0.4), (0.3, 0.5, 0.1, 0.1) and (0.4, 0.1, 0.3,

Table 6. Individual solutions generated by the GA algorithm

Description		Best	Median	Worst
Objective Function	f_1	52.5562	59.6175	72.4155
	f_2	87.4493	53.5941	44.1095
	f_3	87.4107	88.7885	90.6108
	f_4	1.5436	1.5823	1.59962
Decision Variable	x_1	62.35	77.4579	80.4639
	x_2	10.0113	10.2203	11.8286
	x_3	2.17086	2.64723	2.36643
	x_4	4.83728	4.95008	4.97075
HVI		45926.9	17231.4	7160.4

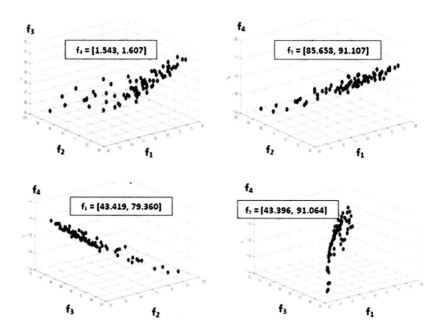

Figure 2. The approximate Pareto frontiers obtained by the GA algorithm

0.2). As for the approximate construction of the Pareto frontier, 81 solutions for various weights were obtained using the GA as shown in Figure 2:

The individual solutions for the DE algorithm and their degrees of dominance are presented in Table 6. The weights (w_1, w_2, w_3, w_4) obtained at the best, median and worst solutions are (0.4, 0.2, 0.3, 0.1), (0.4, 0.2, 0.1, 0.3) and (0.2, 0.2, 0.1, 0.5). The approximation of the Pareto frontier using the DE algorithm is as in Figure 3.

The individual best, median and worst solution solutions for the GSA algorithm and their respective HVI values are as in Table 7. The associated weights (w_1, w_2, w_3, w_4) for the best, median and worst solution are (0.2, 0.3, 0.4, 0.1), (0.1, 0.6, 0.2, 0.1) and (0.1, 0.2, 0.5, 0.2) respectively. The approximate Pareto frontiers contracted by the implementation of the GSA algorithm is shown in Figure 4.

Table 6. Individual solutions generated by the DE algorithm

Description		Best	Median	Worst
Objective Function	f_1	50.2281	45.8468	39.4176
	f_2	137.769	137.19	137.105
	f_3	86.289	85.3638	83.7878
	f_4	1.50073	1.50953	1.52143
Decision Variable	x_1	53.0457	55.5626	59.4985
	x_2	9.39392	9.13418	8.74959
	x_3	2.99849	2.99044	2.99944
	x_4	4.39392	4.13418	3.74959
HVI		79831.04	69296.434	54136.94

Figure 3. The approximate Pareto frontiers obtained by the DE algorithm

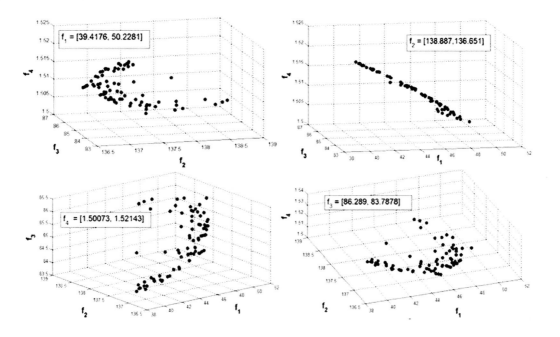

The individual solutions for the PSO algorithm and their individual HVI values are as in Table 8. The associated weights (w_1, w_2, w_3, w_4) for the best, median and worst solution are (0.2, 0.3, 0.4, 0.1), (0.1, 0.6, 0.2, 0.1) and (0.1, 0.2, 0.5, 0.2). The approximate construction of the Pareto frontier was obtained using the PSO algorithm as depicted in Figure 5.

The assessment of quality of the solutions produced by each algorithm was conducted throughout the entire spectrum of solutions that approximate the Pareto frontier and not just individual solutions. The degree of dominance of these solutions produced by each algorithm was computed using the HVI and its values are shown in Figure 6.

Table 7. Individual solutions generated by the GSA algorithm

Description		Best	Median	Worst
Objective Function	f_1	34.6562	25.8028	25.5917
	f_2	205.619	210.839	210.877
	f_3	81.1777	78.8635	78.8113
	f_4	1.46948	1.47897	1.47937
Decision Variable	x_1	52.0015	52.0023	52.0009
	x_2	8.00543	8.00205	8.0007
	x_3	2.49446	1.51268	1.5
	x_4	3.19889	3.00254	3
HVI		62934.52	34353.70	33672.73

Figure 4. The approximate Pareto frontiers obtained by the GSA algorithm

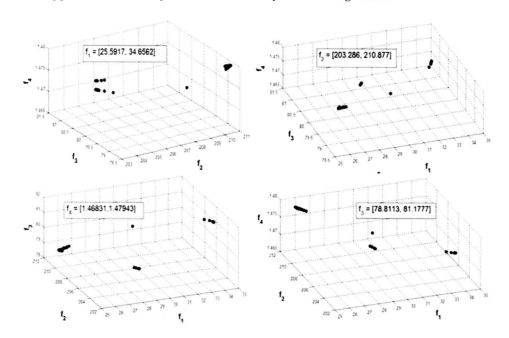

It can be observed that the PSO algorithm produces the most dominant solution followed the DE, GSA and GA algorithms sequentially. The PSO algorithm dominated the DE algorithm by 18.08%. To obtain a better understanding on the two most efficient algorithms (DE & PSO), further analysis on the solution spread characteristics are conducted using the convergence and sigma diversity metrics. These characteristics are shown in Table 9.

In Table 9, it can be observed that the solution spread produced by the PSO algorithm is less diverse as compared to the DE algorithm. As for the convergence aspect, the solution spread produced by the PSO algorithm is more convergent as compared to the DE algorithm. Hence, the solution spread produced by PSO algorithm although convergent lacks diversity as compared to the DE approach. To further boost the diversity of the solutions produced by the PSO algorithm,

Table 8. Individual solutions generated by the PSO algorithm

Description		Best	Median	Worst
Objective Function	f_1	54.3163	60.6924	46.2995
	f_2	148.75	109.761	105.735
	f_3	87.1063	88.5401	86.0016
	f_4	1.49606	1.51546	1.53547
Decision Variable	x_1	52.5239	54.5421	58.2747
	x_2	10.9927	10.598	10.0498
	x_3	2.56221	2.90304	1.73465
	x_4	3.71673	4.83757	4.58633
HVI		99911.41	79753.97	50680.65

Figure 5. The approximate Pareto frontiers obtained by the PSO algorithm

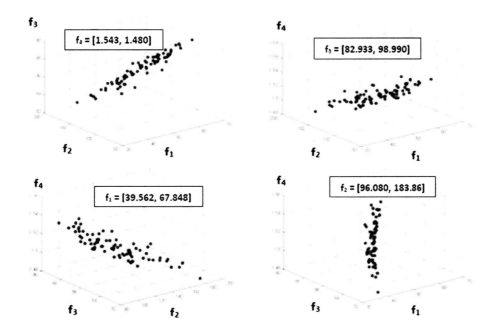

the chaotic component is integrated into the PSO algorithm creating the Ch-PSO algorithm. The Pareto frontier constructed by the solutions of the Ch-PSO algorithm is as shown in Figure 7.

Spread characteristics of the solutions produced the Ch-PSO and PSO algorithms are shown in Table 10. The comparison of the degree of dominance of solution sets produced by the Ch-PSO and PSO algorithms are given in Figure 8.

From Table 10, it can be observed that the Ch-PSO algorithm produces solutions with highly diverse but less convergent spread characteristics as compared to the PSO algorithm. In addition, the Ch-PSO algorithm efficiently approximates the Pareto frontier which is more dominant as compared with the PSO algorithm. Hence, the solutions produced by the Ch-PSO algorithm are more dominant as compared to the PSO algorithm by 17.7%. The execution time for

Figure 6. The Degree of dominance of the solutions produced by each of the algorithms

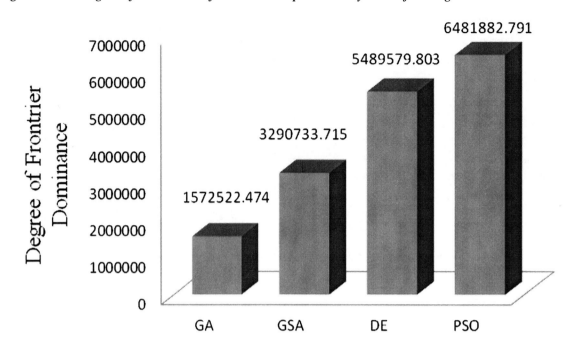

Table 9. Solution spread characteristics of the solutions produced the DE and PSO algorithms

Algorithms	Solution Spread Characteristics	
	Diversity	Convergence
DE	0.6071	0.0673
PSO	0.2222	0.0581

each of the techniques used in this work is shown in Table 11.

In Table 11 it can be observed that the most inefficient algorithm in terms of computational efficiency is the GA algorithm. This is may be due to the difficulty faced by the GA algorithm to search for the optima and hence carries on large number of iterations. The GSA and PSO algorithms performed almost equally efficient. The high algorithmic complexity of the DE algorithm as compared to the PSO and Ch-PSO algorithms causes the DE algorithm to perform computationally inferior to the PSO and Ch-PSO algorithms. Besides, PSO-based algorithms have been known to be computationally very efficient when it comes to execution time as compared to many other algorithms.

One of the setbacks of the weighted sum method employed in this work is that it does not guarantee Pareto optimality (only in the weak sense (Pradyumn, 2007). Besides, scalarization techniques (such as weighted sum as well Normal Boundary Intersection (NBI)) cannot approximate sections of the Pareto frontier that is concave (Zitzler et al., 2008). Although the HVI metric is most effective in benchmarking the dominance of solution sets produced by algorithms, this metric is very dependent on the choice of the nadir point. The HVI metric's only weakness is in this aspect.

Figure 7. The approximate Pareto frontiers obtained by the Ch-PSO algorithm

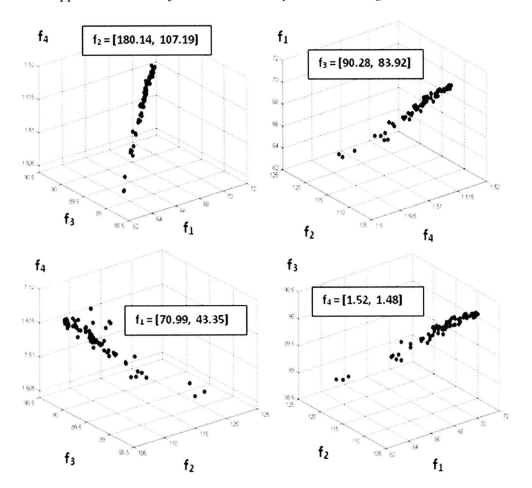

Table 10. Solution spread characteristics of the solutions produced the Ch-PSO and PSO algorithms

Algorithms	Solution Spread Characteristics	
	Diversity	Convergence
Ch-PSO	0.7778	0.0189
PSO	0.2222	0.0581

In this work, the GA, GSA, DE, PSO and Ch-PSO algorithms performed stable computations during the program executions. All Pareto-efficient solutions produced by the algorithms developed in this work were feasible and no constraints were compromised. One of the advantages of using the PSO and Ch-PSO algorithms as compared to the other algorithms in this work is that it produces highly effective results in terms of approximating the Pareto frontier. Since the PSO and Ch-PSO methods are swarm-type algorithms, the diversification of the search space is high and thus resulting in low computational overhead as compared with the GA and DE methods (which are evolutionary-type algorithms). The PSO method can be said to be the second best optimizer as compared to the

Figure 8. Degree of dominance of solution sets produced by the Ch-PSO and PSO algorithms

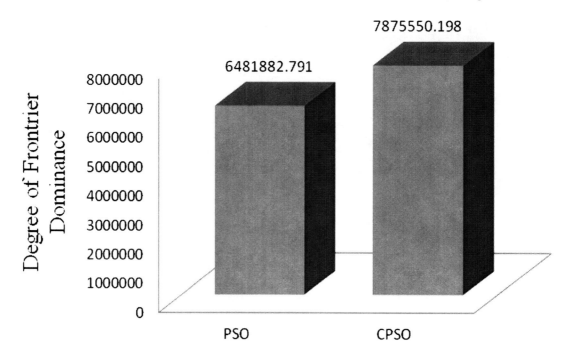

Table 11. Execution time for each algorithm

Algorithm	GA	GSA	DE	PSO	CH-PSO
Execution Time	282.404	21.2	39.08	21.683	35.316

Ch-PSO method. Therefore, with the aid of the information provided by measurement metrics, the diversification augmentations were successfully performed on the PSO algorithm using the chaotic mechanism. The resulting Ch-PSO algorithm was highly effective in producing optimal solutions the construct the approximate Pareto frontier.

CONCLUSION

A new local maximum and a more dominant approximation of the Pareto frontier were obtained using the Ch-PSO method. More Pareto-efficient solution options to the green mould system MO optimization problem were obtained. Besides, using the Ch-PSO algorithm, the solution spread of the frontier was more diversely distributed. This property is much attributed to the chaotic mechanism in the Ch-PSO algorithm. When gauged with the HVI metric, the PSO and CPSO algorithms produced the most dominant approximate of the Pareto frontier as compared to the GA, GSA and the DE methods.

For future works, other meta-heuristic algorithms such as Analytical Programming (AP) (Zelinka, 2002), Hybrid Neuro-GP (Ganesan et al., 2011) and Hybrid Neuro-Swarm (Ganesan et al., 2012) techniques should be applied to the green mould system. The solution sets obtained by these algorithms should be tested and evaluated using the HVI.

REFERENCES

Beume, N., Naujoks, B., & Emmerich, M. (2007). SMS-EMOA: Multiobjective selection based on dominated hypervolume. *European Journal of Operational Research, 181*(3), 1653–1669. doi:10.1016/j.ejor.2006.08.008

Colorni, A., Dorigo, M., & Maniezzo, V. (1991). distributed optimization by ant colonies. In *Proceedings of the first European Conference of Artificial Intelligence.* (pp. 134-142). Paris: Elsevier Publishing.

Das, I., & Dennis, J. E. (1998). Normal-boundary intersection: A new method for generating the Pareto surface in nonlinear multicriteria optimization problems. *SIAM Journal on Optimization, 8*(3), 631–657. doi:10.1137/S1052623496307510

Deb, K., & Jain, S. (2002). *Running performance metrics for evolutionary multiobjective optimization* (KanGAL Report No. 2002004). Kanpur, India: Indian Institute of Technology.

Deb, K., Pratap, A., Agarwal, S., & Meyarivan, T. (2002). A fast and elitist multiobjective genetic algorithm: NSGA-II. *IEEE Transactions on Evolutionary Computation, 6*(2), 182–197. doi:10.1109/4235.996017

Eschenauer, H., Koski, J., & Osyczka, A. (1990). *Multicriteria design optimization.* Berlin: Springer-Verlag. doi:10.1007/978-3-642-48697-5

Fishburn, P. C. (1967). *Additive utilities with incomplete product set: Applications to priorities and assignments.* Baltimore, MD: Operations Research Society of America.

Flake, G. W. (1998). *The computational beauty of nature: Computer explorations of fractals, chaos, complex systems, and adaptation.* Cambridge, MA: MIT Press.

Ganesan, T., Vasant, P., & Elamvazuthi, I. (2011). Optimization of nonlinear geological structure mapping using hybrid neuro-genetic techniques. *Mathematical and Computer Modelling, 54*(11-12), 2913–2922. doi:10.1016/j.mcm.2011.07.012

Ganesan, T., Vasant, P., & Elamvazuthi, I. (2012). Hybrid neuro-swarm optimization approach for design of distributed generation power systems. *Neural Computing & Applications, 23*(1), 105–117. doi:10.1007/s00521-012-0976-4

Jakobson, M. (1981). Absolutely continuous invariant measures for one-parameter families of one-dimensional maps. *Communications in Mathematical Physics, 81,* 39–38. doi:10.1007/BF01941800

Kennedy, J., & Eberhart, R. (1995). Particle swarm optimization. In *IEEE Proceedings of the International Conference on Neural Networks: Perth, Australia,* (pp. 1942-1948).

Lorenz, E. N. (1963). Deterministic nonperiodic flow. *Journal of the Atmospheric Sciences, 20*(2), 130–141. doi:10.1175/1520-0469(1963)020<0130:DNF>2.0.CO;2

Luyben, M. L., & Floudas, C. A. (1994). Analyzing the interaction of design and control. 1. A multiobjective framework and application to binary distillation synthesis. *Computers & Chemical Engineering, 18*(10), 933–969. doi:10.1016/0098-1354(94)E0013-D

Mostaghim, S., & Teich, J. (2003). Strategies for finding good local guides in multiobjective particle swarm optimization. In *IEEE Swarm Intelligence Symposium, Indianapolis, USA,* (pp. 26-33).

Mostaghim, S., & Teich, J. (2005). A new approach on many objective diversity measurement. In *Dagstuhl Seminar Proceedings 04461, Practical Approaches to Multiobjective Optimization,* (pp. 1-15).

Parappagoudar, M. B., Pratihar, D. K., & Datta, G. L. (2007). Non-linear modeling using central composite design to predict green sand mould properties. *Proceedings of the Institution of Mechanical Engineers. Part B, Journal of Engineering Manufacture, 221*, 881–894. doi:10.1243/09544054JEM696

Pradyumn, S. K. (2007). On the normal boundary intersection method for generation of efficient front. In Y. Shi et al. (Eds.), *ICCS 2007, Part I, LNCS 4487* (pp. 310–317). Berlin, Heidelberg: Springer-Verlag.

Rashedi, E., Nezamabadi-pour, H., & Saryazdi, S. (2009). GSA: A gravitational search algorithm. *Information Sciences, 179*, 2232–2248. doi:10.1016/j.ins.2009.03.004

Rosenberg, R. S. (1967). *Simulation of genetic populations with biochemical properties* (Ph.D. thesis). University of Michigan, MI.

Statnikov, R. B., & Matusov, J. B. (1995). *Multicriteria optimization and engineering*. New York: Chapman and Hall. doi:10.1007/978-1-4615-2089-4

Storn, R., & Price, K. V. (1995). *Differential evolution – A simple and efficient adaptive scheme for global optimization over continuous spaces* (ICSI, Technical Report TR-95-012), pp. 1-12.

Surekha, B., Kaushik, L. K., Panduy, A. K., Vundavilli, A. P. R., & Parappagoudar, M. B. (2011). Multiobjective optimization of green sand mould system using evolutionary algorithms. *International Journal of Advanced Manufacturing Technology, 58*, 1–9.

Sushil, K., Satsangi, P. S., & Prajapati, D. R. (2010). Optimization of green sand casting process parameters of a foundry by using Taguchi method. *International Journal of Advanced Manufacturing Technology, 55*, 23–34.

Triantaphyllou, E. (2000). *Multi-criteria decision making: A comparative study*. Dordrecht, The Netherlands: Kluwer Academic Publishers. doi:10.1007/978-1-4757-3157-6

Zelinka, I. (2002). Analytic programming by means of SOMA Algorithm. In *Proc. 8th International Conference on Soft Computing Mendel'02, Brno, Czech Republic*, (pp. 93-101).

Zitzler, E., Knowles, J., & Thiele, L. (2008). Quality assessment of Pareto set approximations. In J. Branke et al. (Eds.), *Multiobjective optimization, LNCS 5252* (pp. 373–404). Berlin, Heidelberg: Springer-Verlag. doi:10.1007/978-3-540-88908-3_14

Zitzler, E., & Thiele, L. (1998). Multiobjective optimization using evolutionary algorithms - A comparative case study. In *Conference on Parallel Problem Solving from Nature (PPSN V)*, (pp. 292–301).

ADDITIONAL READING

Aboosadi, Z. A., Jahanmiri, A. H., & Rahimpour, M. R. (2011). Optimization of tri-reformer reactor to produce synthesis gas for methanol production using differential evolution (DE) method. *Applied Energy, 88*, 2691–2701. doi:10.1016/j.apenergy.2011.02.017

Elamvazuthi, I., Ganesan, T., & Vasant, P. (2011), 'A Comparative Study of HNN and Hybrid HNN-PSO Techniques in the Optimization of Distributed Generation (DG) Power Systems', *International Conference on Advance Computer Science and Information System* 2011 (ICACSI), 195 -199. ISBN: 978-979-1421-11-9

Eschenauer, H., Koski, J., & Osyczka, A. (1990). *Multicriteria Design Optimization*. Berlin: Springer-Verlag. doi:10.1007/978-3-642-48697-5

Ganesan, T., Elamvazuthi, I., Ku Zilati Ku Shaari, and Vasant, P. (2012) 'Swarm intelligence and gravitational search algorithm for multiobjective optimization of synthesis gas production', *Journal of Applied Energy* http://dx.doi.org/10.1016/j.apenergy.2012.09.059.

Ganesan, T., Vasant, P., & Elamvazuthi, I. (2012). *Hybrid Optimization Techniques for Optimization in a Fuzzy Environment in Handbook of Optimization: From Classical to Modern Approach* (Vol. 38). Intelligent Systems Reference Library.

Grosan, C. Oltean., M., & Dumitrescu, D. (2003), 'Performance Metrics For Multiobjective Optimization Evolutionary Algorithms'. In *Proceedings Of Conference On Applied And Industrial Mathematics (Caim), Oradea*.

Holland, J. H. (1992). *'Adaptation in Natural and Artificial Systems: An Introductory Analysis with Applications to Biology, Control and Artificial Intelligence*. USA: MIT Press.

Igel, C., Hansen, N., & Roth, S. (2007). Covariance matrix adaptation for multiobjective optimization. *Evolutionary Computation*, *15*(1), 1–28. doi:10.1162/evco.2007.15.1.1 PMID:17388777

Jirapong, P., & Ongsakul, W. (2008) 'Available Transfer Capability Determination Using Hybrid Evolutionary Algorithm', *International Conference on Power Control and Optimization (PCO)*, 273 – 277.

Koza, J. R. (1992). *Genetic Programming: On the Programming of Computers by means of Natural Selection*. USA: MIT Press.

Masand, B. (1994). *'Optimizing confidence of text classification by evolution of symbolic expressions', Advances in Genetic Programming*. USA: MIT Press.

Mistree, F., Hughes, O. F., & Bras, B. A. (1993). *'The Compromise Decision Support Problem and the Adaptive Linear Programming Algorithm', Structural Optimization: Status and Promise* (pp. 247–286). Washington, D.C.: AIAA.

Mohanty, S. (2006). Multiobjective optimization of synthesis gas production using non-dominated sorting genetic algorithm. *Computers & Chemical Engineering*, *30*, 1019–1025. doi:10.1016/j.compchemeng.2006.01.002

Oakley, H. (1994). Two scientific applications of genetic programming: Stack filters and non-linear equation fitting to chaotic data. In *Advances in Genetic Programming*. USA: MIT Press.

Oplatkova, Z., & Zelinka, I. (2006), Investigations using Artificial Ants using Analytical Programming, *Second International Conference on Computational Intelligence, Robotics, and Autonomous Systems*, Singapore.

Oplatkova, Z., & Zelinka, I. (2007). Creating evolutionary algorithms by means of analytic programming - design of new cost function. In *ECMS 2007* (pp. 271–276). European Council for Modeling and Simulation. doi:10.7148/2007-0271

Pytel, K. (2012), 'The fuzzy Genetic System for Multiobjective Optimization', *Proceedings of the Federated Conference on Computer Science and Information Systems*, 137 - 140.

Rahimpour, M. R., Parvasi, P., & Setoodeh, P. (2009). Dynamic optimization of a novel radial-flow, spherical-bed methanol synthesis reactor in the presence of catalyst deactivation using Differential Evolution (DE) algorithm. *International Journal of Hydrogen Energy*, *34*, 6221–6230. doi:10.1016/j.ijhydene.2009.05.068

Ryan, C., Collins, J., & O'Neill, M. (1998). Grammatical evolution: Evolving programs for an arbitrary language, *Lecture Notes in Computer Science, First European Workshop on Genetic Programming*.

Sandgren, E. (1994), 'Multicriteria design optimization by goal programming'. In Hojjat Adeli, editor, 'Advances in Design Optimization', Chapman & Hall, London, chapter 23, 225–265.

Sankararao, B., & Gupta, S. K. (2007). Multiobjective optimization of an industrial fluidized-bed catalytic cracking unit (FCCU) using two jumping gene adaptations of simulated annealing. *Computers & Chemical Engineering*, *31*, 1496–1515. doi:10.1016/j.compchemeng.2006.12.012

Schott, J. R. (1995), 'Fault tolerant design using single and multicriteria genetic algorithms optimization'. *Master's thesis, Department of Aeronautics and Astronautics, Massachusetts Institute of Technology, Cambridge, MA.*

Varacha, P. (2011). *Neural Network Synthesis via Asynchronous Analytical Programming*. Recent Researches in Neural Networks, Fuzzy Systems, Evolutionary Computing and Automation.

Vasant, P., Ganesan, T., & Elamvazuthi, I. (2012). Improved Tabu Search Recursive Fuzzy Method For Crude Oil Industry. *International Journal of Modeling, Simulation, and Scientific Computing, World Scientific*, *3*(1).

Vasant, P., Ganesan, T., & Elamvazuthi, I. (2012). Hybrid Tabu Search Hopfield Recurrent ANN Fuzzy Technique to the Production Planning Problems: A Case Study of Crude Oil in Refinery Industry [IGI Global]. *International Journal of Manufacturing, Materials, and Mechanical Engineering*, *2*(1), 47–65. doi:10.4018/ijmmme.2012010104

Zelinka, I., & Oplatkova, Z. (2003), Analytic programming – Comparative Study. CIRAS'03, *Second International Conference on Computational Intelligence, Robotics, and Autonomous Systems*, Singapore.

KEY TERMS AND DEFINITIONS

Chaotic Particle Swarm Optimization (Ch-PSO): A novel metaheuristic algorithm that merges ideas from the chaos theory and swarming behavior of organisms to improve the solving capabilities of the standard PSO.

Differential Evolution (DE): A type of metaheuristic algorithm that employs perturbative methods and evolutionary dynamics for solving optimization problems iteratively.

Genetic Algorithm (GA): A type of metaheuristic algorithm that uses concepts from evolutionary biology and genetics to solve optimization problems.

Gravitational Search Algorithms (GSA): A type of metaheuristic algorithm that integrates ideas from Newtonian gravitational laws to search for optimal solutions in the objective space.

Green Mould Sand System: An aggregate of pulverized coal, water, and bentonite clay that is employed as a mould for the casting of metals.

Measurement Metrics: Mathematical metrics used to identify and measure solution characteristics such as degree of convergence, diversity, and dominance.

Metaheuristic: A framework consisting of a class of algorithms employed to find good solutions to optimization problems by iterative improvement of solution quality.

Multiobjective Optimization: Optimization problems that are represented with more than one objective function.

Particle Swarm Optimization (PSO): A type of metaheuristic algorithm that uses concepts from swarming behavior of organisms to search for optimal solutions.

Weighted Sum Approach: A solution approach in multiobjective optimization where the objective functions are aggregated by multiplying them by weights (level of importance) and summing them over.

Chapter 4
Hybrid Evolutionary Optimization Algorithms:
A Case Study in Manufacturing Industry

Pandian Vasant
Universiti Teknologi PETRONAS, Malaysia

ABSTRACT

The novel industrial manufacturing sector inevitably faces problems of uncertainty in various aspects such as raw material availability, human resource availability, processing capability and constraints and limitations imposed by the marketing department. These problems have to be solved by a methodology which takes care of such unexpected information. As the analyst faces this man made chaotic and due to natural disaster problems, the decision maker and the implementer have to work collaboratively with the analyst for taking up a decision on an innovative strategy for implementation. Such complex problems of vagueness and uncertainty can be handled by the hybrid evolutionary intelligence algorithms. In this chapter, a new hybrid evolutionary optimization based methodology using a specific non-linear membership function, named as modified S-curve membership function, is proposed. The modified S-curve membership function is first formulated and its flexibility in taking up vagueness in parameters is established by an analytical approach. This membership function is applied for its useful performance through industrial production problems by employing hybrid evolutionary optimization algorithms. The novelty and the originality of this non-linear S-curve membership function are further established using a real life industrial production planning of an industrial manufacturing sector. The unit produces 8 products using 8 raw materials, mixed in various proportions by 9 different processes under 29 constraints. This complex problem has a cubic non-linear objective function. Comprehensive solutions to a non-linear real world objective function are achieved thus establishing the usefulness of the realistic membership function for decision making in industrial production planning.

DOI: 10.4018/978-1-4666-5836-3.ch004

INTRODUCTION

In the literature, there are several techniques and heuristic/metaheuristic have been used to improve the general efficiency of the evolutionary algorithms. Some of the most commonly used hybrid approaches are summarized as below:

1. Hybridization among several techniques of evolutionary algorithm and another evolutionary algorithm (example: an evolutionary strategy is used to improve the performance of a genetic algorithm)
2. Neural network and evolutionary algorithms
3. Fuzzy logic and evolutionary algorithm
4. Particle swarm optimization (PSO) and evolutionary algorithm
5. Ant colony optimization (ACO) and evolutionary algorithm
6. Bacterial foraging optimization and evolutionary algorithm
7. Evolutionary algorithms incorporating prior knowledge
8. Hybridization between evolutionary algorithm and other heuristic approaches (such as local search, general pattern search, mesh adaptive direct search, tabu search, simulated annealing, hill climbing, dynamic programming, greedy random adaptive search procedure and others).

The combination of different learning and adaptation techniques, to overcome individual limitations and achieve synergetic effects through hybridization or fusion of these techniques, has in recent years contributed to a large number of new hybrid evolutionary systems. Most of these approaches, however, follow an ad hoc design methodology, further justified by success in certain application domains. Due to the lack of a common framework it remains often difficult to compare the various hybrid systems conceptually and evaluate their performance comparatively. There are number of ways to hybridize a conventional evolutionary algorithm for solving optimization problems. Some of them are summarized below (Swain & Morris, 2000).

- The solutions of the initial population of evolutionary algorithm may be created by problem-specific heuristics.
- Some or all the solutions obtained by the evolutionary algorithm may be improved by local search. This kind of algorithms is known as memetic algorithms (Hart, 1998; Moscato, 1999).
- Solutions may be represented in an indirect way and a decoding algorithm maps any genotype to a corresponding phenotypic solution. In this mapping, the decoder can exploit problem-specific characteristics and apply heuristics approaches.
- Variation operators may exploit problem knowledge. For example, in recombination more promising properties of one parent solution may be inherited with higher probabilities than the corresponding properties of the other parent(s). Also mutation may be biased to include in solutions promising properties with higher probabilities than others.

Fuzzy logic controller (FLC) is composed by a knowledge base, that includes the information given by the expert in the form of linguistic control rules, a fuzzification interface, which has the effect of transforming crisp date into fuzzy sets, an inference system, that uses them together with the knowledge base to make inference by means of a reasoning method, and a defuzzification interface, that translates the fuzzy control action thus obtained to a real control action using defuzzification method. Fuzzy logic controllers have been used to design adaptive evolutionary algorithms. The main idea is to use a fuzzy logic controller whose inputs are any combination of evolutionary algorithms performance measures and current control parameter values and whose outputs are evolutionary algorithms control parameter values. Lee and Takagi (1993) proposed the dynamic parametric genetic algorithms (DPGA) that uses a fuzzy logic controller for controlling genetic algorithms parameters. The inputs to the fuzzy logic controller are any combination of genetic algorithms performance measures or current control settings, and outputs may be any of the genetic algorithms control parameters. Hererra and Lozano (1996) reported tightly coupled, uncoupled, and loosely coupled methods for adaptation. Three levels of tightly coupled adaptation may be implemented at the level of individuals, the level of subpopulation and the level of population. In an uncoupled adaptation, a totally separate adaptive mechanism adjusts the performance of evolutionary algorithms. It is to be noted that an uncoupled approach does not rely upon the evolutionary algorithms for the adaptive mechanism. In the loosely coupled method, an evolutionary algorithm is partially used for the adaptive mechanism, for example, either the population or the genetic operators are used in some fashion.

The evolutionary algorithm control parameter settings such as mutation probability (Pm), crossover probability (Pc), and population size (N) are key factors in the determination of the exploitation versus exploration tradeoff (Holland, 1975).

Example: Mutation rates (Pm) may be adapted to prevent premature convergence and to speed up the optimization. The rules that take care of adjusting mutation rates could be formulated as follows:

- If convergent then set Pm = 0.001
- If not convergent then set Pm= 0.8

A difference between global and local search procedures is that global techniques are largely independent on the initial conditions while local methods produce solutions that are strongly dependent on the starting point. Besides, local procedures tend to be coupled to the solution domain.

The genetic algorithms were largely developed for the purpose of performing global searches. The conventional genetic operators were developed with the main purpose of enhancing the algorithm capability of finding global optima basins.

Memetic algorithms (MA) or hybrid genetic algorithm (HGA) denote the association of local and global search operators inside genetic algorithms. This kind of strategy is used by many successful global optimization procedures with the goal of refining the solution of the problem and improving the speed of convergence to the actual optimum point (not only to its vicinity).

Richard Dawkins in his book the "Selfish Gene" has introduced the word meme to denote the idea of a unit of imitation in cultural transmission which is some aspects is analogues to the gene (Dawkins, 1976). The first use of the name memetic algorithms in the literature appeared in the work of Pablo Moscato (1999). On evolution, search, optimization, genetic algorithm and martial arts: Towards memetic algorithms, for denoting algorithms that use some kind of structured information, that is obtained and refined as the algorithm evolves, and is transmitted from one generation to another, for enhance the search. Since then, this idea has gained wide acceptance in the computing community and has been successfully applied in a large class of problems (Lozano, Herrera, Krasnogor, & Molina, 2004; Moscato, 1999).

The main advantage obtained from the use of memetic algorithms is that the space of possible solutions is reduced to a subspace (or a lower-order set) of local optima. The introduction of local search in the traditional genetic algorithms has some computational cost, but this is compensated by the decrease in the search apace that must be explored in order to find the solution (Wanner, Guimarae, Saldanha, Takahashi, & Fleming, 2005).

A new methodology, RBGA-CQA hybrid (Real-Biased Genetic algorithms with Constraints Quadratic Approximation), is presented by Wanner, Guimarae, Saldanha, Takahashi, and Fleming

(2005), deals with nonlinear equality constraint using quadratic approximation technique and the bisection method as a local search operator in the real-biased genetic algorithm. This new local search operator improves the solution because it allows the equality constraint to be reached with increased precision. Furthermore, this operator does not impose any significant additional computational cost on the traditional RBGA if the computational cost is measured by the number of calls of the objective and constraint functions. In this way, the new algorithm is kind of memetic algorithm or hybrid genetic algorithm for equality constrained problems.

Aruldoss and Ebenezer (2005) proposed a sequential quadratic programming (SQP) method for the dynamic economic dispatch problem (DEDP) of generating units considering the valve-point effects. The developed method is a two-phase optimizer. In the first phase, the candidates of evolutionary programming explore the solution space freely. In the second phase, the sequential quadratic programming is invoked when there is an improvement of solution (a feasible solution) during the evolutionary programming run. Thus, the sequential quadratic programming guides evolutionary programming for better performance in the complex solution space. A similar hybrid approach involving evolutionary programming and sequential quadratic programming techniques was proposed by Attaviriyanupap, Tanaka, and Hasegawa (2002), where the evolutionary programming is applied to obtain a near global solution; once the evolutionary programming terminates its procedures, the sequential quadratic programming is applied to obtain final optimal solution.

Hybrid methods which combine evolutionary computation techniques with deterministic procedures for numerical optimization problems have been recently investigated. Papadrakakis, Tsompanakis, and Lagaros (1999) used evolution strategies with the sequential quadratic programming method, while Waagen, Diercks, and McDonnell (1992) combined evolutionary programming with the direct search method of Hooke and Jeeves (1961). The hybrid implementation proposed in Papadrakakis, Tsompanaki,s and Lagaros (1999) was found very successful on shape optimization test examples, while the method proposed in Waagen, Diercks, and McDonnell (1992) was applied to unconstrained mathematical test functions. Myung, Kim, and Fogel (1995) considered a similar to Waagen, Diercks, and McDonnell (1992) approach, but they experimented with constrained mathematical test functions. Myung, Kim, and Fogel (1995) combined a floating-point evolutionary programming technique, with a method-developed by Maa and Shanblatt (1992) applied to the best solution found by the evolutionary programming technique. The second method iterates until the system defined by the combination of the objective function, the constraints and the design variables reach equilibrium.

A characteristic property of the sequential quadratic programming based optimizers is that they capture very fast the right path to the nearest optimum, irrespective of its nature a local or global optimum. However, after locating the area of this optimum it might oscillate until all constraints are satisfied since it is observed that even small constraint violation often slow down the convergence rate of the method. On the other hand evolutionary algorithm proceed with slower rate, due to their random search, but the absence of strict mathematical rule, which govern the convergence rate of the mathematical programming methods, make evolutionary algorithm less vulnerable to local optima and therefore it is likely to converge towards the global optimum in non-convex optimization methods. These two facts give the motivation to combine evolutionary algorithm and mathematical programming methodologies. Between the two evolutionary algorithms examined by Lagaros, Papadrakakis, and Kokossalakis (2002) the genetic algorithms seems to be faster than evolutionary strategies since they do not always operate on the feasible region of the design space as evolutionary algorithms.

However, they are most often found unable to converge to feasible designs.

In order to benefit from the advantages of both methodologies (Evolutionary algorithm and Sequential Quadratic Programming) a hybrid approach is proposed by Lagaros, Papadrakakis, and Kokossalakis (2002), which combines the two methods in an effort in increase the robustness and the computational efficiency of the optimization procedure. The optimization process is divided into two separate phases. During the first phase, the evolutionary algorithm optimizes the objective function. After the termination of this phase, the second phase starts by applying the sequential quadratic programming to the best solution found during the first phase.

The procedure of evolutionary algorithm is first used in order to locate the region where the global optimum lies, and then the sequential quadratic programming activated in order to exploit its higher rate of convergence in the neighborhood of the optimum. The switch from evolutionary algorithm to sequential quadratic programming is performed when evolutionary algorithm reaches the vicinity of an optimum that is considered a good one. This approach appears to be more rational in the general case when more complex and non-convex design problems are to be solved with many local optima. In the case of genetic algorithm-sequential programming, the final design achieved by genetic algorithm can be tolerated to be infeasible since sequential quadratic programming will eventually locate a feasible design (Lagaros, Papadrakakis, and Kokossalakis, 2002).

In this research work, six hybrid methodologies are examined which are based on the combination of genetic algorithm with Line search (LS), general pattern search (GPS), mesh adaptive direct search (MADS); general pattern search with genetic algorithm and line search; mesh adaptive direct search with genetic algorithm and line search.

Ganesh and Punniyamoorthy (2004) proposed a hybrid genetic algorithm – simulated annealing (SA) algorithm for continuous- time aggregate production- planning problems. The motivation behind genetic algorithm- simulated annealing combination is the power of genetic algorithm to work on the solution in a global sense while allowing simulated annealing to locally optimize each individual solution (Ganesh & Punniyamoorthy, 2004). The hybrid algorithm executes in two phases. In the first phase, the genetic algorithm generates the initial solution randomly. The genetic algorithm then operates on the solutions using selection, crossover, mutation and recombination operators to produce new and better solutions. After each generation, the genetic algorithm sends each solution to the simulated annealing (second phase) to be improved. The neighborhood generation scheme used in simulated annealing is a single insertion neighborhood scheme. Once the simulated annealing finished for a solution of genetic algorithm, another genetic algorithm solution is passed on to simulated annealing. This process continues until all solutions of genetic algorithm in one generation are exhausted. Once the simulated annealing is finished for all solutions in one generation of genetic algorithm, the best solutions of population size obtained from simulated annealing are the solution of genetic algorithm for the next generation. The genetic algorithm and simulated annealing exchange continues until the required stopping criterion is achieved (Ganesh & Punniyamoorthy, 2004).

Evolutionary Pattern Search algorithms (EPSAs) are a class of evolutionary algorithms (EAs) that adapt the step size of the mutation operators to guarantee convergence to a stationary point of an objective function, where the gradient is zero, with probability one (Hart, 1998). While, EAs are typically described as methods for global optimization, the convergence analysis for EPSAs does not guarantee that the global optimum is found. However, the constraints imposed on EPSAs to ensure convergence to a stationary point provide insight into the requirements needed to converge to a global optimum. For example, the convergence analysis of pattern search methods highlights the fact that the rate of convergence of EPSAs is likely to decrease as the dimension of the problem increases (Torczon, 1997).

Aside from theoretical considerations, experience with pattern search algorithms suggests that EPSAs can be successfully applied to a wide range of optimization problems (Wright, 1996). For example, Meza and Martinez (1994) applied a pattern search method, parallel directed search, to the global optimization of the conformation energy of a simple chain molecule. Their comparison of parallel directed search to a genetic algorithm and simulated annealing algorithm indicates that parallel directed search can be equally effective at performing global optimization for this problem.

The definition of EPSAs is motivated by the similarity between EAs and pattern search methods (Torczon, 1997). Pattern search methods use an exploratory moves algorithm to conduct a series of moves about the current iterates before identifying a new iterate. In EPSAs, the iteration of the loop indexed by h corresponds to an iteration of an exploratory moves algorithm for a stochastic pattern search algorithm. The restriction on the replacement strategy ensures that the best individual found is used during the search (Hart, 1998).

Hart (1998) experimental evaluation of EPSAs indicates that they can find near-optimal points on some standard global optimization test functions. Their experiment illustrate the different search dynamics of EPSAs and EAs, and they demonstrate the EPSAs can perform nonlocal optimization. Furthermore, the ability for EPSAs to reliably terminate near stationary points offers a practical advantage over other EAs, which are typically stopped by a bound on the number of iteration is specified by the user.

In Hart (1998) experiments, the EPSAs performed about as well as the EAs, which is interesting since EPSAs cannot in general perform global optimization. Although a comparison with adaptive EAs like evolutionary programming is needed, these results suggest that EPSAs are better described as nonlocal optimizers since they perform nonlocal search through the stochastic competition between disparate solutions. In fact, Hart (1998) believe that nonlocal methods like EPSAs could be advantages, even if they ultimately do not perform a global optimization.

There are two ways in which simulated annealing may be combined with an alternative method; either using the alternative method to provide a 'good' initial solution which simulated annealing attempts to improve, or by using simulated annealing to provide a 'good' initial solution as a starting point for the alternative method (Eglese, 1990). In this research work the first approach is adopted as an optimization problem solving method.

The first of these approaches is illustrated by Chams, Hertz, and Werra (1987) for example, when considering graph colouring problems. Johnson, Aragon, McGeogh, and Schevon (1987) also provide some experimental results for the graph partitioning problem. They show that both quality of solution and running time may be improved by the use of a good starting solution. When a good initial solution is used, the initial temperature in the cooling schedule is reduced; otherwise the benefits of the good initial solution will be lost. They also show that starting solutions which take advantage of the special structure of the problem instance being considered seem preferable to those obtained by general heuristics.

The second approach is exemplified by using simulated annealing as a way of obtaining a good initial solution for a branch and bound algorithm or integer programming algorithms (Eglese, 1990).

For many real world optimization problems, the environment is uncertain or fuzzy, leading to dramatic changes in the fitness or objective function values of individual solution. The industrial production planning problems with uncertainty in the profit function is a good example. Optimization under uncertain environment can be handled nicely by hybrid evolutionary techniques (Jong, 1999). The industrial production planning problems can be handled with successfully in an uncertain environment.

Based on the above mention advantages of hybrid evolutionary computation, the following listed objectives are thoroughly investigated in this research work.

- **Provides a new formulation of the industrial production planning problem with the global solution approaches:** Jimenez, Sanchez, Vasant, and Verdegay (2006) have formulated the above problems with quadratic objective function and solved by fuzzy evolutionary approach. But in this research work the author has formulated a new objective function in the form of cubic function. The idea of cubic objective function was obtained from the reference (Lin & Yao, 2002). Lin and Yao (2002) solved a theoretical problem of fuzzy optimal profit with cubic function by using genetic algorithm approach without fuzzy membership function. In this research the cubical objective function with 21 inequality constraints and 8 bound constraints was solved by various hybrid evolutionary optimization techniques with fuzzy modified s-curve membership function.
- **Reports the best known approaches and its results and findings for the real world application problems:** The major approaches which were adopted in this research work are: Hybrid Genetic Algorithms and Line Search, Hybrid Genetic Algorithms and General Pattern Search, Hybrid Genetic Algorithms and Simulated Annealing, Hybrid General Pattern Search and Simulated Annealing, Hybrid Mesh Adaptive Direct Search and Genetic Algorithms and Hybrid Line Search and Simulated Annealing. All the above methods are successfully utilized and solved the real world problems of industrial production planning.
- **Investigate the optimization of the objective function with constraint handling approaches of the problem:** Various methods have been provided in Vasant (2012) on constraint handling techniques. Majority of these techniques are related to penalty function approaches. The main drawback of penalty functions is the careful fine tuning required by the penalty factors, which determine the severity of the penalization. Moreover it's difficult to use penalty function approaches for the large scale problem with many constraints. Based on this limitation, several approaches have been proposed to deal with it and also alternative techniques such as hybrid approaches have been adopted in this research work.
- **Proposes a new form of the hybridization approaches as a global solution procedures:** There are 11 different techniques of hybridization have be utilized in this research work for solving a non linear cubic objective function of industrial production planning problems with 21 constraint and 8 bound constraints represent the eight decision variables. The non linear cubic objective function contains 24 coefficients for the 8 decision variables.
- **Find optimal solution for the objective function respect to decision variables, level of satisfaction, vagueness factor and computational time:** The fuzziness and uncertainty in the technological coefficients of constraints in the industrial production planning problems will be deal by fuzzy membership function of modified s-curve. The optimal profit function depends on the major factors such as vagueness and computational time.
- **Comparison of classical approaches versus hybrid approaches in solving optimization problems of industrial production planning:** There are four non-hybrid approaches such as genetic algorithms; general pattern search, mesh adaptive direct

search and line search have been adopted in solving this problem. On the other hand 11 hybrid techniques have been utilized in solving the similar problem.

The scope of this thesis is to propose novel methods, which would take benefit of newly formed hybrid approaches: being sufficiently general to be used for particular real world problems of industrial production planning, it should still be able to incorporate specific knowledge about this problem.

The final goal is therefore to achieve findings best possible optimal profit for the industrial production planning problems with non linear objective function with 21 constraints and eight decision variables with respect to decision variables, level of satisfaction, vagueness factor and computational time. To perform this task, the methods proposed in this thesis should succeed in finding optimal but realistic solutions for the implementation.

BACKGROUND

Optimization problems occur in many engineering, economics, and scientific research projects, like cost, time, and risk minimization or quality, revenue, and efficiency maximization (Back, Fogel, & Michalewicz, 1997). Thus, the development of general strategies is of great value.

In real-world situations, the objective function f and the constraints g_j are often not analytically treatable or are even not given in closed form, of the function definition are based on a simulation model (Schwefel, 1979).

The traditional approach is such cases is to develop a formal model that resembles the original functions close enough but is solvable by means of traditional mathematical methods such as linear and non-linear programming. This approach most often requires simplifications of the original problem formulation. Thus, an important aspect of mathematical programming lies in the design of the formal model.

No doubt, this approach has proven to be very successful in many applications, but has several drawbacks, which motivated the search for novel approaches, where evolutionary computation is one of the most promising directions. The most severe problem is that, due to over simplifications; the computed solutions do not solve the original problem. Such problems, in the case of simulation models, are often considered unsolvable.

The fundamental difference in the evolutionary computation approach is to adapt the method to the problem at hand. In author's opinion, evolutionary algorithms should not be considered as off-the-peg, ready-to-use algorithms but rather as a general concept which can be tailored to most of the real-world applications that often are beyond solution by means of traditional methods. Once a successful evolutionary computation framework has been developed it can be incrementally adapted to the problem under consideration (Michalewicz, 1993), to changes of the requirements of the research project, to modifications of the model, and the change of hardware resources.

In concluding this section, referring to the research field of computational intelligence (Bezdek, 1994) and the applications of hybrid evolutionary computation to other main fields of computational intelligence, namely fuzzy logic and simulated annealing. An overview of the utilization and hybridization of genetic algorithms with simulated annealing is given in (Wah & Chen, 2003). Similarly, both the rule base and membership functions of fuzzy systems can be optimized by hybrid evolutionary algorithms, typically yielding improvements of the performance of the fuzzy system (Karr, 1991). The interaction of computational intelligence techniques and hybridization with other methods such as evolutionary computation, general pattern search, mesh adaptive direct search and local optimization techniques certainly opens a new direction of research toward hybrid systems that exhibit problem solving capabilities those of naturally intelligent systems in the future. Hybrid evolutionary algorithms, seen as a technique to evolve machine intelligence (Fogel, 1995), are one of the mandatory prerequisites for achieving this goal by means of algorithms principles that are already working quite successfully in natural evolution (Dennett, 1995).

Based on the above arguments, the author is convinced that it is just a beginning to understand and to exploit the full potential of hybrid evolutionary computation. Concerning basic research as well as practical application to challenging industrial production planning problems, hybrid evolutionary algorithms offer a wide range of promising further investigation, and it is exciting to observe the future development of the field.

DEVELOPMENT OF PRODUCTION PLANNING MODEL

The S-curve membership function is a particular case of the logistic function with specific values of B, C and α. This logistic function as given by Equation (1) and depicted in Figure 1 is indicated as S-shaped membership function by Gonguen (1969) and Zadeh (1971).

The logistic membership function for the FLP problem is defined as:

$$f(x) = \begin{cases} 1 & x < x_L \\ \dfrac{B}{1 + Ce^{\alpha x}} & x_L < x < x_U \\ 0 & x > x_U \end{cases} \quad (1)$$

where f(x) is the degree of membership function of a specific parameter value x, such that 0 < f(x) < 1. The parameter x is considered to be a member of the related fuzzy set ; x_L and x_U are respectively the lower boundary and upper boundary for the fuzzy parameter x. B and C are constants and the parameter α > 0 determines the shapes of

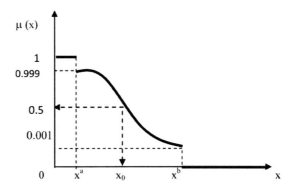

Figure 1. S-Curve membership function

membership function. The larger the value of α, the more is its vagueness.

The definition of a modified S-curve membership functions as follows:

$$\mu(x) = \begin{cases} 1 & x < x^a \\ 0.999 & x = x^a \\ \dfrac{B}{1 + Ce^{\alpha x}} & x^a < x < x^b \\ 0.001 & x = x^b \\ 0 & x > x^b \end{cases} \quad (2)$$

where μ is the degree of membership function.

Figure 1 shows the S-curve . Equation (2) is similar to Equation (1) except that the membership function is redefined as 0.001 ≤ μ (x) ≤ 0.999. This range is selected because in manufacturing system the work force need not be always 100% of the requirement. At the same time the work force will not be 0%. Therefore there is a range between x^0 and x^1 with $0.001 \leq \mu(x) \leq 0.999$.

We rescale the x axis as $x^a = 0$ and $x^b = 1$ in order to find the values of B, C and α. Novakowska (1977) has performed such a rescaling in his work of social sciences.

The values of B, C and α are obtained from Equation (2) as

$$B = 0.999 (1 + C) \quad (3)$$

$$\frac{B}{1+Ce^{\alpha}} = 0.001 \qquad (4)$$

By substituting Equation (3) into Equation (4):

$$\frac{0.999(1+C)}{1+Ce^{\alpha}} = 0.001 \qquad (5)$$

Rearranging Equation (5)

$$\alpha = \ln\frac{1}{0.001}\left(\frac{0.998}{C}+0.999\right) \qquad (6)$$

Since, B and α depend on C, we require one more condition to get the values for B, C and α

Let, when $x_0 = \dfrac{x^a + x^b}{2}$, $\mu(x_0) = 0.5$; Therefore

$$\frac{B}{1+Ce^{\frac{\alpha}{2}}} = 0.5 \qquad (7)$$

and hence

$$\alpha = 2\ln\left(\frac{2B-1}{C}\right) \qquad (8)$$

Substituting Equation (3) and Equation (6) in to equation (8), we obtain

$$2\ln\left(\frac{2(0.999)(1+C)-1}{C}\right) = \ln\frac{1}{0.001}\left(\frac{0.998}{C}+0.999\right) \qquad (9)$$

Rearranging equation (2.27) yields

$$\left(0.998 + 1.998C\right)^2 = C\left(998 + 999C\right) \qquad (10)$$

Solving equation (2.28):

$$C = \frac{-994.011992 \pm \sqrt{988059.8402 + 3964.127776}}{1990.015992} \qquad (11)$$

Since C has to be positive, Equation (11) gives $C = 0.001001001$ and from Equation (3) and (8), B = 1 and α = 13.81350956.

This section outlines an introduction to real-world industrial problem for product-mix selection involving eight variables and 21 constraints with fuzzy technological coefficients and thereafter, a formulation for an optimization approach to solve the problem. This problem occurs in production planning in which a decision maker plays a pivotal role in making decision under fuzzy environment. Decision-maker should be aware of his/her level of satisfaction as well as degree of fuzziness while making the product-mix decision. Thus, a thorough analysis performed on a modified S-curve membership function for the fuzziness patterns and fuzzy sensitivity solution found from the various optimization methodologies (Bhattacharya & Vasant, 2007; Bhattacharya, Vasant, & Susanto, 2007; Vasant & Kale, 2007; Vasant & Barsoum, 2006). An evolutionary algorithm is proposed to capture multiple non dominated solutions in a single run of the algorithm. Results obtained have been compared with the well-known various hybrid evolutionary algorithms.

It is well known that optimization problems arise in a variety of situations. Particularly interesting are those concerning management problems as decision makers usually state their data in a vague way: "high profits," "low cost," "average revenue," etc. Because of this vagueness, managers prefer to have not just one solution but a set of them, so that the most suitable solution can be applied according to the state of existing decision of the production process at a given time and without increasing delay. In these situations fuzzy optimization is an ideal methodology, since it allows us to represent the underlying uncertainty of the optimization problem, while finding optimal solutions that reflect such uncertainty and then applying them to possible instances, once the uncertainty has been solved. This allows us to obtain a model of the behavior of the solutions based on the uncertainty of the optimization problem.

Fuzzy constrained optimization problems have been extensively studied since the seventies. In the linear case, the first approaches to solve the so-called fuzzy linear programming problem appeared in Bellman and Zadeh (1970). Since then, important contributions solving different linear models have been made and these models have been the subject of a substantial amount of work. In the nonlinear case (Ramik & Vlach, 2002) the situation is quite different, as there is a wide variety of specific and both practically and theoretically relevant nonlinear problems, with each having a different solution method.

In this chapter, a real-life industrial problem for product mix selection involving 21 constraints and eight bound constraints has been considered. This problem occurs in production planning in which a decision-maker plays a pivotal role in making decision under a highly fuzzy environment (Vasant, Bhattacharya, Sarkar, & Mukherjee, 2007; Vasant, Barsoum, Kahraman, & Dimirovski, 2007). Decision maker should be aware of his/her level-of satisfaction as well as degree of fuzziness while making the product mix decision. Thus, we have analyzed using the sigmoidal membership function, the fuzziness patterns and fuzzy sensitivity of the solution. Vasant (2006) considered a linear case of the problem and solved by using a linear programming iterative method, which is repeatedly applied for different degrees of satisfaction values. In this research, a nonlinear case of the problem is considered and proposed a various optimization approaches in order to capture solutions for different levels of satisfaction with a single and multiple run of the algorithm. This various optimization approaches has been proposed by Liang (2008); Sanchez, Jimenez, and Vasant (2007); Turabieh, Sheta, and Vasant (2007); Bhattacharya, Abraham, Vasant, and Grosan (2007); Bhattacharya, Vasant, Sarkar, and Mukherjee (2006); and Jiménez, Gómez-Skarmeta, and Sánchez (2004) within a soft computing optimization general context. The detail on the nonlinear case study will be provided in the following section.

Due to limitations in resources for manufacturing a product and the need to satisfy certain conditions in manufacturing and demand, a problem of fuzziness occurs in industrial systems. This problem occurs also in chocolate manufacturing when deciding a mixed selection of raw materials to produce varieties of products. This is referred to as the product-mix selection problem (Tabucanon, 1996).

There are a number of products to be manufactured by mixing different raw materials and using several varieties of processing. There are limitations in resources of raw materials and facility usage for the varieties of processing. The raw materials and facilities usage required for manufacturing each product are expressed by means of fuzzy coefficients. There are also some constraints imposed by marketing department such as product-mix requirement, main product line requirement and lower and upper limit of demand for each product. It is necessary to obtain maximum profit with certain degree of satisfaction of the decision-maker.

The firm Chocoman Inc. manufactures 8 different kinds of chocolate products. Input variables x_i represents the amount of manufactured product in 10^3 units.

The function to maximize is the total profit obtained calculated as the summation of profit obtained with each product and taken into account their coefficients. Table 1 shows the profit coefficients (c_i), (d_i) and (e_i) for each product i.

There are 8 raw materials to be mixed in different proportions and 9 processes (facilities) to be utilized. Therefore, there are 17 constraints with fuzzy coefficients separated in two sets such as raw material availability and facility capacity. These constraints are inevitable for each material and facility that is based on the material consump-

Table 1. Profit Coefficients c_i, d_i and e_i (Profit function in US $ per 10^3 units)

Product (x_i)	Synonym	c_i	d_i	e_i
x_1 = Milk chocolate, 250g	MC 250	c_1 = 180	d_1 = 0.18	e_1 = 0.01
x_2 = Milk chocolate, 100g	MC 100	c_2 = 83	d_2 = 0.16	e_2 = 0.13
x_3 = Crunchy chocolate, 250g	CC 250	c_3 = 153	d_3 = 0.15	e_3 = 0.14
x_4 = Crunchy chocolate, 100g	CC 100	c_4 = 72	d_4 = 0.14	e_4 = 0.12
x_5 = Chocolate with nuts, 250g	CN 250	c_5 = 130	d_5 = 0.13	e_5 = 0.15
x_6 = Chocolate with nuts, 100g	CN 100	c_6 = 70	d_6 = 0.14	e_6 = 0.17
x_7 = Chocolate candy	CANDY	c_7 = 208	d_7 = 0.21	e_7 = 0.18
x_8 = Chocolate wafer	WAFER	c_8 = 83	d_8 = 0.17	e_8 = 0.16

Table 2. Raw material and Facility usage required (per 10^3 units) ($\tilde{a}_{ij} = [a^l_{ij}, a^h_{ij}]$) and Availability ($b_j$)

Material or Facility	MC 250	MC 100	CC 250	CC100	CN250	CN100	Candy	Wafer
Cocoa (kg)	[66, 109]	[26, 44]	[56,9]	[22,37]	[37,62]	[15,25]	[45, 75]	[9, 21]
Milk (kg)	[47, 78]	[19, 31]	[37,6]	[15,25]	[37,62]	[15,25]	[22, 37]	[9, 21]
Nuts (kg)	[0, 0]	[0, 0]	[28,4]	[11,19]	[56,94]	[22,37]	[0, 0]	[0, 0]
Cons. sugar (kg)	[75, 125]	[30, 50]	[66,109]	[26,44]	[56,94]	[22,37]	[157,262]	[18,30]
Flour (kg)	[0, 0]	[0, 0]	[0, 0]	[0, 0]	[0, 0]	[0, 0]	[0, 0]	[54,90]
Alum. foil (ft²)	[375,625]	[0, 0]	[375,625]	[0, 0]	[0, 0]	[0, 0]	[0, 0]	[187,312]
Paper (ft²)	[337,562]	[0, 0]	[337,563]	[0, 0]	[337,562]	[0, 0]	[0, 0]	[0, 0]
Plastic (ft²)	[45, 75]	[95, 150]	[45, 75]	[90,150]	[45,75]	[90, 150]	[1200,200]	[187,312]
Cooking(ton-hours)	[0.4, 0.6]	[0.1, 0.2]	[0.3, 0.5]	[0.1, 0.2]	[0.3,0.4]	[0.1, 0.2]	[0.4, 0.7]	[0.1,0.12]
Mixing (ton-hours)	[0, 0]	[0, 0]	[0.1, 0.2]	[0.04,0.07]	[0.2, 0.3]	[0.07, 0.12]	[0, 0]	[0, 0]
Forming(ton-hours)	[0.6, 0.9]	[0.2, 0.4]	[0.6, 0.9]	[0.2, 0.4]	[0.6, 0.9]	[0.2, 0.4]	[0.7, 1.1]	[0.3, 0.4]
Grinding(ton-hours)	[0, 0]	[0, 0]	[0.2, 0.3]	[0.07, 0.12]	[0, 0]	[0, 0]	[0, 0]	[0, 0]
Wafer making (ton-hours)	[0, 0]	[0, 0]	[0, 0]	[0, 0]	[0, 0]	[0, 0]	[0, 0]	[0.2, 0.4]
Cutting (hours)	[0.07,0.2]	[0.07,0.12]	[0.07,0.12]	[0.07, 0.12]	[0.07, 0.12]	[0.07, 0.12]	[0.15, 0.25]	[0, 0]
Packaging1 (hours)	[0.2, 0.3]	[0, 0]	[0.2, 0.3]	[0, 0]	[0.2, 0.3]	[0, 0]	[0, 0]	[0, 0]
Packaging2 (hours)	[0.04,0.6]	[0.2, 0.4]	[0.04, 0.06]	[0.2, 0.4]	[0.04, 0.06]	[0.2, 0.4]	[1.9, 3.1]	[0.1, 0.2]
Labour (hours)	[0.2, 0.4]	[0.2, 0.4]	[0.2, 0.4]	[0.2, 0.4]	[0.2, 0.4]	[0.2, 0.4]	[1.9, 3.1]	[1.9, 3.1]

tion, facility usage and the resource availability. Table 2 shows fuzzy coefficients $\tilde{a}_{ij}$ represented by (a^l_{ij}, a^h_{ij}) for required materials and facility usage j for manufacturing each product i and non fuzzy coefficients b_j for availability of material or facility j is given in Table 4.

Additionally, the following constraints were established by the sales department of Chocoman Inc.:

1. Main product line requirement. The total sales from candy and wafer products should not exceed 15% of the total revenues from

Table 3. Demand (u_k) and Revenues/Sales (r_k) in US $ per 10^3 units

Product (x_k)	Synonym	Demand (u_k)	Revenues/Sales (r_k)
x_1 = Milk chocolate, 250g	MC 250	u_1 = 500	r_1 = 375
x_2 = Milk chocolate, 100g	MC 100	u_2 = 800	r_2 = 150
x_3 = Crunchy chocolate, 250g	CC 250	u_3 = 400	r_3 = 400
x_4 = Crunchy chocolate, 100g	CC 100	u_4 = 600	r_4 = 160
x_5 = Chocolate with nuts, 250g	CN 250	u_5 = 300	r_5 = 420
x_6 = Chocolate with nuts, 100g	CN 100	u_6 = 500	r_6 = 175
x_7 = Chocolate candy	CANDY	u_7 = 200	r_7 = 400
x_8 = Chocolate wafer	WAFER	u_8 = 400	r_8 = 150

Table 4. Raw material availability (b_j)

Material or Facility	Availability
Cocoa (kg)	100000
Milk (kg)	120000
Nuts (kg)	60000
Cons. sugar (kg)	200000
Flour (kg)	20000
Alum. foil (ft^2)	500000
Paper (ft^2)	500000
Plastic (ft^2)	500000
Cooking (ton-hours)	1000
Mixing (ton-hours)	200
Forming (ton-hours)	1500
Grinding (ton-hours)	200
Wafer making (ton-hours)	100
Cutting (hours)	400
Packaging 1 (hours)	400
Packaging 2 (hours)	1200
Labour (hours)	1000

the chocolate bar products. Table 3 show the values of sales/revenues (r_i) for each product i.

2. Product mix requirements. Large-sized products (250 g) of each type should not exceed 60% of the small-sized product (100 g).

Finally, the lower limit of demand for each product i is 0 in all cases, while the upper limit (u_i) is shown in Table 3.

Membership Function for Technological Coefficients

The modified S-curve membership function proposed by Vasant (2006) is considered. For a value x, the degree of satisfaction $\mu_{\tilde{a}_{ij}}(x)$ for fuzzy coefficient $\tilde{a}_{ij}$ is given by the membership function given in Equation (12).

$$\mu_{\tilde{a}_{ij}}(x) = \begin{cases} 1.000 & x < a_{ij}^l \\ 0.999 & x = a_{ij}^l \\ \dfrac{B}{1 + Ce^{\alpha\left(\frac{x - a_{ij}^l}{a_{ij}^h - a_{ij}^l}\right)}} & a_{ij}^l < x < a_{ij}^h \\ 0.0001 & x = a_{ij}^h \\ 0.000 & x > a_{ij}^h \end{cases}$$

(12)

Given a level of satisfaction value μ, the crisp value $a_{ij}|_\mu$ for fuzzy coefficient $\tilde{a}_{ij}$ can be calculated using (13).

$$\tilde{a}_{ij}|_\mu = a_{ij}^l + \left(\frac{a_{ij}^h - a_{ij}^l}{\alpha}\right)\ln\frac{1}{C}\left(\frac{B}{\mu} - 1\right) \quad (13)$$

The value α determines the shape of the membership function, while B and C values can be calculated from α, given in Equations (14) and (15).

$$C = -\frac{0.998}{(0.999 - 0.001\,e^{\alpha})} \quad (14)$$

$$B = 0.999(1 + C) \quad (15)$$

If we wish that for a level of satisfaction value μ = 0.5, the crisp value $a_{ij}|_{0.5}$ is in the middle of the interval $[a_{ij}^l, a_{ij}^h]$, that is:

$$\tilde{a}_{ij}|_{0.5} = \frac{a_{ij}^l + a_{ij}^h}{2} \quad (16)$$

then, α = 13.81350956 (Vasant, 2006).

Problem Formulation

Optimization techniques are primarily used in production planning problems in order to achieve optimal profit, which maximizes certain objective function by satisfying a number of constraints. The first step in an optimal production planning problems is to formulate the underlying nonlinear programming (NLP) problem by writing the mathematical functions relating to the objective and constraints.

Given a degree of satisfaction value μ, the fuzzy constrained optimization problem can be formulated (Jiménez, Cadenas, Sánchez, Gómez-Skarmeta, & Verdegay, 2006; Vasant, 2006) as the non linear constrained optimization problem shown below.

$$\text{Maximize} \sum_{i=1}^{8} (c_i x_i - d_i x_i^2 - e_i x_i^3)$$

Subject to:

$$\sum_{i=1}^{8}\left[a_{ij}^l + \left(\frac{a_{ij}^h - a_{ij}^l}{\alpha}\right) \ln\frac{1}{C}\left(\frac{B}{\mu} - 1\right)\right] x_i - b_j \leq 0,$$
$$j = 1, 2, \ldots, 17 \quad (17)$$

$$\sum_{i=7}^{8} r_i x_i - 0.15 \sum_{i=1}^{6} r_i x_i \leq 0$$

$$x_1 - 0.6 x_2 \leq 0$$

$$x_3 = 0.6 x_4 \leq 0$$

$$x_5 - 0.6 x_6 \leq 0$$

$$0 \leq x_i \leq u_i, \quad i = 1, 2, \ldots, 8$$

In the above non-linear programming problem, the variable vector x represents a set of variables x_i, i = 1, 2,…, 8. The above optimization problem contains eight continuous variables and 21 inequality constraints. A test point x_i satisfying constrains is called feasible, if not infeasible. The set satisfying constrains is called the feasible domain. The aim of the optimization is to maximize the total production profit for the industrial production planning problems. The formulation of the new non-linear cubic function for this particular problem has been refereed to Lin (2007) and Chaudari (2007). The cubic objective function (Equation 17) has 24 coefficients for eight decision variables. The left hand side of the second equation of (17) represents the fuzzy technical coefficient. This problem considered one of the most challenging problems in the research area of industrial production planning.

Solutions and Recommendations

These sections discuss the computational results of hybrid optimization techniques such as hybrid genetic algorithms with general pattern search, hybrid genetic algorithms with simulated annealing, hybrid mesh adaptive direct search with genetic algorithms and hybrid line search with genetic algorithm have be reported. All these methods are utilized in solving industrial production planning problems.

Hybrid Genetic Algorithms and General Pattern Search

The hybrid genetic algorithm and general pattern search is a combination of genetic algorithms and general pattern search approach. Below is the algorithm for this hybrid approach.

Step 1: Initial population: Generate initial population randomly.
Step 2: Genetic operators:
  Selection: stochastic uniform
  Crossover: arithmetic crossover operator
  Mutation: Adaptive feasible mutation operator
Step 3: Evaluation: Best offspring and fitness function.
Step 4: Stop condition: If a pre-defined maximum generation number, time limit, fitness limit reached or an optimal solution is located during genetic search process, then stop; otherwise, go to Step 2.
Step 5: Continue with general pattern techniques.
Step 6: End

Simulation Results

Figure 2 provides the best optimal solution for the fitness function respect to level of satisfaction at $\alpha = 13.813$.

Table 5 reports the solution for the best fitness function and the decision variables at $\alpha = 13.813$. CPU time for running the simulation for $\alpha = 13.813$ is 16 minutes. The best optimal solution for the fitness function is 180776.1 at $\gamma = 0.1$.

Figure 3 explains the detail solution for the decision variables versus level of satisfaction. CPU time for $\alpha = 13.813$ at $\gamma = 0.99$ is 78.375 seconds.

The solution for the decision variable x_8 is unrealistic since for the majority values of γ, x_8 is zero. Therefore, this solution needs some improvement. For this purpose, the following hybrid methods investigated thoroughly.

Fitness values for various values of α from 1 to 41 given in Figure 4. The optimal value for fitness functions occurs at $\alpha = 9$ and $\gamma = 0.99$ is 197980.93. Total CPU time for the simulation of Figure 5.8 is 2 hours 18 minutes.

Figure 5 provides the holistic view of 3D mesh plot solution for various values of α, and γ respect to level of satisfaction. CPU time for 3D mesh plot is 2 hours 52 minutes.

The main contribution of hybrid genetic algorithms with general pattern search is that, it is able to locate the best near global optimal solution for the fitness function. On the other hand, it is unable to provide a reasonable computational CPU time for the fitness function values. Furthermore, this method fails to compute the non-zero solution for the x_8 decision variable at $\alpha = 13.813$. Therefore, further analysis and tests are required for other hybrid optimization methods in order to overcome this drawback. It is clear that this method skips the nonzero values of x_8.

Hybrid Genetic Algorithms and Simulated Annealing (HGASA)

In this section, integration of genetic algorithms and simulated annealing approaches applied in solving the industrial production planning problems. Below is the algorithm for simulated annealing.

Figure 2. Fitness value vs. level of satisfaction

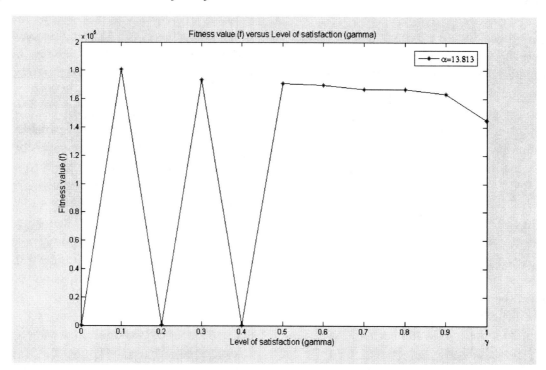

Table 5. Optimal value for fitness function

γ	x_1	x_2	x_3	x_4	x_5	x_6	x_7	x_8	f
0.001	0.62	0.80	0.94	0.00	0.21	0.24	0.00	0.00	366.4
0.1	346.7	580.5	276.2	460.4	148.7	247.9	189.2	4.54	180776.1
0.2	0.43	0.90	0.73	0.58	0.04	0.68	0.57	0.00	475.5
0.3	314.8	577.5	291.0	485.7	104.2	173.7	175.9	5.34	173329.7
0.4	0.00	0.00	0.00	0.61	0.51	1.69	0.59	0.00	351.2
0.5	278.2	466.2	282.0	521.1	156.5	260.8	164.2	0.00	170805.1
0.6	280.6	497.6	269.3	493.6	146.5	244.2	161.4	0.00	169598.1
0.7	258.1	602.6	260.3	472.7	128.5	214.2	153.6	0.00	166739.9
0.8	305.2	512.4	238.8	398.0	136.8	228.1	159.4	0.00	166566.4
0.9	257.7	444.3	269.2	450.5	146.1	243.6	148.2	0.00	163208.3
0.99	283.8	566.7	153.1	255.2	106.4	177.3	123.5	0.00	144803.8

Figure 3. Decision variables vs. level of satisfaction

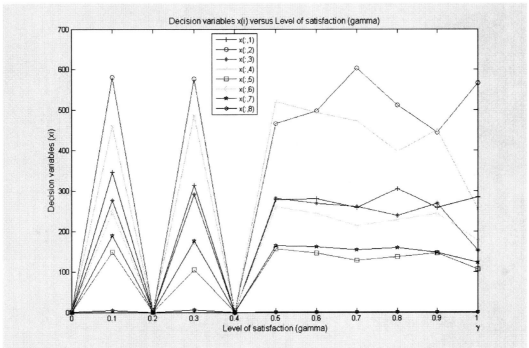

Figure 4. Fitness value vs. level of satisfaction

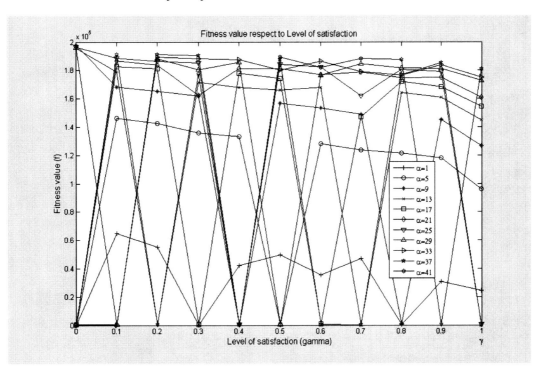

Figure 5. 3D Mesh plot for fitness value, α and γ

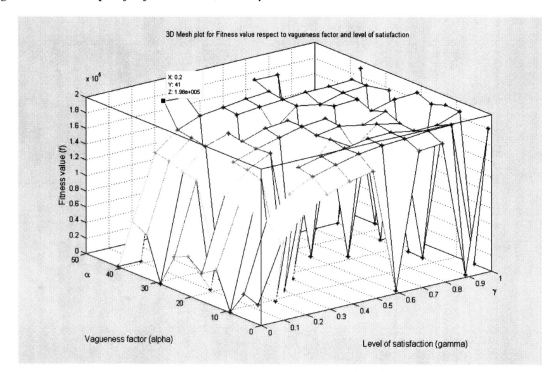

Outline of the Algorithm and Parameter Setting

The following is an outline of the steps performed for the simulated annealing algorithms:

1. The algorithm begins by randomly generating a new point. The distance of the new point from the current point, or the extent of the search, determined by a probability distribution with a scale proportional to the current temperature.
2. The algorithm determines whether the new point is better or worse than the current point. If the new point is better than the current point, it becomes the next point. If the new point is worse than the current point, the algorithm may still make it the next point. Simulated annealing accepts a worse point based on an acceptance probability. Threshold acceptance accepts a worse point if the objective function is raised by less than a fixed threshold.
3. The algorithm systematically lowers the temperature and (for threshold acceptance) the threshold, storing the best point found so far.
4. Reannealing is performed after a certain number of points (ReannealInterval) are accepted by the solver. Reannealing raises the temperature in each dimension, depending on sensitivity information. The search is resumed with the new temperature values.
5. The algorithm stops when the average change in the objective function is very small, or when any other stopping criteria are met.

Stopping Conditions for the Algorithm

The simulated annealing algorithms use the following conditions to determine when to stop:

- **Tol Fun:** The algorithm runs until the average change in value of the objective function in StallIter Lim iterations is less than Tol Fun. The default value is 1e-6.
- **Max Iter:** The algorithm stops if the number of iterations exceeds this maximum number of iterations. One can specify the maximum number of iterations as a positive integer or Inf. Inf is the default.
- **Max Fun Eval** specifies the maximum number of evaluations of the objective function. The algorithm stops if the number of function evaluations exceeds the maximum number of function evaluations. The allowed maximum is 3000*number of variables.
- **Time Limit** specifies the maximum time in seconds the algorithm runs before stopping.
- **Objective Limit:** The algorithm stops if the best objective function value is less than or equal to the value of Objective Limit.

Parameter Settings

The following is the typical commands.

- **Annealing Fcn:** @annealing fast
- **TemperatureFcn:** @temperature exp
- **Acceptance Fcn:** @acceptance sa
- **TolFun:** 1.0000e-006
- **StallIter Limit:** '500*number of variables'
- **Max Fun Evals:** '3000*number of variables'
- **Time Limit:** Infinite
- **Max Iter:** Infinite
- **Objective Limit:** Infinite
- **InitialTemperature:** 100
- **ReannealInterval:** 100
- **Data Type:** 'double'

Computational and Simulation Results for Hybrid Genetic and Simulated Annealing Algorithms

Figure 6 provides the outcome for the best optimal fitness value respect to level of satisfaction at $\alpha = 13.813$.

The best optimal fitness value f is 176057.7. Total CPU simulation time for this value at $\gamma = 0.99$ is 4.30 seconds. CPU time for running GA is 2.1488 seconds and for SA is 2.1491 seconds at $\gamma = 0.99$.

Figure 7 is the results of simulation for feasible solution for decision variables respect to level of satisfaction. CPU time for running GA and SA is 4.5931 s and 4.5936 s respectively at $\gamma = 0.99$. Total CPU time for $\gamma = 0.001$ to $\gamma = 0.99$ is 48.24 s.

Table 6 reports a significant realistic and feasibility solution for all the eight decision variables. The best optimal realistic and feasible non-zero solution obtained for x_8 decision variable for $\gamma = 1$ to $\gamma = 0.99$. This is a significant performance of this method in this optimization process. Total CPU time for GA and SA at $\gamma = 0.99$ is 9.1868 s. The improved fitness function value now is 195828.7 at $\gamma = 0.99$.

Figure 8 provides the detail investigation on the relationship between the eight decision variables and fitness function value.

Total CPU time for the simulation is 45.39 s. CPU time for running GA and SA is 1.8211s and 1.8214 s respectively. The results for the eight decision variables in the Figure 5.12 reflect the realistic solutions in the practical world.

Further investigation carried out to enhance the best optimal solution for the fitness function value. Figure 9 depicts the simulation results for $\gamma = 1$ to $\gamma = 41$. Total CPU time for the computations is 8 minutes 25.78 seconds. CPU time for running GA and SA at $\gamma = 0.99$ and $\alpha = 41$ is 5.5209 and 5.5212 respectively. The best optimal solution for the fitness function is 198601.7 at $\alpha = 17$ and $\gamma = 0.99$.

Figure 6. Fitness value vs. level of satisfaction

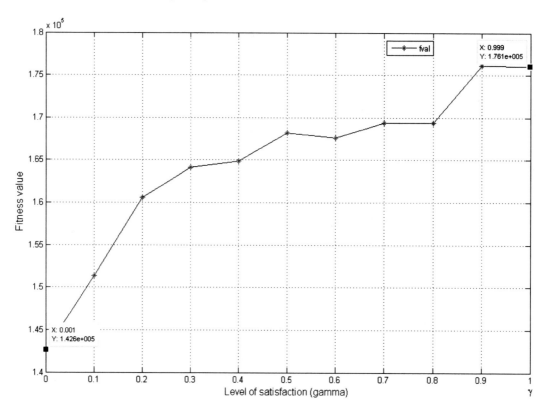

Figure 7. Decision variables vs. level of satisfaction

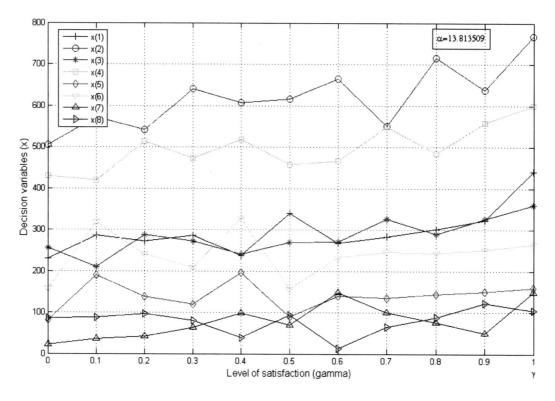

Table 6 Optimal value for fitness function (f)

γ	x_1	x_2	x_3	x_4	x_5	x_6	x_7	x_8	f
0.001	230.4	504.8	256.0	429.9	81.5	158.4	22.3	85.8	138507.0
0.1	286.3	572.3	210.2	419.4	190.5	317.6	36.6	87.9	155100.9
0.2	273.0	541.2	287.6	513.5	138.6	240.1	41.3	96.4	158323.2
0.3	285.9	640.0	272.1	472.0	120.3	207.9	63.8	80.2	161210.7
0.4	236.8	606.8	240.8	518.4	196.0	327.3	98.4	39.3	163769.9
0.5	340.4	616.8	269.6	457.6	90.00	157.2	69.9	94.2	161944.9
0.6	267.8	665.4	270.6	266.5	139.9	233.4	148.2	13.4	170991.2
0.7	282.5	549.7	326.6	549.0	134.6	246.0	100.2	64.2	169956.3
0.8	301.8	715.9	289.4	484.9	144.5	242.8	76.3	87.6	170203.5
0.9	323.5	637.1	326.2	558.3	150.4	250.9	49.2	122.8	171209.0
0.99	440.5	767.5	359.0	599.4	159.0	265.2	149.0	104.6	195828.7

Figure 8. Fitness value vs. decision variables

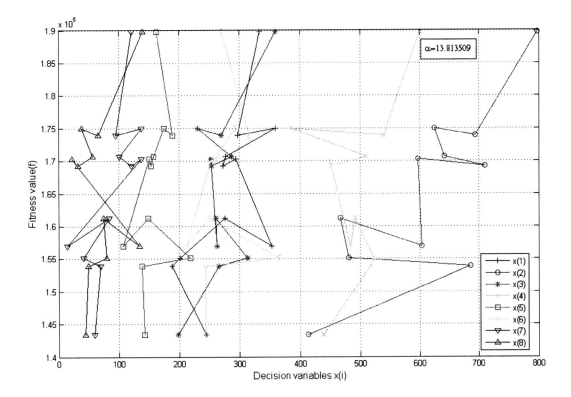

Figure 9. Fitness value vs. level of satisfaction

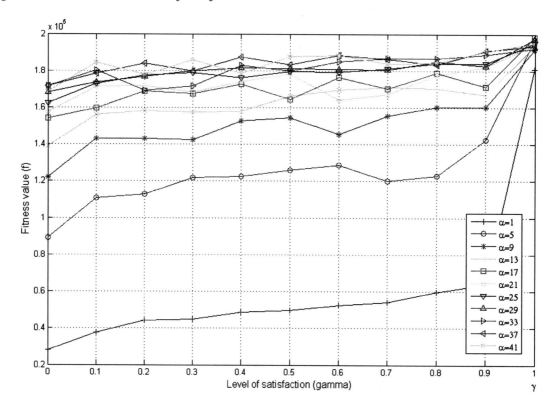

The mesh 3D plot in Figure 10 provides a holistic solution for the fitness function respect to decision variables and level of satisfaction. The best optimal fitness function value is 199171.28 occurs at α = 13 and γ = 0.99. Total CPU time is 4 minutes 29.28 seconds. CPU time for running GA and SA at α = 29 and γ = 0.99 is 3.081 and 3.013 respectively.

Figure 11 depicts the scenario where by the GA solution trapped and stuck at local optima. By using SA algorithm, the local optima solution removed and near global optimal value found for the fitness function.

The best near global optimal for the fitness found by the simulation results for hybrid GA and SA techniques. The significant contribution for this method is on the realistic, feasible and quality solutions for decision variables especially for x_8. There was a tremendous improvement in the results for this method. The best part of this method is on the reasonable computational CPU time in running GA and SA algorithms. The major contribution of population-based method of GA is on the search of good initial solution and the robust convergence optimal solution. On the other hand, SA performed as a reasonably good optimizer. The author believes that the major contribution of SA in these case studies is very much highly appreciated.

Hybrid Mesh Adaptive Direct Search and Genetic Algorithms (HMADSGA)

In this section, computational and simulation results are presented for the performance of the fitness function, decision variables and CPU time of the proposed hybridization method of MADS and GA. Vasant (2013) provides MADS algorithms and parameter settings. Vasant (2012) provides GA algorithms and parameter settings.

Figure 10. 3D Mesh plot for fitness value f, decision variables x and α

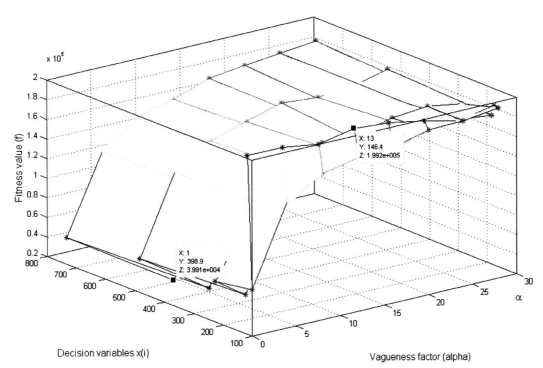

Figure 11. Vagueness factor vs. level of satisfaction

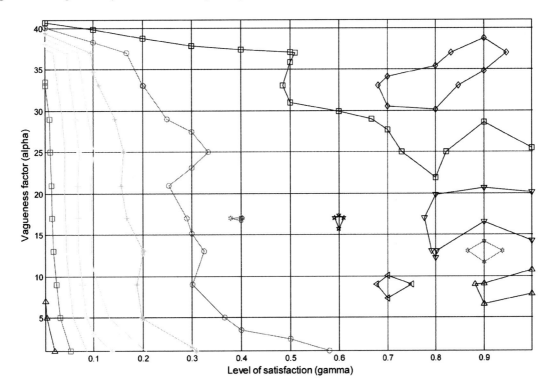

A thorough investigation performed for the quality of the best fitness function solution, decision variables and computational CPU time.

Simulation Results

Figure 12 depicts the best fitness function solution for $\alpha = 13.813$ and $\gamma = 0.001$ to $\gamma = 0.99$. The best optimal fitness function value is 197364.2 obtained at $\gamma = 0.99$ after three iteration with 8251 number of function evaluations. CPU time for running MADS and GA is 39.427 s and 39.4275 s respectively. Total CPU time for running at $\alpha = 13.813$ and $\gamma = 0.001$ to $\gamma = 0.99$ is 8m 49.32 s.

Table 7 reports feasible optimal solution for eight decision variables with γ and fitness function values.

The feasible solutions in Table 7 for the eight decision variables reflect the real world situation for the industrial production planning problems. The best optimal solution for the fitness function value at $\gamma = 0.99$ is less than fitness function value in the MADS method alone. This is due to MADS stopped the optimization process earlier than GA searching process. Furthermore the CPU time for this hybrid approach is almost double than MADS alone. The major drawback of this method is, it is suffering from slow convergence and it is wandering around optimal solution if high accuracy needed.

From the above explanation, it suspected that further experiment be carried on for several of α in order to investigate improved optimal solution for the fitness function value. Figure 13 depicts the simulation results for $\alpha = 1$ to $\alpha = 41$.

The best optimal fitness function value is 199813.94 obtained at $\gamma = 0.99$ and $\alpha = 1$. CPU time for running MADS and GA is 6.1055 s and 6.1059 s respectively for $\gamma = 0.99$ and $\alpha = 41$. Number of iterations and function evaluations is 3 and 6001 respectively. Total CPU time for $\alpha = 1$ to $\alpha = 41$ is 1 h 54 m 48.87 s.

Table 8 reports the best optimal fitness function values for $\alpha = 1$ to $\alpha = 41$ at $\gamma = 0.99$.

From the Table 8, it is concluded that this hybrid approach reaches the best optimal fitness value 199813.94 at $\alpha = 1$. This is the major significant contribution of MADS and GA hybrid approach in this case studies. On the other hand, MADS alone reaches optimal fitness value 198002 at $\alpha = 1$ and $\gamma = 0.99$.

Further experiment carried out for the investigation of simulation results for the fitness function respect to α and γ. Figure 14 depicts the results.

Table 7. Optimal value for fitness function

γ	x_1	x_2	x_3	x_4	x_5	x_6	x_7	x_8	f
0.001	255.5	545.2	102.3	341.9	92.4	233.5	66.5	41.8	141916.7
0.1	286.3	644.3	213.6	459.1	125.0	208.4	67.3	65.6	156521.4
0.2	270.8	540.7	285.9	488.6	114.4	192.3	109.4	41.0	163081.0
0.3	249.0	496.2	269.2	522.5	188.9	314.8	99.4	41.6	164043.3
0.4	259.7	669.3	237.4	408.7	185.8	309.6	114.3	30.6	166244.4
0.5	326.3	664.2	185.0	425.8	161.8	269.7	133.8	22.3	167783.4
0.6	352.8	588.7	284.1	473.5	91.4	169.4	94.8	77.0	166901.8
0.7	300.7	598.5	242.5	566.8	187.7	312.9	72.0	78.2	167212.6
0.8	322.8	574.7	284.4	479.2	161.3	268.9	123.0	51.2	175329.0
0.9	368.3	653.2	240.2	402.1	167.3	278.8	157.2	30.8	178822.2
0.99	444.0	740.4	358.9	598.2	149.3	248.8	178.0	81.0	197364.2

Hybrid Evolutionary Optimization Algorithms

Figure 12. Fitness value vs. γ

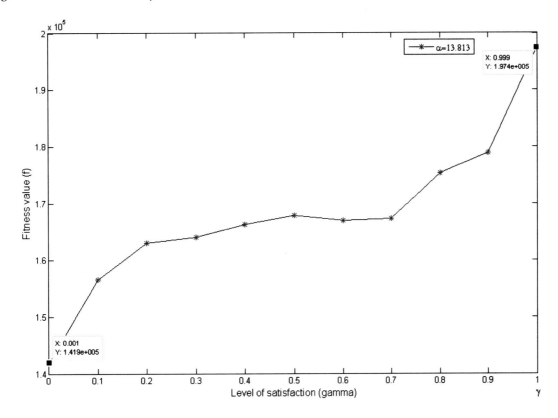

Figure 13. Fitness value vs. γ

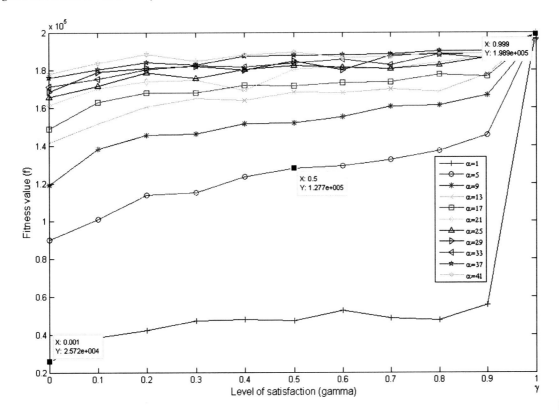

Figure 14. Fitness function vs. α and γ

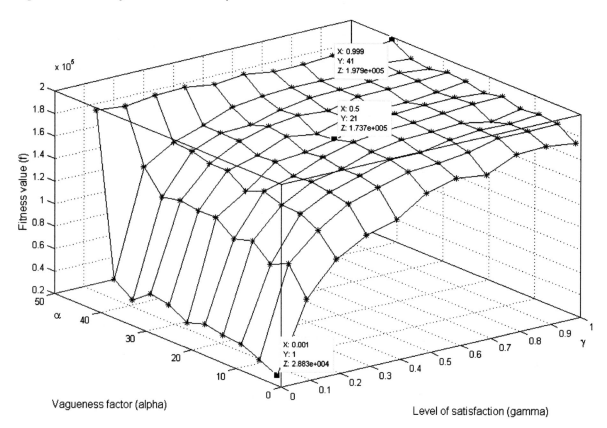

Table 8.. Optimal value for Fitness Function at γ = 0.99

α	f
1	199813.94
5	199171.89
9	197168.54
13	199357.83
17	197303.77
21	195918.66
25	198955.51
29	197928.65
33	198738.09
37	198915.82
41	196138.06

Total CPU time for this simulation is 1 h 45m 17.79 s. Optimal fitness function value at α = 41 and γ = 0.99 is 197902.8. CPU time running GPS and GA for this result is 83.3417 s and 83.3422 s respectively.

The findings of the computational and simulation results reveal that the significant contribution of MADS and GA hybrid approach is in finding a reasonable quality solution for the feasible decision variables and fitness function value. The incorporation of GA into MADS exhibit far more superior speedy solution in CPU computational time compare to HGPSGA techniques. On the other hand, its major drawback is on the longer computational CPU time in running MADS and GA algorithms. This is due to the role of GA in this optimization process as searching technique for the best initial solution and MADS played as an optimizer. Nevertheless, there is a very strong possibility of improving this draw back with other additional hybrid optimizations techniques.

Hybrid Line Search and Genetic Algorithms (HLSGA)

In this section, the simulation and computational results on the hybridization techniques of merging LS and GA presented for the industrial production planning problems is propounded. LS techniques utilized for the searching of best initial points to start the optimization process, and GA techniques are used for finding the best global close to the optimal solutions. GAs are in a class of biologically motivated optimization methods that evolve a population of individuals where individuals who are more fit have a higher possibility of surviving into subsequent generations. Line search method is a type of gradient-based method that uses derivative of objective function and constraints for the continuous functions. Its' suggested that this hybrid method would be a good candidate to find best global near optimal value for the fitness function and feasible solution for the decision variables as well as a reasonable computational CPU time.

Simulation and Computational Analysis

The hybrid algorithm for LS and GA in the program is as follows:

Step 1: Start Line search (Vasant, 2012)
Step 2: Genetic algorithms (Vasant, 2013)
Step 3: End

Parameter setting for GA:

Population type: double vector
Population size: 20
Elite count: 2
Crossover fraction: 0.8
Migration direction: forward
Generation: 100
Time limit: Infinite
Fitness limit: Infinite
Creation function: uniform
Fitness scaling: rank
Selection function: stochastic uniform
Crossover: Scattered
Mutation: Gaussian

Figure 15 depicts the simulation result for the optimal solution of objective function at $\alpha = 13.813$. The optimal value for objective function is 200116.4 at $\gamma = 0.99$.

Table 9 indicates the feasible solution for the decision variables with $\gamma = 0.001$ to $\gamma = 0.99$ and objective function values at $\alpha = 13.813$.

Form Table 9, it is observed that the average CPU time running LS and GA is 0.3443 s and 0.3446 s respectively. Even though the best optimal objective function value same as the best objective function value for LS method alone but the superiority of hybrid LS and GA method lies on the computational CPU time supremacy. Total CPU time for running LS and GA for $\gamma = 0.001$ to $\gamma = 0.99$ is 3.7874 s and 3.7907 s respectively. This CPU time extremely lower compare to LS method alone and Hybrid GA with LS method. This is the major contribution of novel techniques of GA in helping LS to achieve the best optimal objective function value of 200116.4 with average CPU time 0.3446 s. However, the major drawback of HLSGA techniques is on the inability of obtaining non-zero solution for the decision variable x_8 for $\gamma = 0.001$ to $\gamma = 0.90$. There is a possibility of obtaining non-zero solution for the decision variable x_8 at $\gamma = 41$.

Figure 16 depicts the objective function values versus feasible solution of decision variables at $\alpha = 12.813$. Unfortunately, there is no any improvement in the feasible values for decision variables compare to LS method (Vasant, 2012) alone and hybrid GA with LS (Vasant, 2013). This is the major setback of incorporating GA with LS in this hybridization approach. Moreover, similar

Figure 15. Objective value vs. γ

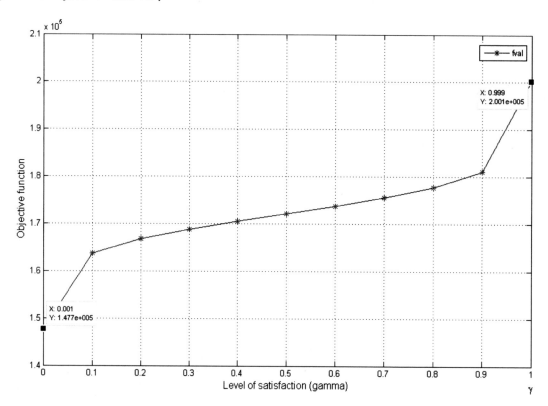

Table 9. Optimal value for Objective Function

γ	x_1	x_2	x_3	x_4	x_5	x_6	x_7	x_8	f	$LS_t(s)$	$GA_t(s)$
0.001	246.8	411.4	205.7	342.8	134.3	223.9	120.6	0.00	147712.9	2.8205	2.8210
0.1	284.8	474.7	239.2	398.7	148.8	248.0	150.4	0.00	163731.6	0.5864	0.5869
0.2	292.5	487.6	246.1	410.1	151.8	253.0	156.7	0.00	166763.3	0.0303	0.0306
0.3	297.9	496.5	250.8	418.0	153.8	256.4	161.2	0.00	168830.5	0.0267	0.0269
0.4	302.5	504.1	254.8	424.7	155.6	259.3	165.1	0.00	170548.7	0.0286	0.0289
0.5	306.8	511.7	258.6	431.0	157.2	262.1	168.8	0.00	172146.4	0.0795	0.0797
0.6	311.2	518.6	262.5	437.5	158.9	264.9	172.6	0.00	173763.9	0.0264	0.0267
0.7	316.1	526.9	266.9	444.8	160.8	268.1	176.9	0.00	175548.3	0.0559	0.0562
0.8	322.4	537.4	272.4	454.1	163.3	272.1	182.4	0.00	177754.9	0.0277	0.0279
0.9	332.3	553.9	281.2	468.6	167.1	278.5	191.3	0.00	181132.7	0.0679	0.0682
0.99	414.3	690.6	354.0	590.0	200.0	333.4	200.0	54.5	200116.4	0.0375	0.0377

$LS_t(s)$: CPU time for LS technique and $GA_t(s)$: CPU time for GA technique

Figure 16. Objective value vs. decision variables

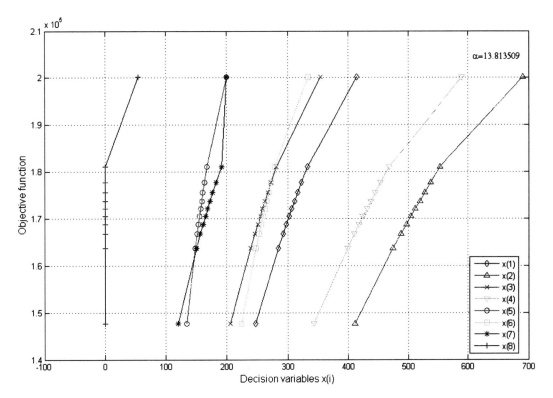

Figure 17. Objective value vs. α and γ

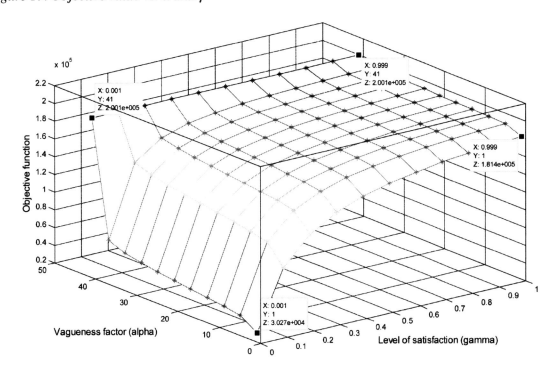

tragedy occurs for the hybrid GA with LS approach (Vasant, 2013). These techniques fail to produce a productive solution for the decision variable x_8 while they are capable to produce a reasonable best solution for x_2 and x_4 feasible decision variables.

Figure 17 depicts the simulation and computational results for objective function respect to α and γ via 3D mesh plot. Total CPU time for running this result is 7.61 seconds. This CPU time is extremely low compare to CPU time for LS (alone) techniques and hybrid GA with LS techniques. Computational efficiency is one of the novel characteristic of GA techniques when it incorporated in the hybridization process. This is one of major achievement in these research case studies.

Table 10 describe the superiority of computational efficiency in finding the best optimal objective function respect to $\alpha = 1$ to $\alpha = 41$ at $\gamma = 0.99$ respect to CPU time. This lowest CPU time reveal the major significant contribution of superb techniques of GA in these research findings. The findings also indicate that GA is the best quality approach as far as CPU time concern in this research work.

The strength of GA in this hybridization process lies in great contribution of computational efficiency of CPU time. This clearly indicated in Table 10. The average CPU time for running LS and GA for $\alpha = 1$ to $\alpha = 41$ at $\gamma = 0.99$ is 0.0553 s and 0.0555 s respectively. In fact, GA has contributed for his own CPU running time as well as great help in CPU running time for LS techniques. The major contribution of LS is in finding the benchmark solution for the best global near optimal value for the cubic objective function of industrial production planning problems. On the other hand, the great contribution of GA is on the quality of computational efficiency and robust convergence computing CPU time in this particular hybridization approach.

FUTURE RESEARCH DIRECTIONS

The integration of GA, SA, LS, MADS and GPS with other emerging technology such ant colony optimization (ACO), particle swarm optimization (PCO) and artificial immune system (AIS) could be another challenging research areas. The combination of these emerging technologies may not only involve GA and LS as a helper to these three, but could result in the emerging technologies being able to assist GA and LS applications.

Table 10. Best CPU time for objective function at $\gamma = 0.99$

α	f	LS CPU time (s)	GA CPU time (s)
1	200116.44	0.03104	0.03129
5	200116.44	0.05659	0.05685
9	200116.44	0.02970	0.02995
13	200116.44	0.15890	0.15917
17	200116.44	0.01922	0.01947
21	200116.44	0.02525	0.02550
25	200116.44	0.02510	0.02535
29	200116.44	0.18363	0.18389
33	200116.44	0.02599	0.02625
37	200116.44	0.02660	0.02685
41	200116.44	0.02620	0.02645

Different combinations may offer us a fruitful result in intelligent optimization systems.

Overall, the knowledge generated from hybrid evolutionary and heuristic optimization over the last three decades has now become mature. The prospect of applying hybrid intelligent optimization techniques for practical applications is overwhelming. A considerable growth in the application of hybrid intelligent optimization, particularly in the field of industrial engineering, anticipated in the near future.

To the best of author's knowledge, this research work is among the first to apply hybrid line search with GA, and hybrid GA with MADS and GPS to the industrial production planning problems. Further, in this research work there are no any global solutions available at this moment. Therefore, further work includes solving other types of nonlinear and complex multi-objective problems arise in real world situation, by considering other meta-heuristic techniques such as hybrid Tabu search with ant colony and particle swarm optimization. In particular, construction-planning problems, product mix problem, inventory models (discount, demand and variable replenishment), design of electrical networks, mechanical components, and facility location in optimization formulated in crisp, fuzzy or fuzzy-stochastic environment.

CONCLUSION

In conclusion, the hybrid optimization techniques outperformed the traditional and non-hybrid optimization techniques as per indicated in the intelligent performance analysis. The insight information obtained from this analysis will be very useful for the decision maker and implementer in their wise and strategic decision-making process under the uncertain and turbulence environment. The main advantage and significant contribution of this analysis is on the well-informed choices and options to the decision maker and implementer in their successful decision making process.

The novel innovative of hybridization of soft computing techniques such as GA and SA with classical optimization techniques such as GPS, MADS and LS provide a great future especially in solving complex problems of industrial engineering and other real world applied design optimization engineering problems.

REFERENCES

Aruldoss, A. V., & Ebenezer, J. A. (2005). A modified hybrid EP-SQP approach for dynamic dispatch with valve-point effect. *International Journal of Electrical Power & Energy Systems*, *27*(8), 594–601. doi:10.1016/j.ijepes.2005.06.006

Attaviriyanupap, K. H., Tanaka, E., & Hasegawa, J. (2002). A hybrid EP – SQP for dynamic economic dispatch with nonsmooth incremental fuel cost function. *IEEE Transactions on Power Systems*, *17*(2), 411–416. doi:10.1109/TPWRS.2002.1007911

Back, T., Fogel, D. B., & Michalewicz, Z. (1997). *Handbook of evolutionary computation*. New York: Oxford University Press and Institute of Physics. doi:10.1887/0750308958

Bellman, R. E., & Zadeh, L. A. (1970). Decision making in a fuzzy environment. *Management Science*, *17*, 141–164. doi:10.1287/mnsc.17.4.B141

Bezdek, J. C. (1994). What is computational intelligence. In J. M. Zurada, R. J. Marks, & C. J. Robinson (Eds.), *Computational intelligence: Imitating life* (pp. 1–12). New York: IEEE Press.

Bhattacharya, A., Abraham, A., Vasant, P., & Grosan, C. (2007). Meta-learning evolutionary artificial neural network for selecting FMS under disparate level-of-satisfaction of decision maker. *International Journal of Innovative Computing, Information, & Control*, *3*(1), 131–140.

Bhattacharya, A., & Vasant, P. (2007). Soft-sensing of level of satisfaction TOC product-mix decision heuristic using robust fuzzy-LP. *European Journal of Operational Research, 177*(1), 55–70. doi:10.1016/j.ejor.2005.11.017

Bhattacharya, A., Vasant, P., Sarkar, B., & Mukherjee, S. K. (2006). A fully fuzzified, intelligent theory-of-constraints product-mix decision. *International Journal of Production Research, 46*(3), 789–815. doi:10.1080/00207540600823187

Bhattacharya, A., Vasant, P., & Susanto, S. (2007). Simulating theory of constraint problem with a novel fuzzy compromise linear programming model. In A. Elsheikh, A. T. Al Ajeeli, & E. M. Abu-Taieh (Eds.), *Simulation and modeling: Current technologies and applications* (pp. 307–336). Hershey, PA: IGI Publishing. doi:10.4018/978-1-59904-198-8.ch011

Chams, M., Hertz, A., & Werra, D. (1987). Some experiments with simulated annealing for coloring graphs. *European Journal of Operational Research, 32*, 260–266. doi:10.1016/S0377-2217(87)80148-0

Dawkins, R. (1976). *The selfish gene*. Oxford: Oxford Press.

Dennett, D. C. (1995). *Darwin's dangerous idea*. New York: Touchstone.

Eglese, R. W. (1990). Simulated annealing: A tool for operational research. *European Journal of Operational Research, 46*, 271–281. doi:10.1016/0377-2217(90)90001-R

Fogel, D. B. (1995). *Evolutionary computation: Toward a new philosophy of machine intelligence*. Piscataway, NJ: IEEE Press. doi:10.1109/ICEC.1995.489143

Goguen, J. A. (1969). The logic of inexact concepts. *Syntheses, 19*, 325–373. doi:10.1007/BF00485654

Hart, W. E. (1998). On the application of evolutionary pattern search algorithms. In *Proceedings of the 7th International Conference on Evolutionary Programming* (pp. 301-312). Berlin: Springer.

Hererra, F., & Lozano, M. (1996). Adaptation of genetic algorithm parameters based on fuzzy logic controllers. In F. Hererra, & J. L. Verdegay (Eds.), *Genetic algorithms and soft computing* (pp. 95–125).

Holland, J. H. (1975). *Adaptation in natural and artificial systems*. Ann Arber, MI: University of Michigan Press.

Jiménez, F., Gómez-Skarmeta, A. F., & Sánchez, G. (2004). Nonlinear optimization with fuzzy constraints by multi-objective evolutionary algorithms. *Computational Intelligence. Theory and Applications: Advances in Soft Computing, 33*, 713–722.

Jimenez, F., Sanchez, G., Vasant, P., & Verdegay, J. (2006). A multi-objective evolutionary approach for fuzzy optimization in production planning. In *Proceedings of IEEE International Conference on Systems, Man, and Cybernetics* (pp. 3120-3125). USA: IEEE Press.

Johnson, D. S., Aragon, C. R., McGeogh, L. A., & Schevon, C. (1987). Optimization by simulated annealing: An experimental evaluation, Part 1. *Operations Research, 37*, 865–892. doi:10.1287/opre.37.6.865

Karr, C. L. (1991). Genetic algorithms for fuzzy controllers. *Artificial Intelligence Expert, 6*(2), 27–33.

Lagaros, N. D., Papadrakakis, M., & Kokossalakis, G. (2002). Structural optimization using evolutionary algorithms. *Computers & Structures, 80*, 571–589. doi:10.1016/S0045-7949(02)00027-5

Lee, M. A., & Takagi, H. (1993). Dynamic control of genetic algorithms using fuzzy logic techniques. In S. Forrest (Ed.), *Proceedings of the 5th International Conference on Genetic Algorithms* (pp. 76-83). San Mateo, CA: Morgan Kaufmmann.

Liang, T. F. (2008). Interactive multi-objective transportation planning decisions using fuzzy linear programming. *Asia Pacific Journal of Operational Research, 25*(1), 11–31. doi:10.1142/S0217595908001602

Lin, F. T., & Yao, J. S. (2002). Applying genetic algorithms to solve the fuzzy optimal profit problem. *Journal of Information Science and Engineering, 18*, 563–580.

Lozano, M., Herrera, F., Krasnogor, N., & Molina, D. (2004). Real coded memetic algorithms with crossover hill climbing. *Evolutionary Computational Journal, 12*(3), 273–302. doi:10.1162/1063656041774983 PMID:15355602

Maa, C., & Shanblatt, M. (1992). A two phase optimization neural network. *IEEE Transactions on Neural Networks, 3*(6), 1003–1009. doi:10.1109/72.165602 PMID:18276497

Michalewicz, Z. (1993). A hierarchy of evolution programs: An experimental study. *Evolutionary Computation, 1*(1), 51–76. doi:10.1162/evco.1993.1.1.51

Moscato, P. (1999). Memetic algorithms: A short algorithms. In D. Corne (Ed.), *New ideas in optimization* (pp. 219–234).

Myung, H., Kim, J. H., & Fogel, D. (1995). Preliminary investigation into a two-stage method of evolutionary optimization on constrained problems. In J. R. MacDonnell, R. G. Reynolds, & D. B. Fogel (Eds.), *Proceedings of the 4th Annual Conference on Evolutionary Programming* (pp. 449-463). Cambridge, MA: MIT Press.

Nowakowska, N. (1977). Methodological problems of measurement of fuzzy concepts in the social sciences. *Behavioral Science, 22*, 107–115. doi:10.1002/bs.3830220205

Papadrakakis, M., Tsompanakis, Y., & Lagaros, N. D. (1999). Structural shape optimization using evolution strategies. *Engineering Optimization, 31*, 515–540. doi:10.1080/03052159908941385

Ramik, J., & Vlach, M. (2002). Fuzzy mathematical programming: A unified approach based on fuzzy relations. *Fuzzy Optimization and Decision Making, 1*, 335–346. doi:10.1023/A:1020978428453

Ramik, J., & Vlach, M. (2002). Fuzzy mathematical programming: A unified approach based on fuzzy relations. *Fuzzy Optimization and Decision Making, 1*, 335–346. doi:10.1023/A:1020978428453

Sanchez, G., Jimenez, F., & Vasant, P. (2007). Fuzzy optimization with multi-objective evolutionary algorithms: A case study. In *Proceedings of the 2007 IEEE Symposium on Computational Intelligence in Multi-criteria Decision Making* (pp. 58-64). Honolulu, Hawaii.

Schwefel, H. P. (1979). Direct search for optimal parameters within simulation models. In *Proceedings of the 12th Annual Simulation Symposium* (pp.91-102), Tampa, Florida.

Schwefel, H. P. (1979). Direct search for optimal parameters within simulation models. In *Proceedings of the 12th Annual Simulation Symposium* (pp.91-102), Tampa, Florida.

Swain, A. K., & Morris, A. S. (2000). A novel hybrid evolutionary programming method for function optimization. In *Proceedings of the Congress on Evolutionary Computation* (pp. 1369-1376).

Tabucanon, T. T. (1996). Multi objective programming for industrial engineers. In M. Avriel, & B. Golany (Eds.), *Mathematical programming for industrial engineers* (pp. 487–542). New York: Marcel Dekker, Inc.

Torczon, V. (1997). On the convergence of pattern search methods. *SIAM Journal on Optimization*, 7, 1–25. doi:10.1137/S1052623493250780

Turabieh, H., Sheta, A., & Vasant, P. (2007). Hybrid optimization genetic algorithm (HOGA) with interactive evolution to solve constraint optimization problems for production systems. *International Journal of Computational Science*, 1(4), 395–406.

Vasant, P. (2006). Fuzzy production planning and its application to decision making. *Journal of Intelligent Manufacturing*, 17(1), 5–12. doi:10.1007/s10845-005-5509-x

Vasant, P. (2012). A novel hybrid genetic algorithms and pattern search techniques for industrial production planning. *International Journal of Modeling, Simulation, and Scientific Computing*, 3(4). DOI No: 10.1142/S1793962312500201

Vasant, P. (2013). Hybrid optimization techniques for industrial production planning. In Z. Li, & A. Al-Ahmari (Eds.), *Formal methods in manufacturing systems: Recent advances* (pp. 84–111). Hershey, PA: Engineering Science Reference. doi:10.4018/978-1-4666-4034-4.ch005

Vasant, P., Barsoum, N., Kahraman, C., & Dimirovski, G. (2007). Application of fuzzy optimization in forecasting and planning of construction industry. In D. Vrakas, & I. Vlahavas (Eds.), *Artificial intelligence for advanced problem solving techniques* (pp. 254–265). Hershey, PA: IGI Publishing.

Vasant, P., & Barsoum, N. N. (2006). Fuzzy optimization of units products in mix- product selection problem using FLP approach. *Soft Computing. A Fusion of Foundations. Methodologies and Applications*, 10, 144–151.

Vasant, P., Bhattacharya, A., Sarkar, B., & Mukherjee, S. K. (2007). Detection of level of satisfaction and fuzziness patterns for MCDM model with modified flexible S-curve MF. *Applied Soft Computing*, 7, 1044–1054. doi:10.1016/j.asoc.2006.10.005

Vasant, P., & Kale, H. (2007). Introduction to fuzzy logic and fuzzy linear programming. In A. Frederick, & P. Humphreys (Eds.), *Encyclopedia of decision making and decision support technologies* (pp. 1–15). Hershey, PA: IGI Publishing.

Waagen, D., Diercks, P., & McDonnell, J. (1992). The stochastic direction set algorithm: A hybrid techniques for finding function extreme. In D. B. Fogel and W. Atmar (Eds.), *Proceedings of the 1st Annual Conference on Evolutionary Programming* (pp. 35-42). Evolutionary Programming Society.

Wah, B. W., & Chen, Y. (2003). Hybrid evolutionary and annealing algorithms for nonlinear discrete constrained optimization. *International Journal of Computational Intelligence and Applications*, 3(4), 331–355. doi:10.1142/S1469026803001063

Wanner, E. F., Guimarae, F. G., Saldanha, R. R., Takahashi, R. H., & Fleming, P. J. (2005). Constraint quadratic approximation operator for treating equality constraints with genetic algorithms. In *Proceedings of the 2005 IEEE Congress on Evolutionary Computation* (pp. 2255-2262).

Zadeh, L. A. (1971). Similarity relations and fuzzy orderings. *Information Sciences*, 3, 177–206. doi:10.1016/S0020-0255(71)80005-1

ADDITIONAL READING

Angelova, M., Atanassov, K., & Pencheva, T. (2012). Intuitionistic fuzzy estimations of purposeful model parameters genesis, IS'2012 – 2012, *6th IEEE International Conference Intelligent Systems, Proceedings*, art. no. 6335217, pp. 206-211.

Angelova, M., Atanassov, K., & Pencheva, T. (2012). Purposeful model parameters genesis in simple genetic algorithms. *Computers & Mathematics with Applications (Oxford, England), 64*(3), 221–228. doi:10.1016/j.camwa.2012.01.047

Angelova, M., Melo-Pinto, P., & Pencheva, T. (2012). Modified simple genetic algorithms improving convergence time for the purposes of fermentation process parameter identification. *WSEAS Transactions on Systems, 11*(7), 256–267.

Angelova, M., & Pencheva, T. (2012). Algorithms improving convergence time in parameter identification of fed-batch cultivation. *Comptes Rendus de L'Academie Bulgare des Sciences, 65*(3), 299–306.

Bhattacharya, A., & Vasant, P. (2009). Soft-sensing of level of satisfaction in TOC product-mix decision heuristic using robust fuzzy-LP. *European Journal of Operational Research, 177*(1), 55–70. doi:10.1016/j.ejor.2005.11.017

Cebi, S., Kahraman, C., & Kaya, I. (2012). Soft Computing and Computational Intelligent Techniques in the Evaluation of Emerging Energy Technologies. In P. Vasant, N. Barsoum, & J. Webb (Eds.), *Innovation in Power, Control, and Optimization: Emerging Energy Technologies* (pp. 164–197). Hershey, PA: Engineering Science Reference.

Chang, C.-T. (2010). An approximation approach for representing S-shaped membership functions. *IEEE Transactions on Fuzzy Systems, 18* (2), art. no. 5411779, 412-424.

Dostál, P. (2013). The Use of Soft Computing for Optimization in Business, Economics, and Finance. In P. Vasant (Ed.), *Meta-Heuristics Optimization Algorithms in Engineering, Business, Economics, and Finance* (pp. 41–86). Hershey, PA: Information Science Reference.

Elamvazuthi, I., Ganesan, T., & Vasant. P. A comparative study of HNN and Hybrid HNN-PSO techniques in the optimization of distributed generation power systems. In *Proceedings of the 2011 International Conference on Advanced Computer Science and Information Systems* (ICACSIS'11), Jakarta, Indonesia, 2011, pp. 195-199. ISBN 978-979-1421-11-9.

Elamvazuthi, I., Ganesan, T., Vasant, P., & Webb, J. F. (2009). Application of a fuzzy programming technique to production planning in the textile industry. *International Journal of Computer Science and Information Security, 6*(3), 238–243.

Elamvazuthi, I., Vasant, P., & Ganesan, T. (2010). Fuzzy Linear Programming using Modified Logistic Membership Function [IREACO]. *International Review of Automatic Control, 3*(4), 370–377.

Elamvazuthi, I., Vasant, P., & Ganesan, T. (2012). Integration of Fuzzy Logic Techniques into DSS for Profitability Quantification in a Manufacturing Environment. In M. Khan, & A. Ansari (Eds.), *Handbook of Research on Industrial Informatics and Manufacturing Intelligence: Innovations and Solutions* (pp. 171-192). doi: doi:10.4018/978-1-4666-0294-6. ch007.

Feng, X., & Yuan, G. (2011). Optimizing two-stage fuzzy multi-product multi-period production planning problem. *Information, 14*(6), 1879–1893.

Ganesan, T., & Vasant, P., P, & Elamvazuthi, I. (2011). Solving engineering optimization problems with KKT Hopfield Neural Networks. *International Review of Mechanical Engineering, 7*(7), 1333–1339.

Ganesan, T., Vasant, P., & Elamvazuthi, I. (2012). Hybrid PSO approach for solving non-convex optimization problems. *Archives of Control Sciences, 22*(1), 5–23. doi:10.2478/v10170-011-0014-2

Hasuike, T., & Ishii, H. (2009). Product mix problems considering several probabilistic conditions and flexibility of constraints. *Computers & Industrial Engineering, 56*(3), 918–936. doi:10.1016/j.cie.2008.09.006

Jimenez, F., Sanchez, G., & Vasant. P. (2013). A multi-objective evolutionary approach for fuzzy optimization in production planning, *Journal of Intelligent and Fuzzy Systems*, Pre-Press, March 2013, DOI: 10.3233/IFS-130651.

Leng, K., & Chen, X. (2012). A genetic algorithm approach for TOC-based supply chain coordination. *Applied Mathematics and Information Sciences, 6*(3), 767–774.

Liang, T.-F., Cheng, H.-W., Chen, P.-Y., & Shen, K.-H. (2011). Application of fuzzy sets to aggregate production planning with multiproducts and multitime periods. *IEEE Transactions on Fuzzy Systems*, 19 (3), art. no. 5713250, pp. 465-477.

Madronero, M. D., Peidro, D., & Vasant, P. (2010). Vendor selection problem by using an interactive fuzzy multi-objective approach with modified s-curve membership functions. *Computers & Mathematics with Applications (Oxford, England), 60*, 1038–1048. doi:10.1016/j.camwa.2010.03.060

Mawengkang, H., Guno, M., Hartama, D., Siregar, A. S., Adam, H. A., & Alfina, O. (2012). An Improved Direct Search Approach for Solving Mixed-Integer Nonlinear Programming Problems, Global *Journal Technology and Optimization*, 3 (1). ISSN: 2229-8711.

Peidro, D., & Vasant, P. (2011). Transportation planning with modified s-curve membership functions using an interactive fuzzy multi-objective approach. *Applied Soft Computing, 11*, 2656–2663. doi:10.1016/j.asoc.2010.10.014

Pinto, G., Israelí, U., Ainbinder, I., & Rabinowitz, G. (2010). GA for the Resource Sharing and Scheduling Problem, *Global Journal Technology and Optimization*, 1, ISSN 1985-9406.

Senvar, O., Turanoglu, E., & Kahraman, C. (2013). Usage of Metaheuristics in Engineering: A Literature Review. In P. Vasant (Ed.), *Meta-Heuristics Optimization Algorithms in Engineering, Business, Economics, and Finance* (pp. 484–528). Hershey, PA: Information Science Reference.

Svancara, J., Kralova, Z., & Blaho, M. (2012). Optimization of HMLV manufacturing systems using genetic algorithm and simulation. *International Review on Modelling and Simulations, 5*(1), 482–488.

Tsoulos, I. G., & Vasant, P. (2009). AIP Conference Proceedings: Vol. 1159. *Product mix selection using an evolutionary technique* (pp. 240–247). doi:10.1063/1.3223936

Vasant, P. (2013). Hybrid Optimization Techniques for Industrial Production Planning. In Z. Li, & A. Al-Ahmari (Eds.), *Formal Methods in Manufacturing Systems: Recent Advances* (pp. 84–111). Hershey, PA: Engineering Science Reference. doi:10.4018/978-1-4666-4034-4.ch005

Vasant, P., & Barsoum, N. (2009). Hybrid genetic algorithms and line search method for industrial production planning with non-linear fitness function. *Engineering Applications of Artificial Intelligence, 22*(4-5), 767–777. doi:10.1016/j.engappai.2009.03.010

Vasant, P., & Barsoum, N. (2010). Hybrid pattern search and simulated annealing for fuzzy production planning problems. *Computers & Mathematics with Applications (Oxford, England), 60*(4), 1058–1067. doi:10.1016/j.camwa.2010.03.063

Vasant, P., Elamvazuthi, I., Ganesan, T., & Webb, J. F. (2010). Iterative fuzzy optimization approach for crude oil refinery industry. *Scientific Annals of Computer Science, 8*(2), 262–280.

Vasant, P., Ganesan, T., Elamvazuthi, I., & Webb, J. F. (2011). Fuzzy linear programming for the production planning: The case of Textile Firm. *International Review on Modelling and Simulations*, *4*(2), 961–970.

Zelinka, I., Vasant, P., & Barsoum, N. (Eds.). (2013). *Power, Control and Optimization, Series: Lecture Notes in Electrical Engineering* (Vol. 239). Springer. doi:10.1007/978-3-319-00206-4

KEY TERMS AND DEFINITIONS

Crossover Operation: Generates offspring from randomly selected pairs of individuals within the mating pool by exchanging segments of the chromosome strings from the parents.

Evolutionary Pattern Search Algorithms (EPSAs): A class of evolutionary algorithms (EAs) that adapt the step size of the mutation operators to guarantee convergence to a stationary point of an objective function, where the gradient is zero, with probability one.

Exploitation: The process using information gathered from previously visited points in the search space to determine which places might be profitable to visit next.

Exploration: The process of visiting entirely new regions of a search space, to see if anything promising may be found there.

Genetic Algorithm: A global stochastic method based on the mechanism of nature selection and evolutionary genetics.

Hybrid Evolutionary Methods: The combination of evolutionary computation techniques with deterministic procedures for numerical optimization problems.

Pattern Search Method: The technique uses an exploratory moves algorithm to conduct a series of exploratory moves about the current iterates before identifying a new iterate.

Chapter 5
A Framework for the Modelling and Optimisation of a Lean Assembly System Design with Multiple Objectives

Atiya Al-Zuheri
University of South Australia, Australia & Ministry of Science and Technology, Iraq

Lee Luong
University of South Australia, Australia

Ke Xing
University of South Australia, Australia

ABSTRACT

The newest assembly system is lean assembly, which is specifically designed to respond quickly and economically to the fluctuating nature of the market demands. Successful designs for these systems must be capable of satisfying the strategic objectives of a management in manufacturing company. An example of such systems is the so-called walking worker assembly line WWAL, in which each cross-trained worker travels along the line to carry out all tasks required to complete a job. Design approaches for this system have not been investigated in depth both of significant role in manual assembly process design; productivity and ergonomics. Therefore these approaches have had a limited success in actual applications. This chapter presents an innovative and integrated framework which offers significant potential improvement for productivity and ergonomics requirements in WWAL design. It establishes a systematic approach clearly demonstrating the implementation of a developed framework based on the simultaneous application of mathematical and meta- heuristic techniques.

DOI: 10.4018/978-1-4666-5836-3.ch005

1. INTRODUCTION

The newest assembly system is lean system, which is specifically designed to respond quickly and economically to the fluctuating nature of the market demands (Katayama & Bennett, 1996). One of the cornerstones of lean assembly is the advantage of work-force flexibility in systems where the workers are able to switch tasks rapidly enabling the line to respond quickly to changing production demands. This results in improved system efficiency in the form of higher throughput, less work in process (WIP), and shorter cycle times without significant additional investment in equipment and labour (Downey, 1992). One advantageous form of flexible force-work is the use of cross-trained workers to perform all assembly operations required on a product under manufacture. As such, they are capable of shifting their capacity to where it is needed (Hopp et al., 2004). An example of such systems is the so-called walking worker assembly line (WWAL), in which each cross-trained worker walks down the line carrying out each assembly task at each workstation as scheduled. Thereby, each walking worker completes the assembly of a product in its entirety from start to end. Figure 1 illustrates the concept of WWAL, where a walking worker completes the product assembly process at the last workstation and then moves back to the first workstation to begin the assembly of a new product. At most workstations the nature of the assembly process in manual assembly systems requires manual tasks performed by the worker. Therefore, the optimal design should consider both productivity and ergonomics aspects of the system. To deal with such optimisation problems, robust modelling and optimisation techniques are required, particularly when human-centric-manufacturing process relationships in the problem increase, as in manual assembly operations in WWAL. As a result of the exhaustive literature review in research of Al-Zuheri (2013), it has become clear currently there are no design frameworks which include the required techniques for a demonstration of this. Therefore, this chapter describes the development of a novel framework, developed to optimise designs of WWAL. To clearly demonstrate the development of the framework approach step by step, this chapter proceeds as follows: firstly, existing research about WWAL provides no complete experimental or empirical modelling or optimisation techniques in relation to overall system performance. Rather, a massive body of literature has focused on modelling and optimisation in order to find solutions for many fixed-worker (FWAL) design problems in traditional assembly lines. Therefore, these are reviewed as the theoretical foundation for a new approach. On this basis, the capabilities and limitations of the techniques proposed in this literature are analysed. As a result, an appropriate framework mechanism is proposed for comprehensively addressing all essential characteristics of an efficient design of WWAL which profoundly impacts productivity and ergonomics. Based on such a mechanism, a methodically structured framework is developed to achieve comprehensiveness and optimality in

Figure 1. Layout of walking worker line (Wang et al., 2005)

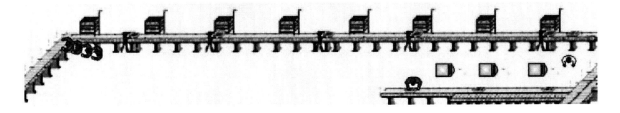

the design scenarios for WWAL. Finally, following the development of the framework, three basic techniques; mathematical modelling, genetic algorithms (GA) and desirability function, are adopted in this research to enable modelling and optimisation design of WWAL are introduced and their applications in the proposed framework are discussed.

2. EVALUATION OF ALTERNATIVE TECHNIQUES TO MODELLING AND OPTIMISATION OF MANUAL ASSEMBLY SYSTEM DESIGN

Optimisation technique applications for manual assembly system design processes are vital for companies to be capable of responding quickly and economically to current rapidly changing scenarios in manufacturing industries (Feyzbakhsh & Matsui, 1999). In practice, the use of optimisation methods in finding the optimal design of a manual assembly system involves determining the best combination of design variables to optimise performance measures considering various objectives (Kim, Kim, & Kim, 1996). Optimisation of design variables in flexible production systems like WWAL, require two stages to undertake (Anderson & Ferris, 1994; Smed, Johnsson, Johtela, & Nevalainen, 1999):

1. Modelling of input, design variables and objectives relationship; and
2. Determination of an optimal combination of those variables that could achieve higher performance without violating the imposed constraints.

An optimisation model is displayed in Figure 2. Output of the modelling process is used by an optimisation technique to provide feedback on progress of the search for the optimal solution. This in turn determines further input to the modelling process (Carson & Maria, 1997).

Over the years, modelling and optimisation techniques for designing manual assembly systems underwent substantial development and expansion. A cursory look at the literature focused on the issues inherent the design of manual assembly systems, and displays comprehensive uses for modelling and optimisation techniques in order to determine optimal or near-optimal designing condition(s) with respect to numerous objective criteria (Solot & Vliet, 1994). In the next subsection, an attempt has been made to evaluate existing and frequently used modelling and optimisation techniques, specific to designing of manual assembly systems. Critical appraisal of these techniques identifies key issues needing to be addressed whilst carrying out optimising designs of walking worker assembly lines.

Figure 2. Conceptual schematic of modelling and optimisation process

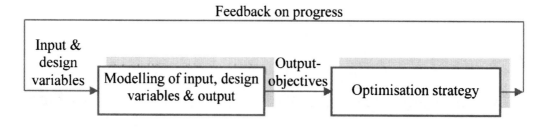

Figure 3. Classification of modelling techniques in manual assembly system design problems

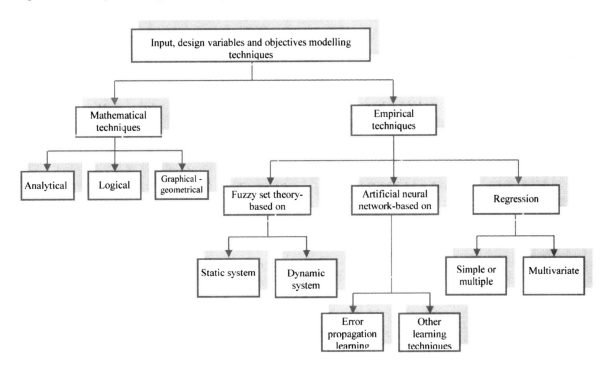

2.1 Modelling Techniques of Inputs, Design Variables and Objectives Relationship

A crucial step towards optimal designs for manual assembly systems is to understand the functional relationship between structural and other operational elements in the performance of a given system. This understanding may be achieved by developing an explicit model that formulates this relationship. Modelling techniques to manual assembly system problems may be classified into two categories in terms of their formulation; mathematical (mechanistic) and empirical (Box & Hunter, 1987). The first is a process-based mathematical model which integrates the different processes (mechanisms) involved whilst the empirical model refers to any kind of modelling based entirely on statistical data, rather than on mathematically describable relationships between variables of the system modelled (Byrne & Bakir, 1999). Figure 3 provides a general classification of different inputs, design variables and outputs objectives modelling techniques, in manual assembly systems design.

Both types of modelling techniques may work satisfactorily in different situations; there are constraints, assumptions and shortcomings, limiting the use of a specific technique. The usefulness, applications, and limitations of these techniques are explained below.

2.1.1 Mathematical Techniques

In general, the planning and design phase of flexible assembly systems involves evaluating numerous alternative system configurations. In the preliminary design phase, the designer first employs a set of methods that allow quick and inexpensive analysis of alternative designs. Mathematical (analytical) models are one of these methods. Using mathematical models, the designer would be able to understand the relationship among the operating parameters and the effect that changing those values has on system performance (Kamath, Suri, & Sanders, 1988). In the case of mathematical techniques, the modelling process is an abstraction of a real system in terms of algebraic equations or differential equations. As far as logic is taken as a part of the process; the mathematical models may include logical, graphical and geometrical models also. For all types of mathematical models, there are formal analytical solution procedures available for optimisation (Byrne & Bakir, 1999). Mathematical models are considered from the earlier versions of operational research techniques that are used in modelling manual assembly system design problems. Extensive research has used mathematical models to solve manual assembly system design problems. A sample from that research is represented by studies of Dashchenko and Loladze (1991), Hillier and So (1996), Martin (1994), and Solot and Vliet (1994).

Due to the sources of randomness in manufacturing systems such as processing or assembly times at each workstation and skills of workers who perform these tasks, most of these systems are stochastic rather than deterministic (Wang, 2005). Among specialists, it is widely accepted that use of mathematical modelling techniques is not sufficient to describe a system with random behaviour (Wang, 2005; Wang, Lassalle, Mileham, & Owen, 2009). This is because mathematical models do not consider the stochastic nature of the system, based on many simplifying assumptions and provide a limited number of system performance measures (Hsieh, 2002). Consequently, accuracy often becomes a major problem for system optimisation when using mathematical models (Wang, 2005; Wang, Lassalle, Mileham, & Owen, 2009).

2.1.2 Empirical Techniques

The list of applications for the empirical modelling techniques of manual assembly systems is endless. This category mainly includes simulation, statistical regression (Montgomery, Peck, & Vining, 2001), artificial neural network (Fu, 1994), and fuzzy set theory (Klir, Clair, & Yuan 1997; Ross, 2010).

2.1.2.1 Simulation Modelling

Simulation modelling has emerged as a powerful tool for the optimisation of complex manufacturing systems that are characterised by stochastic operating environments (Bulgak & Sanders, 1990; Jayaraman & Gunal, 1997; Siebers, 2004). Currently, simulation modelling is considered as the most common technique used behind optimisation (Shafer & Smunt, 2004). Components of the simulation model try to represent, with varying degrees of accuracy, the actual operations of the real components of the system. In simulation, the flow of entities through the system is controlled by logical rules derived from the operating rules, which are associated with underlying assumptions. Like other manufacturing systems, the simulation model of the manual assembly system is used to obtain performance measure values for different combination scenarios of design variables.

There is much published work in simulation modelling-based optimisation of manual assembly systems. Examples from this research are in studies by Boër, El-Chaar, Imperio, and Avai (1991), Chan and Smith (1993), Battini, Faccio, Persona, and Sgarbossa (2007), Jayaram, Kim, DeChenne, Lyons, Palmer, and Mitsui (2007), Kung and Changchit (1991), Lin and Cochran (1987), Longo, Mirabelli, and Papoff (2006), Nomura and Takakuwa (2006), Ooi (2005), Spieckermann, Gutenschwager, Heinzel, and Voß (2000), and Yoshimura, Yoshida, KonishI, Izui, Nishiwaki, InamorIz, Nomura, Mitsuyuki, Kawaguchi, and Inagaki (2006).

However, a simulation modelling technique often requires more effort and higher costs to obtain solutions than a mathematical one. Also, the accuracy of the model is totally dependent its quality and the skill of the modeller, therefore it is sometimes hard to accurately interpret the simulation results. Because of these limitations, it seems, it is difficult to adopt simulation modelling as the sole modelling technique for manual assembly systems.

2.1.2.2 Statistical Regression Modelling

Several applications of regression equation-based modelling in manual assembly systems design are reported in the literature (Ben-Gal & Bukchin, 2002; Bulgak, Tarakc, & Verter, 1999; Seijo-Vidal & Bartolomei-Suarez, 2010; Singholi, Chhabra, & Bagai, 2010). Statistical regression may implement a statistical model effectively; this modelling technique is especially useful just when the relationship between the decision variables and measures of the system is almost linear. Non-linear complex relationships are often unsuitable for modelling using this technique. A statistical regression model is valid in predicting relationship(s) such as linear, quadratic, higher-order-polynomial, and exponential relationship(s) between output(s), and input decision variable(s), only if there exist prior assumptions regarding these functional relationships. This technique cannot be used to predicate a cause and effect relationship, only as an aid to confirm the cause-effect relationship. Also, for each output-predicted from a particular input decision variable, the regression line variation needs to be mutually independent, normally distributed, and having constant variance (Montgomery, Peck, & Vining, 2001).

2.1.2.3 Artificial Neural Network (ANN)-Based Modelling

Due to the complexity of worker–machine or workstation modelling in manual assembly systems, handling the complex input-output relationship of manual assembly line design problems effectively are achievable when using ANN. This is because of the precision and capability of ANN in relation to capturing unknown nonlinear relationships between inputs and outputs, as well as avoiding both the mathematical complexities and prior assumptions on the functional form of the relationship between them (e.g. linear, quadratic, higher order polynomial and exponential).

All the above mentioned attributes of ANN motivates many researchers to use it as an attractive alternative choice for modelling diverse applications in science and engineering (Antari, Chabaa, & Zeroual, 2011). Owing to a multi-variable, dynamic, non-linear estimator, ANN solves the problems of modelling through self-learning and self-organisation (Fu, 1994). The intelligence of neural networks in modelling depend mainly on a constitutive unit called the "artificial neuron," and driving process knowledge is usually referred to the input and output data set (Shtub & Zimerman, 1993). Performance evaluation and validating of ANN models along with its potential for applications in manufacturing system designs have been discussed extensively by many researchers (Coit, Jackson, & Smith, 1998; Feng & Wang, 2004; Zang & Huang, 1995). In relation to modelling problems of manual assembly systems design, several applications of ANN have been reported in the literature (Sukthomya & Tannock, 2005; Zang & Huang, 1995).

It should be noted there are certain assumptions, constraints, and limitations inherent in applying ANN techniques. They are considered as an alternative choice when the empirical polynomial models provided by regression techniques are inappropriate. In addition, ANN techniques are not without drawbacks. These drawbacks are:

- For non-linear relationships, the interpretation of model parameters may not be accurate;
- The size of the data set that is available for training plays an important role in determining the accuracy level for the predicted model. Although by increasing the training data set gives better accuracy, on the other hand it increases time to train the network. Also, a small size of training data set even gives a high level of predicating accuracy,

- it is potentially unable to identify other data sets; and
- The capability of ANN to identify influential observations, outliers, and significance of various predictors is uncertain. This is because of the uncertainty existing in the limited convergence of approaches used in predictive models of ANN, and basically the setting of convergence criteria is depends on prior understandings gained from earlier applications. In regard to manufacturing systems design, there exist no rules that can be used to select the appropriate type of a particular ANN technique for any typical system.

2.1.2.4 Fuzzy Set Theory-Based Modelling

In addition to a great variety of applications in engineering design reviewed by different authors (Klir, Clair, & Yuan, 1997; Zimmermann, 1996), this technique may be also used in the modelling of manual assembly system design problems. Uncertainties are inherent with manufacturing systems (e.g. manual assembly systems), such as the uncertainties of operation time, walking time, quality, failure of the assembly system and changes to product structure and demand etc. Unlike other modelling techniques, fuzzy set theory admits the existence of such uncertainties due to vagueness or imprecision of human cognitive processes (referred to as 'fuzzy uncertainty'), not just due to randomness alone (Zimmermann, 1996).

Fuzzy set theory-based modelling is considered a preferable choice of technique especially when defining the objective function and decision variables of problems where subjective knowledge or opinion(s) of a process expert(s) is a key issue in setting both the objectives and decision variables (Gottwald 1993). In literature, there are some applications of fuzzy set theory-based modelling of manual assembly system design problems. Ramnath, Elanchezhian, and Kesavan (2010) proposed a fuzzy based simulation (FBS) model to address the suitability of lean kitting assembly for an assembly line of a leading two wheeler manufacturer. Hui, Chan, Yeung, and Ng (2002) have introduced a fuzzy logic model to determine the number of workers required in hybrid assembly lines in apparel manufacturing. Zha and Lim (2003) adopted a fuzzy logic structure in conjunction with neural networks when modelling for human workplace designs and the simulation of manual assembly workstations. Dependence on developed rules of fuzzy set theory-based modelling techniques is based on process expert(s) knowledge, and their prior experiences and opinion(s). This is considered to be one of the main limitations of this technique in modelling manual assembly systems design problems. Because it cannot deal efficiently with the discrete attributes of those systems and in particular the dynamic changes of underlying manual assembly operations using this technique. It also does not always give definitive answers on how to express analytical models of manual assembly processes. In addition, it is difficult to program the fuzzy part.

2.2 OPTIMISATION TECHNIQUES FOR DETERMINATION OF OPTIMAL COMBINATION OF DESIGN VARIABLES

Worker reaction time, fatigue effects, and a wide product variety cause more complexity in manual assembly systems design. This complexity, however, directly affects outcomes for the optimal or near-optimal combination of design parameters, which is considered to be a critical and difficult task for researchers and practitioners (Rekiek, De Lit, & Delchambre, 2000), and is often guided by heuristics or trial-and-error approaches. Hence, the design of manual assembly systems with optimal performance can be formulated as an optimisation problem. This problem also includes the need to confront discrete and continuous parameter spaces, differentiable as well as non-differentiable objective functions or measure(s). The choice of the suitable optimisation technique to solve this problem is based on; (i) design parameters that determine the model of the system, (ii) the objective function(s) that quantifies the system performance with respect to the design variables with or without constraint(s). Diverse techniques have been used by researchers to determine the best combination of design variables so as to optimise a performance criterion for manual assembly systems. As displayed in Figure 4, optimisation techniques can be roughly classified into two categories: conventional techniques and non-conventional techniques. In general, conventional optimisation techniques attempt to provide a local optimal solution; therefore, these techniques are also called local search techniques. On the other hand, non-conventional techniques are based on extrinsic models or pre-established objective functions. Thus, it is only an approximation in the attempt to provide a near-optimal combination of design parameters (conventional optimisation techniques). In turn, conventional technique categories may be broadly classified into two categories as well. The first one involves experimental techniques that include statistical designs of experiments, such as Taguchi method, and response surface design methodology (RSM). The second category is called iterative mathematical search techniques. Linear programming (LP), non-linear programming (NLP), and dynamic programming (DP) algorithms are included in this category. Non-conventional meta-heuristic search-based techniques, which are sufficiently general and extensively used by researchers in recent works, are based on simulated annealing (Kirkpatrick, Gelatt, & Vecchi, 1983), Tabu search (Glover, 1989), and genetic algorithm (GA) (Goldberg, 1989a). A critical evaluation of each of these techniques is introduced in the next subsection.

Figure 4. Available optimisation techniques to determine optimal combination of design variables of manual assembly systems

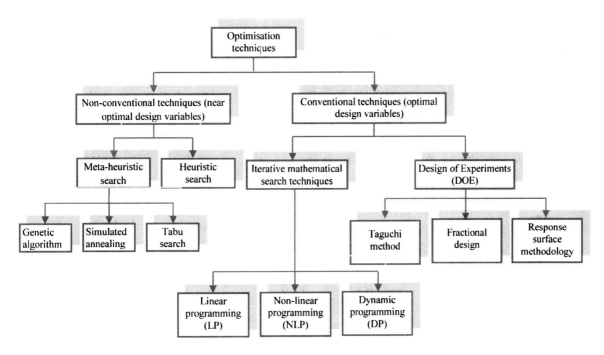

2.2.1 Conventional Techniques

2.2.1.1 Taguchi Method

The statistical techniques which are employed by Professor Genichi Taguchi to improve the quality of products and processes, attracted many researchers to use these techniques as an invaluable contribution to quality engineering (Kalagnanam & Diwekar, 1997; Nair, Abraham, MacKay, Nelder, Box, Phadke, Kacker, Sacks, Welch, Lorenzen, Shoemaker, Tsui, Lucas, Taguchi, Myers, Vining, & Wu, 1992). These techniques are based on the use of classical fractional (or factorial) designs, Taguchi's orthogonal arrays (OAs) and robust designs in designing a system in such a manner, that the performance is insensitive to variation due to the uncertainty represented by uncontrolled or noise factors. Due to its distribution-free and orthogonal array design (Ross, 1995; Tjahjono, Ball, Ladbrook, & Kay, 2009), this method is used to determine the combination of design variables for the optimum design process, with the least number of experiments and consequently it reduces the time and resources needed to determine those variables (Unal & Dean, 1991).

A common class of applications has focused on optimising manual assembly systems design through the Taguchi method. Besseris (2008) proposed a methodology for solving multi-objective optimisation problem in electronic assembly operations by employing Taguchi methods and a non-parametric statistical technique. Ding, Ceglarek, and Shi (2002) presented a methodology for evaluating and benchmarking design configurations of multi-station assembly processes through the Taguchi method; sensitivity-based design evaluation. Ismail, Haniff, Deros, Rani, Makhbul, and Makhtar (2010) considered the effects of environmental factors such as luminance, humidity, wet bulb temperature on worker productivity. The Taguchi method was used in that research to determine the sequence of dominant factors playing significant role in worker productivity at particular workstations at an automotive components assembly factory. Tjahjono, Ball, Ladbrook, and Kay (2009) applied simulation and the Taguchi method to derive a set of principles that can be used by engineers and practitioners to design manual assembly lines. D'Angelo, Gastaldi, and Levialdi (1998) focused on designing an optimal configuration of a flexible manufacturing system with particular reference to printed circuit board assembly. The Taguchi method was combined with response surface methodology, to assess the morphology of the response surface representing performance cost of system configuration.

Although the Taguchi method contributes towards robust design optimisation and introduces innovative techniques to find best possible design and performance of the system, there are some limitations and weaknesses associated with this method. Below is summary of these limitations.

- The Taguchi orthogonal array method does not account for the total number of experiments required by the design process. Consequently, many important interacting factors are not considered using this method which hinders finding the optimal solution (Box & Hunter, 1987; Nair et al., 1992);
- While the Taguchi method results in reducing the extremely time consuming number and the cost of experimentations, it does not exactly indicate what design variables have the highest effect on the performance measures (Box & Hunter, 1987; Nair et al., 1992);
- It is difficult to maintain a certain level of confidence when conducting simultaneous optimisation of multiple measures using this method. This is because of a lack of use in conjunction with the controversial S/N ratio in the simultaneous optimisation of mean quantities and standard deviations in system performance, for given signal and control factors. On the other hand,

other techniques have been suggested for tackling similar optimisation problems such as data transformation and the use of dual-response surface technique (Myers, Walter, & Carter, 1973; Tang, Goh, Yam, & Yoap, 2006), and Lambda plot (Gunter, 1988), are available in literature and can accurately provide a better solution than the Taguchi method;

- The functional relationships expressing the complexity of the manual assembly system may not be quadratic in nature; hence, use of quadratic loss function in this method as a universal expression to approximate functional relationships is unconvincing (Mukherjee & Ray, 2006);
- In Taguchi's optimisation procedure method, there is no reliance upon intrinsic empirical or mechanistic modelling during experimentation. As such there is always a need for greater in-depth knowledge of key design parameters that have the most effect on the system performance (Gunter, 1988);
- Phadke (1989) demonstrated that the optimisation of multiple objective problems using the Taguchi method is purely based on critical and subjective process knowledge.

2.2.1.2 Response Surface Design Methodology (RSM)

Response surface methodology (RSM) (Box & Wilson, 1951; Hinkelmann & Kempthorne, 2008) is comprised of mathematical, statistical and optimisation techniques for exploring optimal operating conditions through 'design of experiment' (DOE). Typically, this involves doing several experiments, establishing the relationship between the measure(s) of a process with its input decision taken from experimental data in order to achieve the objective of maximising or minimising the properties of the given measures. The RSM optimisation procedure begins with building the first regression model and searching for the optimal solution through multivariate regression analysis. After reaching the vicinity of the optimal point, second order models are employed (Khuri & Mukhopadhyay, 2010). In recent years RSM has been prevalent in optimisation studies. Some examples of RSM applications have been performed for optimising manual assembly systems. RSM requires a smaller number of simulation experiments in comparison to many gradient based methods; therefore it is frequently used to optimise stochastic simulation models (Neddermeijer, Oortmarssen, Piersma, & Dekker, 2000). Ben-Gal and Bukchin (2002) have used RSM to optimise designs for manual workstation from the perspective of productivity and ergonomics. In a similar way, optimising assembly cell design was the main purpose in a study by Spedding, Souza, Lee, and Lee (1998). In which they presented a two-phase approach based on experimentation with simulation models for the design and analysis of assembly cells. Response Surface Methodology (RSM) was used to find the optimal number of workstations and workers to be used in the paint shop line of an automotive factory (Allen & Yu, 2007). On the other hand, some reviewers have indicated the limitations of response surface methodology. Carlyle, Montgomery, and Runger (2000) and Wu (2009) indicated that the application of mathematical iterative search algorithms, and heuristic or meta-heuristic search techniques in an optimisation problem, gave high accuracy results in comparison with RSM applications, especially when problems have highly nonlinear, multi-modal, objective functions. In addition, they claimed that it is extremely difficult to use RSM to solve problems which have polynomial complexities. The presence of multiple objectives further increases the complexity of the problem. The applicability of RSM in engineering design has been well recognised by Castillo and Semple (2000), but they found that this optimisation technique performed efficiently when the number of measures did not exceed three. If the optimisation problem had more than three measures, the response surface methodology gave an indefinite saddle function in the quadratic response surface model.

2.2.1.3 Iterative Mathematical Search Techniques

The mathematical search technique provides a unified framework for the optimisation of problems of sequential decision-making given the presence of uncertainty inherent in manual assembly systems. For a variety of optimality criteria, these problems can be solved by dynamic programming. The main strength of this technique is that fairly general stochastic and linear/nonlinear dynamics systems can be considered (Farias, 2002). Early research in the optimisation problems of manual assembly system designs applied mainly mathematical search techniques (Wong, Mok, & Leung, 2006). These techniques included linear integer programming, mixed integer programming (Hiller & Liebermann, 1999), quadratic assignment (Koopmans & Beckmann, 1957) and graph theoretic (Caccetta & Kusumah, 2001). The selection of the appropriate search technique can be used to generate a sequence of improving approximate solutions for a class of problems; it strongly depends on the specific problem type that is to be solved. When using the techniques mentioned, constructing an actual physical model of manual assembly systems gives limits to the validation of developed optimisation technique results.

A specific implementation of linear integer programming in any general manual assembly system design problem is dependent on the existing linear functions of both objective function(s) and constraint equation(s). A common search algorithm in linear integer programming is simplex.

Manual assembly systems have been recognised as highly complex manufacturing systems where assembly processes become increasingly complex due to an increase in product variety, differently skilled manual workers, and the dynamic behaviour of workers (Wang, Koc, & Nagi, 2005). This complexity impacts system performance in terms of productivity, ergonomics and quality. Thus, LP techniques cannot adequately address the complexity and non-linear nature of manual assembly systems. However, due to difficulty in handling the direction of complex flows in optimisation designs, multi-model functions based on multiple nonlinear measure functions are formulated and solved based on NLP solution techniques, such as mixed integer programming, quadratic assignment problem and graph theoretic.

Table 1 summarises the studies which are used these mathematical search technique to solve manual assembly system design problems.

The desirability function approach which was proposed by Derringer and Suich (1980) is a powerful technique for multiple measure optimisation and ranks well in industry use. This technique was earlier described Harrington (1965). It is based on translating functions into a common scale [0, 1], combining them using the arithmetic

Table 1. The research used mathematical techniques for modelling of manual assembly systems

Techniques	The research
Linear programming	Aase, Olson, and Schniederjans (2004), AlGeddawy and ElMaraghy (2010), Gadidov and Kurtoglu (2004), Lee and Stecke (1995), Ooi (2005), Tirpak (2008), and Wilhelm (2000)
Mixed integer programming	Eguia, Lozano, Racero, and Guerrero (2011), Klampfl, Gusikhin, and Rossi (2006), Luo, Li, Tu, Xue, and Tang (2011), Rave and Álvarez (2011)
Quadratic assignment	Kouvelis, Chiang, and Kiran (1992), Lan and Kang (2006), and Xu and Liang (2006)
Graph theoretic	Kim and Kim (1995)

mean and then optimising the overall measure. The equations may represent model predictions or analytic equations.

The generalised distance technique used to simultaneously optimise multiple measures has been developed by Khuri and Conlon (1981). It may be considered as a two-stage process. The first stage obtains individual optima of the calculated measures over the experimental region. The second stage obtains the optimum by minimising the distance function, the distance from the ideal optimum.

When a large feasible region of solutions exists and multiple measures are to be investigated, the optimising designs of manual assembly systems it might not be suitable to use both desirability function and generalised distance techniques.

Iterative mathematical search techniques work well in optimisation formulations, but for various reasons, their successful application is not always straightforward:

- Often mathematical iterative search techniques are formulated to solve design problems with only single aspects, in a sense design variables are determined by optimising one objective function while real problem design of manual assembly systems has several desired objectives; and
- Although most of the iterative mathematical search techniques are adequate enough to be successful in dealing with processes involving a sequence of decisions, and are flexible for applications in various optimisation problems, translating a decision problem into a mathematical programming formulation requires a high degree of skill in practice (Wang, 2005).

To address dynamics and non-linearity of manual assembly systems, heuristic and meta-heuristic techniques may provide alternative near-optimal design parameter(s), which are efficient and reasonably acceptable for implementation by designers and practitioners rather than searching for exact optimal design parameter(s), based on iterative mathematical search techniques.

2.2.2 Non-Conventional Techniques

2.2.2.1 Heuristic Search Techniques

Many optimisation problems are too complex to solve using conventional optimisation techniques. This complexity is because of existing high-dimensional search spaces with many local optima. In such circumstances, suitable heuristic search techniques are the most effective in finding good, although not necessary optimal solutions (Gavrilas, 2010). Even so, they provide the framework into which acceptable solution(s) at a reasonable computational cost can be found (Caserta & Voß, 2010). Researchers and practitioners prefer to use certain techniques within a reasonable time frame to find adequate solutions (not perfectly accurate, but good quality approximations to exact solutions) rather than exactly optimal ones, as it may be impossible to generate an exact optimal solution, particularly for higher dimension and multimodal search space problems.

Heuristic search techniques such as evolutionary algorithms have been applied to the design of manual assembly systems. Sirovetnukul and Chutima (2010) developed an algorithm named COIN to solve multi-objective worker allocation problems for single and mixed-model assembly lines with manually operated machines in several fixed U-shaped layouts. Also Battini, Faccio, Ferrari, Persona, and Sgarbossa (2007) used an evolutionary algorithm to optimise the configuration of a mixed-model assembly system permitting a significant improvement in task repeatability and job enlargement for workers, by reducing load and set-up times to the minimum, even for automated equipment (Pierreval, Caux, Paris, & Viguier, 2003).

2.2.2.2 Meta-Heuristic Search Techniques

Due to the need of heuristic techniques for the high number of function evaluation, especially for high dimensional and multi-model optimisation problems where such need cannot be afforded, these techniques are inefficient in the numerical search process for optimisation (Raidl, 2006). The most popular techniques able to successfully cope with the above problems are meta-heuristic (Gavrilas, 2010). These techniques are proven to be significantly useful and practical for approximately solving complex optimisation problems (Raidl, 2006). Meta-heuristic techniques require no special knowledge of the optimisation problem which is considered to be their most recommending feature (Blum & Rolli, 2003). This group of techniques include genetic algorithm (GA), simulated annealing (SA), and Tabu search (TS). A brief review of these three techniques along with their applications and limitations in field of optimising design of manual assembly systems will be presented in the next subsections.

1. **Genetic Algorithm (GA):** Genetic algorithm (GA) is a subclass of the general structure of evolutionary computations. The work of Goldberg (1994), Holland (1992), Konak, Coit, and Smith (2006) are based on principles of selection and evolution, using three basic operators: selection, crossover and mutation. The work of genetic algorithms can be explained by the following steps; (i) randomly creating of group of individuals (solutions) and placing them in a population, (ii) the programmer or the end-user then allocates an evaluation fitness function or scores to the individuals based on their task performance, (iii) depending on fitness basis, two individuals are then chosen. The individuals with high fitness having a higher chance of being selected, (iv) these individuals reproduce, creating one or more offspring which are then subsequently, randomly mutated, (v) dependant on the requirements of the programmer, this continues until a suitable solution has been found or a certain number of generations have passed. GA is well-suited to accommodate single and multi-objective optimisation problems. In comparison with mathematical search techniques, GA has a number of advantages (Goldberg, 1989). Some of its advantages include the following; (i) an implementation of GA does not require the continuity or convexity of the design space, because it is not based on gradient-based information, (ii) a structural genetic algorithm gives us the possibility of exploring a large search space, using probabilistic transition rules rather than deterministic rules, and consequently, the chance of reaching the global optimal solution significantly increases, (iii) since the genetic algorithm execution technique works throughout the solution space, rather than as a single solution as in conventional techniques, the chance of avoiding local optimality is increased, (iv) multi-dimensional, non-differential, non-continuous, and even non-parametrical problems can be solved using genetic algorithms. This is because the execution of GA does not rely on the existence and continuity of derivatives or other auxiliary information. A large amount of literature exists on the optimisation of manual assembly systems using GA. A number of researchers have used GA to optimise structural design (Banerjee, Zhou, & Montreuil, 1997; Dikos, Nelson, Tirpak, & Wang, 1997; Lee, Khoo, & Yin, 2000; Manziniy, Gamberiy, Regattieriy, & Personaz, 2004) or for operational design (Feyzbakhsh & Matsui, 1999; Ho & Ji, 2005; Wong, Kwong, Mok, Ip, & Chan, 2005) or even both (Rekiek, Lit, & Delchambre, 2002). Although genetic algorithms have been proven to efficiently search complex solution spaces in many situations, there are

a few limitations inherent in this technique. These limitations included; (i) convergence of the GA is not guaranteed, (ii) attaining near-optimal solutions requires a significant amount of computer execution time, and the convergence speed of GA may be not fast, (iii) identifying fitness function must be carefully considered. In many cases, if the fitness function is identified poorly or imprecisely, the genetic algorithm may not be capable of finding a solution to the problem, or may end up solving the wrong problem, (iv) since no universal rule exists, genetic algorithm parameters such as population size, number of generations to be evaluated, crossover probability, mutation probability, and string length, must be chosen with care.

2. **Simulated Annealing:** Simulated annealing (SA) was a concept inspired by modelling and simulating the physical process of annealing solids; it was developed in the 1980's based on the Metropolis algorithm (Kirkpatrick, Gelatt, & Vecchi, 1983). The SA is a stochastic search technique performing well in combinatorial process optimisation problems. In the SA process, the solution to the problem is equivalent to the state of solids, and the cost of the solution is equivalent to the energy of the solids. This technique moves in the search space towards worst and better neighbouring solutions with certain probabilistic measures to escape from the local minima in a multimodal measure function. Thus, in comparison with local search techniques, finding the desirable solution using SA is not strongly dependent upon the choice of initial solution. Therefore, it is considered an effective and robust technique (Aarts & Korst, 1989). A comprehensive investigation of manufacturing layout problem literature includes using SA. Singh and Sharma (2006) presented the most recent survey of layout papers on SA based facilities. Some recent research has covered the use of SA in the attempt to solve optimisation in manual assembly system design. Wang, Wu, and Liu (2001) applied SA to search for the optimal configuration of cellular assembly system. Although SA is appreciated for its ability to solve complex and intractable problems and has ease of implementation, it has three main limitations: (i) it cannot guarantee optimality of its solutions because the repeatability of the near-optimal solution obtained by SA with the same initial solution is not guaranteed, (ii) the changes to solutions are typically local, (iii) pursues just one state solution at a time.

3. **Tabu Search (TS):** It is a local search-based technique Introduced by Glover (1989, 1990a, 1990b). The search optimisation strategy in this technique is based on improving an initial feasible solution which is obtained from a random feasible combination of decision variables, and avoids returning to the solution space previously visited by maintaining Tabu lists that store successive feasible solutions. Tabu lists allow some solutions to be revisited, at some point and after a number of steps. New solutions may be generated in different ways, by adding a new solution to a complete Tabu list and removing the oldest based on a FIFO principle (first in – first out). The Tabu search technique has been applied to provide optimal and desirable solutions to combinatorial optimisation problems. Some of the applications in the literature involve optimising designs for manual assembly systems using TS, as in the research of Martín-del-Campo, François, and Morales (2002). This technique is still actively researched, and is continuing to evolve and improve (Singh & Sharma, 2006). TS suffers from some restrictions, similar to SA, this technique cannot guarantee that the convergence for multi-model objective functions can be achieved in a finite number of steps. The selection of the

Tabu list size can have an important influence on determining the global solution to the problem. More particularly, a list of a small size may result in wasteful revisit of the same combinations of design variables; whilst a list of a long size may result in an exponential increase of CPU time required to verify the Tabu status of candidate combinations of design variables.

2.3 Analysis of Research Progress in Previous Techniques

After reviewing the modelling and optimisation techniques of manual assembly system designs described above, it was found that there exist no common foundations or principles for selecting the best one and judge the performance of the different alternative techniques in any manual assembly system design optimisation problem. However, it was recognised by Baines and Kay (2002), Wang and Chatwin (2005), and Borenstein, Becker, and Santos (1999) that the optimising design of flexible assembly systems remains as a challenging problem because of the following reasons:

- The extensive interactions between available facilities for production (such as labour, tools, fixtures, information, products and assembly workstations);
- Uncertainties in production demands (such as production schedules); and
- The presence of randomness (such as variability in task completion time).

Additional to the above mentioned reasons, and based on a specific application and other design considerations in the system, the number of measure characteristics to be optimised simultaneously may vary. Hence, the modelling and optimisation process becomes more complex especially where there exists a non-linear, constrained multiple measure optimisation problem with multi-model distribution of measure characteristics.

3 GUIDELINES FOR SELECTING A FRAMEWORK APPROACH TO OPTIMISING DESIGN OF MANUAL ASSEMBLY SYSTEMS

In perspective of the above, and dependant on the capabilities and limitations of the reviewed modelling and optimising techniques, this research proposes a number of guidelines that should be taken into consideration when developing an integrated model capable of handling designs for optimising manual assembly systems. The guidelines are following:

- The model ought to combine all considerations (decisions) including technical, economic, ergonomics and strategic ones, throughout assembly system design;
- Because of the 'socio-technical' nature of manual assembly (equipment——technical and human——social), the assemblage of naturally related variables used to develop the model, should accurately identify interacting processes between these variables, to clarify the scale of the comprehensive nature of the system;
- The model should elucidate the complexity of manual assembly systems which simultaneously embody all or most of the variables that have an impact on system performance;
- The optimisation technique of the model must be able to handle non-linear and discrete system models and multi-objective problems; and
- The proposed performance measures used to evaluate the system play a crucial role in determining the selection of the most appropriate optimisation technique.

As stated at the end of section 1, a lack of appropriate methodologies capable of assisting evaluation and analysis during WWAL design in terms of its real performance, especially in areas of productivity and ergonomics, are an important factor. Therefore, this system has yet to be widely adopted in the industrial environment. Taking into account the proposed criteria above here follows a complete description of the framework developed to manage this situation, describing its phases and logic.

4 DEVELOPMENT OF A FRAMEWORK FOR THE OPTIMISING DESIGN OF WWAL

4.1 General Framework Structure

This research proposes a framework combining modelling and optimising techniques to effectively design a manual assembly system, in particular walking worker assembly line. This framework for modelling and optimising the design of walking worker assembly line (MODWWAL) is used to tackle the requirements of productivity and ergonomics. Utilising one unifying framework for WWAL design is more effective than tackling each requirement separately without any loss of solution quality.

The schematic of this framework is outlined in Figure 5. The framework developed here consists of the following two phases:

- Mathematical modelling phase: develop mathematical model formulas to map the process input parameter relationships with the performance measures for system design; worker productivity and ergonomics.
- Genetic algorithms (GA) optimisation phase: GA is utilised to obtain the optimal values of performance measures with corresponding optimal process design variable combinations.

4.2 Framework Tools and Techniques

The techniques of the MODWWAL framework were selected for the target system design because they satisfy most of the criteria mentioned above. The MODWWAL framework techniques and the reasons for their selection are described in the following subsections.

4.2.1 The Mathematical Model

The mathematical model of this research is a combination of two individual models: the productivity model and the ergonomics model. In turn, the ergonomics model is also a combination of two individual models: a physiological model and a biomechanical model. The mathematical model may be considered as the system definition. Such a description is extremely valuable for two reasons. Firstly, it provides an understanding of how changes to the design factors can significantly affect the system performance. Secondly, it allows the study of a wider variety of "interesting" architectures for the system since it is conveniently implementable, requiring no further development for other architectures (Lam, Wilton, Leong, & Luk, 2008). In this research, the mathematical model is used to mathematically describe some processes of WWAL and to predict the performance measures proposed for use for the system optimisation objectives. The resultant model provides the basic mathematical input required to formulate the process of optimisation required to determine the optimal settings for the design variables that may contribute to the overall optimisation of WWAL design in productivity and ergonomics measures. For the purpose of formulating the mathematical model it is recognised that there are certain assumptions inherent in these models, most of these assumptions are consistent with the actual application of WWAL. Simulation modelling is a more appropriate and accurate method capable of representing the complex static structure, and dynamic behaviour of manufacturing systems

Figure 5. General framework structure for modelling and optimisation of WWAL design problem

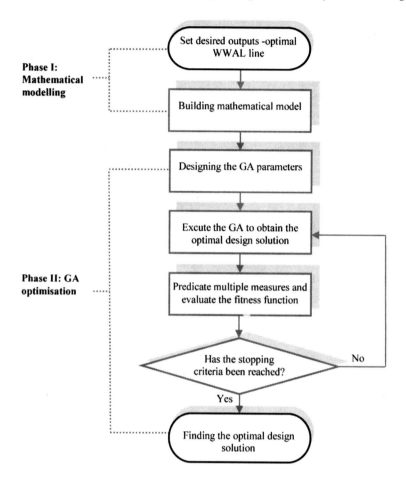

(Borenstein, 2000). Accordingly, the simulation model is used here simply to confirm that the mathematical model of WWAL is correct and the model representation of the system behaviour has "reasonable" accuracy for the intended purpose of the model. To summarise, in this research, the overall model of the total system is a mathematical model and the simulation model is used in the validation process of that model.

4.2.2 The Desirability Function

In this research, the required measures of the WWAL design are that of productivity and ergonomics measures. Some of these measures must be kept small, while other must be kept large. Therefore, these output measures result in a multi-objective optimisation problem. Because these four output measures have to be minimised and maximised respectively, they can convert the multi-objective functions into a single objective using two strategies. The first strategy is that individual performance measures are transformed into individual desirability functions with values between zero and one (Harrington, 1965; Castillo & Montgomery, 1996). In modified desirability functions, every response (y_i) is assigned an individual desirability function (d_i) where the range of value varies between zero and one. With $(d_i) = 1$, the measure (y_i) is at the maximum desirable limit, and if the measure is outside the acceptable lower limit, $(d_i) = 0$. In the second strategy, individual desirability functions of measures are then transformed into one objective function using the geometric means of the individual desirability functions. The outcome from the second strategy is the overall desirability function $(\breve{D}_k)$. To combine it with a genetic algorithm the overall desirability function is modified, according to research of Ortiz and Simpson (2002). The modified overall desirability function will act as the fitness function to GA. The basic idea of using the technique above is that the desirability function serves as evaluation criteria for the solutions produced; using a genetic algorithm in this research it is the following:

- The regions of search are expanded to include infeasible solutions and gain some insight into internal structure of good solution strings and memorises paths that could lead to good solutions; and
- Feasible and infeasible solutions may be easily distinguished from each other, and consequently will not allow infeasible solutions to have the same desirability value as feasible solutions.

4.2.3 Genetic Algorithm

Genetic algorithm was first pioneered by John Holland (1975). The GA is an adaptive heuristic search algorithm based on the evolutionary ideas of natural selection and natural genetics (Ang & Sivakumar, 2007). GA is chosen to perform the optimisation for two main reasons (Noorossana, Tajbakhsh, & Saghaei, 2009):

- Due to the necessity of response surfaces for computing the gradient and direction of improvement, deterministic gradient-based algorithms such as Conjugate Gradient (CG), cannot be used to create the next search steps; and
- GA is known for improved accuracy when compared to traditional methods of optimising highly nonlinear, multi-modal and heavily constrained functions.

In addition to the above reasons, GA is widely known to reduce the time and effort required to provide the global optimal solution. This stage of research involves using a GA to optimise the overall (or total) desirability value thus obtaining the optimal design variable values and the corresponded multiple measures (productivity and ergonomics) from the possible solutions space. When implementing a GA to solve an optimisation problem, there are three major design decisions to be considered. Firstly, a representation for candidate solutions must be encoded in the GA chromosome. Secondly, a fitness function must be specified to evaluate the quality of each solution. Finally GA run operators must be specified. Both design variables and input are used to create the solutions for the GA approach. The solution of GA represents a vector of design variables. A possible solution represents a chromosome. The accuracy and levels of design variables are determined based on the system characteristics. GA then evaluates the quality of each individual solution (or chromosome) using a fitness function. For the problem under consideration, the overall desirability function represents the value of the fitness function of the GA.

After an initial set of solutions (population) is randomly generated, GA evolves through basic operators; mate selection, crossover and mutation. In mate selection, a pair of feasible solutions from a randomly generated so-called 'initial population' is selected to produce a new solution (offspring) for the next generation. A process of selection is carried out based on the evaluation of final stage composite desirability function/fitness function values for each feasible solution. Those two solutions which have higher composite desirability function in that generation will be chosen to produce new offspring. Crossover is the basic of the genetic operators. It consists of swapping parts (gens) of the pair of new solutions (offspring) between solutions to generate new solution pairs which may be superior to the first pair in the current generation. The operation of mutation creates a new solution by the mutation of one-gene from one solution to another according to mutation probability. It should be noted by using the crossover and mutation operations, new populations are created. A matlab in (C++) can be used in writing of fitness function and other calculations and coding in genetic algorithm (Noorossana, Tajbakhsh, & Saghaei, 2009).

4.3 Implementation of the Developed WWAL Approach

The MODWWAL framework may be summarised in the following phases:

4.3.1 Phase I: Mathematical Modelling

In this research, the main objective of mathematical WWAL design modelling is to reveal the underlying relationships between independent design decision variables and the dependent output measures, with special emphasis on productivity and ergonomics measures. This objective is met by developing ordinary differential formulas describing the interaction of workers with work conditions, job content, system configuration, demand schedules and operating costs of production facilities. In addition to assumptions, these formulas are based on a number of technical studies that develop the requisite operational and structural inputs. The following steps require incorporation into the modelling phase:

Step 1: State the assumptions on which the model will be based.
Step 2: Completely describe the design variables (control factors), input parameters (non-factors) and performance measures of the process design of WWAL to be used in the model.
Step 3: Use the assumptions (from Step 1) to derive mathematical equations relating the parameters and variables (from Step 2).
Step 4: Determine the objectives and that the line is designed to accomplish them.
Step 5: Set the constraints that must be satisfied by the optimal solution, which achieves the objectives mentioned in the previous step.
Step 6: The developed mathematical model requires testing and validation under different conditions. Validating the mathematical model of a structural dynamic system (i.e. WWAL) is usually achieved by comparing predictions from the model to results measured from experimental data (results from simulation model or real life experiments).

4.3.2 Phase II: GA Optimisation

In this research, optimisations refer to finding the values of design variables corresponding to and providing the maximum or minimum of desired objectives in WWAL design, as specified in the next section. Meta-heuristic technique, specifically "GA" was chosen for application to the research problem. The GA technique execution in this work consists of five major steps:

Step 1: The fitness function is the overall desirability function according to Step 3 on phase II.
Step 2: Initialisation of GA parameters, choice of population size N, crossover, mutation and other parameters, and choice of a maximum allowable generation number t_{max} and set $t = 0$.
Step 3: Generation of initial feasible random population.
Step 4: Calculation of the fitness of each individual in the population. In this research, the performance measures are productivity measures and ergonomics measures. Accordingly, it is multi-objective optimisation problem. One of the methods for combining multiple objective functions into a scalar

fitness function (single objective) was the desirability function approach (Harrington, 1965). The following steps are established to calculate the fitness function:
- Specification for the relative weights and relative importance of each performance measure using the analytic hierarchy process (Saaty, 1990) or dependent upon on the desire. However, this is not a simple task and impacts the optimal design variables settings:
- Calculation of the individual desirability functions for each performance measure based on:
  - Optimisation process design constraints and objectives criteria; and
  - The relative weight of that measure.
- Derivation of the overall desirability function from the geometric mean of individual desirability values, whilst considering the relative importance assigned subjectively and respectively to each performance measure. The overall desirability function will act as the fitness function to GA.

Step 5: Perform of GA operations including selection, crossover, and mutation. This creates a new solution set, potentially better than the previous generation. This completes one generation.

Step 6: Set $t = t + 1$. If $t > t_{max}$ or other termination criteria is satisfied, then the process is terminated. Otherwise, the process is repeated from Step 4.

Step 7: The optimal level combination of design variables is found.

4.4 Approach Input and Output

The WWAL design approach (inputs) is used to formalise the requirements (outputs) that the approach must accommodate. Here, inputs and outputs of the problem are presented as follows: select the basic data to be manipulated by the approach (input) and the design variables to be supplied by the approach (output). Both the inputs and the outputs of WWAL design problems are divided into two classes; operational and structural.

Basic Data

It is assumed that the products to be assembled and the assembly workstations (including human workers) have already been selected. The basic data is classified into two groups, as follows:

1. *Operational data:* operation time at each workstation, weight of product, anthropometric data of worker, labour cost, metabolic energy requirements, shift time and variability in tasks.
2. *Structural data:* sequence of job, and process environment.

Design Variables

Taking the basic data mentioned, as input parameters, the problem is in finding values for the following eight design variables for the best WWAL design able to provide improved ergonomics conditions whilst also improving worker productivity:

Operational Variables

1. Skill levels of workers
2. Number of slow workers
3. Number of workers at the line
4. Walking speed

Structural Variables

5. Distance between workstations
6. Number of workstations
7. Layout shape design
8. Floor surface roughness

4.5 Objectives and Overall Project Plan of the Approach

The WWAL should be designed with capability for a "level of improvement" for design variables that will result in the best system design, where the opportunity for greater impact on labour productivity and ergonomics is expected to be greatest. There are four design objectives to consider in WWAL design.

Productivity Objectives

1. Enhanced utilisation of labour by minimising the possibility of blocking rate between one worker and the next due to unevenly skilled workers, different working speeds and / or individual abilities. These factors can lead to in-process waiting time in front of the bottleneck workstation resulting in a decreased production capacity, therefore WWAL should be designed in order to minimise in-process waiting time; and
2. Reducing the operating design cost by minimising labour costs for the shift time required to meet the required production rate.

Ergonomic Objectives

3. Lowering the amount of effort spent on the task (minimising metabolic energy consumption); and
4. Resulting in maximising the mechanical exposure of workers' bodies through assembly work.

While each objective is unique, they share the main operational objectives when designing manual assembly systems; to improve worker-hour productivity and reduce line operating costs (Bukchin, 1998).

4.6 The Approach Constraints

The MODWWAL framework is a design approach that selects the best design of WWAL based on setting selected objectives (as mentioned above) and constraints. Therefore, the best design must satisfy both the objectives and constraints. The following constraints identify the requirements to be met in a set of feasible design solutions of WWAL, before considering one of them as a superior choice that meets the previous objectives:

1. **Production requirements:** the expected design candidate for the solution has a production rate equal to or greater than that of a traditional assembly line (FWAL). When considering economic factors in operating manufacturing assemblies (Ritchie, Dewar, & Simmons, 1999), productivity rarely needs to be sacrificed in order to attain other measures. Hence, if this requirement is not met, why would any adopt this system?
2. **Design operating cost:** total costs associated with the assembly line design solution, specifically, the sum of costs associated with direct labour requirement must be lower than that of FWAL. This cost reduction can result in significant product cost savings and consequently retains a competitive advantage in increasingly competitive markets.
3. **Physical worker capacity:** the workers selected in the design of the system must be physically able to perform all work-related tasks required of them. In this context, it is desirable to adopt ergonomically sound system design which matches the worker's capacity to the job's demands. If the worker's capacity is neglected when designing and analysing of manual assembly systems, they remain exposed to the major risk of work related musculoskeletal disorders WRMDs working under those conditions (Wang, Tang, & Loua, 2009).

5 CONCLUSION

This chapter has presented a new framework for an optimising design of WWAL. The development of that framework known as MODWWAL was based mainly on the comprehensive review of the techniques of modelling and optimising for manual assembly systems design. A general classification was used as framework for describing the major modelling and optimisation techniques that have emerged in this field. Main features, strengths and limitations of these techniques as major solution methods are discussed through this review. Moreover, the dynamic nature of the WWAL problem was also discussed to highlight some of the limitations of existing techniques for modelling and optimising. The framework developed here consists of two phases: mathematical modelling phase and a genetic algorithms GA optimisation phase. In the first phase, a mathematical model is developed to map the design variables and input parameters relationships with performance measures. In the second phase, GA is utilised to obtain the optimal values of performance measures with corresponding optimal design variables combinations.

The framework attempts to provide a foundation for WWAL system design which may contribute to more applications for this system within industry.

REFERENCES

Aarts, E., & Korst, J. (1989). *Simulated annealing and boltzmann machines: A stochastic approach to combinatorial optimization and neural computing*. New York, NY: John Wiley & Sons Inc.

Aase, G., John, R. O., & Schniederjans, M. J. (2004). U-shaped assembly line layouts and their impact on labor productivity: An experimental study. *European Journal of Operational Research*, 156(3), 698–711. doi:10.1016/S0377-2217(03)00148-6

Al Geddawy, T., & El Maraghy, H. (2010). Design of single assembly line for the delayed differentiation of product variants. *Flexible Services and Manufacturing Journal*, 22(3-4), 163–182. doi:10.1007/s10696-011-9074-7

Al-Zuheri, A. (2013). *Modelling and optimisation of walking worker assembly line for productivity and ergonomics improvement* (PhD thesis). University of South Australia, Australia.

Allen, T., & Yu, L. (2007). Paintshop production line optimization using response surface methodology. In *Proceedings of the 2007 Winter Simulation Conference* (pp.1667-1672). Washington, DC: IEEE Press Piscataway.

Anderson, E. J., & Ferris, M. C. (1994). Genetic algorithms for combinatorial optimisation: The assembly line balancing problem. *Journal on Computing*, 6(2), 161–173.

Ang, A. T. H., & Sivakumar, A. I. (2007). Online multiobjective single machine dynamic scheduling with sequence-dependent setups using simulation-based genetic algorithm with desirability function. In *Proceedings of the 2007 Winter Simulation Conference* (pp.1828-1834). Washington, DC: IEEE Press Piscataway.

Antari, J., Samira, C., & Zeroual, A. (2011). Modeling non linear real processes with ANN techniques. In *Proceedings of the 2011 International Conference on Multimedia Computing and Systems* (pp.1-5). Ouarzazate, Morocco: IEEE Press.

Baines, T. S., & Kay, J. M. (2002). Human performance modelling as an aid in the process of manufacturing system design: A pilot study. *International Journal of Production Research*, *40*(10), 2321–2334. doi:10.1080/00207540210128198

Banerjee, P., Zhou, Y., & Montreuil, B. (1997). Genetically assisted optimization of cell layout and material flow path skeleton. *IIE Transactions*, *29*(4), 277–291. doi:10.1080/07408179708966334

Battini, D., Faccio, M., Ferrari, E., Persona, A., & Sgarbossa, F. (2007). Design configuration for a mixed-model assembly system in case of low product demand. *International Journal of Advanced Manufacturing Technology*, *34*(1-2), 188–200. doi:10.1007/s00170-006-0576-5

Ben-Gal, I., & Bukchin, J. (2002). The ergonomic design of workstations using virtual manufacturing and response surface methodology. *IE Transactions*, *34*(4), 375–391. doi:10.1080/07408170208928877

Besseris, G. J. (2008). Multi-response optimisation using Taguchi method and super ranking concept. *Journal of Manufacturing Technology Management*, *19*(8), 1015–1029. doi:10.1108/17410380810911763

Blum, C., & Rolli, A. (2003). Metaheuristics in combinatorial optimization: Overview and conceptual comparison. *ACM Computing Surveys*, *35*(3), 268–308. doi:10.1145/937503.937505

Boër, C. R., El-Chaar, J., Imperio, E., & Avai, A. (1991). Criteria for optimum layout design of assembly systems. *CIRP Annals - Manufacturing Technology*, *40*(1), 415-418.

Borenstein, D. (2000). Implementation of an object-oriented tool for the simulation of manufacturing systems and its application to study the effects of flexibility. *International Journal of Production Research*, *38*(9), 2125–2152. doi:10.1080/002075400188537

Borenstein, D., Becker, J. L., & Santos, E. R. (1999). A systemic and integrated approach to flexible manufacturing systems design. *Integrated Manufacturing Systems*, *10*(1), 6–14. doi:10.1108/09576069910370639

Box, G. E. P., & Hunter, J. S. (1987). *Empirical model-building and response surface*. New York: John Wiley.

Box, G. E. P., & Wilson, K. B. (1951). On the experimental attainment of optimum conditions. *Journal of the Royal Statistical Society. Series A (General)*, *13*(1), 1–45.

Bukchin, J. (1998). A comparative study of performance measures for throughput of a mixed model assembly line in a JIT environment. *International Journal of Production Research*, *36*(10), 2669–2685. doi:10.1080/002075498192427

Bulgak, A. A., & Sanders, J. L. (1990). An analytical assembly systems stations performance model for assembly systems with automatic inspection and repair loops. *Computers & Industrial Engineering*, *18*(3), 373–380. doi:10.1016/0360-8352(90)90059-U

Bulgak, A. A., Tarakc, Y., & Verter, V. (1999). Robust design of asynchronous flexible assembly systems. *International Journal of Production Research*, *4*(3), 3169–3184. doi:10.1080/002075499190220

Byrne, M. D., & Bakir, M. A. (1999). Production planning using a hybrid simulation-analytical approach. *International Journal of Production Economics*, *59*(1-3), 305–311. doi:10.1016/S0925-5273(98)00104-2

Caccetta, L., & Kusumah, Y. S. (2001). *Graph theoretic based heuristics for the facility layout design problems*. Retrieved from orsnz.org.nz

Carlyle, W. M., Montgomery, D. C., & Runger, G. C. (2000). Optimization problem and method in quality control and improvement. *Journal of Quality Technology, 32*(1), 1–17.

Carson, Y., & Maria, A. (1997). Simulation optimization: Methods and applications. In *Proceedings of the 29th conference on Winter Simulation* (pp. 118-126). Washington, DC, USA: IEEE Computer Society.

Caserta, M., & Voß, S. (2010). Metaheuristics: Intelligent problem solving. In V. Maniezzo (Ed.), *Matheuristics* (pp. 1–38). New York: Springer US.

Castillo, E. D., & Montgomery, D. C. (1996). Modified desirability functions for multiple response optimization. *Journal of Quality Technology, 28*(3), 337–345.

Castillo, E. D., & Semple, J. (2000). Optimization problem and method in quality control and improvement. *Journal of Quality Technology, 32*(1), 20–23.

Chan, F. T. S., & Smith, A. M. (1993). Simulation approach to assembly line modification: A case study. *Journal of Manufacturing Systems, 12*(3), 239–245. doi:10.1016/0278-6125(93)90334-P

Coit, D., Jackson, B. T., & Smith, A. E. (1998). Static neural network process models: Considerations and case studies. *International Journal of Production Economics, 36*(11), 2953–2967. doi:10.1080/002075498192229

D'Angelo, A., Gastaldi, M., & Levialdi, N. (1998). Performance analysis of a flexible manufacturing system: A statistical approach. *International Journal of Production Economics, 56-57*, 47–59. doi:10.1016/S0925-5273(96)00115-6

Dashchenko, A. I., & Loladze, T. N. (1991). Choice of optimal configurations for flexible (readjustible) assembly lines by purposeful search. *CIRP Annals - Manufacturing Technology, 40*(1), 13-16.

Derringer, G. C., & Suich, R. (1980). Simultaneous optimization of several response variables. *Journal of Quality Technology, 12*(4), 214–219.

Dikos, A., NeIson, P. C., Tirpak, T. M., & Wang, W. (1997). Optimization of high-mix printed circuit card assembly using genetic algorithms. *Annals of Operations Research, 75*(0), 303–324. doi:10.1023/A:1018919815515

Ding, Y., Ceglarek, D., & Shi, J. (2002). Design evaluation of multi-station assembly processes by using state space approach. *Journal of Mechanical Design, 124*(3), 408–418. doi:10.1115/1.1485744

Eguia, I., Lozano, S. R. J., & Guerrero, F. (2011). A methodological approach for designing and sequencing product families in reconfigurable disassembly systems. *Journal of Industrial Engineering and Management, 4*(3), 418–435. doi:10.3926/jiem.2011.v4n3.p418-435

Farias, D. P. D. (2002). *The linear programming approach to approximate dynamic programming: Theory and application* (Doctoral dissertation). Stanford University, Palo Alto, California.

Feng, C.-X. J., & Wang, X.-F. (2004). Data mining techniques applied to predictive modeling of the knurling process. *IIE Transactions, 36*(3), 253–263. doi:10.1080/07408170490274214

Feyzbakhsh, S. A., & Matsui, M. (1999). Adam—Eve-like genetic algorithm: A methodology for optimal design of a simple flexible assembly system. *Computers & Industrial Engineering, 36*(2), 233–258. doi:10.1016/S0360-8352(99)00131-X

Fu, L. (1994). *Neural networks in computer intelligence*. New York: McGraw Hill.

Gadidov, R., & Wilhelm, W. (2000). A cutting plane approach for the single-product assembly system design problem. *International Journal of Production Research, 38*(8), 1731–1754. doi:10.1080/002075400188564

Gavrilas, M. (2010). Heuristic and metaheuristic optimization techniques with application to power systems. In the *Proceedings of 12th WSEAS International Conference on Mathematical Methods and Computational Techniques in Electrical Engineering* (pp. 95-103). Timisoara, Romania: WSEAS Press.

Glover, F. (1989). Tabu search: Part 1. *ORSA Journal on Computing*, *1*(3), 190–206. doi:10.1287/ijoc.1.3.190

Glover, F. (1990a). Tabu search-Part II. *ORSA Journal on Computing*, *2*(1), 4–32. doi:10.1287/ijoc.2.1.4

Glover, F. (1990b). Tabu search: A tutorial. *Interfaces*, *20*, 74–94. doi:10.1287/inte.20.4.74

Goldberg, D. E. (1989). *Genetic algorithms in search, optimization and machine learning*. Boston, MA: Longman.

Goldberg, D. E. (1994). Genetic and evolutionary algorithms come of age. *Communications of the ACM*, *37*(3), 113–119. doi:10.1145/175247.175259

Gottwald, S. (1993). *Fuzzy sets and fuzzy logic: The foundations of application—From a mathematical point of view*. Wiesbaden, Germany: Vieweg & Sohn Verlagsgesellschaft mbH. doi:10.1007/978-3-322-86812-1

Gunter, B. (1988). Signal-to-noise ratios, performance criteria, and transformations [Discussion]. *Technometrics*, *30*(1), 32–35. doi:10.2307/1270316

Harrington, E. C. J. (1965). The desirability function. *Industrial Quality Control*, *21*(10), 494–498.

Hiller, F. S., & Liebermann, G. J. (1999). *Operations research*. New Delhi, India: CBS Publications and Distributions.

Hillier, F. S., & So, K. C. (1996). On the simultaneous optimization of server and work allocations in production line systems with variable processing times. *Operations Research*, *44*(3), 435–443. doi:10.1287/opre.44.3.435

Hinkelmann, K., & Kempthorne, O. (2008). *Design and analysis of experiments*. Hoboken, NJ: John Wiley and Sons.

Ho, W., & Ji, P. (2005). PCB assembly line assignment: A genetic algorithm approach. *Journal of Manufacturing Technology Management*, *16*(6), 682–692. doi:10.1108/17410380510609519

Holland, J. H. (1992). *Adaptation in natural and artificial systems*. Cambridge: The MIT Press.

Hsieh, S.-J. (2002). Hybrid analytic and simulation models for assembly line design and production planning. *Simulation Modelling Practice and Theory*, *10*(1-2), 87–108. doi:10.1016/S1569-190X(02)00063-1

Hui, P. C.-L., Chan, K. C. C., Yeung, K. W., & Ng, F. S.-F. (2002). Fuzzy operator allocation for balance control of assembly lines in apparel manufacturing. *IEEE Transactions on Engineering Management*, *49*(2), 173–180. doi:10.1109/TEM.2002.1010885

Ismail, A. R., Haniff, M. H. M., Deros, B. M., Rani, M. R. A., Makhbul, Z. K. M., & Makhtar, N. K. (2010). The optimization of environmental factors at manual assembly workstation by using Taguchi method. *Journal of Applied Sciences*, *10*(13), 1293–1299. doi:10.3923/jas.2010.1293.1299

Jayaram, S., Jayaram, U., Kim, Y. J., DeChenne, C., Lyons, K. W., Palmer, C., & Mitsui, T. (2007). Industry case studies in the use of immersive virtual assembly. *Virtual Reality (Waltham Cross)*, *11*(4), 217–228. doi:10.1007/s10055-007-0070-x

Jayaraman, A., & Gunal, A. K. (1997). Applications of discrete event simulation in the design of automotive powertrain manufacturing systems. In the *Proceedings of 29th Conference on Winter Simulation* (pp. 758–764). Atlanta, GA: IEEE Computer Society Washington.

Kalagnanam, J. R., & Diwekar, U. M. (1997). An efficient sampling technique for off-line quality control. *Technometrics*, *39*(3), 308–319. doi:10.1080/00401706.1997.10485122

Kamath, M., Suri, R., & Sanders, J. L. (1988). Analytical performance models for closed-loop flexible assembly systems. *International Journal of Flexible Manufacturing Systems*, *1*(1), 51–84. doi:10.1007/BF00713159

Katayama, H., & Bennett, D. (1996). Lean production in a changing competitive world: A Japanese perspective. *International Journal of Operations & Production Management*, *16*(2), 8–23. doi:10.1108/01443579610109811

Khuri, A. I., & Conlon, M. (1981). Simultaneous optimization of multiple responses represented by polynomial regression functions. *Technometrics*, *23*(4), 363–375. doi:10.1080/00401706.1981.10487681

Khuri, A. I., & Mukhopadhyay, S. (2010). Part I. The foundational years: 1951-1975. *WIREs Computational Statistics*, *2*, 128–149. doi:10.1002/wics.73

Kim, J.-Y., & Kim, Y.-D. (1995). Graph theoretic heuristics for unequal-sized facility layout problems. *Omega*, *23*(4), 391–401. doi:10.1016/0305-0483(95)00016-H

Kim, Y. K., Kim, Y. J., & Kim, Y. (1996). Genetic algorithms for assembly line balancing with various objectives. *Computers & Industrial Engineering*, *30*(3), 397–409. doi:10.1016/0360-8352(96)00009-5

Kirkpatrick, S., Gelatt, C. D., & Vecchi, M. P. (1983). Optimization by simulated annealing. *Science*, *220*(4598), 671–680. doi:10.1126/science.220.4598.671 PMID:17813860

Klampfl, E., Gusikhin, O., & Rossi, G. (2006). Optimization of workcell layouts in a mixed-model assembly line environment. *International Journal of Flexible Manufacturing Systems*, *17*(4), 277–299.

Klir, G. J., Clair, U. H. S., & Yuan, B. (1997). *Fuzzy set theory: Foundations and applications*. Upper Saddle River, NJ: Prentice Hall.

Konak, A., Coit, D. W., & Smith, A. E. (2006). Multi-objective optimization using genetic algorithms: A tutorial. *Reliability Engineering & System Safety*, *91*(9), 992–1007. doi:10.1016/j.ress.2005.11.018

Koopmans, T. C., & Beckmann, M. (1957). Assignment problems and the location of economic activities. *Econometrica*, *25*(1), 53–76. doi:10.2307/1907742

Kouvelis, P., Chiang, W.-C., & Kiran, A. S. (1992). A survey of layout issues in flexible manufacturing systems. *Omega*, *20*(3), 375–390. doi:10.1016/0305-0483(92)90042-6

Kung, H.-K., & Changchit, C. (1991). Just-in-time simulation model of a PCB assembly line. *Computers & Industrial Engineering*, *20*(1), 17–26. doi:10.1016/0360-8352(91)90036-6

Kurtoglu, A. (2004). Flexibility analysis of two assembly lines. *Robotics and Computer-integrated Manufacturing*, *20*(3), 247–253. doi:10.1016/j.rcim.2003.10.011

Lam, A., Wilton, S. J. E., Leong, P., & Luk, W. (2008). An analytical model describing the relationships between logic architecture and FPGA density. In *the Proceedings of the International Conference on Field Programmable Logic and Applications* (pp. 221-226). Heidelberg, Germany: IEEE.

Lan, C.-H., & Kang, C.-J. (2006). Constrained spatial layout and simultaneous production evaluation for a production system. *Concurrent Engineering, 14*(2), 111–120. doi:10.1177/1063293X06065529

Lee, H. F., & Stecke, K. E. (1995). *An integrated design support method for flexible assembly systems* (Working paper no. 681-d). University of Michigan, Ann Arbor, MI.

Lee, S. G., Khoo, L. P., & Yin, X. F. (2000). Optimising an assembly line through simulation augmented by genetic algorithms. *International Journal of Advanced Manufacturing Technology, 16*(3), 220–228. doi:10.1007/s001700050031

Lin, L., & Cochran, D. K. (1987). Optimization of a complex flow line for printed circuit board fabrication by computer simulation. *Journal of Manufacturing Systems, 6*(1), 47–57. doi:10.1016/0278-6125(87)90049-5

Longo, F. G. M., & Papoff, E. (2006). Effective design of an assembly line using modeling & simulation. In the *Proceedings of the 38th Conference on Winter Simulation* (pp. 1893-1898). Monterey, California: Winter Simulation Conference.

Luo, X., Li, W., Tu, Y., Xue, D., & Tang, J. (2011). Operator allocation planning for reconfigurable production line in one-of-a-kind production. *International Journal of Production Research, 49*(3), 689–705. doi:10.1080/00207540903555486

Manziniy, R., Gamberiy, M., Regattieriy, A., & Personaz, A. (2004). Framework for designing a flexible cellular assembly system. *International Journal of Production Research, 42*(17), 3505–3528. doi:10.1080/00207540410001696023

Martin, G. E. (1994). Optimal design of production lines. *International Journal of Production Research, 32*(5), 989–1000. doi:10.1080/00207549408956983

Martín-del-Campo, C., François, J. L., & Morales, L. B. (2002). BWR fuel assembly axial design optimization using Tabu search. *Nuclear Science and Engineering, 142*(1), 107–115.

Mitchell, M. (1996). *An introduction to genetic algorithms*. Cambridge, UK: The MIT Press.

Montgomery, D. C., Peck, E. A., & Vining, G. G. (2001). *Introduction to linear regression analysis*. New York, NY: Wiley.

Mukherjee, I., & Ray, P. K. (2006). A review of optimization techniques in metal cutting processes. *Computers & Industrial Engineering, 50*(1-2), 15–34. doi:10.1016/j.cie.2005.10.001

Myers, R. H., Walter, H., & Carter, J. R. (1973). Response surface techniques for dual response systems. *Technometrics, 15*(2), 301–317. doi:10.1080/00401706.1973.10489044

Nair, V. N., Abraham, B., MacKay, J., & Nelder, J. A., Box, Ge., Phadke, M. S., … Wu, C. F. J. (1992). Taguchi's parameter design: A panel discussion. *Technometrics, 34*(2), 127–161. doi:10.1080/00401706.1992.10484904

Neddermeijer, H. G., van Oortmarssen, G. J., Piersma, N., & Dekker, R. (2000). A framework for response surface methodology for simulation optimization. In the *Proceedings 32nd Conference on Winter Simulation* (pp. 129–136). Orlando, FL: IEEE.

Nomura, J., & Takakuwa, S. (2006). Optimization of a number of containers for assembly lines: The fixed-course pick-up system. *International Journal of Simulation Modelling, 5*(4), 155–166. doi:10.2507/IJSIMM05(4)3.066

Noorossana, R., Tajbakhsh, S. D., & Saghaei, A. (2009). An artificial neural network approach to multiple-response optimization. *International Journal of Advanced Manufacturing Technology*, *40*(11), 1227–1238. doi:10.1007/s00170-008-1423-7

Ooi, K. T. (2005). Design optimization of a rolling piston compressor for refrigerators. *Applied Thermal Engineering*, *25*(5), 813–829. doi:10.1016/j.applthermaleng.2004.07.017

Ortiz, F. Jr, & Simpson, J. R. (2002). *A genetic algorithm with a modified desirability function approach to multiple response optimization*. Florida: College of Engineering.

Phadke, M. S. (1989). *Quality engineering using robust design*. Englewood Cliffs, NJ: Prentice Hall.

Pierreval, H., C. C., Paris, J. L., & Viguier, F. (2003). Evolutionary approaches to the design and organization of manufacturing systems. *Computers & Industrial Engineering*, *44*(3), 339–364. doi:10.1016/S0360-8352(02)00195-X

Raidl, G. R. (2006). A unified view on hybrid metaheuristics. In F. Almeida (Ed.), *Hybrid metaheuristics* (pp. 1–12). Heidelberg, Germany: Springer Berlin Heidelberg. doi:10.1007/11890584_1

Ramnath, B. V., Elanchezhian, C., & Kesavan, R. (2010). Suitability assessment of lean kitting assembly through fuzzy based simulation model. *International Journal of Computers and Applications*, *4*(1), 25–31. doi:10.5120/795-1129

Rave, J. I. P., & Álvarez, G. P. J. (2011). Application of mixed-integer linear programming in a car seats assembling process. *Pesquisa Operacional*, *31*(3), 593–610. doi:10.1590/S0101-74382011000300011

Rekiek, B., De Lit, P., & Delchambre, A. (2000). Designing mixed-product assembly lines. *IEEE Transactions on Robotics and Automation*, *16*(3), 268–280. doi:10.1109/70.850645

Rekiek, B., De Lit, P., & Delchambre, A. (2002). Hybrid assembly line design and user's preferences. *International Journal of Production Research*, *40*(5), 1095–1111. doi:10.1080/00207540110116264

Ritchie, M., Dewar, R., & Simmons, J. (1999). The generation and practical use of plans for manual assembly using immersive virtual reality. *Proceedings of the Institution of Mechanical Engineers. Part B, Journal of Engineering Manufacture*, *213*(5), 461–474. doi:10.1243/0954405991516930

Ross, P. J. (1995). *Taguchi techniques for quality engineering*. New York, NY: McGraw-Hill Professional.

Ross, T. J. (2010). *Fuzzy logic with engineering applications*. Chicester, UK: Wiley-Blackwell. doi:10.1002/9781119994374

Saaty, T. L. (1990). How to make a decision: The analytic hierarchy process. *European Journal of Operational Research*, *48*(1), 9–26. doi:10.1016/0377-2217(90)90057-I

Seijo-Vidal, R. L., & Bartolomei-Suarez, S. M. (2010). Testing line optimization based on mathematical modeling from the metamodels obtained from a simulation. In the *Proceedings of the 2010 Winter Simulation Conference* (pp. 1739–1749). Baltimore: IEEE.

Shafer, S. M., & Smunt, T. L. (2004). Empirical simulation studies in operations management: Context, trends, and research opportunities. *Journal of Operations Management*, *22*(4), 345–354. doi:10.1016/j.jom.2004.05.002

Shtub, A., & Zimerman, Y. (1993). A neural-network-based approach for estimating the cost of assembly systems. *International Journal of Production Economics*, *93*(3), 189–207. doi:10.1016/0925-5273(93)90068-V

Siebers, P.-O. (2004). *The impact of human performance variation on the accuracy of manufacturing system simulation models* (PhD thesis). Cranfield University, UK.

Singh, S. P., & Sharma, R. R. K. (2006). A review of different approaches to the facility layout problems. *International Journal of Advanced Manufacturing Technology, 30*(5), 425–433. doi:10.1007/s00170-005-0087-9

Singholi, A., Chhabra, D., & Bagai, S. (2010). Performance evaluation and design of flexible manufacturing system: A case study. *Global Journal of Enterprise Information System, 2*(1), 24–34.

Sirovetnukul, R., & Chutima, P. (2010). The impact of walking time on U-shaped assembly line worker allocation problems. *English Journal, 14*(2), 53–78.

Smed, J., Johnsson, M., Johtela, T., & Nevalainen, O. (1999). *Techniques and applications of production planning in electronics manufacturing systems (Technical Report)*. Turku Centre for Computer Science.

Solot, P., & van Vliet, M. (1994). Analytical models for FMS design optimization: A survey. *International Journal of Flexible Manufacturing Systems, 6*(3), 209–233. doi:10.1007/BF01328812

Spedding, T. A., De Souza, R., Lee, S. S. G., & Lee, W. L. (1998). Optimizing the configuration of a keyboard assembly cell. *International Journal of Production Research, 36*(8), 2131–2144. doi:10.1080/002075498192814

Spieckermann, S., Gutenschwager, K., Heinzel, H., & Voß, S. (2000). Simulation-based optimization in the automotive industry - A case study on body shop design. *Simulation, 75*(5), 276–286.

Sukthomya, W., & Tannock, J. D. T. (2005). Taguchi experimental design for manufacturing process optimisation using historical data and a neural network process model. *International Journal of Quality & Reliability Management, 22*(5), 485–502. doi:10.1108/02656710510598393

Tang, L. C., Goh, T. N., Yam, H. S., & Yoap, T. (2006). A unified approach for dual response surface optimization. *Journal of Quality Technology, 34*(4), 37–52.

Tirpak, T. M. (2008). Developing and deploying electronics assembly line optimization tools: A Motorola case study. *Decision Making in Manufacturing and Services, 2*(1-2), 63–78.

Tjahjono, B., Ball, P., Ladbrook, J., & Kay, J. (2009). Assembly line design principles using six sigma and simulation. In *the Proceedings of the 2009 Winter Simulation Conference* (pp. 3066- 3076). Austin, TX: IEEE.

Unal, R., & Dean, E. B. (1991). Taguchi approach to design optimization for quality and cost: An overview. In the *Proceedings of 13th Annual Conference of the International Society of Parametric Estimators* (pp. 1-9). New Orleans, LA: NASA Technical Documents.

Wang, A., Koc, B., & Nagi, R. (2005). Complex assembly variant design in agile manufacturing. Part I: System architecture and assembly modeling methodology. *IIE Transactions, 37*(1), 1–15. doi:10.1080/07408170590516764

Wang, J. (2005). *A review of operations research applications in workforce planning and potential modeling of military training. (Land operations division: Systems sciences laboratory report)*. Australian Government Department of Defense.

Wang, Q., & Chatwin, C. R. (2005). Key issues and developments in modelling and simulation-based methodologies for manufacturing systems analysis, design and performance evaluation. *International Journal of Advanced Manufacturing Technology, 25*(11-12), 1254–1265. doi:10.1007/s00170-003-1957-7

Wang, Q., Lassalle, S., Mileham, A. R., & Owen, G. W. (2009). Analysis of a linear walking worker line using a combination of computer simulation and mathematical modeling approaches. *Journal of Manufacturing Systems, 28*(2-3), 64–70. doi:10.1016/j.jmsy.2009.12.001

Wang, T. Y., Wu, K. B., & Liu, Y. W. (2001). A simulated annealing algorithm for facility layout problems under variable demand in cellular manufacturing systems. *Computers in Industry, 46*(2), 181–188. doi:10.1016/S0166-3615(01)00107-5

Wang, X., Tang, D., & Loua, P. (2009). An ergonomic assembly workstation design using axiomatic design theory. In the *Proceedings of the 16th ISPE International Conference on Concurrent Engineering* (pp. 403-412). London, UK: Springer London.

Wong, C. K. K., Mok, P. Y., Ip, W. H., & Chan, C. K. (2005). Optimization of manual fabric-cutting process in apparel manufacture using genetic algorithms. *International Journal of Advanced Manufacturing Technology, 27*(1-2), 152–158. doi:10.1007/s00170-004-2161-0

Wong, P. Y. M., & Leung, S. Y. S. (2006). Developing a genetic optimisation approach to balance an apparel assembly line. *International Journal of Advanced Manufacturing Technology, 28*(3-4), 387–394. doi:10.1007/s00170-004-2350-x

Wu, F.-C. (2009). Robust design of nonlinear multiple dynamic quality characteristics. *Computers & Industrial Engineering, 56*(4), 1328–1332. doi:10.1016/j.cie.2008.08.001

Xu, Z., & Liang, M. (2006). Integrated planning for product module selection and assembly line design / reconfiguration. *International Journal of Production Research, 44*(11), 2091–2117. doi:10.1080/00207540500357146

Yoshimura, M., Yoshida, S., Konish, I., Izui, Y., Nishiwaki, K., & Inamor, S. et al. (2006). A rapid analysis method for production line design. *International Journal of Production Research, 44*(6), 1171–1192. doi:10.1080/00207540500336355

Zha, X. F., & Lim, S. Y. E. (2003). Intelligent design and planning of manual assembly workstations: A neuro-fuzzy approach. *Computers & Industrial Engineering, 44*(4), 611–632. doi:10.1016/S0360-8352(02)00238-3

Zhang, C. H., & Huang, S. H. (1995). Application of neural network in manufacturing—A state of art survey. *International Journal of Production Research, 33*(3), 705–728. doi:10.1080/00207549508930175

Zimmermann, H. J. (1996). *Fuzzy set theory and its applications*. Boston, MA: Kluwer. doi:10.1007/978-94-015-8702-0

KEY TERMS AND DEFINITIONS

Assembly Operation: A set of assembly tasks involving the joining of a number of parts as inputs with a subassembly or the final products as output.

Assembly Workstations: Designated workspace along the work flow line at which one or more assembly tasks are performed by a worker.

Framework: A methodological design approach for the system that can enhance the performance within specified goals.

Lean Assembly Line: An assembly line that considers the expenditure of available resources for any goal other that creation of value for the end customer to be wasteful, and thus a target of elimination of all the wastes.

Manual Assembly Line: A set of assembly workstations in which the majority of the assembly tasks are manually performed by the workers.

Modelling: The representation (mostly mathematically) of process, concept, or operation of a system.

Optimisation: A technique used to find optimal system or design from available range of alternatives. The optimal one is supposed be the most cost-effective or higher performance could be achieved without violating restrictions.

Worker: An operator (female or male) at the workstation.

Section 3
Smart Manufacturing Enabling Technologies

Chapter 6
Design of Anti-Metallic RFID for Applications in Smart Manufacturing

Bo Tao
Huazhong University of Science and Technology, China

Hu Sun
Huazhong University of Science and Technology, China

Jixuan Zhu
Huazhong University of Science and Technology, China

Zhouping Yin
Huazhong University of Science and Technology, China

ABSTRACT

Anti-metallic passive RFID tags play a key role in manufacturing automation systems adopting RFID techniques, such as manufacturing tool management, logistics and process control. A novel long range passive anti-metallic RFID tag fabrication method is proposed in this chapter, in which a multi-strip High Impendence Surface (HIS) with a feeding loop is designed as the antenna radiator. Firstly, the bandwidth enhancement methods for passive RFID tags based on micro strips are discussed. Then, a RFID tag design based on multi-strip antenna is proposed and its radiation efficiency is analyzed. After that, some key parameters of the RFID antenna proposed are optimized from the viewpoint of radiation efficiency and impedance match performance. Targeted for manufacturing plants with heavy metallic interfering, the proposed RFID tag can significantly enhance the radiation efficiency to improve the reading range as well as the bandwidth. Finally, some RFID tag prototypes are fabricated and tested to verify their performance and applicability against metallic environment, and the experimental results show that these fabricated RFID tags have outstanding reading performance and can be widely used in manufacturing plant full of heave metallic interfering.

DOI: 10.4018/978-1-4666-5836-3.ch006

1. INTRODUCTION

With the trend of consumer personalization and demand diversification, an increasing number of manufacturing enterprises are employing the informationization of Manufacturing Execution System (MES) to realize the seamless integration between the resource management systems and the real-world production environment, which is achieved by the collection and synchronization of real manufacturing data (Dai et al., 2012; Liu et al., 2012). Traditionally, barcode techniques have been mainly used for MES data acquisition. However, a barcode can easily get scuffed, damaged or wrinkled, and the reading is unreliable in dirty environment. Also, due to the difficulties to store data or identify moving objects, a barcode cannot acquire varying production data correctly and efficiently (Chao et al., 2007; Gaukler, 2011). Recently, a new electronic tag technology, i.e. RFID (Radio Frequency Identification), has been introduced to manufacturing automation due to its advantages such as long recognition distance, fast reading speed, large storage and programmable memory. Providing a non-contact solution for automatic identification, RFID has been increasingly used for online data acquisition and process control in the modern manufacturing system (DiGiampaolo et al., 2012; Ngai et al., 2008; Saad et al., 2011; Shih-Kang et al., 2010; Zhou et al., 2007). Nowadays, many researchers have applied RFID techniques for accurate and on-line decision-making in production management, allowing for closed-loop manufacturing systems (Bottani et al., 2009; Guo et al., 2009; Kim et al., 2007).

A typical RFID system is composed of a RFID reader and a passive RFID tag attached on items or pallets, among which the RFID tag is the key component to determine the maximum reading range (Rao et al., 2005). In a manufacturing logistic system, supervision of forklifts and pallets get into or out of warehouse requires RFID tags that have a reading range more than 5 meters. Therefore, long range UHF (Ultra High Frequency) RFID tags are needed. However, when a RFID tag is used in manufacturing plant, metal environment will severely affects its radiation efficiency. Especially, when the RFID tag is mounted on or near to metal objects, the reading/writing performance of the RFID tag will be dramatically degraded due to the change of the antenna parameters (frequency, efficiency, bandwidth, radiation pattern, and input impedance) (Dobkin et al., 2005; Prothro et al., 2006). For example, the reading range of an ideal passive UHF RFID tag will sharply drop from 10 m to less than 0.5 m in air when it is placed close to a metallic object. The reason for the performance degradation is mainly because that the phase of the impinging wave is reversed by metal surface, resulting in destructive interference with the wave emitted in the other directions (Sievenpiper, 1999).

There are several kinds of approaches to design RFID tags where metallic objects are involved. In recent years, the studies of anti-metallic RFID tags have attracted much attention. Among them, the inverted-F antenna (IFA), planar inverted-F antenna (PIFA), and patch-type antenna structure, are the typical ones. However, the sizes of these antennas tend to be very large and the reading range is only about 4 m (Ukkonen, Engels, Sydänheimo, & Kivikoski, 2004; Sievenpiper, 1999; Ukkonen, Sydanheirno, & Kivikoski, 2004). The miniature tag antenna structure with a compact artificial magnetic conduct (AMC) substrate can reduce the size of a RFID tag antenna, but the max reading range is still less than 5 m (Wu, 2011; Yu et al, 2006). Another approach to avoid the metallic interfering is to add a high impedance surface (HIS) between the tag antenna and the metal surface, also known as artificial magnetic conductor, which reflect incident wave in certain frequency range with no phase shift (Kim et al., 2008; Kim et al., 2012). Unlike the above approaches, the new idea is to directly use HIS as a radiator in a RFID tag antenna, which allows for small footprints (Kuo et al., 2010; Chen et al., 2009; Chen et al., 2008; Yang et al., 2011). However, when the RFID chip is directly attached to the HIS units (Chen et al., 2009;

Chen et al., 2008), the radiation efficiency of the antenna is very low in order to achieve impedance match between chip and antenna. Therefore, novel RFID tags featured with long reading ranges and small sizes as well as compatibility with metallic environment are desired for applications of RFID in real-world manufacturing plant.

In this chapter, a new design of passive anti-metallic RFID tag with long reading range is proposed for manufacturing plant with heavy metallic interfering. The antenna of the proposed solution uses a multi-strip HIS with a feeding loop as radiator to improve the RFID radiation efficiency. A transmission line model is presented to analyze the RFID antenna, and full wave simulation is carried out to optimize the key parameters of the antenna. Some prototypes of the RFID tag proposed are fabricated and tested, and it is shown that the reading range reaches up to 9 meters when they are attached on metallic objects.

2. DESIGN OF WIDEBAND RFID TAG USING MICRO STRIP ANTENNA

Micro strip antennas are widely used in RFID applications because of their small profile planar feature. Almost all of the antennas used in RFID readers are micro strip antennas. Usually, the micro strip antennas for hand-hold readers have a height from 3mm to 5mm. Unlike hand-hold readers, desktop RFID readers generally use multi-antenna configurations to achieve maximum throughput or reading rate. So, Right handed or left handed circular polarized antenna arrays, which have a height from 20mm to 100mm, are applied in these circumstances. Moreover, Micro strip antennas are used in RFID tag design, because of their reliability when placed on or near to metal, water or used in other unpredictable environments. When placed on water or metal, their electromagnetic performance would not change substantially. This characteristic makes micro strip tag antennas much more reliable than dipole antennas. However, micro strip tag antennas still have two drawbacks: low radiating efficiency and small bandwidth. Besides the radiating mechanism of micro strip antennas, these two drawbacks partly come from the dimension restriction of RFID tags. For the convenience of manufacturing or batch reading/writing, RFID tags are usually linear polarized. Moreover, RFID tags are usually required to be small so that they can be used for small tools or other objects, so most RFID tags are designed to be thin and slim, and usually have a dimension of about 20mm×100mm, and a height less that 5mm. A micro strip antenna with this dimension usually has a radiating efficiency of about 20%, and its bandwidth is less than 25MHz. However, an etched RFID dipole antenna, having the radiating efficiency over 90% and the bandwidth over 100MHz can be easily achieved.

2.1. Basic Micro Strip Antennas for RFID Tag

2.1.1. Half Wavelength Micro Strip Antenna

A typical micro strip RFID tag antenna consists of a half wavelength radiating patch on one side of a substrate and a ground plane on the other side of the substrate. As shown in Figure 1, the radiating patch is connected to one port of a RFID chip through a micro strip line, which acts as impedance matching circuit. The other port of the RFID chip is connected to the ground plane by a shorting pin. The half wavelength radiating patch comprises non-radiating edges which have a length of L, and radiating edges which have a width of W.

The radiating edges W will radiate EM waves to space, and lead to an equivalent extensive slot in the ends of the radiating patch, which can be represented by ΔL. For effective radiating in certain frequency, the effective length of the radiating patch must equal to the effective half wave length of that frequency, that is,

Figure 1. Basic micro strip antenna configuration

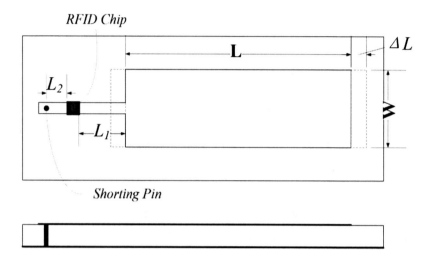

$$L_e = L + 2\Delta L = \frac{\lambda_{eff}}{2} \quad (1)$$

$$\lambda_{eff} = \lambda / \sqrt{\varepsilon_{eff}}$$

Where, ε_{eff} is the effective dielectric constant of the patch, λ is the wave length in free space. The extensive slot length ΔL can be determined by (Kirschning et al., 1981),

$$\Delta L / h = (\xi_1 \xi_2 \xi_3 / \xi_4) \quad (2)$$

With

$$\xi_1 = 0.434907 \frac{\varepsilon_{eff}^{0.81} + 0.26}{\varepsilon_{eff}^{0.81} - 0.189} \cdot \frac{(W/h)^{0.8544} + 0.236}{(W/h)^{0.8544} + 0.87}$$

$$\xi_2 = 1 + \frac{(W/h)^{0.371}}{2.358\varepsilon_r + 1}$$

$$\xi_3 = 1 + \frac{0.5274 \arctan[0.084(W/h)^{1.9413/\xi_2}]}{\varepsilon_{eff}^{0.9236}}$$

$$\xi_4 = 1 + 0.0377 \arctan[0.067(W/h)^{1.456}]$$
$$\times \{6 - 5\exp(0.036(1-\varepsilon_r))\}$$

$$\xi_5 = 1 - 0.218\exp(-7.5W/h)$$

(3)

Where, h is the height of the substrate, ε_r is the relative dielectric constant of the substrate.

Transmission line theory can be used to calculate the input impedance or impedance bandwidth of the micro strip antenna, as shown in Figure 2. The admittance Y_s represents the admittance caused by the radiating edge or the extensive slot ΔL, where,

$$Y_s = G_s + jB_s \quad (4)$$

As demonstrated by Pues (1984), the conductance G_s and susceptance B_s can be calculated by:

$$G_s = \frac{1}{\pi \eta_0} \left\{ \left(\omega \operatorname{Si}(\omega) + \frac{\sin \omega}{\omega} + \cos \omega - 2\right)\left(1 - \frac{s^2}{24}\right) \right.$$
$$\left. + \frac{s^2}{12}\left(\frac{1}{3} + \frac{\cos \omega}{\omega^2} - \frac{\sin \omega}{\omega^3}\right) \right\}$$

(5)

And,

$$B_s = Y_0 \tan(\beta \Delta L) \quad (6)$$

Figure 2. Transmission line model

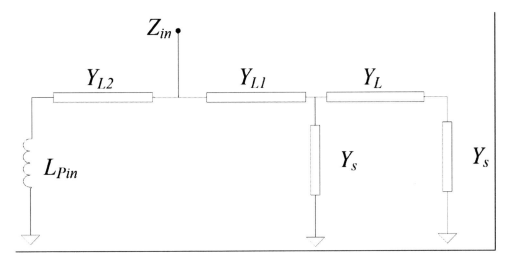

Herein, $\omega = k_0 w_e$, $s = k_0 \Delta l$, η_0 is the free space wave impedance, k_0 is the free space wave number, β is the phase constant, w_e is the effective width of the radiating patch which can be determined by Dearnley et al. (1989). Y_0 is the characteristic admittance of the radiating patch.

The inductance of the shorting pin can be determined according to Chen et al. (2010), where,

$$L_{stub} = \frac{\eta_0 h}{2\pi c} \ln\left[\frac{4c}{\zeta 2\pi f d_{pin} \sqrt{\varepsilon_r}}\right], \text{ and}$$
$$\zeta = 1.781072 \quad (7)$$

Once the parameters mentioned above are determined, the characteristic admittances of all transmission lines can be calculated. The input impedance of the half wavelength micro strip antenna can be analytically calculated.

Except using extensive transmission line as impedance match circuit, another impedance match method is changing the feeding position, the configuration and transmission line model of which are shown in Figure 3. The effective length of the radiating patch is,

$$L_e = L' + \Delta L' + L'' + \Delta L'' \quad (8)$$

The extensive radiating slots $\Delta L'$ and $\Delta L''$ should be calculated respectively.

The impedance matching methods can be generally explained by Figure 4, which approximately presents the impedance changing from one end of the radiating patch to the chip bonding position. The impedance match method shown in Figure 1 uses an extending transmission line to convert antenna impedance near the conjugate match impedance, which is presented by the solid line of Figure 4. However, for the impedance match method shown in Figure 3 cuts an insert L' to feed the radiation patch within it. The impedance in the feeding point is inductive, the conjugate match between antenna and RFID chip can be obtained by modifying the insert length L', and this process is presented by the dashed line of Figure 4.

2.1.2. Quarter Wavelength Micro Strip Antenna with One-End Shorted

The configuration and transmission line model of a micro strip antenna with one end shorted are respectively shown in Figure 5 and Figure 6. With one end of the radiating patch shorted to

Figure 3. Another configuration of half wavelength micro strip antenna and its transmission line model

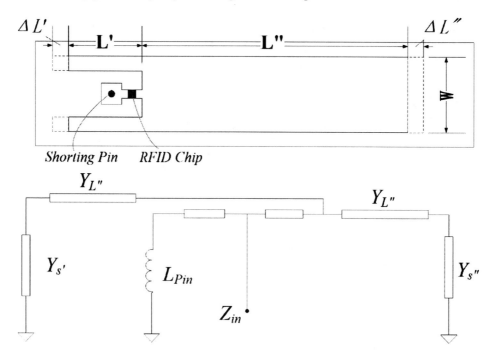

Figure 4. Smith chart of the two impedance match methods

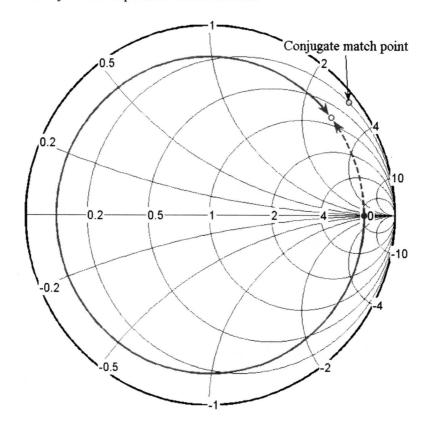

Figure 5. Quarter wave micro strip antenna with one end shorted

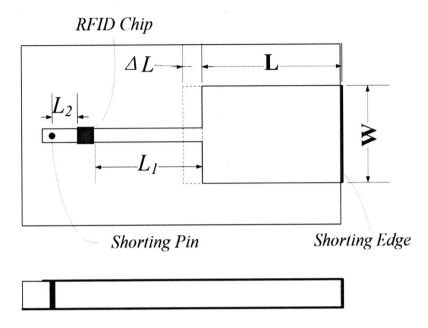

Figure 6. Transmission line model of a micro strip antenna with one end

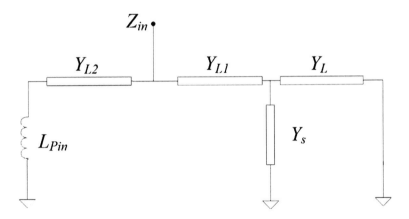

the ground plane the profile of the whole antenna will be much more compacted, compared to the half wavelength micro strip antenna. The analysis method is as same as the half wave length micro strip antenna.

The main problem of this configuration is that its radiating efficiency would be worse. A better configuration and its transmission line model is shown in Figure 7. The two radiating edges are opposite to each other. For the purpose of saving place, two symmetric short circuit stubs are used to match the antenna and chip impedance. Since the whole structure is symmetric, only anglicizing the left/right side of it can get the input impedance of the antenna.

Figure 7. Symmetric configuration for quarter wave micro strip antenna and its transmission line model

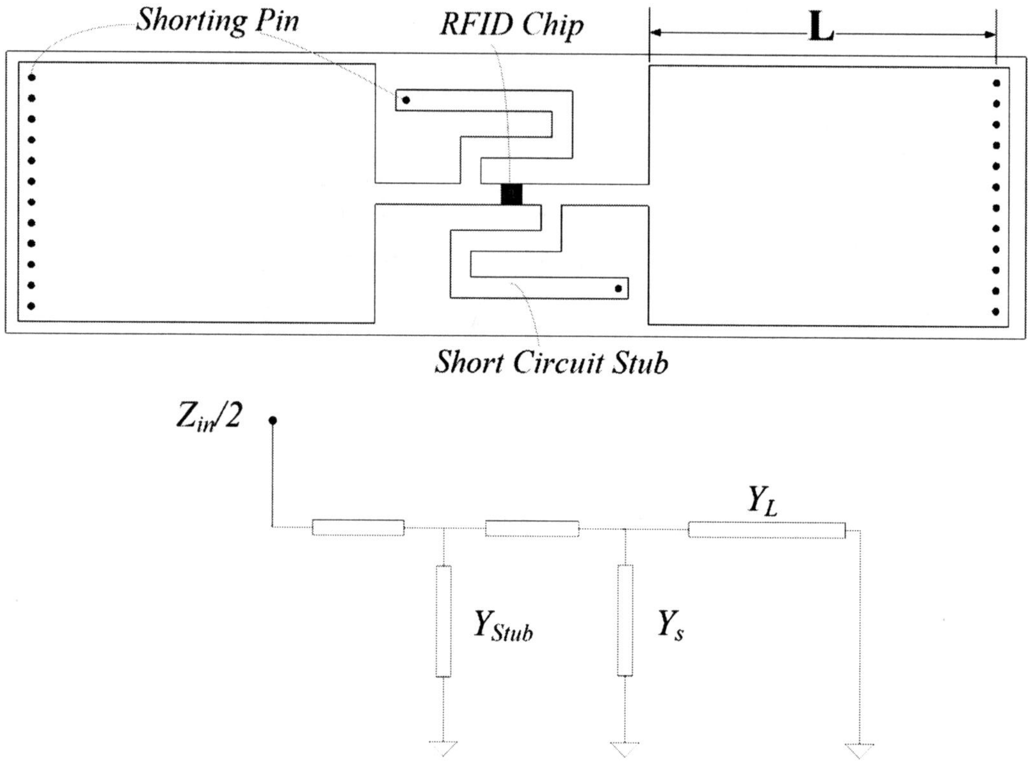

2.2. Bandwidth Enhancement of Micro Strip Antenna

The bandwidth of an antenna can be discussed in two aspects: radiating bandwidth and impedance bandwidth. The radiating bandwidth is defined as follow, in certain frequency band the radiation pattern is almost the same, or the radiating efficiency will not degrade more than 3dB. The impedance bandwidth is defined that the transmission coefficient between the RFID chip and the antenna will not degrade more than 3dB. There are three approaches to improve the bandwidth of a micro strip antenna: adding parasitic patch, opening slots in the radiating patch, and using wideband impedance mating network. The first two approaches mainly improve the radiating bandwidth of the antenna, and the impedance bandwidth will be increased simultaneously. The third approach just increases the impedance bandwidth.

2.2.1. Adding Parasitic Patch

Adding parasitic patch is the simplest approach to improve the bandwidth of a micro strip antenna. As shown in Figure 8, two additional radiating patches are added parallel to the main patch along its non-radiating edge. The resonant frequencies of the parasitic patches are higher or lower than the resonant frequency of the main patch. When the operating frequency exceeds the resonant frequency of the main patch, the current coupled to the parasitic patches would radiate effectively at the higher or lower frequency. Slight modification of the impedance match lines should be made to match the RFID chip impedance. As shown in Figure 9, the bandwidth of the parasitic patch

Figure 8. A half wavelength micro strip antenna having a parasitic patches

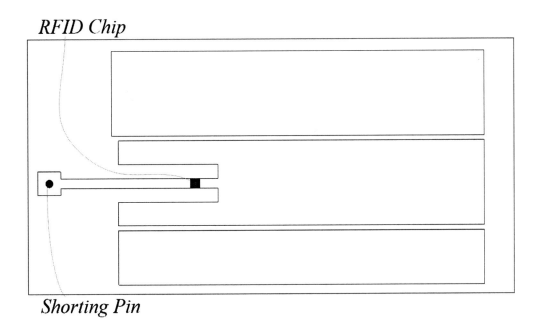

Figure 9. Impedance and power reflection coefficient of the parasitic patch antenna

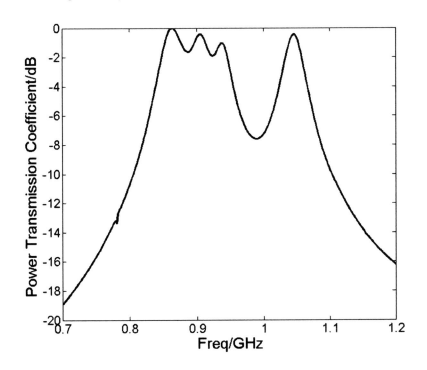

Figure 10. Configuration for stacked parasitic patch antenna

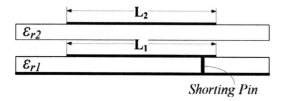

antenna has a bandwidth of over 100Mhz. And an example of this approach is also reported in Tentzeris (2010). The drawback of the parasitic patch antenna is apparently that the dimension of which will be increased by more than 3 or 4 times. That is not acceptable in most occasions. This drawback can be avoided by stack configuration of parasitic patches, which is shown in Figure 10. Stacked configuration can keep the RFID tag's dimension small, but it will suffer from high manufacture cost and large thickness.

2.2.2. Opening Slots in the Radiating Patch

By properly design slots in the radiating patch of micro strip antenna, a high frequency and a lower frequency mode can be excited in one radiating patch, which will make the micro strip antenna to have at least two resonant frequencies. The impedance bandwidth of the antenna configuration shown from Figure 11 to Figure 13 has a bandwidth of over 100MHz, which is comparable to wideband dipole antennas.

2.2.3. Wideband Impedance Match Network

By capacitive feeding the radiating patch, wideband impedance match network can be achieved to improve the bandwidth of micro strip antenna of RFID tags. Ref. (Hirvonen et al., 2006; Son et al., 2011; Son et al., 2006) apply this strategy to get a bandwidth of near 50MHz or dual band characteristic. Figure 12 shows the basic configuration of a micro strip antenna having a capacitive feeding impedance match network. And the transmission line model is presented in Figure 14.

By examine the impedance of Z_1 and Z_2, the effect of capacitive feeding to the antenna input impedance. As shown in Figure 1, The dashed line is the impedance at Z_1 (one end of the radiating patch) in the frequency range from 0.7GHz to 1.2GHz, while the solid line is the impedance at Z_2 in the same frequency range. The impedance at Z_2 concentrates in a small area, which means a broad band impedance bandwidth near the center of the area. The antenna input impedance is presented in Figure 16 After the transmission line L_1 and L_2, the antenna input impedance is shifted near the conjugate match impedance of the RFID chip (whose impedance is 10-140j). According to Figure 17, the capacitive feed micro strip antenna presented in Figure 14 achieves a total bandwidth of 50MHz, and covering the RFID frequency band of 860MHz and 920MHz.

Another configuration for the above capacitive feed micro strip antenna is shown in Figure 19. And this configuration is much more compact. The input impedance of it is shown in Figure 20, which has an impedance bandwidth of more than 50MHz, and covers the RFID frequency band of 860MHz and 915MHz. As reported in Hirvonen et al. (2006), the half wavelength radiating patch of Figure 14 or Figure 19 can be replaced by quarter wavelength patch, which would make the whole antenna even smaller.

3. DESIGN OF ANTI-METALLIC RFID TAG USING MULTI-STRIP HIS

The standard frequency bands assigned to UHF RFID application are 860-960MHz. A UHF RFID tag is mostly designed to work at 915MHz. Usually, two key issues should be overcome during the RFID tag design: one is the impedance match between the RFID chip and antenna, which guar-

Figure 11. U-slot loaded micro strip antenna by Zhang et al. (2008)

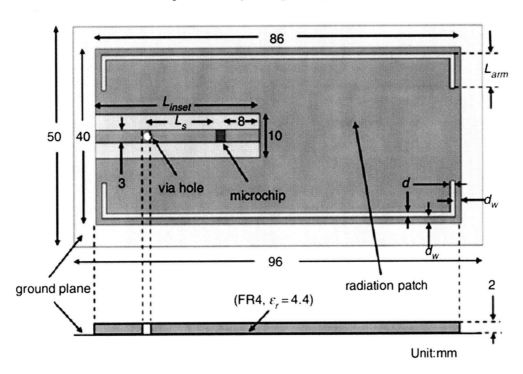

Figure 12. Slotted micro strip antenna by Huang et al. (2009)

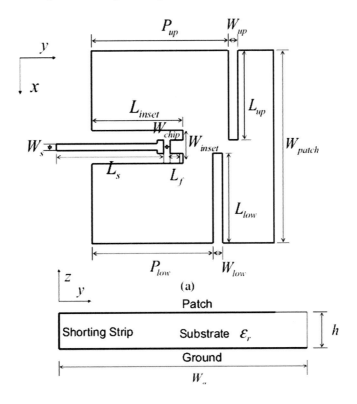

Figure 13. Reactive loaded micro strip antenna by Lai et al. (2010)

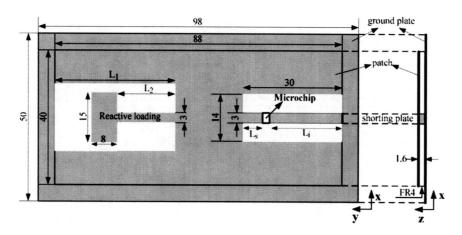

Figure 14. Capacitive feed micro strip antenna

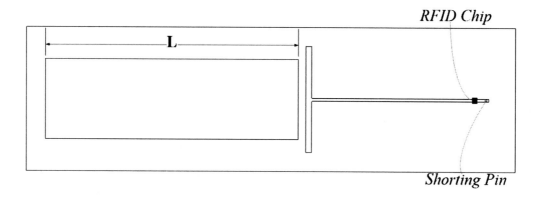

Figure 15. Transmission line model of the capacitive feed micro strip antenna

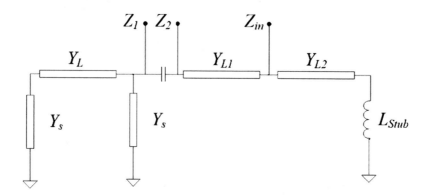

Figure 16. Z_1 and Z_2 in smith chart

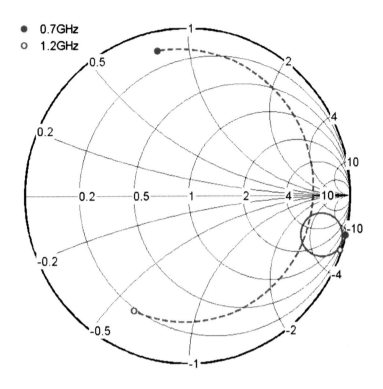

Figure 17. Antenna input impedance Z_{in} shown in smith chart

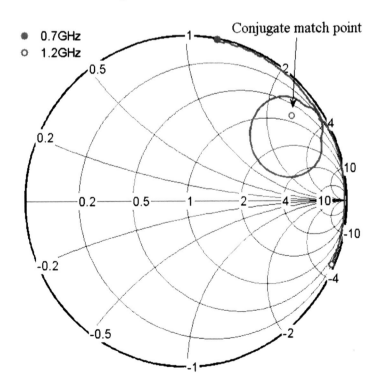

Figure 18. Conjugate match factor of the capacitive feed micro strip antenna

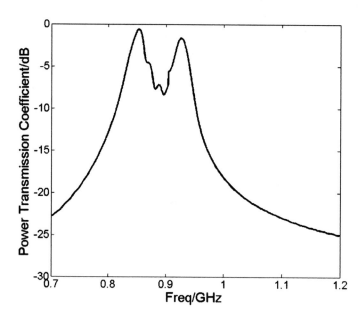

Figure 19. Compact configuration for the above capacitive feed micro strip antenna

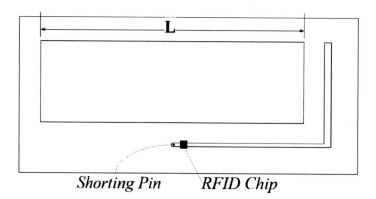

antees efficient power transmission between chip and antenna. The other is the radiation efficiency or the antenna gain, which determines the power radiated by the tag antenna or received by the antenna of the RFID reader. Therefore, a RFID antenna usually contains two parts: radiator and impedance matching network.

3.1 Influence of Metallic Environment on RFID Tag

When a plane wave is impinging on a conductor, as shown in Figure 21, the electric field in air environment has the following form:

$$\vec{E}_i(z) = \hat{E}_0 e^{-j\beta_1 z} \vec{a}_x, \qquad (9)$$

Figure 20. The input impedance of the compact antenna

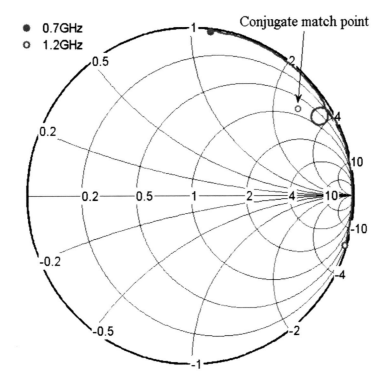

Figure 21. A plane wave impinging on a conductor

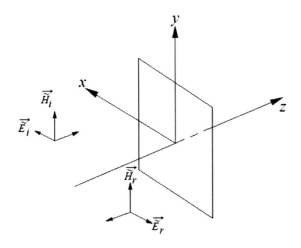

$$z_s = \frac{E_x}{H_y}. \quad (10)$$

For a finite conductor, z_s has positive real and imaginary parts and both approach 0. So E_x at the conductor surface has a value approaching 0.

The impedance of a metal surface is equivalent to a small inductance. When z_s inductive, TEM surface wave can occur on the interface between air and the conductor (Sievenpiper, 1999). With the addition of boundary condition of electric field of Maxwell equation, the reflected electric field can be approximately described by equation (3). i.e.

$$\vec{E}_r(z) = -\hat{E}_0 e^{j\beta_1 z}\vec{a}_x + \hat{E}_s e^{-j\beta_r x + \alpha_r z}\vec{a}_z. \quad (11)$$

where, β_1 is the phase constant of the transverse electromagnetic (TEM) wave propagating in air environment. The surface impedance of the conductor is given by:

Then the total electric field in the air is

$$\vec{\bar{E}}_a(z) = -j2\hat{E}_0 \sin(\beta_1 z)\vec{a}_x + \hat{E}_s e^{-j\beta_r x + \alpha_r z}\vec{a}_z. \quad (12)$$

Herein, β_r and α_r are the phase constant and decay constant of TEM surface wave respectively. Both of them are related to the characteristics of materials on both sides of the interface. According to equation (11) and (12), the electric field in x direction of the reflected wave has a phase shift of 180 degree, which prevent the RFID tag placing near a metal object to obtain electromagnetic energy. The simulation result is shown in Figure 22. According to the simulation result, decay constant α_r has a value of about 55.

When a high impedance surface is adopted to replace the conductor, z_s of high impedance surface is large, which means $H_y \to 0$. So the total electric field in the air can be approximately described by:

$$\vec{\bar{E}}_a(z) = 2\hat{E}_0 \cos(\beta_1 z)\vec{a}_x + \hat{E}_s e^{-j\beta_r x + \alpha_r z}\vec{a}_z. \quad (13)$$

The magnetic field in y direction of the reflected wave has a phase shift of 180 degree. The surface wave is restrained, and the electric field is enhanced.

Generally, there are two key problems need to be overcome first during RFID tag design. One is the impedance match between the RFID chip and its attached antenna, which provides efficient power transmission between chip and antenna. The other is the radiation efficiency or the antenna gain, which indicates the energy radiated into space. So, a RFID antenna usually contains two parts: radiator and impedance matching network. By directly applying the HIS as radiator other than using HIS as ground plane, the thickness of RFID tag can be reduced.

3.2 Antenna Design of Anti-Metallic RFID

The electric field near the conductor surface is approximately zero, when electromagnetic wave impinges on a conductor such as metallic objects. This is because the reflected electric field near the conductor surface has reverse phase comparing with the impinging electric field. When an antenna such as dipole is placed near or on the metal surface, the electric field of the antenna will be canceled by the reflected electric field of the metal surface, so the radiation efficiency of the antenna is very low for normal operation. The relationship between the electric field and the magnetic field in a surface can be characterized as surface impedance. The magnitude of surface impedance is calculated by:

$$z_{mag} = \frac{|E_{tangent}|}{|H_{tangent}|}. \quad (14)$$

The surface impedance near a metal surface is approximately zero, because the value of electric field is very small. High electric field can exist only when the surface impedance is large enough, and an antenna will radiate well near a high impedance surface.

The principle structure of the proposed RFID tag based on multi-strip HIS is shown in Figure 23. Herein, W_R denotes HIS total width, L_R denotes HIS total length, G denotes HIS gap between a pair of opposite strips, L_{strip} denotes the strip length from the shorting pin, w and g denotes strip width and gap between a pair of parallel strips respectively, W_L and L_L denotes feeding loop width and length respectively.

A multi-strip HIS is used herein as the radiator and a feeding loop is used to feed the HIS. Each strip is a quarter-wave-length transmission line with one shorting end. According to the transmission line theory, the middle area of the

Figure 22. Simulation result of a plane wave impinging on a) perfect conductor, b) high impedance ground

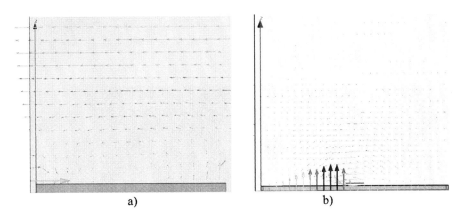

Figure 23. Structure of the proposed RFID tag

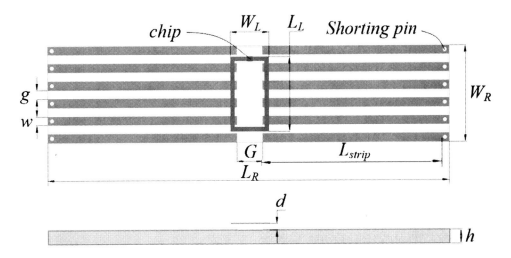

multi-strip upper surface forms a high impedance surface area, where the electric field strength can be large. Hence, the strip length can be determined by:

$$L_{strip} = \frac{\lambda_{strip}}{4} - \Delta l - l_{pin\_eff}. \quad (15)$$

λ_{strip} is the electromagnetic wave length along the strip line of the multi-strip HIS, and it is defined by:

$$\lambda_{strip} = \frac{c}{f\sqrt{\varepsilon_{eff}}}. \quad (16)$$

Δl is the effective open end extension of the strip which can be determined by Pues et al. (1984), and l_{pin_eff} is the effective length of the shorting pin. According to full wave simulations, l_{pin_eff} can be evaluated by:

$$l_{pin\_eff} = 2h\sqrt{\frac{\varepsilon_r}{\varepsilon_{eff}}}. \quad (17)$$

Figure 24. Surface impedance of the multi-strip HIS at 915MHz

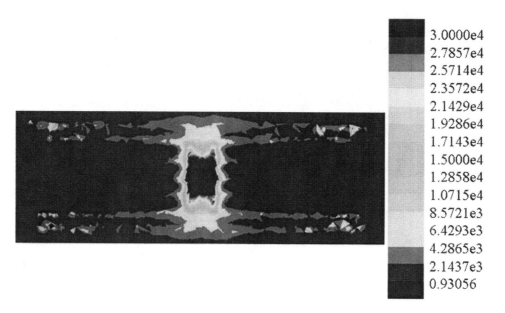

Herein, h is the thickness of the antenna substrate, c is the light speed, f is frequency, ε_r and ε_{eff} are the relative permittivity of the substrate and the effective permittivity of the strip line respectively.

The open end of the multi-strip transmission line will form a radiating slot, which enables the multi-strip HIS structure as a radiator. The radiating slot has a width of Δl. A feeding loop can efficiently feed the multi-strip HIS in the high impedance surface area, and achieve better impedance match. The simulated result of the surface impedance near the multi-strip HIS upper surface is shown in Figure 24.

Compared with the antenna applying single HIS unit (Yang et al., 2011), of which the current is only concentrated upon the long sides of HIS unit, the surface current of multi-strip HIS will be constrained along each strip, and the radiation efficiency of multi-strip HIS will be much better. The current distribution of the proposed solution is simulated, as shown in Figure 25.

Figure 26 shows a comparison of radiation efficiency of multi-strip HIS against single HIS unit, both of which are placed on ideally infinite metal plate and have a width of 22mm. The single HIS unit has a length of 90mm and the multi-strip HIS has a length of 94mm, to make both of them to resonant around 915MHz. Apparently, The maximum radiation efficiency of the multi-strip HIS is at least 13% more than that of the single HIS unit.

3.3 Antenna Analysis Using Transmission Line Model

A transmission line model, as shown in Figure 27, is proposed to analyze the input impedance of the proposed RFID antenna. The proposed antenna is symmetrical, so the symmetry plane can be a virtual ground for analysis. The dashed box in Figure 27 shows the model for one strip of the multi-strip HIS structure. The strip with prior known w can be effectively modeled with a transmission line of characteristic admittance Y_0 and propagation constant γ. The radiation slot admittance Y_s is composed of conductance and susceptance, which are related to the power radiated from the slot and the stored energy, given by:

Figure 25. Current distribution at 915MHz of a) single HIS unit antenna and b) multi-strip HIS antenna

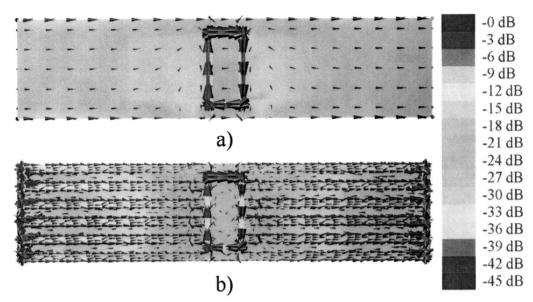

Figure 26. Radiation efficiency of single HIS unit antenna and multi-strip HIS antenna

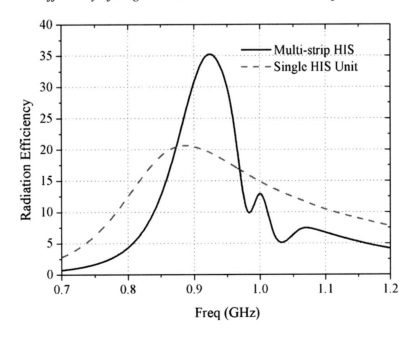

Figure 27. Transmission line model for the multi-strip HIS antenna

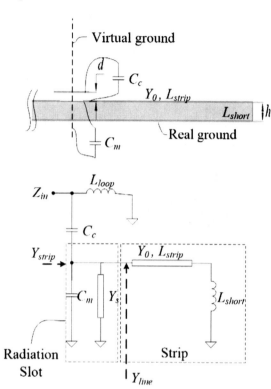

$$Y_s = G_s + jB_s. \qquad (18)$$

As demonstrated in Pues et al. (1984), the conductance G_s and susceptance B_s can be calculated by:

$$G_s = \frac{1}{\pi\eta_0}\left\{\left(\omega \operatorname{Si}(\omega) + \frac{\sin\omega}{\omega} + \cos\omega - 2\right)\left(1 - \frac{s^2}{24}\right) + \frac{s^2}{12}\left(\frac{1}{3} + \frac{\cos\omega}{\omega^2} - \frac{\sin\omega}{\omega^3}\right)\right\} \qquad (19)$$

and

$$B_s = Y_0 \tan(\beta\Delta l). \qquad (20)$$

Herein, $\omega = k_0 w_e$, $s = k_0 \Delta l$, η_0 is the wave impedance in free space, k_0 is free space wave number, β is phase constant of the propagation constant γ, w_e is the effective width of the strip, and Δl is open end extension of the strip defined above. So the transmission line admittance Y_{line} at the open end can be determined by:

$$Y_{line} = Y_0 \frac{Y_{short} + Y_0 \tanh(\gamma L_{strip})}{Y_0 + Y_{short} \tanh(\gamma L_{strip})}. \qquad (21)$$

Herein, L_{strip} is the length from the shorting pin to the strip open end, Y_{short} is the admittance of the shorting pin, and L_{short} is the inductance of the shorting pin. L_{strip} is defined by equation (16), Y_{short} and L_{short} can be calculated by Dearnley et al. (1989):

$$Y_{short} = \frac{1}{j2\pi f L_{short}}, \qquad (22)$$

and

$$L_{short} = \frac{\eta_0 h}{2\pi c} \ln\left[\frac{4c}{\zeta 2\pi f d_{pin}\sqrt{\varepsilon_r}}\right]. \qquad (23)$$

Herein, ζ is the correction coefficient, which is determined experimentally. d_{pin} is the diameter of the shorting pin, which should be at least 0.6mm considering the fabrication convenience. The mutual coupling capacitance C_m between two adjacent radiation slots and inductance L_{loop} of the feeding loop can be calculated by:

$$C_m = \frac{\varepsilon w}{2\pi}\left\{\ln\left[0.25 + \left(\frac{h}{G}\right)^2\right] + \frac{G}{h}\tan^{-1}\left(\frac{2h}{g}\right)\right\}, \qquad (24)$$

and

$$L_{loop} = m \times 0.4(W_L + L_L)\ln\left[\frac{2W_L L_L}{w_{loop}(W_L + L_L)}\right](\mu H). \quad (25)$$

Herein, w_{loop} is the line width of the loop, ε is the permittivity of substrate, and m is the correction coefficient of the loop inductance. Coupling capacitance C_c between feeding loop and the multi-strip can be calculated as parallel plate capacitors:

$$C_c = n\varepsilon_0 \frac{N_C \times (w_{loop} \times w)}{d}, \quad (26)$$

where, ε_0 is permittivity of free space. n is the correction coefficient considering that the electric field line is not ideally perpendicular to the parallel plate. N_C is the number of parallel plates. For the proposed design as shown in Figure 23, N_C is 8. Then, the total admittance of one strip can be calculated as:

$$Y_{strip} = Y_s + Y_{line} + j2\pi f C_m. \quad (27)$$

Herein, Y_s, Y_{line} and C_m are defined by equation (18), (21) and (24). If there are N_{Strip} pairs of symmetrical strips (For the proposed design as shown in Figure 23, it is 6), the input impedance Z_{in} of the proposed multi-strip HIS antenna is defined as:

Figure 28. Tuning results of the transmission line model

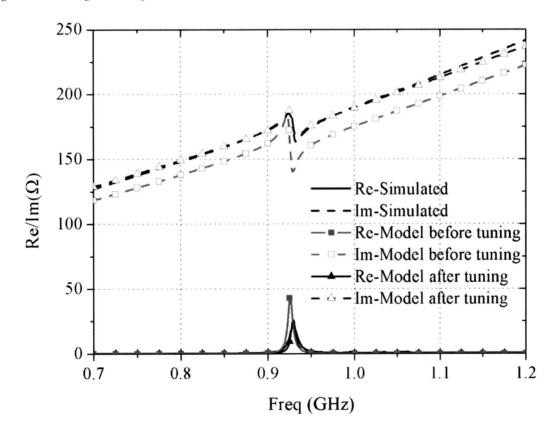

$$Z_{in} = \left(\frac{1}{j2\pi f L_{loop}} + \left(\frac{2}{N_{Strip} Y_{strip}} + \frac{1}{j2\pi f C_c} \right)^{-1} \right)^{-1}. \quad (28)$$

Considering the complex coupling factors, m and n are usually determined experimentally. After comparing with the full wave simulation results, $m = 1.1$ and $n = 0.7$ are determined to tune the transmission line model in the proposed design. After tuning, both real part and imaginary part of the antenna input impedance match well with the full wave simulation, as shown in Figure 28.

3.4 Key Parameters Study of Antenna

In general, the reading range of a RFID tag is dominantly determined by its antenna's radiation efficiency. In this section, a MOM (Method of Moment) simulation tool is adopted to study the influence of some key parameters of the proposed RFID tag on its performance according to the antenna's radiation efficiency and input impedance, in which an ideally infinite metal ground is used to simulate the real metal object.

3.4.1. Radiation Efficiency Optimization

Firstly, the influence of HIS total width W_R on radiation efficiency is studied. As shown in Figure 29, the maximum radiation efficiency increases remarkably with the increment of W_R. This is maybe because larger W_R will produce a larger radiating slot. Hence, the total width of multi-strip HIS should be as large as possible while meeting to the requirement of RFID tag size.

Another key parameter relating to the radiation efficiency is the number of strips. When W_R is given, the number of strips is determined by strip width w and strip gap g. Usually, $w = g$. As shown in Figure 30, the maximum radiation efficiency decreases monotonously when w increases. Considering the fabrication convenience of the strip itself and its shorting pin (as aforementioned, the diameter of the shorting pin d_{pin} should be at least 0.6mm), w should be larger than 2mm in practice.

According to the transmission line analysis, HIS gap G affects the mutual coupling capacitance C_m, so the resonant frequency of the proposed antenna can be slightly adjusted by changing the value of G. Also, G has an influence on radiation efficiency. As shown in Figure 31, for the given $G = 2mm, 4mm, 6mm, 8mm$, both the maximum radiation efficiency and resonant frequency of the antenna increase with the increment of G. Hence, if the goal resonant frequency is near 915MHz, the optimum G is 6mm.

3.4.2. Impedance Match Optimization

The impedance match performance between the RFID chip and antenna can be evaluated by their power transmission coefficient, given by:

$$\tau = \frac{4R_c R_{in}}{|Z_c + Z_{in}|}, 0 \leq \tau \leq 1. \quad (29)$$

Herein, is the chip impedance and is the antenna input impedance defined by equation (28). The power transmission between chip and antenna will be lossless, when conjugate match between them is achieved. Z_c is determined by the RFID chip manufacturing process and is prior-known, so the impedance match performance is directly determined by antenna input impedance.

In order to study the influence of the loop shape on the antenna input impedance, the sum length of W_L and L_L should keep 26mm to make sure there will be conjugate match in the frequency range near 915MHz, if the RFID chip is

Figure 29. Maximum radiation efficiency for different HIS total width W_R

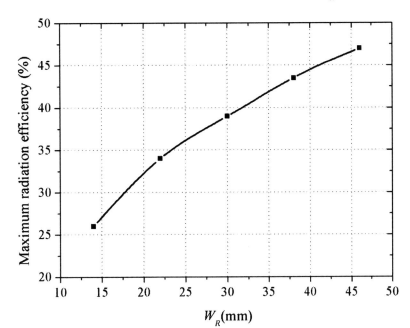

Figure 30. Maximum Radiation efficiency for different strip width w

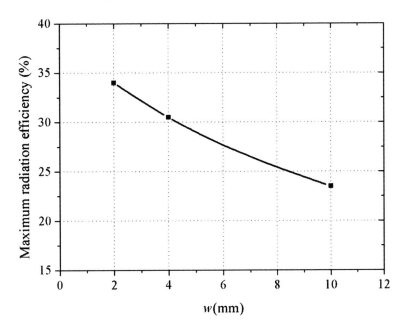

Figure 31. Radiation efficiency for different strip gap G

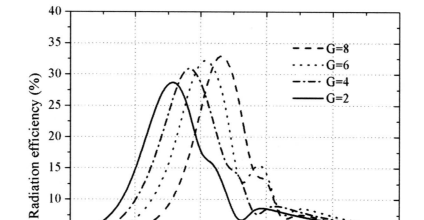

Figure 32. Real and imaginary part of antenna impedance for different loop length L_L

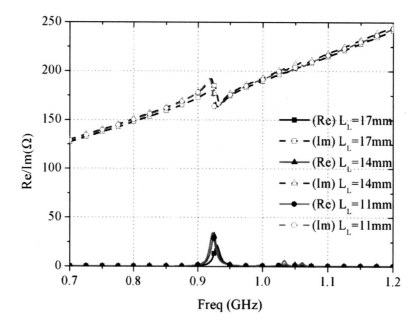

NXP G2XM that has an input impedance of (22-j195) Ω at 915MHz according to equation (25). As shown in Figure 32, it can be found that the input impedance curves are nearly same when L_L is 11mm and 14mm respectively, and the curve shifts slightly when $L_L = 17mm$. In fact, resonant frequency shifting is mainly caused by the parallel plate area, which changes the value of C_c. If the parallel plate area remains nearly same, the resonant frequency will keep constant. In other words, there is no remarkable influence on the resonant frequency of antenna when L_L varies from 11mm to 17mm. Moreover, in the case of $L_L = 17\,mm$, the amplitude of the input impedance imaginary part fluctuates in a small range, it means a larger impedance bandwidth. Therefore, in the given range, $L_L = 17\,mm$ is desirable.

Another key parameter for antenna input impedance is the distance of feeding loop surface and multi-strip HIS, denoted by d. According to the transmission line model, d affects the coupling capacitance C_c, which will apparently affect the antenna input impedance. For three given d, the input impedance curves are plotted together, as shown in Figure 33. From Figure 33, it can be found that the input impedance curve when $d = 0.5mm$ deviates far from the other two cases and behaves an abnormal fluctuation in 0.9~1GHz frequency range. It implies that the there is a coupling effect between feeding loop and radiator, and the mutual coupling capacitance C_m might be affected when d is too small. Usually, d should be larger than 1mm in practice.

In practice, the ground plane size of a RFID antenna is always finite. Therefore, some simulations with finite ground are also carried out to study the influence of ground plane on the resonant frequency of antennas. For a comparison purpose, same simulation configurations and structure parameters, as listed in table 1, are used in the simulations except for the ground plane. The finite ground plane used in the simulations is a 250mm-by-250mm aluminum plate. As shown in Figure 34, in the case using finite ground plane aforementioned, the input impedance curve slightly shift left, compared with that using ideal infinite ground plane. Hence, considering the influence of real ground plane and materials, the calculation frequency used to estimate the structure parameters should be a little larger than the goal frequency in practice when beginning to design a real RFID tag using the method proposed in this study. For example, if the goal frequency of a required anti-metallic RFID tag is 915MHz, the calculation frequency should be 925MHz or so.

4. PROTOTYPE FABRICATION AND TEST

In this section, fifteen real RFID tag prototypes in total are fabricated and tested to verify the method proposed, as shown in Figure 35. The size of the RFID tag should be less than 25mm width and 100mm length to meet application requirements. The RFID chips used in the prototypes is NXP G2XM that has an input impedance of (22-j195) Ω at 915MHz and the threshold input power is about -15dBm. The implemented antenna is fabricated on a PTFE substrate, of which the thickness is 3 millimeters, the relative permittivity is 3.5, and the dielectric loss tangent is 0.008. The feeding loop is fabricated on 50um PET substrate, of which the relative permittivity is near 1. The copper thickness is 17um.

According to the analysis in section III, W_R and L_R are determined to be 22mm and 96mm respectively. A little space is remained outside each HIS edge for future packaging and the shorting pin is fabricated 1mm apart from each strip edge. L_L, W_L, G, g, w, are also determined according to the optimization analysis in section III, as listed in Table 1. Other parameters, namely d_{pin}, d, h are determined according to

Figure 33. Real and imaginary part of antenna impedance for different distance d

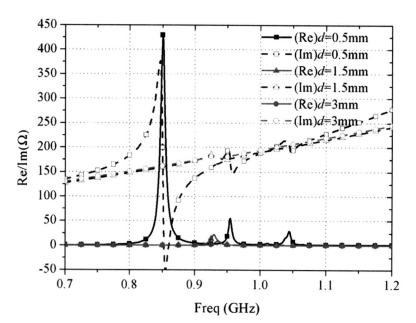

Figure 34. Input impedance curve for Simulation and real testing

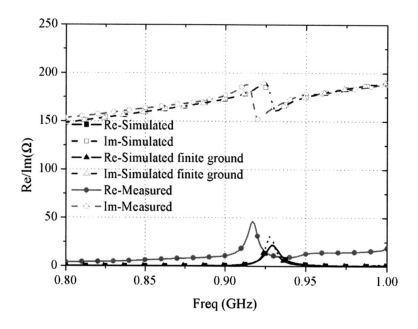

Design of Anti-Metallic RFID for Applications in Smart Manufacturing

Table 1. Structure parameters used for RFID tags fabrication

Parameters	Value (mm)	Parameters	Value (mm)
W_L	9	g	2
L_L	17	w	2
W_R	22	G	6
L_R	96	d	1.5
L_{strip}	43.7	h	3
d_{pin}	0.6	N_{Strip}	6

Figure 35. The prototype of the proposed anti-metallic RFID tag

manufacturing practice, as listed in Table 1. Putting the parameters given herein into equation (15), the resultant strip length L_{strip} is 43.7mm.

The input impedance of the prototype is tested using the method introduced in Kirschning et al. (1981), and the test result is compared with the simulation results using ideal infinite ground and finite ground, as shown in Figure 34. It can be found that three curves are very close to each other. The input impedance curve of the prototype only slightly shift left and the resonant frequency is 914.8MHz while the resonant frequencies calculated are 925.7MHz and 924.6MHz respectively for the cases using ideal infinite ground and finite ground. Hence, it can be concluded that the characteristic of the RFID tags fabricated accord with the theoretic analysis result very well and the method proposed in this paper is effective to guide the RFID tag design.

Reading range is a key parameter to characterize the performance of an anti-metallic RFID tag. The maximum reading range of a RFID tag can be calculated by:

$$R = \frac{\lambda_0}{4\pi} \sqrt{\frac{P_{Reader}}{P_{Chip}} \cdot G_{Reader} \cdot G_{Tag} \cdot \tau} \qquad (30)$$

Herein, is the electromagnetic wave length, at 915MHz, it is 32.8cm. P_{Reader} and G_{Reader} are the output power and antenna gain of the RFID reader. In the tests, a commercial RFID reader is used, and its P_{Reader} and G_{Reader} are 30dBm and 5dBi respectively. P_{Chip} is the minimum activation power of the RFID chip. For NXP G2XM chip, it is -15dBm. G_{Tag} is the maximum gain of the tag antenna, and τ is the power transmission coefficient.

The antenna gain and power transmission coefficient of the RFID tag prototype are calculated and plotted in Figure 36 and Figure 37 respectively. It is found that the maximum gain of the tag prototype placed on a 250mm by 250mm aluminum plate is about 2.5dBi, and the maximum power transmission coefficient is about -1.5dB. So the theoretic reading range calculated using equation (30) is 9.4m.

Moreover, four different metal plates, namely aluminum, zinc, copper, steel, are selected to test the applicability of the tags fabricated against different metallic environment. These plates have same size of 250mm width, 250mm length and 5mm thickness. In each case, five randomly selected tags are tested and the mean reading range is recorded, as shown in Table 2. For a comparison purpose, the reading range of the tags without any metallic interfering is also tested using the same method. From Table 2, it is found that the tags behave a good reading performance in each metallic case, and their reading ranges are all far longer than that without metallic interfering. The

Figure 36. Antenna gain of the RFID tag prototype

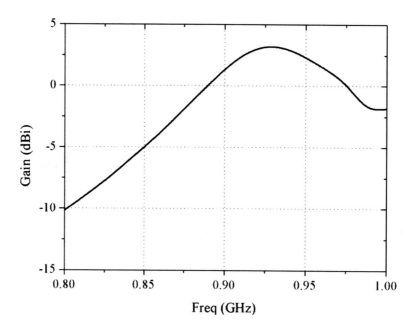

Figure 37 Power transmission coefficient of the RFID tag prototype

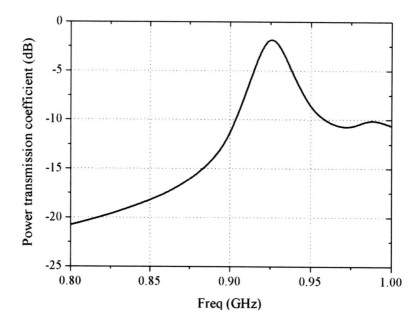

metallic material with better conductivity has a longer reading range. Specifically, in the case of aluminum plate, the real reading range of the tags is 8.8m, which accords well with the theoretic result 9.4m with a relative difference 6.4%. This result is also compared with some published results in Yang et al. (2011). The RFID tag size is 22mm width and 83mm length in Yang et al. (2011) as

Table 2. Reading range

Metal material	Mean Range (m)
None	4.5
Zinc	7.9
Copper	9
Steel	8
Aluminum	8.8

nearly equivalent as our prototype, whose reading range is only 5m. Hence, it can be concluded that the method proposed in this paper is a perfect solution to fabricate long range anti-metallic RFID tags.

5. CONCLUSION

A long range passive anti-metallic RFID tag used in manufacturing plant with heavy metallic interfering is proposed in this paper. The antenna of the proposed tag uses a multi-strip HIS as the radiator and a feeding loop to feed the multi-strip HIS. A transmission line model is presented to analysis the antenna's input impedance, and it is found that the radiation efficiency of the tag antenna can be significantly improved after adopting multi-strip HIS proposed. This will help to improve the reading range of RFID tags when they are used in the metallic environment. The antenna parameters are studied and optimized according to its radiation efficiency using full wave simulation. Some RFID tag prototypes are fabricated and tested in different metallic environment to verify their performance and applicability against environment. The reading range reaches up to 8.8m when the prototype is placed on Aluminum plate, better than the published result. Hence, the method proposed is a perfect solution to fabricate the anti-metallic RFID tag, which can be widely used in manufacturing logistics management system full of heave metallic interfering.

ACKNOWLEDGMENT

This work is supported by the National Science Foundation of China under grant 91023033, the National Fundamental Research Program of China under Grant 2013CB035803 and 2011CB013003, and the New Century Excellent Talents Plan of Education Ministry under grant NCET-10-0414.

REFERENCES

Bottani, E., & Bertolini, M. (2009). Technical and economic aspect of RFID implementation for asset tracking. *International Journal of RF Technologies: Research and Applications, 1*(3), 169–193. doi:10.1080/17545730903159034

Chao, C. C., Yang, J. M., & Jen, W. Y. (2007). Determining technology trends and forecasts of RFID by a historical review and bibliometric analysis from 1991 to 2005. *Technovation, 27*(5), 268–279. doi:10.1016/j.technovation.2006.09.003

Chen, H., & Tsao, Y. (2010). Broadband capacitively coupled patch antenna for RFID tag mountable on metallic objects. *Antennas and Wireless Propagation Magazine, 9*, 489–492. doi:10.1109/LAWP.2010.2050854

Chen, S. L., & Lin, K. H. (2008). A slim RFID tag antenna design for metallic object applications. *IEEE Antennas and Wireless Propagation Letters, 7*, 729–732. doi:10.1109/LAWP.2008.2009473

Chen, S. L., Lin, K. H., & Mittra, R. (2009). A low profile RFID tag designed for metallic objects. In *Proceedings of Microwave Conference, Singapore*, 226-228.

Dai, Q. Y., Zhong, R. Y., Huang, G. Q., Qu, T., Zhang, T., & Luo, T. Y. (2012). Radio frequency identification-enabled real-time manufacturing execution system: A case study in an automotive part manufacturer. *International Journal of Computer Integrated Manufacturing, 25*(1), 51–56. doi:10.1080/0951192X.2011.562546

Dearnley, R. W., & Barel, A. F. (1989). A broad-band transmission line model for a rectangular microstrip antenna. *IEEE Transactions on Antennas and Propagation, 37*(1), 6–15. doi:10.1109/8.192158

DiGiampaolo, E., & Martinelli, F. (2012). A passive UHF-rfid system for the localization of an indoor autonomous vehicle. *IEEE Transactions on Industrial Electronics, 9*(10), 3961–3970. doi:10.1109/TIE.2011.2173091

Dobkin, D. M., & Weigand, S. M. (2005). Environmental effects on RFID tag antennas. *IEEE MTT-S International Microwave Symposium Digest,* 135-138.

Gaukler, G. M. (2011). Item-level RFID in a retail supply chain with stock-out-based substitution. *IEEE Trans. Industrial Informatics, 7*(2), 362–370. doi:10.1109/TII.2010.2068305

Guo, Z. X., Wong, W. K., Leung, S. Y. S., & Fan, J. T. (2009). Intelligent production control decision support system for flexible assembly lines. *Expert Systems with Applications, 36*(3), 4268–4277. doi:10.1016/j.eswa.2008.03.023

Hirvonen, M., & Jaakkola, K. (2006). Dual-band platform tolerant antennas for radio-frequency identification. *IEEE Transactions on Antennas and Propagation, 54*(9), 2632–2637. doi:10.1109/TAP.2006.880726

Huang, J. Z., Yang, P. H., Chew, W. C., & Ye, T. T. (2009). A compact broadband patch antenna for UHF RFID tags. In *Proceedings of 2009 Asia Pacific Microwave Conference,* 1044–1047.

Kim, C., Nam, S. Y., Park, D. J., Park, I., & Hyun, T. Y. (2007). Product control system using RFID tag information and data mining. *Lecture Notes in Computer Science, 4412,* 100–109. doi:10.1007/978-3-540-71789-8_11

Kim, D., & Yeo, J. (2008). Low-profile RFID tag antenna using compact AMC substrate for metallic objects. *IEEE Antennas and Wireless Propagation Letters, 7,* 718–720. doi:10.1109/LAWP.2008.2000813

Kim, D., & Yeo, J. (2012). Dual-band long range passive RFID tag antenna using an AMC ground plane. *IEEE Transactions on Antennas and Propagation, 60*(6), 2620–2626. doi:10.1109/TAP.2012.2194638

Kirschning, M., Jansen, R. H., & Koster, N. H. L. (1981). Accurate model for open end effect of microstrip lines. *Electronics Letters, 17*(3), 123–124. doi:10.1049/el:19810088

Kuo, S. K., Chen, S. L., & Lin, C. T. (2010). Design and development of RFID label for steel coil. *IEEE Transactions on Industrial Electronics, 57*(6), 2180–2186. doi:10.1109/TIE.2009.2034174

Kuo, S. K., & Liao, L. G. (2010). an analytic model for impedance calculation of an RFID metal tag. *IEEE Antennas and Wireless Propagation Letters, 9,* 603–607. doi:10.1109/LAWP.2010.2053511

Lai, M., & Li, R. (2010). A low-profile broadband RFID tag antenna for metallic objects. In *Proceedings of 2010 International Conference on Microwave and Millimeter Wave Technology,* 1891–1893.

Liu, W. N., Zheng, L. J., Sun, D. H., Liao, X. Y., Zhao, M., & Su, J. M. et al. (2012). RFID-enabled real-time production management system for Loncin motorcycle assembly line. *International Journal of Computer Integrated Manufacturing, 25*(1), 86–99. doi:10.1080/0951192X.2010.523846

Mo, L., Zhang, H., & Zhou, H. (2008). Broadband UHF RFID tag antenna with a pair of U slots mountable on metallic objects. *Electronics Letters, 44*(20), 5–6. doi:10.1049/el:20089813

Ngai, E. W. T., Moon, K. K. L., Riggins, F. J., & Yi, C. Y. (2008). RFID research: An academic literature review (1995-2005) and future research directions. *International Journal of Production Economics, 112*(2), 510–520. doi:10.1016/j.ijpe.2007.05.004

Prothro, J. T., Durgin, G. D., & Griffin, J. D. (2006). The effects of a metal ground plane on RFID tag antennas. *IEEE Antennas and Propagation Society International Symposium, Albuquerque,* 3241-3244.

Pues, H., & Van, A. (1984). Accurate transmission-line model for the rectangular microstrip antenna. *IEE Proceedings H: Microwaves Optics and Antennas, 131*(6), 334-340.

Rao, K., Nikitin, P. V., & Lam, S. F. (2005). Antenna design for UHF RFID tags: A review and a practical application. *IEEE Transactions on Antennas and Propagation, 53*(12), 3870–3876. doi:10.1109/TAP.2005.859919

Saad, S. S., & Nakad, Z. S. (2011). A stand-alone RFID indoor positioning system using passive tags. *IEEE Transactions on Industrial Electronics, 58*(5), 1961–1970. doi:10.1109/TIE.2010.2055774

Sievenpiper, D. F. (1999). *High-impedance electromagnetic surfaces* (Ph.D. dissertation). University of California, Los Angeles, CA.

Son, H., Choi, G., & Pyo, C. (2006). Design of wideband RFID tag antenna for metallic surfaces. *Electronics Letters, 42*(5), 2–3. doi:10.1049/el:20064323

Son, H., & Jeong, S. (2011). Wideband RFID Tag Antenna for Metallic Surfaces Using Proximity-Coupled Feed. *IEEE Antennas and Wireless Propagation Letters, 10*, 377–380. doi:10.1109/LAWP.2011.2148151

Tentzeris, M. M. (2010). Low-profile broadband RFID tag antennas mountable on metallic objects. *IEEE Antennas and Propagation Society International Symposium,* 1–4.

Ukkonen, L., Engels, D., Sydänheimo, L., & Kivikoski, M. (2004). Planar wire-type inverted-F RFID tag antenna mountable on metallic objects. *IEEE Antennas and Propagation Society International Symposium, 1,* 101-104

Ukkonen, L., Sydanheirno, L., & Kivikoski, M. (2004). A novel tag design using inverted-F antenna for radio frequency identification of metallic objects. *IEEE/Sarnoff Symposium on Advances in Wired and Wireless Communication,* 91-94.

Wu, J. Y., Li, J. X., Cui, X. S., & Mao, L. H. (2011). miniaturized dual-band patch antenna mounted on metallic plates for RFID passive tag. In *Proceedings of International Conference on Control, Automation and Systems Engineering (CASE),* Singapore, 1-4.

Yang, P. H., Li, Y., Jiang, L. J., Chew, W. C., & Ye, T. T. (2011). Compact metallic RFID tag antennas with a loop-fed method. *IEEE Transactions on Antennas and Propagation, 59*(12), 4454–4462. doi:10.1109/TAP.2011.2165484

Yu, B., Kim, S. J., Jung, B., Harackiewicz, F. J., & Lee, B. (2006). RFID tag antenna using two-shorted microstrip patches mountable on metallic objects. *Microwave and Optical Technology Letters, 49*(2), 414–416. doi:10.1002/mop.22159

Zhou, S. Q., Ling, W. Q., & Peng, Z. X. (2007). An RFID based remote monitoring system for enterprise internal production management. *International Journal of Advanced Manufacturing Technology, 33*(7-8), 837–844. doi:10.1007/s00170-006-0506-6

KEY TERMS AND DEFINITIONS

High Impendence Surface (HIS): Surfaces that reflect the impinging electromagnetic waves in-phase, which are widely used in microwave and antenna engineering such as filters, substrates with suppressed surface-wave propagation, or artificial magnetic conductors.

Radio Frequency Identification (RFID): The wireless non-contact use of radio-frequency electromagnetic fields to transfer data for the purposes of automatically identifying and tracking tags attached to objects. The tags contain electronically stored information. Some tags are powered and read at short ranges (a few meters) via magnetic fields.

Transmission Line Model: In communications and electronic engineering, a radio-frequency transmission line is a specialized cable or other structure designed to carry alternating current of radio frequency, that is, currents with a frequency high enough that their wave nature must be taken into account.

Chapter 7
Towards Smart Manufacturing Techniques Using Incremental Sheet Forming

J.B. Sá de Farias
University of Aveiro, Portugal

S. Marabuto
University of Aveiro, Portugal

M.A.B.E. Martins
University of Aveiro, Portugal

J.A.F Ferreira
University of Aveiro, Portugal

A. Andrade Campos
University of Aveiro, Portugal

R.J. Alves de Sousa
University of Aveiro, Portugal

ABSTRACT

The current world's economical crisis raised the necessity from the industry to produce components cheaper and faster. In this sense, the importance of smart manufacturing techniques, proper articulation between CAD/CAM techniques and integrated design and assessment becomes critical. The Single Point Incremental Forming (SPIF) process represents a breakpoint with traditional forming processes, and possibly a new era in the small batches production or customized parts, being already used by automotive industry for light components. While classical stamping processes need a punch, a die, a holder and a press, in the SPIF process the final geometry is achieved incrementally through the action of a punch with a spherical head. Since the blank is clamped at the edges, there is no need to employ a die with the shape of the final part. However, this process must be further improved in terms of speed and dimensional accuracy. Because the process is cheap and easy to implement, it is currently the subject of intensive experimental and numerical research, but yet not deeply understood. This chapter gives an overview on the techniques currently being employed to optimize the process feasibility.

DOI: 10.4018/978-1-4666-5836-3.ch007

1. INTRODUCTION

Over the last centuries, sheet metal components have been produced using different tools and techniques. A method universally applied is stamping, which uses dies and punches, specifically manufactured according to the shape and dimensions of the component. However, despite being widespread, it presents technological hindrances, such as large energy costs and very high investment, which make this process very expensive. Because of the high cost of punches and dies, this method is only suitable for mass production, where the cost can be shared with a large number of parts.

On the other hand, the recent diversification in customer demand has resulted in the downsizing of the production lots. Because of this reduction the cost of manufacturing the tools needed to be reduced. This necessity gave origin to the intensification in the development of new production methods for a small lot. One of those methods is to create an incremental deformation in the sheet metal using a simple tool. The idea of Incremental Sheet Forming (ISF) with a single tool was initially patented by Leszak in 1967 (Leszak, 1967). This method has later become very attractive, due to the advance of manufacturing technology in the fields of Numerical Control and Automation. With the massification of computer or numerical control of manufacturing systems, incremental forming can now be fully automated and tends to be more available to general public. One of the great advantages of this method is ability to use small/simple tools that create a deformation along a defined path. The tool can be a simple hammer, a laser beam, or a water jet. One of the techniques that are receiving great attention by the scientific community and from the industry is the Single Point Incremental Forming (SPIF).

2. SCOPE AND MOTIVATION

As referred, Incremental Forming Processes have been gaining a growing interest both on academia and production industry. It is a low cost sheet forming process – mainly applied to metals – once it doesn't require the use of dies and punches as in press sheet forming. Doing so, it is still a versatile option in terms of geometrical possibilities of final parts. Several applications have been found on biomedical industry (implants) and usable prototypes, which can represent considerable savings on RD costs (Ambrogio et al., 2005). However, the low production rate and relevant geometrical inaccuracy are impairing a deeper implementation into industrial environment.

Several studies are being carried out by many research groups from all over the world, covering both numerical and experimental aspects (Aerens et al., 2009; Tanaka et al., 1999). Nevertheless, there is still much to explore and understand on the field of Incremental Forming Processes and several questions remain open.

In this work, focus is more given on how these techniques can work together in order to optimize the process, make it more fast and feasible, and finally easier to implement in industrial environment. Summing up, the scope of the present survey is to highlight how computer aided design and manufacturing, computational mechanics, optimization processes and experimental techniques can act together to push up processes boundaries and enlarge its potential towards a smart manufacturing approach.

3. ABOUT SINGLE POINT INCREMENTAL FORMING

3.1 Description

Single Point Incremental Forming is a manufacturing process that has the capability of producing both axisymmetric and asymmetric

Figure 1. Single point incremental forming main components

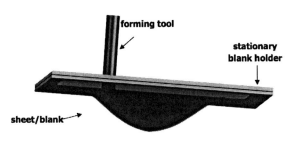

Figure 2. Path followed by the forming tool

parts. It uses simple tools, usually a cylindrical metal tool with a flat or spherical tip. It also does not require the use of dies with the shape of the piece being produced. This eliminates the use of presses, because the forces involved in the process tend to be much smaller. Figure 1 shows the components involved.

The deformation is performed by applying pressure on the surface of the sheet, forcing it to deform gradually. The metal sheet is usually restrained by the blankholder to avoid displacement and the flow of material into the forming area. Therefore, all the forming operations are made at the expense of sheet thinning phenomenon. Blank thickness may be expressed depending on the wall angle, by means of the Sine Law (1).

$$T_f = T_i \sin \alpha \qquad (1)$$

where,

T_f: sheet's final thickness;
T_i: sheet's initial thickness;
α: wall angle.

The forming tool is controlled by CNC software and the path is predefined by a CAD/CAM toolpath processor. Figure 2 illustrates an exemplificative path taken by the forming tool.

The main advantages and drawbacks of SPIF are summarized as follows (Jeswiet et al., 2005).

Advantages:

- Useable parts can be formed directly from CAD data, requiring a minimum of specialized tooling. These can be either prototype parts or small batches;
- The process does not require positive or negative dies, hence it is dieless. However, it does need a backing plate to create a clear angle transition in the sheet surface and improve accuracy through bending mechanisms;
- Changes in part design sizes can be easily and quickly accommodated, due to ISF's high flexibility;
- Producing ready-to use sheet metal prototypes is normally difficult, but this process made it easier;
- The deformation mechanism based on bending-stretch and the incremental nature of the process contributes to increased formability, making it easier to deform even sheets produced from low formability metals;
- A conventional CNC milling machine or lathe can be used for this process;
- The size of the part is limited only by the size of the machine. Forces do not increase significantly because forming mechanisms in ISF have localized scope area, and regardless of the total size of the part to be

manufactured, the contact zone remains the same;
- The surface finish of the part can be adjusted according to other process parameters;
- The operation is stable and relatively noise free.

Disadvantages:
- Forming time is much longer than other alternative sheet forming processes, such as deep drawing;
- The process is limited to small size batch production;
- The forming of vertical wall cannot be done in one step due to the excessive sheet thinning, thus requiring a multi-step approach to the process (Skjoedt et al., 2010; Verbert et al., 2008) ;
- A Springback phenomenon occurs during the whole production phase, affecting geometrical and dimensional accuracy.

3.2 Applications

Single Point Incremental Forming has a vast field of applications, such as aerospace, automobile, home appliances industry, marine industry and even fields such as medicine and food processing. It is a very versatile technology that can handle different kinds of materials, such as steel, aluminum, composite and polymeric materials. Figure 3 shows examples of the versatility in geometries that can be achieved using SPIF.

One of the major application areas of SPIF is in Rapid Prototyping because it has the capability of attaining functional parts. Given that, the automotive industry had its attention focused on this process because the field of application in areas like prototypes and replacement parts were vast and profitable.

Computer Aided Design and Manufacturing must be effectively integrated for successful operations on SPIF. Figure 4 shows a CAD face, the corresponding generated toolpath and the final part in a work made at the University of Aveiro.

Another area where SPIF can have a great impact is in the medical industry, particularly in the manufacture of prostheses and braces. Using Reverse Engineering (RE) it is possible to reproduce high accurate models that mimic parts of a human body or support them.

For example, a cranial implant can be easily produced using SPIF (Duflou et al., 2005b). The only necessary input is the CAD model, which can be obtained by a surface acquisition of the patient's skull. Then, this geometry is replicated in a metallic shell. An example can be seen in Figure 5.

Figure 3. Different geometries achieved by SPIF (Jeswiet et al., 2005)

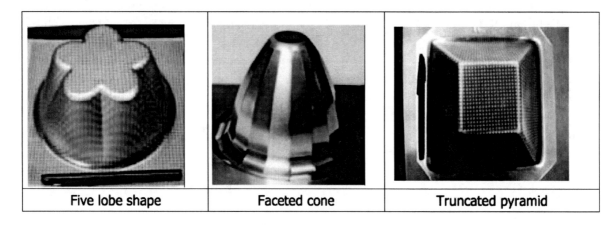

Figure 4. From the initial CAD to the final part in SPIF

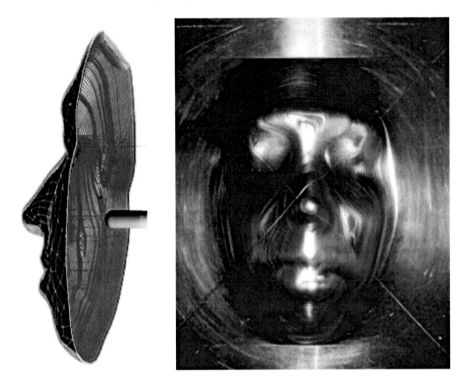

Figure 5. Cranial implant obtained by incremental forming

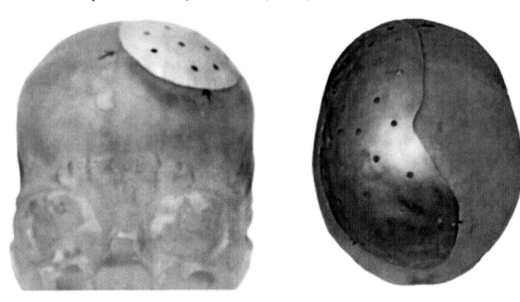

Another example can be found in Ambrogio et al. (2005). The process began by using a non contact inspection, laser scanning, to obtain the morphological and dimensional information of the body part to manufacture in the form of a cloud of points. From this information, surfaces were made and ultimately a CAD model was built. Finally, via a CAD/CAM application, an ISO file was created and sent to a numerical control machine that carried out the Incremental Forming operation. Figure 6 illustrates the different stages required to manufacture an ankle support.

This type of product is ideal for the SPIF process due to its high degree of customization, and because the price of the product does not represent a major vector of competitiveness, a characteristic that is common throughout the biomedical component industry.

With the development of this process, areas like biomechanics can benefit itself through of great enhancement. Nowadays, biomechanics lack tools to produce or transform with accuracy the necessary materials into usable products. With the development of the process and new SPIF machines this obstacle can be overcame.

3.3 SPIF Parameters

3.3.1 Forming Forces

One of the major characteristics that define a deformation process is the forming forces involved in it. According to Allwood et al. (2005), for SPIF process the forces involved in the process can be predicted using an approximate calculation by a theoretical model. They divided their analyses in two tool loading situations represented in Figure 7, (a) the tool travels normally to a flat sheet, causing a hemispherical indentation of the sheet; (b) during deformation, the tool moves horizontally 'around' the existing deformed area of sheet, creating a one-sided groove.

From the analyses of the previous diagrams the vertical force, F_v and horizontal force, F_h can be estimated by the following equations:

$$F_v = \pi \cdot r \cdot t \cdot \sigma_y \cdot \sin\alpha \qquad (2)$$

$$F_h = r \cdot t \cdot \sigma_y \cdot \left(\sin\alpha + 1 - \cos(\alpha)\right) \qquad (3)$$

Figure 6. The process of manufacturing an ankle support (Ambrogio et al., 2005)

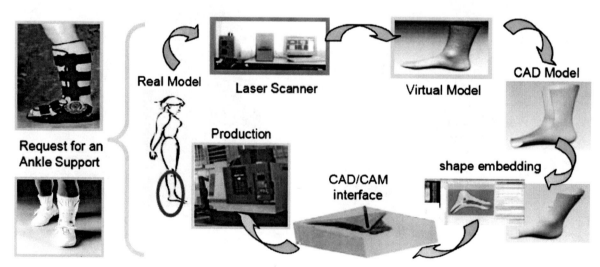

Figure 7. Two tool load cases (Allwood et al., 2005)

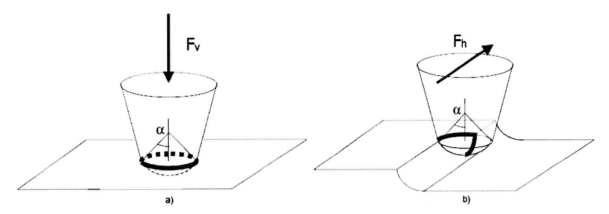

where,

r : tool radius;

t : initial blank thickness;

σ_y : yield strength;

α : cone interior angle.

This theoretical model gives a rough assessment of the forming forces involved in ISF. Various other authors developed experimental and numerical simulations in order to estimate the forces involved in the process. However, the parameters between each analysis are significantly different. Aspects like different materials, blank thickness, tool diameter, step size and others were combined by the authors. Table 1 shows some of the forces predicted and measured by different research groups.

3.3.2 Friction at the Tool/Sheet Interface

One of the most relevant parameters in SPIF process is the interaction between the tool and the blank. This interaction can occur in different ways:

- The tool slides over the blank without rotation, but with a significant amount of friction.
- The tool rolls over the blank having spindle speed and with some friction.
- The tool rolls over the blank with free rotation and with low friction.

Table 1. Measured forces by different authors

Author	Maximum force (kN)	Method
Allwood et al. (2005)	13 (vertical), 6.5 (horizontal)	Theoretical model
Duflou et al. (2005)	1.46	Experimental
Rauch et al. (2009)	0.9	Experimental
Jackson et al. (2008)	3	Experimental
Durante et al. (2008)	2	Experimental
Bouffioux et al. (2007)	1.3	Simulation
Decultot et al. (2011)	12	Experimental
	14	Simulation

Figure 8. Coated carbide tools (left), spherical tip tools (center), flat tip tools (right), from ref. Jeswiet et al., 2005

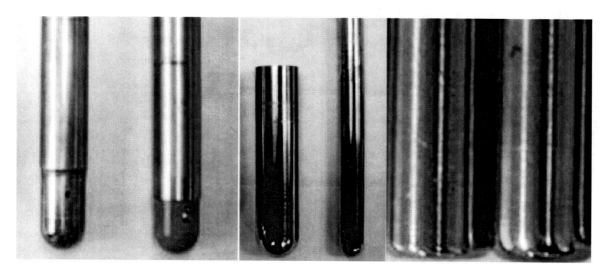

The control of the spindle speed allows the control of the friction between the tool and the blank during forming process. Friction levels should be maintained low because friction generates heat, causing surface degradation to occur. According to Kim and Park (2002), the appearance of friction at the tool/sheet interface would result in the improvement of formability and delay of fracture occurrence. However, they also concluded that if the friction increases too much, the sheet would fracture.

3.3.3 Tools Size and Geometry

The tools normally used for SPIF have a spherical or a flat tip. This assures a continuous point contact between the sheet and forming tool. A wide range of tool diameters can be used, from small diameters such as 6 *mm*, up to a large tool with 100 *mm* in diameter. These last are normally used for the manufacturing of large parts such as car's hoods and when the forming forces increase because of the large contact angle involved.

3.3.4 Vertical Step Size

Surface roughness is a major concern in a final product. Therefore, the quality of the product must be ensured from the beginning of the process. In SPIF the major factor in determining surface roughness is the incremental step size, Δ_z. The surface roughness tends to increase when the distance between two successive indentations increases, phenomenon caused by the displacement of the tool on the blank. A representation of that occurrence is shown in Figure 9.

An experiment conducted by Jeswiet et al. (2005) shows different surface roughness for different step sizes. The profiles presented in Figure 10 were obtained by white light interferometry and are 3.6 *mm* x 4.6 *mm*. The tool diameter is 12.5 *mm*. This is a SPIF characteristic that needs to receive some attention, since the market is very demanding in terms of quality, and because quality is related to the vertical increment, it becomes imperative to use a small increment. Unfortunately a reduction in the step size can represent a significant increase in production time. Therefore, a compromise must be achieved between quality and production time.

Figure 9. Two different approaches to measure step size

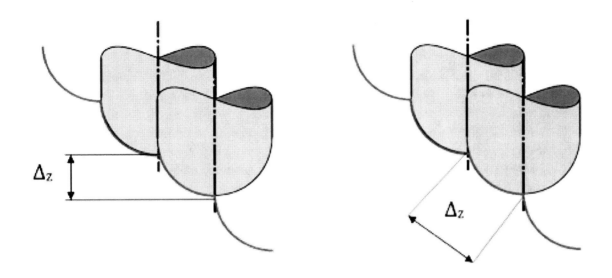

Figure 10. 3D Surface roughness for four different step sizes, Δz

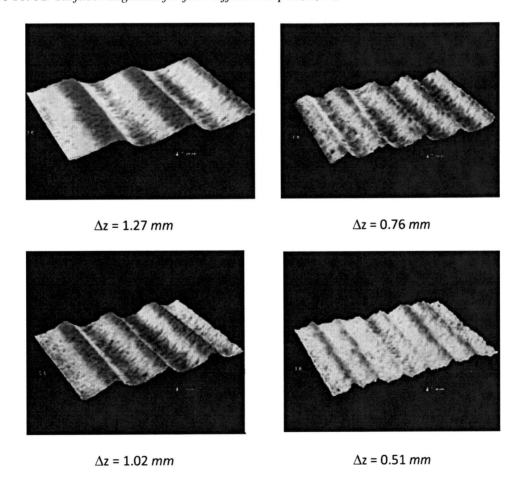

Δz = 1.27 mm   Δz = 0.76 mm

Δz = 1.02 mm   Δz = 0.51 mm

3.3.5 Materials

With the evolution of technology to process materials, there is a follow up response in creating new materials. With the rise of aerospace and biomedical applications, new materials have been created and others have seen their characteristics improved. In order to develop SPIF process, a number of materials have been tested. The more common used material is aluminum mainly because the machines where the tests are conducted have limitation in the maximum applicable force. Steel is target material for this process but do to the limitation mentioned above, only very thin sheets can the deformed. Table 2 resumes some of the experiments made and the results observed while testing different materials.

3.3.6 Formability

One very interesting feature of SPIF is the increase in the material formability (Figure 11). Although the scientific community agrees that there is an increase of formability, compared with other processes like stamping or deep drawing, it is still not clear how the mode of deformation influences the formability of the process. This subject is still an open discussion in the scientific community.

Some authors state that the deformation occurs by stretching, combined with bending, which promotes increased formability (Emmens et al., 2008). Other explanations are based on the presence of through thickness shear strain which leaded to the developments of extended Marciniak-Kuszinski models (Aerens et al., 2009).

Table 2. Tested materials and their characteristics

Author	Material	Application	Characteristics
Ham and Jeswiet (2006)	AA – 5754 AA – 6451 AA- 5182	experimental study	Thickness: 0,8 – 1,5 mm
Jeswiet et al. (2005)	AA – 3003 O	experimental study	Thickness: 0,93 – 2,1 mm
Micari et al. (2007)	AA 1050 O	experimental study	Thickness: 1,2 mm
Tanaka et al. (2005)	pure titanium	denture plate	Main difficulties in the production of this part were the surface quality, needing to find optimal combination between feed rate and lubrication
Hussain et al. (2008)	pure titanium	experimental study	Proper tool, good lubricant and lubrication method were required
Jackson et al. (2008)	sandwich panels	aircraft interiors, car body panels, architecture panels saving weight, absorbing sound, vibrations and impact, and isolating thermically	–
Franzen et al. (2009)	PVC (Polyvinyl Chloride)	complex polymer sheet components with very high depths	–
Le et al. (2008)	polypropylene (PP)	different geometries	–
Ji et al. (2008)	Magnesium AZ31	structural applications	Warm temperatures

Figure 11. Typical enhanced formability curve for SPIF process (Centeno et al., 2012)

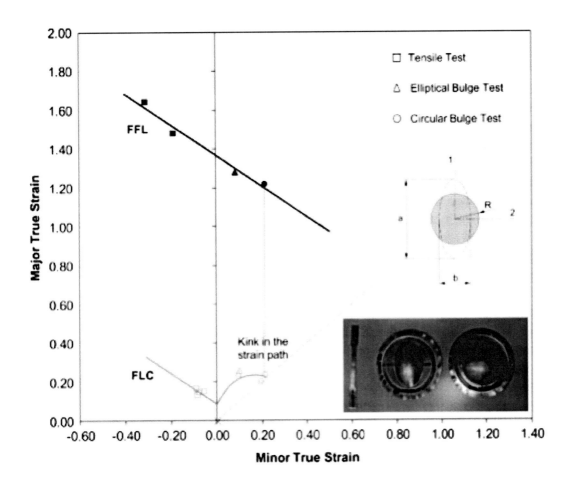

3.3.7 Toolpath

Toolpath is the process parameter than have direct influence of the all previously referred parameters. Doing so, speak on Smart Manufacturing for SPIF means to act directly on the toolpath itself. Doing so, section 4 will be exclusively devoted to analyze toolpath optimization.

4. TOOLPATH OPTIMIZATION

As referred in the previous section, toolpath strategies and their parameters represent the main optimization basis for ISF. These can be divided into two main aspects: the strategy itself, as being derived from machining processes, or specifically designed accounting ISF's demands; and the relevant parameters' adjustment within these toolpaths.

Several methods have been developed over the last decade to improve SPIF's accuracy, to control and predict formability as well as to minimize surface roughness on the final parts. Some of them comprise on-line monitoring and control of the process, correction algorithms, tuning variables and noise control just to name a few.

Although there are many procedures and methods to control ASPIF process' parameters, none of them is effective in what regards production time. Finite element analyses approaches provide accurate arrangement of parameters to produce a given part. However these are extremely time consuming when using try-out methodologies that may take several days processing a solution for a part that takes few minutes to manufacture. Consequently, researchers have been seeking for new approaches that may act as an optimizer in order to make ASPIF a competitive technology in prototyping industrial applications.

Examples of studies on the influence of parameters in incremental sheet forming, such as forming speeds, sheet thickness, maximum depth etc., may be found in the works carried out by Duflou et al. (2003, 2011), Ham and Jeswiet (2007), Jeswiet et al. (2005), Emmens and Boogaard (2008), Kim and Park (2002), Franzen et al. (2009), and Ambrogio et al. (2003, 2006), among others. Major research on forming strategies and toolpath optimization was performed by Filice et al. (2006) or Skjoedt et al. (2010).

However, these works were focused on only one or two single process optimization parameters. More relevant is the fact that some works are related to performing ASPIF with adapted CNC milling machines which may limit the range of materials and thicknesses to be processed, given its limited payload (Marabuto et al., 2011).

The use of Design of Experiments (DOE) techniques in incremental forming processes was already tried by Ambrogio et al. (2004) using an experimental campaign. However, the work has only included a preliminary analysis for the prediction of the formability and its detection of failure insurgence. Additionally, the use of numerical analysis of the technological process (FEM simulation), coupled with numerical optimization techniques was not yet studied. This is certainly an innovative and challengeable approach based on the capabilities that finite elements have to accurately simulate SPIF, despite long CPU times (Sena, 2011).

It is the general opinion within the scientific community that this process has presented some unpredicted challenges and difficulties in achieving an industrially acceptable level of operation: nowadays performing ISF in industry requires a deep knowledge of the state-of-the-art regarding this technology, and it is a know-how detained by very few people in the world.

Nevertheless, ISF has evolved to meet some cutting-edge smart manufacturing approaches, which in a short-term can lead to a successful industrial dissemination. Nowadays, the tool trajectory is considered as a process variable in SPIF, whereas usually it is directly derived from knowledge within the machining field. Below are described some traditional machining techniques applied to SPIF, namely spiral, helical and z-level toolpath strategies. Furthermore some correction algorithms are presented.

4.1 Z Constant Tool Path

This type of toolpath is adapted from milling machining 3D roughing strategy, where the final surface is vertically sliced to 2D contours performed at the same z level. It was one of the first strategies applied to ISF, due to its' easiness of implementation. The majority of research teams started testing z constant toolpaths to produce their first parts in ISF and assess their setup (Guzmán et al., 2012, Rauch et al., 2009).

One of the main advantages of this technique is the easiness of generating numerically controlled codes as well as short control programs that can be interpreted by any milling machine. However, this strategy often leads to visible stretch marks and a coarse surface finishing. These problems can be minimized by using smaller vertical increments, but at the expense of larger forming times. State of-the-art in ISF evolved towards the use of more complex, yet more suited toolpaths.

4.2 Spiral Toolpath Optimization

As stated before, the first tool path strategies designed for ISF were derived from machining

Figure 12. Constant level toolpath

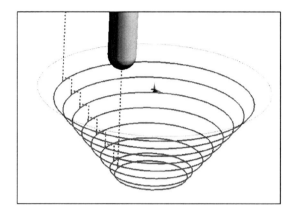

Figure 13. Spiral toolpath strategy derived from machining

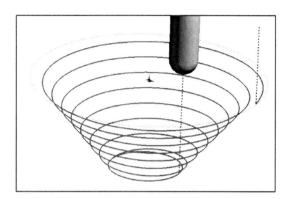

operations. The process itself can be regarded as a fusion between conventional stamping and milling machining: because it is performed over a sheet metal blank and, at a first glance, has many setup features in common with milling (Filice et al., 2006).

Toolpath strategy selection has a dominant influence on sheet thickness distribution. Several strategies were tested to both homogenize thickness variation, which penalizes formability and reliability in the final part, and minimize thinning phenomena. It should be noticed the thinning phenomena leads frequently to sheet tearing and destruction of the piece, endowing a loss of robustness to ISF.

Aiming to overcome the above mentioned drawbacks, four manufacturing strategies were defined by Manco et al. (2009) for axisymmetric profiles.

1. The first one consists on what is traditionally known as the conventional negative tool path (from outer to inner and from higher to deeper): Single Slope (SS).
2. The Incremental Slope (IS) strategy was designed to evaluate the contribution of the additional material in avoiding localized thinning phenomena, assuming sheet volume constancy. To accomplish that, it starts accounting more material than what is strictly required, and part of the blank usually not involved in the manufacturing step is now deformed.
3. Wall Slope (WS) strategy intends to balance hardening and stretching phenomena, by restoring the steady-state conditions after the transient response. This is accomplished by splitting the total depth of the piece into three steps, each with both increasing depth and wall angle, also fractionated. Each subsequent step repeatedly works the same portion of material, testing it's hardening.
4. Decremental Slope (DS) strategy artificially increases the wall angle in order to minimize the drawbacks of excessive stretching in the last loops of the manufacturing process. For a given wall angle (alpha), forming occurs with a corrected angle ($\alpha+10°$), returning to an overcompensated forming angle ($\alpha-10°$) to achieve the intended geometric profile.

In ISF it is possible to obtain the same final shape inducing different thickness distributions, just by altering the toolpath strategy, and thus enhance formability. The tendency regarding spiral toolpaths relies on deforming as much material as possible, from a theoretical point of view, or to deform the material according to guidelines which are 'not conventional'.

Figure 14. Toolpath strategies, according to Manco et al. (2009)

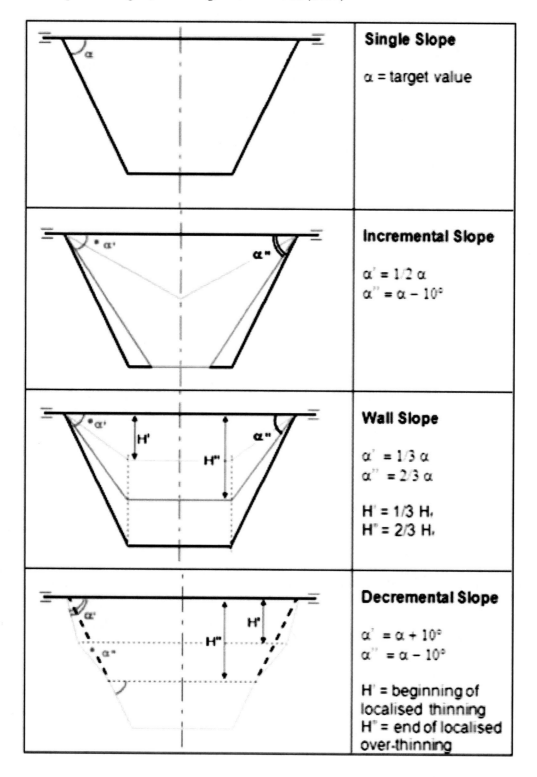

4.3 Constant Feed Tool Path

Another toolpath strategy was proposed by Skjoedt et al. (2010) for ISF. It is derived from the profile milling or contour milling machining strategy, used to obtain complex surfaces, and it consists in setting up the tool feed ratio equal in all three directions: Rolling direction, tangential direction, and the one orthogonally defined by the previous two. Starting from a helical toolpath, a dedicated program uses the coordinates from the profile milling code and converts them into one having continuous feed in all three directions.

Starting from a spiral or helical tool path containing information regarding the part's geometry, the next step is to divide it into layers with the same thickness as the step size. The coordinates from the profile tool path act as the input for a program which computes an adapted geometrical helical tool path, having the same increment or feed in the three main directions. The constant feed in three directions strategy can be a problem for some CNC milling machines, because they are not prepared to operate at the high feed rates required in SPIF.

Some geometry composed by dramatically different surfaces may need to use the helical program on each section separately and then combine the two helical outputs, such as mufflers. This methodology results in a working tool path with only one non-helical layer where the two helical outputs are linked.

This method has proven to be successful in removing completely surface scarring in pyramid shaped parts and cones. In parts composed by very different surfaces, the strategy was unable to remove all the visible to the naked eye defects. However, with this method were observed some significant improvements in the overall quality of the parts.

4.4 Design to Improve Accuracy

One of the major drawbacks of ISF relies on the lack of accuracy both geometric and dimensional in the final parts obtained. Actually many strategies were proposed in order to minimize the geometrical error involved in SPIF, although it is clear that many efforts must be spent in order to ensure a higher precision. Only in this way, incremental forming could represent a valid alternative to conventional processes.

The springback phenomenon is particularly relevant since the sheet is simply clamped in a frame, and free to bend when the punch action is relaxed and the part is trimmed. According to Micari et al. (2007), strategies to minimize geometrical and dimensional deviations can be divided in two main categories: through alterations in the basic SPIF tooling and setup, such as the use of different types of support; and developing optimized tool path trajectories that minimize the geometrical error after springback, unclamping and trimming.

Some researchers focused on accuracy, and their experiments highlighted significant improvements to ISF; on dimensional accuracy, when a smaller tool diameter and a lower tool depth are utilized (Ambrogio et al., 2004); the geometrical error reduces when the blank thickness increases, at forming simple truncated cone shapes, even without any backing. Accuracy then depends not only on the thickness but mainly on the punch diameter versus blank thickness ratio. In the same work, it was also concluded that accuracy depends on the relative position of the forming area to the clamping system. More specifically, the distance between the forming area and the clamping frame must be the smallest, in order to minimize the bending effect in the blank during the tool's forming cycles (Ambrogio et al., 2003). In opposition of what could be assumed, geometry complexity can increase accuracy. For instance, parts having corners, double curvature or, in general, part discontinuities limit undesired blank movements during forming, and increase process accuracy (Ambrogio et al., 2003).

The continuous search for accuracy improvements led to create effective methodologies for error detecting, especially designed for ISF. In their work, Filice and their team (2006) defined two main kinds of error control methods: (i) out of machine control and (ii) on-machine control.

Out of machine control methodologies do not affect the forming process while is occurring. Instead, they assess parameters on setup phase and after forming. Nowadays, three main systems are used:

Laser scanners, which can be is very fast and don't require any contact with the sheet surface. However, these parts need to be subjected to a surface treatment in order to ensure the opacity required by the laser beam.

Mechanical probe, which can acquire the coordinates of points in space, and based on them reconstruct a three dimensional surface. This surface can now be compared with the CAD geometry and computed deviations. The major drawback of this system is the time required to collect the points.

Coordinate Measuring Machine systems. This type of equipments acquires readings in six degrees of freedom and displays the data in mathematical form. These machines may be manually controlled by an operator or it may be computer controlled. Measurements are defined by a probe attached to the third moving axis. Probes may be mechanical, optical, laser, or white light, just to name a few. These machines ensure a very high precision but, on the other hand, they are expensive, require the right positioning of the part in the cube and, finally, the plotting process is sometimes not very fast.

On-machine control methodology relies on inspecting the process directly on the mandrel, acquiring coordinates and subsequent information regarding surface in forming, transforming the equipment into a sort of hybrid milling-CMM machine, who plots the actual geometry of parts obtained through SPIF. These types of measuring systems were initially designed for accounting tool wear in milling and turning, and therefore provide the adequate compensation. The measured geometry can be used by the machine controller in order to develop a corrective action. Nowadays, these systems are being used to carry out an adaptive "in process" control of the geometry where properly developed corrective strategies are implemented. The control of ISF, in particular the on-machine control has proven to be an effective tool to improve the overall quality of the process. Some research resulted in the implementation of an on-machine control based on forming force behavior and patterns to predict sheet failure and excessive thinning. It basically consists in an algorithm which assesses force detection and monitoring to provide automatic modification of the current part's program, updating some user-defined parameters and thus correcting the process in order to avoid failure and to produce sound parts (Filice et al., 2006).

4.5 Alternative Approaches

Besides the above mentioned, over the years were presented some very interesting and innovative strategies to improve accuracy in ISF. The most part were based on the use of different types of support, thus complicating the basic SPIF tooling and didn't have a successful implementation. However, are briefly presented some of the most interesting ideas published within the scope.

Tanaka and his co-workers (2005, 1999) designed a flexible support under the blank made in rubber that it is acted as a counter-tool. They concluded that the springback was reduced, and the sheet metal formability was improved due to a hydrostatic stress state generated in the blank between the punch and the elastic die, minimizing the risk of fracture.

Chen et al. (2005) proposed another approach to perform ISF on large parts: sectional multipoint forming technology. It consists in the use of several counter punches to help bending the sheet while a

Figure 15. Force components in ISF

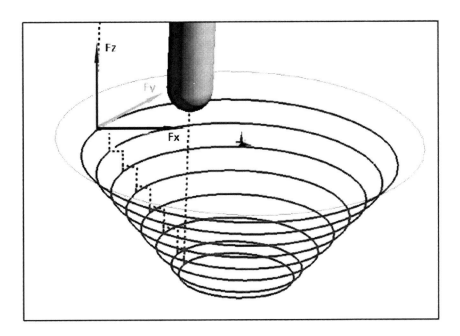

punch forms the upper side of the sheet. This set of counter punches was vertically controlled to fit the counter shape of the geometry to be formed, thus forming an interactive counter die which reduces springback phenomenon. Although this technique enables the manufacturing of complex shapes, it has to be performed in properly designed machines able to control at the same time several counter punches, making ISF very expensive and, above all, it cannot be carried out on simple machines, such as CNC milling machines. Furthermore, the development of part programs for this application represents a critical task, because a limited experience in this field is today available.

The use of counter pressure is another technique that derives from the same concept but acts as a combination of hydroforming and incremental sheet forming. In this case, the blank support effect is obtained using pressurized water jets (Iseki, 2001; Shim et al., 2001). The drawback is the loss of flexibility associated to hydroforming technique and its setup requirements as well as the adjustment of the water pressure to ensure the desired formability since a too high pressure may cause undesired sheet bulging that reduces itself the part precision and a too low pressure may not affect as a counter support.

4.6 Monitoring Forming Forces

One of the most determinant aspects to control the ISF process is measuring and monitoring the forces between the punch and the sheet. Some studies covered the magnitude and behavior while others tried to find patterns characteristic of catastrophic phenomena such as sheet failure and critical thinning. Notice that since the beginning of SPIF studies that evaluating forces has taken a dominant role in optimizing incremental sheet forming (Duflou et al., 2005a).

Duflou et al. (2007) described one of the first studies carried out to evaluate forces in ISF and their relationship between the most influential process' parameters. In this work, they formed six cones made from different sheet thicknesses and measured the three components of the forming force, F_x, F_y and F_z, as illustrated in figure below. The forming forces were evaluated according to the maximum value reached during a test, the peak forming force (F_p), and the average value after it's stabilization, the 'settled' force (F_s), reached over the bulk of the time taken to form a partial force sum.

The main conclusions drawn in this work, regarding the influence of forces in ISF and their interaction with the most relevant parameters, are summarized below:

1. **Influence of Step Size:** As the vertical step size increases it is verified that force magnitude also rises in a linear trend, and a directly proportional behavior can be derived between the magnitude of force and the depth of the vertical step.
2. **Influence of Tool Diameter:** An increase in tool diameter also causes an increase in the force amplitude required for forming, which is valid for both the peak forces (F_p) and the mean force after the peak (F_s).
3. **Influence of Wall Angle:** As the part wall becomes steeper, the magnitude of the force gradually increases. Furthermore, with the increasing of the wall angle, the difference between F_p and F_s rises. It is believed that this forms an indication that the maximum wall angle that can be achieved by a conventional single point incremental forming tool path is being approached.
4. **Influence of Sheet Thickness:** The resultant of the forces increases considerably with increasing sheet thickness. It can be concluded that sheet thickness is a dominant factor determining the force magnitude required for ISF.
5. **Influence of Lubrication:** When forming occurs without any lubricant, there is a slight increase on the peak force, often followed by a sudden drop due to catastrophic failure. Actually, a part that is about to fail shows this pronounced peak and subsequent drop in force magnitude. However, the part develops tears before reaching a minimum force level and slowly increasing again. This is due to excessive friction at the tool-sheet interface and the presence of lubricant is quite important to avoid wear on both the working material and the forming tool regardless of the type of lubricant used. Furthermore, when the process is performed with reduced friction, the surface quality is improved.

So, when vertical step size, tool diameter, wall angle or sheet thickness are increased, the forming forces will also increase. From all the tested parameters, vertical step size has the least significant impact and can therefore be increased to achieve better production times.

Also, Filice et al. (2006) carried out a study to analyze forming forces in ISF, namely force peak and trends, also stating that an accurate detection and online analysis of forces can provide important information on material behavior and on damage prediction. The main idea of implementing an on-line control on ISF is having a method able to obtain reliable information on damage insurgence and to control the process's parameters in order to avoid sheet metal failure and its related consequences in terms of costs and time. Since that, forming force started to be considered a "spy" variable of the process' conditions. The tangential component of the forming force was then evaluated, in particular its behavior, in order to find patterns that could predict ISF instability and explain deformation mechanisms.

In the same study, force trend was accurately investigated at the varying of the main process parameters, including sheet thickness, allowing draw some conclusions, which represent the technological base for the further implemented correction algorithm. It was verified:

- a strong sensitivity of the forming load to the variation of the process' parameters;
- a strong correlation between the force load peak and the sheet thickness;
- an increasing the tool depth step, the forming load increases;
- that the tool diameter (D_p) strongly influences the forming load;
- an increasing the wall inclination angle (α), the force peak increases.

Some of the above described behavior can easily be explained. For example, since higher strains are locally imposed on the sheet, the increasing of the tool depth step causes the forming load to increase. In addition, when a larger thinning is imposed to the sheet metal, increases material opposition to deformation, and it can explain why force peak increases with wall angle.

The experiments allowed defining three main classes of force behavior:

1. Steady state force trends;
2. Polynomial force trends;
3. Monotonically decreasing force trends.

The steady state behavior is obtained when forming forces achieve equilibrium conditions. This occurs when sheet deformation and, consequently, thinning reduce the load, still maintaining material strain hardening which results in a load increasing, and reaching dynamic equilibrium. On the other side, polynomial trend curves are typically obtained for severe but not dangerous values of the wall inclination angle.

Finally, monotonically decreasing trends are typically obtained for processes which yield to failure. After the peak, thinning becomes the dominant phenomenon while material strain hardening is not able to stabilize the process; the forming forces suddenly drop up to material failure. Having these results, it can be defined a threshold between safe forming conditions, the steady trend, and instability; therefore it is possible to conclude that the force gradient after the peak strongly depends on the experiment parameters and can be effectively recognized as an index of the process stability. However, when material work hardening compensates for sheet thinning, the stability in ISF can be recovered and steady conditions are reached. If on the contrary, thinning prevails, the process becomes instable and the forces drop up to failure. It is due to this phenomenon that the gradient of the tangential force can be considered a "spy variable," able to assess ISF stability.

After reaching the above stated conclusions, Duflou et al. (2007) carried out an experimental campaign, based on a proper Design of Experiments (DOE), to obtain the correlations between the main process parameters and the forming forces trends, and in addition, determine the critical values for the ISF design. Having in consideration that the force gradient after the peak is highly influenced by sheet thickness and wall inclination angle, as well as by the punch diameter and, at last, by the tool depth step or vertical increment, they were now able to design a control model, set-up the control strategy.

The force peak, defined as K, was evaluated using a response surface statistical analysis highlighted that a quadratic model well describes the K-variable evolution with respect to the input data. The suitability of the model was evaluated through the analysis of the residual normal probability and it was concluded that the model predicts adequately force gradient depending on the mentioned process parameters. So, measuring forces in ISF is a relatively affordable technique, which can be used to control and monitor the process on-line, as well as provide for automatic correction of the parameters, avoiding failure. This strategy has potential to represent a powerful tool for industrial applications, since it is easy to program, update some user-defined parameters, and prevent material collapse.

5 MACHINE DESIGNS FOR INCREMENTAL FORMING

To perform Single Point Incremental Forming there are a few indispensable requirements. One of them is the need for a CNC-controlled three axis machine. Currently there are two ways to produce parts using Single Point Incremental Forming, adapted and purpose-built machines. These machines may vary in type, but there are only a few purpose-built models available, mainly because this is a recent technology and there are only a few companies that invest in developing a SPIF machine. So the most common is the use of adapted machinery. The characteristics that have to be accounted for SPIF work are speed, output rate, stiffness and payload. High speeds, large working volumes, adequate stiffness and high load capacity are favorable. The types of machines that can be used to perform incremental forming are:

- CNC milling machines;
- Robots;
- Purpose built machines;
- Stewart platforms and Hexapods.

5.1 CNC Milling Machines

The more common in SPIF process is the use of three-axis CNC-controlled milling machines, which are adapted for the process (Figure 17). According to Allwood et al. (2005) these machines are very attractive because adapting a milling machine to perform ISF does not require many additional alterations and costs. Unfortunately they also have some disadvantages; firstly because "milling machines are generally not designed for high loads tangential to the spindle, there is a danger that the machine will be damaged during incremental sheet forming operations," as highlighted in the blue area of Figure 16. Therefore, the use of these systems is limited to softer materials, such as aluminum alloys, that normally involves

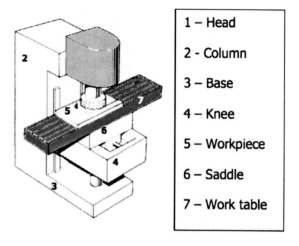

Figure 16. Milling machine

1 – Head
2 – Column
3 – Base
4 – Knee
5 – Workpiece
6 – Saddle
7 – Work table

forces applied to the tool much smaller than the forces needed for materials, such as steel. Secondly, "CNC machines do not generally provide instrumentation for measurement of three-axis forces at the tool tip." Third reason is that "the worktable of a CNC machine is generally solid and has limited width, so there is limited access to the reverse side of the workpiece" (Jeswiet et al., 2005), and also limits the work volume.

5.2 Robots

Another technology that has been used/adapted in SPIF has been the robots with serial kinematics (Figure 18). These systems are attractive because they have high flexibility and mobility which enables the manufacture of parts with high complexity and work at high speeds. The final position of the tool is the result of the combination of the different joint movements, which can occur at the same time. Another advantage is that they allow the tool to operate in the best angle configuration. This is an important aspect when dealing with complex shapes because it allows to nullify efforts that otherwise would be hazardous to the tool, and to the machine itself. It also prevents the degradation of the metal sheet and improves accessibility.

Figure 17. SPIF adapted milling machines (Left, Jeswiet et al., 2005), (right, Kopac et al., 2005)

The disadvantages, just like the milling machines, are that they have low stiffness and low admissible forces at the tip of the tools. Therefore, the use of these machines is limited to softer materials or else they would require investment in stiffening the joints. Another disadvantage is the lower precision. Because it's a serial configuration, the positioning errors of the individual axes tend to sum and this result in higher geometric errors when comparing with other machines. An example of these systems can be seen in Figure 19.

5.3 Purpose Built Machines

The purpose built or dedicated machines, unlike adapted machines, are designed for incremental forming process, and try to have simultaneously high rigidity and high flexibility. These systems allow the manufacturing of parts with complex geometries, while maintaining high accuracy and good surface finish. There are very few models on the market, and a few currently being studied and proposed.

Figure 18. Articulated robot arm

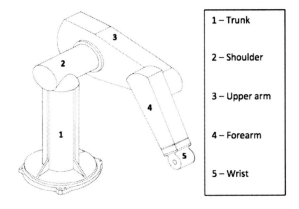

5.3.1 Amino Dieless NC Machine

A Japanese company named Amino Corporation developed a series of commercial models that work with the Single and Two Point Incremental Forming (Amino et al., 2002). Their technology is based on a complete package that allows going straight from data to finished metal parts with only minimal soft tooling. According to the information available, the Dieless NC Machine can use a wide variety of materials, including mild steels, aluminum, titanium and even perforated steel mesh, with thickness ranging from 0.1 *mm*

Figure 19. Robot performing Single Point Incremental Forming (Duflou et al., 2005a)

Figure 20. Amino's Corporation, Dieless NC machine (Amino et al., 2002)

Table 3. Amino's Corporation models of the Dieless NC machine.

		DLNC-RA	DLNC-RB	DLNC-PA	DLNC-PB	DLNC-PC	DLNC-PD
Max. blank size (*mm*)		400x400	600x600	1100x900	1600x1300	2100x1450	2600x1830
Max. forming size (*mm*)		300x300	500x500	1000x800	1500x1200	2000x1300	2500x1750
Max. forming depth (*mm*)		150	250	300	400	500	600
Stroke (*mm*)	X axis	330	550	1100	1600	2100	2600
	Y axis	330	550	900	1300	1450	1900
	Z axis	200	300	350	450	550	650
Max. work holder size (*mm*)		700x750	1000x950	1300x1100	1800x1500	2300x1650	2800x2030
Forming capacity thickness (*mm*)	Steel (CR)	0.6~1.6	0.6~1.6	0.6~2.3	0.6~3.2	0.6~3.2	0.6~3.2
	Stainless	0.5~1.0	0.5~1.0	0.5~1.5	0.5~2.0	0.5~2.0	0.5~2.0
	Aluminum	0.5~3.0	0.5~3.0	0.5~4.0	0.5~5.0	0.5~5.0	0.5~5.0
Forming speed (m/min)	X, Y axis	30.0	30.0	60.0	60.0	60.0	60.0
	Z axis	310.0	10.0	10.0	10.0	10.0	10.0
AC servo motor power (kW)	X axis	0.9	1.4	5.9	8.2	10.0	16.0
	Y axis	0.9	1.4	4.5	4.5	4.5	7.0
	Z axis	0.9	0.9	1.0	1.0	1.0	3.0
Machine weight (ton)		3.0	5.0	6.0	8.0	10.0	(18)

to 4.0 *mm*, depending on the model. The Dieless NC machine can be seen in Figure 20.

Amino Corporation offers standard sized models of their Dieless NC Machine. Two of the models, DLNC-RA and DLNC-RB, are designated as research-use and have smaller tables and slower in-table traverse speeds. The other four models, DLNC-PA, DLNC-PB, DLNC-PC and DLNC-PD, are primarily designed for industrial use. The specifications for all the models are shown in Table 3.

5.3.2 Cambridge ISF

In 2004, at Cambridge University, the Cambridge ISF was developed with the objective of performing SPIF operations (Figure 21).

According to Allwood et al. (2005), the development of this machine was centered on the distinctive requirement to maintain a proper accuracy while in the presence of high horizontal and vertical tool forces. With that intent and to avoid excessive moment loads on any of the system's bearings, the vertical axis was maintained as close to the workpiece as possible. The specifications of the Cambridge ISF are summarized in Table 4.

The workpiece is restrained in a channel section frame, and may be clamped over a backing plate. The frame is mounted on six 10 *kN* load cells organized to provide a six degrees-of freedom constraint without any moment loads on the load cells.

5.3.3 Tricept HP1

According to Callegari et al. (2006) the Tricept HP1 is a hybrid machine that uses a spherical wrist with serial kinematics (Roll-Pitch-Roll) mounted on a mobile platform with parallel kinematics. It can handle six degrees of freedom. The apparatus of the Tricept HP1 can be seen in Figure 22.

Callegari et al. (2006) defined this machine as having a predominantly parallel nature. That

Figure 21. Cambridge ISF machine (Allwood et al., 2005)

Table 4. Cambridge SPIF Machine specifications

Parameter	Measurement
Workpiece (active area)	300 x 300 (*mm*)
Material	Up to 1,6 (*mm*) mild steel
Vertical force	<13 (*kN*)
Horizontal force	<6,5 (*kN*)
Tool tip speed	<40 (*mm/s*)
Tool tip radius	5, 10, 15 (*mm*)
Maximum cone angle	67,5°
Maximum vertical axis travel	100 (*mm*)

allowed high stiffness, proper accuracy, and fine repeatability levels, even under heavy loads. They also explained that the stresses placed on it are transferred to the robot chassis using the three legs, which are mainly loaded with normal stresses. Also, positioning errors of the individual axes tend to counterbalance, rather than summing as in the case of serial configurations. The structure of the Tricept HP1 is built with a thick metal sheet to ensure stiffness, and allows both the robot as the supporting frame where the blankholder is mounted to be held down. The robot can apply a maximum thrust of *15 kN* within an area of *2000 mm x 600 mm*, with repeatability up to than *0.03 mm*. The forming tool is held by a pneumatic gripper that subsequently is attached to the robot flange. The blankholder and the blank are held together by an array of bolts.

Drawbacks of this machine reside mainly in the workspace's complex shape do to its limited agility (Figure 23). Therefore, a simulation tool was implemented in order to allow the machine kinematics off-line analyses and evaluate beforehand the viability of a certain operation.

5.3.4 SPIF-A Machine

Exception made to dedicated Single Point Incremental Forming (*SPIF*) machines like the ones developed by Amino Corporation (Amino et al., 2002) or by Julian Allwood's group at Cambridge University (Allwood et al., 2005), most of SPIF

Figure 22. The hybrid robot Tricept HP1, (Callegari et al., 2006)

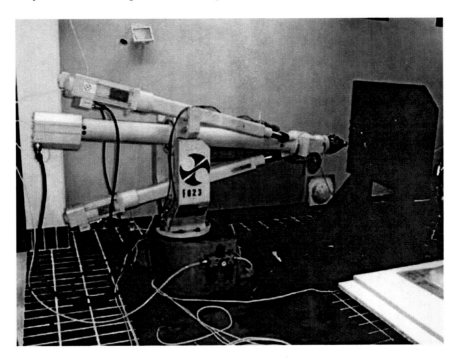

Figure 23. Tricept HP1 workspace's shape. Top view, (left), front view (middle), side view (right), (Callegari et al., 2006)

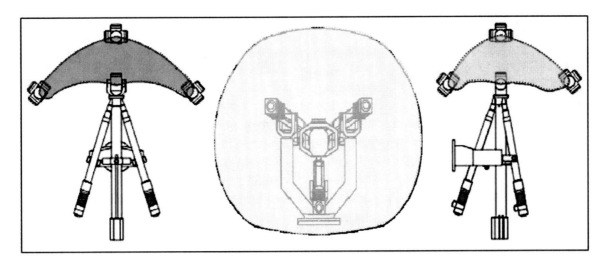

Figure 24. The 6 degree of freedom SPIF machine prototype built in the University of Aveiro

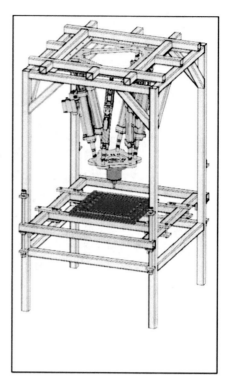

experimental research is being nowadays carried out using adapted CNC milling machines.

As an alternative to adapted machine tools, several authors have implemented the use of robotic serial manipulators to perform SPIF (Meier et al., 2005). This solution becomes attractive due to its high flexibility given by six axes mobility, allowing the most convenient punch positioning at the surface sheet metal. It also makes possible the combination of multiple stages of production in a single manufacturing cell using a single robot. The SPIF process with robotic arms is performed in a very similar way to the one described for milling machines: a 3D CAD model of the part is sliced into horizontal layers in the CAM software and converted into tool paths of the punch.

Another recent approach was to develop and build a machine with six degrees of freedom (Marabuto et al., 2011). The SPIF-A machine joins the distinctive features from CNC machines as high stiffness and flexibility from serial kinematic robots with five axes available for the tool path, Figure 24.

This project was driven by the need of forming harder materials and with higher thicknesses, which demands larger admissible forming loads. As an innovative prototype, SPIF-A was designed intending to extend the process' range of application and reproduce case studies presented in literature. Hence, it was targeted to achieve a work area of 500 by 500 *mm*, and a vertical travel of 400 *mm*. The ground base for the chosen enlarged work area enveloped is to guarantee the possibility of producing not only academic benchmark pieces, but also real industrial components. Nevertheless, the final machine prototype contemplates the possibility to modify the forming table size or to provide it with an additional translational motion.

Table 5. SPIF-A's features

Kinematic system	Parallel with 6 DOF
Punch/sheet interaction	Passive or un-driven tool
Admissible loads: - axial, compressive - bending, lateral	13*kN* 6,5*kN*
Work area	500x500*mm²*
Vertical travel	400 *mm*

6. CONCLUSION

Contrary to traditional press forming, incremental forming can be easily configured for preliminary experimental testing, specially adapting CNC milling machines, an apparatus easy to find in most of manufacturing labs. Even if one considers finite element numerical simulations, research has been showing too long CPU times to simulate forming of simple parts, in the order of days, which makes ISF quite unique in the sense that is more affordable to run experiments than simulate the process.

The referred experimental implementation easiness led to (especially after 2000's) a quick grow on research carried out in the field. Evolution was quite positive and much of the process parameters and peculiarities are much better understood now. It became undeniable that ISF is very appealing to produce ready-to-use (and mechanically reliable) prototypes, replacement batches, customized products and biomedical implants.

Some major aspects are impairing a proper widespread and dissemination of the use of this technique in industrial environment. The geometrical accuracy of formed parts is still not adequate for the requirements of the parts to be produced. In addition, adapted milling machines have low load capacities and are not able to form high strength materials or thick sheets. Third, purpose built machines are too expensive or exclusively academic. Nevertheless, more options are appearing in terms of hardware and parallel kinematics may be the answer to join cost effectiveness, flexibility and stiffness. The recent onset of optimization techniques and design of experiments may represent another step forward on the process development.

In this survey, a glance over the research carried out over the last years and also on the most important conclusions obtained so far, aims to point the directions and trends to follow to enhance SPIF competitiveness. From all parameters, toolpath is the most important, and easy to optimize, making crucial the development of efficient smart manufacturing techniques, joining CAD and CAM techniques and establishing a solid framework of integrated design, modeling and assessment for incremental forming techniques.

REFERENCES

Aerens, R., Eyckens, P., Van Bael, A., & Duflou, J. R. (2009). Force prediction for single point incremental forming deduced from experimental and FEM observations. *International Journal of Advanced Manufacturing Technology*. doi: doi:10.1007/s00170-009-2160-2

Allwood, J. M., Houghton, N. E., & Jackson, K. P. (2005). The design of an incremental forming machine. In *Proceedings of the 11th Conference on Sheet Metal*, Erlangen, (pp. 471-478).

Allwood, J. M., King, G., & Duflou, J. R. (2004). Structured search for applications of the incremental sheet forming process by product segmentation. *IMECH E Proceedings Part B. Journal of Engineering and Manufacture, 219*(B2), 239–244. doi:10.1243/095440505X8145

Ambrogio, G., Costantino, I., De Napoli, L., Filice, L., & Muzzupappa, M. (2004). Influence of some relevant process parameters on the dimensional accuracy in incremental forming: A numerical and experimental investigation. *Int. J. Mater. Process. Technol., 153–154*, 501–507. doi:10.1016/j.jmatprotec.2004.04.139

Ambrogio, G., De Napoli, L., Filice, L., Gagliardi, F., & Muzzupappa, M. (2005). Application of incremental forming process for high customized medical product manufacturing. *Journal of Materials Processing Technology, 162*, 156–162. doi:10.1016/j.jmatprotec.2005.02.148

Ambrogio, G., Filice, L., De Napoli, L., & Muzzupappa, M. (2003). Analysis of the influence of some process parameters on the dimensional accuracy in incremental forming by using a reverse engineering technique. In *Proceedings of the AED 2003 Conference*, Prague.

Ambrogio, G., Filice, L., & Micari, F. De Napol,i L., & Muzzupappa, M. (2006) Some considerations on the precision of incrementally formed double-curvature sheet components. In *Proceedings of the ESAFORM Conference*, Glasgow, UK.

Ambrogio, G., Filice, L., & Micari, F. (2006). A force measuring based strategy for failure prevention in incremental forming. *Journal of Materials Processing Technology, 177*, 413–416. doi:10.1016/j.jmatprotec.2006.04.076

Amino, H., Lu, Y., Maki, T., Osawa, S., & Fukuda, K. (2002). Dieless NC forming, prototype of automotive service parts. In *Proceedings of the 2nd International Conference on Rapid Prototyping and Manufacturing*, Beijing 2002.

Azaouzi, M., & Lebaal, N. (2012). Tool path optimization for single point incremental sheet forming using response surface method. *Simulation Modelling Practice and Theory, 24*, 49–58. doi:10.1016/j.simpat.2012.01.008

Bambach, M., Hirt, G., & Ames, J. (2005). Quantitative validation of FEM simulations for incremental sheet forming using optical deformation measurement. *Advanced Materials Research, 6-8*, 509–516. doi:10.4028/www.scientific.net/AMR.6-8.509

Behera, A., Vanhove, H., Lauwers, B., & Duflou, J. (2011). Accuracy improvement in single point incremental forming through systematic study of feature interactions. *Key Engineering Materials, 473*, 881–888. doi:10.4028/www.scientific.net/KEM.473.881

Bouffioux, C., Eyckens, P., Henrard, C., Aerens, R., Van Bael, A., Sol, H., et al. (2007). Identification of material parameters to predict single point incremental forming forces. In *Proceedings of IDDRG Conference*, Gyor, 2007.

Callegari, M., Amodio, D., Ceretti, E., & Giardini, C. (2006). *Industrial robotics: Programming, simulation and applications*. Germany: Pro Literatur Verlag.

Centeno, G., Silva, M. B., Cristino, V. A. M., Vallellano, C., & Martins, P. A. F. (2012). Hole-flanging by incremental sheet forming. *International Journal of Machine Tools & Manufacture, 59*, 46–54. doi:10.1016/j.ijmachtools.2012.03.007

Ceretti, E., Giardini, C., & Attanasio, A. (2004). Experimental and simulative results in sheet incremental forming on CNC machines. *Journal of Materials Processing Technology, 152*, 176–184. doi:10.1016/j.jmatprotec.2004.03.024

Chen, J. J., Li, M. Z., Liu, W., & Wang, C. T. (2005). Sectional multipoint forming technology for large-size sheet metal. *J. Adv. Manuf. Technol., 25*, 935–939. doi:10.1007/s00170-003-1924-3

Decultot, N., Velay, V., Robert, L., Bernhart, G., & Massoni, E. (2008). *Behaviour modeling of aluminium alloy sheet for single point incremental forming*. Paper presented at the 11th Esaform Conference on Material Forming, Lyon.

Dejardin, S., Thibaud, S., & Gelin, J. C. (2009). Experimental investigations and numerical analysis for improving knowledge of incremental sheet forming process for sheet metal parts. *Journal of Materials Processing Technology*. doi:doi:10.1016/j.jmatprotec.2009.09.025

Duflou, J., Kellens, K., & Dewulf, W. (2011). Unit process impact assessment for discrete part manufacturing: A state of the art. *CIRP Journal of Manufacturing Science and Technology, 4*(2), 129–135. doi:10.1016/j.cirpj.2011.01.008

Duflou, J., Tuncol, Y., Szekeres, A., & Vanherck, P. (2007). Experimental study on force measurements for single point incremental forming. *Journal of Materials Processing Technology, 189*, 65–72. doi:10.1016/j.jmatprotec.2007.01.005

Duflou, J., Verbert, B., Belkassem, J., Gu, Sol, H., Henrard, C., & Habraken, A. M. (2008). Process window enhancement for single point incremental forming through multi-step toolpaths. *CIRP Annals Manufacturing Technology, 57*, 253-256.

Duflou, J. R., Lauwers, B., & Verbert, J. (2005b). Medical application of single point incremental forming: Cranial plate manufacturing. In *Proceedings of the 2005 VRAP Conference,* Leiria, Portugal. (pp. 161-164).

Duflou, J. R., Sol, H., Van Bael, A., & Habraken, A. M. (2003). *Description of the SeMPeR projet (Sheet Metal oriented Prototyping and Rapid manufacturing).* SBO-project financed by the IWT institute, 2003.

Duflou, J. R., Szekeres, A., & VanHerck, A. (2005a). force measurements for single point incremental forming and experimental study. *Journal of Advanced Materials Research, 6-8*, 441–448. doi:10.4028/www.scientific.net/AMR.6-8.441

Durante, M., Formisano, A., Langella, A., & Minutolo, F. (2008). The influence of tool rotation on an incremental forming process. *Journal of Materials Processing Technology*, 2008.

Emmens, W. C., & van dn Boogaard, A. H. (2008). Tensile tests with bending: A mechanism for incremental forming. In *Proceedings of the 11th ESAFORM conference on material forming,* Lyon, France.

Filice, L., Ambrogio, G., & Micari, F. (2006). *On-line control of single point incremental forming operations through punch force monitoring. CIRP Annals - Manufacturing Technology, 55*(1), 245–248.

Filice, L., Fratini, L., & Micari, F. (2002). Analysis of material formability in incremental forming. *Annals of the CIRP, 51*, 199–202. doi:10.1016/S0007-8506(07)61499-1

Franzen, V., Kwiatkowski, L., Martins, P. A. F., & Tekkaya, A. E. (2009). Single point incremental forming of PVC. *Journal of Materials Processing Technology, 209*, 462–469. doi:10.1016/j.jmatprotec.2008.02.013

Fratini, L., Ambrogio, G., Di Lorenzo, R., Filice, L., & Micari, F. (2004). Influence of mechanical properties of the sheet material on formability in single point incremental forming. *Journal of Materials Processing Technology, 153–154*, 501–507.

Guzmán, C., Gu, J., Duflou, J., Vanhove, H., Flores, P. A., & Habraken, A. M. (2012). Study of the geometrical inaccuracy on a SPIF two-slope pyramid by finite element simulations. *International Journal of Solids and Structures, 49*, 3594–3604. doi:10.1016/j.ijsolstr.2012.07.016

Ham, M., & Jeswiet, J. (2007). Forming limit curves in single point incremental forming. *Annals of the CIRP, 56*.

Henrard, C., Bouffioux, C., Eyckens, P., Soly, H., Duflou, J., & Van Houtte, P. et al. (2011). Forming forces in single point incremental forming: Prediction by finite element simulations, validation and sensitivity. *Computational Mechanics, 47*(5), 573–590. doi:10.1007/s00466-010-0563-4

Hussain, G., & Gao, L. (2007). A novel method to test the thinning limits of sheet metals in negative incremental forming. *International Journal of Machine Tools & Manufacture, 47*, 419–435. doi:10.1016/j.ijmachtools.2006.06.015

Hussain, G., Gao, L., Hayat, N., Cui, Z., Pang, Y. C., & Dar, N. U. (2008). Tool and lubrication for negative incremental forming of a commercially pure titanium sheet. *Journal of Materials Processing Technology, 203*, 193–201. doi:10.1016/j.jmatprotec.2007.10.043

Iseki, H. (2001). Flexible and incremental bulging of sheet metal using high-speed water jet. *JSME International Journal, Series C, 44*(2), 486–493. doi:10.1299/jsmec.44.486

Iseki, H., & Kumon, H. (1994). Forming limit of incremental sheet metal stretch forming using spherical rollers. *J. JSTP, 35*, 1336.

Jackson, K., & Allwood, J. (2009). The mechanics of incremental sheet forming. *Journal of Materials Processing Technology, 209*, 1158–1174. doi:10.1016/j.jmatprotec.2008.03.025

Jackson, K. P., Allwood, J. M., & Landert, M. (2008). Incremental forming of sandwich panels. *Journal of Materials Processing Technology, 204*, 290–303. doi:10.1016/j.jmatprotec.2007.11.117

Jeswiet, J., Micari, F., Hirt, G., Bramley, A., Duflou, J., & Allwood, J. (2005). Asymmetric single point incremental forming of sheet metal. *CIRP Annals Manufacturing Technology, 54*(2), 88–114. doi:10.1016/S0007-8506(07)60021-3

Ji, Y. H., & Park, J. J. (2008). Formability of magnesium AZ31 sheet in the incremental forming at warm temperature. *Journal of Materials Processing Technology, 201*, 254–358. doi:10.1016/j.jmatprotec.2007.11.206

Kim, T. J., & Yang, D. Y. (2000). Improvement of formability for the incremental sheet metal forming process. *International Journal of Mechanical Sciences, 42*(7), 1271–1286. doi:10.1016/S0020-7403(99)00047-8

Kim, Y. H., & Park, J. J. (2002). Effect of process parameters on formability in incremental forming of sheet metal. *Journal of Materials Processing Technology*, 130–131.

Kopac, J., & Kampus, Z. (2005). Incremental sheet metal forming on CNC milling machine-tool. *Journal of Materials Processing Technology*, 622–628. doi:10.1016/j.jmatprotec.2005.02.160

Lamminen, L., Tuominen, T., & Kivivuori, S. (2005). Incremental sheet forming with and industrial robot – Forming limits and their effects on component design. In *Proceedings of 3rd International Conference on Advanced Materials Processing,* Finland, (pp. 331).

Le, V. S., Ghiotti, A., & Lucchetta, G. (2008). *Preliminary studies on single point incremental forming for thermoplastic materials.* Paper presented at the 11th ESAFORM 2008 Conference on Material Forming, Lyon, France.

Leszak, E. (1967). *Apparatus and process for incremental dieless* forming (Patent US3342051A1, published 1967-09-19).

Manco, L., Filice, L., & Ambrogio, G. (2009). Analysis of the thickness distribution varying tool trajectory in single-point incremental forming. *Proc. IMechE Vol. 224 Part B: J. Engineering Manufacture.* DOI: 10.1177/09544054JEM1958

Marabuto, S. R., Afonso, D., Ferreira, J. A. F., Melo, R. Q., Martins, M., & Alves de Sousa, R. J. (2011). Finding the best machine for SPIF operations. A brief discussion. *Key Engineering Materials, 473*, 861–868. doi:10.4028/www.scientific.net/KEM.473.861

Meier, H., Buff, B., Laurischkat, R., & Smukala, V. (2009). Increasing the part accuracy in dieless robot-based incremental sheet metal forming. *CIRP Annals – Manufacturing Technology, 58.* Doi:10.1016/j.cirp.2009.03.056

Meier, H., Dewald, O., & Zhang, J. (2005). Development of a robot-based sheet metal forming process. *Steel Research International, 76*(2-3), 167–170.

Micari, F., Ambrogio, G., & Filipe, L. (2007). Shape and dimensional accuracy in single point incremental forming: State of the art and future trends. *Journal of Materials Processing Technology, 191*, 390–395. doi:10.1016/j.jmatprotec.2007.03.066

Rauch, M., Hascoet, J. Y., Hamann, J. C., & Plennel, Y. (2009). Tool path programming optimization for incremental sheet forming applications. *Computer Aided Design*. doi:10.1016/j.cad.2009.06.006

Schafer, T., & Schraft, R. D. (2004). *Incremental sheet forming by industrial robots using a hammering tool*. Paper presented at the 10th European Forum on Rapid Prototyping, Association Française de Prototypage Rapid.

Sena, J. I. V., Alves de Sousa, R. J., & Valente, R. A. F. (2011). On the use of EAS solid-shell formulations in the numerical simulation of incremental forming processes. *Engineering Computations, 28*, 287–313. doi:10.1108/02644401111118150

Shim, M., & Park, J. (2001). The formability of aluminum sheet in incremental forming. *Journal of Materials Processing Technology, 113*, 654–658. doi:10.1016/S0924-0136(01)00679-3

Shim, M. S., & Park, J. J. (2001). Deformation characteristics in sheet metal forming with small ball. *Trans. Mater. Process. J. JSTP, 113*, 654. doi:10.1016/S0924-0136(01)00679-3

Skjoedt, M., Silva, M. B., Martins, P. A. F., & Bay, N. (2010). Strategies and limits in multi-stage single-point incremental forming. *The Journal of Strain Analysis for Engineering Design, 45*(1), 33–44. doi:10.1243/03093247JSA574

Tanaka, S., Nakamura, T., & Hayakawa, K. (1999). Incremental sheet metal forming using elastic tools. In *Proceeding of the Sixth International Conference of Technology of Plasticity*, Nuremberg, (pp. 1477–1482).

Tanaka, S., Nakamura, T., Hayakawa, K., Nakamura, H., & Motomura, K. (2005). Incremental sheet metal forming process for pure titanium denture plate. In *Proceedings of the 8th International Conference on Technology of Plasticity* (pp. 135-136).

Verbert, J., Belkassem, B., Henrard, C., Habraken, A. M., Gu, J., & Sol, H. et al. (2008). Multi-Step toolpath approach to overcome forming limitations in single point incremental forming. *International Journal of Material Forming, 1*, 1203–1206. doi:10.1007/s12289-008-0157-2

KEY TERMS AND DEFINITIONS

CAD/CAM: Computer Aided Design and Manufacturing. Comprehends the entire process of modeling a part in a computer program and the manufacturing in a CNC machine.

CNC: Computer Numerically Controlled. Machine tools that operate controlled by a sequence of instructions given in ISO or G-format.

SPIF: Single Point Incremental Forming. A die-less forming process where the part is formed by the action of a tool with spherical tip that performs successive contours to shape the desired geometry.

Chapter 8
Software Development Tools to Automate CAD/CAM Systems

N. A. Fountas
School of Pedagogical and Technological Education (ASPETE), Greece

A. A. Krimpenis
School of Pedagogical and Technological Education (ASPETE), Greece

N. M. Vaxevanidis
School of Pedagogical and Technological Education (ASPETE), Greece

ABSTRACT

In today's modern manufacturing, software automation is crucial element for leveraging novel methodologies and integrate various engineering software environments such Computer aided design (CAD), Computer aided process planning (CAPP), or Computer aided manufacturing (CAM) with programming modules with a common and a comprehensive interface; thus creating solutions to cope with repetitive tasks or allow argument passing for data exchange. This chapter discusses several approaches concerning engineering software automation and customization by employing programming methods. The main focus is given to design, process planning and manufacturing since these phases are of paramount importance when it comes to product lifecycle management. For this reason, case studies concerning software automation and problem definition for the aforementioned platforms are presented mentioning the benefits of programming when guided by successful computational thinking and problem mapping.

1. INTRODUCTION

Current manufacturing challenges impose the need for innovative solutions to facilitate production process by minimizing time while simultaneously maintaining high quality. Modern industrial manufacturing systems involve CAD, CAPP, CAM and several other engineering software modules which provide flexibility during a product's lifecycle management. Due to the fact that software vendors intend to satisfy all aspects of industrial domains, they provide large suits of software options which may confuse end users and result to longer processing time.

Software automation allows the creation of programming entities capable of overcoming

DOI: 10.4018/978-1-4666-5836-3.ch008

existing limitations occur when applying generic solutions. By building complete automated modules using programming, software systems may be rapidly customized to either fit specific industrial tasks or collaborate with other applications so as to take advantage of the data exchange capability.

To capture such needs, engineering software vendors provide access to their software through the establishment of opened Application Program Interfaces (APIs). That is, users and software developers may create their own macros and add-in programs by taking advantage of available automation objects and routines through available programming languages such as C++ and Visual Basic and JavaScript.

Advancements of software automation solutions are employed mainly to product design, finite element analysis and process planning phases. Notable contributions include the development of programming functions, macro commands and scripts which perform automated design of machine elements such as cutting tool profiles (Ćuković, et al., 2010), work holding / fixturing devices (Farhan, et al., 2012) and product geometries along with their special features (Lamarche & Rivest, 2007 ; Wayzode & Wankhade, 2013; Wayzode & Tupkar, 2013). Wikström (2011) proposed a methodology for combining three dimensional geometry models of various fuel tank systems assembled in combat aircrafts. The methodology calculates fuel surface location and center of gravity and is based on the orientation of the fuel acceleration vector and the amount of fuel. Basic steps involve the automated volume analysis. For each orientation the fuel body is discretized to a predetermined number of volumes and numerical values for each volume, fuel surface location and center of gravity location are automatically recorder. On their efforts to predict surface errors when machining thin walled parts, R. Izamshah et al. (2011) needed to automate simulations for modeling solids, material removal processes and structural analysis operation, through the use of macros and Visual Basic. Reddy and Brioso (2011) developed a methodology where tedious and repetitive processes are automated while designing an industrial robot lower arm. Further on, their model is structurally evaluated in ANSYS®. Finite element analysis process is automated by the use of the programming languages Python and JavaScript. Furthermore a user interface is created using Microsoft Excel with Visual Basic.

Aiming at rapidly producing setup sheets for the CNC machining of industrial products research efforts have also been conducted to automate the process planning stage. Harik et al., (2008) developed an entire platform to facilitate process planning for aeronautical parts using the main application development framework namely *CAA-RADE* of *Dassault Systèmes*® CATIA®. *CAA-RADE* is a C++ based development environment native for CATIA® and provides a low-level access to most of CATIA's features allowing thus; the creation of embedded commands, toolbars, dialogs etc. To overcome limitations found employing this development platform, Fountas (2008) moved towards the automated process planning by creating support functions using Visual Basic as the main programming application. Jeba Singh and Jebaraj (2005) presented an automatic environment within a commercial CAD/CAM software to generate optimal process plans for determined objectives with respect to factory environment for modeled components. In their work, data exchange is achieved between a feature-based design system and spreadsheet software applications, through Visual Basic programming. Ispas et al., (2007) applied C++ programming application to automate specific utilities for the formulation of a solution for improving the machining precision by the machine tool's error minimization in terms of geometrical errors, kinematic errors, static/dynamic load errors, thermal, etc. Deb et al., (2011), developed a methodology of machining operations selection using Artificial Neural Networks (ANNs). Process planning is conducted by a modular expert system fully integrated with an automated data extraction system which obtains data from CAD to feed process planning modules in a fully automated environment. Krimpenis (2008) conducted an extensive research in the area of machining optimization using Artificial Intelligence (AI). In his work, a special programming module was developed to automate tasks needed for argument passing and parameter evaluations among a CAM system and a Genetic Algorithm (GA). Remaining volume, machining time and material uniformity left for finishing were treated as quality objectives for a rough machining optimization problem.

2. SOFTWARE DEVELOPMENT

In general, software development is the process where work practices, computer tools and programming techniques are employed to develop software. Software development involves computer programming. A typical software should meet demands such as friendly user interface, low cost usage, low processing time and low numbers of defects. Various programming elements are usually met when trying to build software. These are the classes, the modules, the methods, etc, whilst each of these elements may either stand alone under a proper graphical user interface (GUI) to formulate software, or being existed to a unique project thus; supporting operations and programming functions required.

Classes found in a project operate as templates for programming objects; whereas modules provide the ability to organize and type coding functions to be later called and execute certain commands. Methods allow the customization of programming objects used within a code through the change of their behavior. That is, an object may be utilized under different aspects of its contribution to a computer program; thus resulting to different outputs depending on the method of which is programmed.

Programming elements mentioned above, are provided in typical platforms and applications used for the development of specialized software codes. Apart from these applications, several other technologies are existed to facilitate software development such as macro programming, scripting technology and development of dynamic link libraries (dlls). The following subsection describes these program tools.

2.1. Program Development Applications

Program Development Applications (PDEs) provide entire suites of programming environments regarding the language used. Programming languages are artificial languages designed to allow argument passing to machines or computers and create programs that will precisely express algorithms. Most known PDEs are Visual Basic, Visual C, C++, Java, etc. Technical computing applications are also provide tools for program development. Such applications are *MathWorks® MATLAB®* and *WOLFRAM Research® Mathematica®*.

2.2. Special Program Development Frameworks

Special program development frameworks allow the establishment of embedded commands, toolbars, dialogs, etc, for an application. Such frameworks incorporate interoperating objects with built-in capabilities delivered as an entire resource to client applications. Program development frameworks may share interfaces as long as they declared as public; whilst private interfaces are only operated by the framework from once they have been built. As a representative example of a special program development framework, *Dassault Systèmes® CAA-RADE®* is mentioned.

2.3. Macro Programming

Macros are programs written in code and are used to automate repetitive tasks. They are frequently used throughout the world of engineering. Typical and commercially available CAM software packages available in the market offer several options to formulate macros written mainly in VB Script language. Thus, macros are typically created by a list or a sequence of several scripts. Apart from their object automation capabilities, macros are used to save computational time and reduce the possibility of human errors by automating manual processes. Further on, macros can improve programming efficiency, standardization, expanding abilities and facilitation of streamlining procedures.

2.4. Scripting Technology

A script is a sequence of programming instructions interpreted or executed by another program. Scripts may be developed using several programming languages like VB-VBA, Java, C++, etc. Moreover, should a software application exposes its objects for automation, sequential tasks which run manually may be recorded as macro commands and then stored in the form of scripts. Hence, operations performed manually by users can be written in a simple script able to be executed whenever its name is entered as a command by a programming module. Scripting is in general easier and faster to code in than the more structured compiled languages mentioned above.

2.5. Dynamic Link Libraries (dlls)

Dynamic link libraries (dlls) contain resources and codes which can be executed by several software applications. Since dlls are application modules; they can execute specific commands or functions required. When it comes to engineering software, dlls provide the ability to compile outputs generated by a specialized software (i.e. CAD) and called from a computing environment such as MATLAB® to analyze data.

2.6. Internal and External Software Development

Two programming modes one may adopt to develop software; internal and external programming. Internal programming mode provides the ability to take advantage of an application's properties (objects, methods and classes) in order to develop code, customize a function or build an "in-house" application. Major prerequisite to develop code "in-house" is the openness of architecture in terms of a specific application's program interface (API). Open API is a trend that most of engineering software vendors are employ due to the benefit of collaborative services through system integration. Major shortcoming of this approach, is that limitations occur owing to a specific language an application may utilize; that is, the language to be adopted for program development should be the very same.

External programming is performed using common programming application environments such as the aforementioned ones. External developing software for engineering applications imposes the need of communication support; that is, to activate an application's type libraries so as to declare objects, methods and properties involved.

3. PROGRAMMING FOR ENGINEERING SOFTWARE CUSTOMIZATION

3.1. Research Goal

As fundamental components of product lifecycle, design; process planning and manufacturing may require time consuming trial and error experimentations to support production and reach the desired status. On the research efforts to shorten processing data time while expanding engineering software capabilities, this work aims at contributing through presenting special programming functions developed to automate various tasks of product lifecycle management. This section presents novel programming strategies to automate and customize parts of engineering software from conceptual design to manufacturing. Through the presentation of case studies fall mainly in the areas of computer aided design (CAD), computer aided process planning (CAPP) and computer aided manufacturing (CAM), complete guidelines and basics are provided to establish well integrated and fast responded programs capable of handling engineering data and executing application commands. In first, automatic computer aided design is presented by applying a numerical method to automatically compute mathematical formulas and extract data for 3D CAD modeling. Computer aided process planning (CAPP) follows next with a case-oriented and semi-automated application to support manufacturing setup preparations. As far as computer aided manufacturing (CAM) is concerned, a set of programming modules capable of automatically set machining parameters subjected to specific quality characteristics; is presented. Finally, an application that manages to produce efficient and accurate CNC code for 3 axis sculptured surface machining is demonstrated.

3.2. Automated CAD

It is a common industrial task that CAD models should be created by groups of point coordinates especially when it comes to curved (free form) surfaces in which their representations are based on advanced mathematical equations. Designs of airfoils, turbine blades, or centrifugal impellers are representative cases for such tasks.

Should computer aided design phase fall under one of the above mentioned cases; numerical methods are applied to first solve a given mathematical equation to obtain the necessary number of points for curve interpolation and then implement programming techniques to call a CAD platform's routines for producing the model. This section aims to represent the methodologies applied to automatically produce CAD models with the

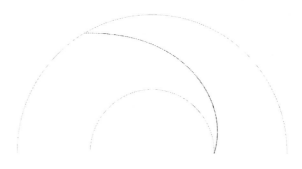

Figure 1. The curve approximated by the circular interpolation approach with the predefined chord error

aid of technical computing software and typical CAD systems.

3.2.1. Numerical Program Development for Automated CAD Data Generation

A mathematical program is developed to facilitate the computation and generation of XY point coordinates from which a centrifugal impeller's blades can be designed. The program numerically solves the respective mathematical formula which is actually a common differential equation widely applied to such components (Chorin & Marsden, 1979). The number of necessary points needed to interpolate the curve is produced under a predetermined chordal deviation from the ideal curve and are then approximated by adopting circular interpolation. Thereby, small circular segments pass through each point to sequentially connect them and formulate the impeller's blade center line (see Figure 1)

A curve in its general polar form can be written as

$$\gamma(r) = [(cos\varphi(r), sin\varphi(r)] \quad (1)$$

with

$$\gamma(r) \in \Re^2, r \in [r_1, r_2].$$

A centrifugal impeller blade can be described by applying the Simpson's rule in polar form, as:

$$d\varphi = \frac{dr}{r\tan(\beta(r))} \Rightarrow \varphi(r) = \frac{180}{\pi}\left(\int_{r_1}^{r_2}\frac{1}{r\tan(\beta(r))}\right) \quad (2)$$

where '$\beta(r)$' is a linear distribution of the outer angle of fluid flow 'β' and it is given by:

$$\beta(r) = \beta_1 + (r_2 - r_1)\frac{\beta_2 - \beta_1}{r_2 - r_1} \quad (3)$$

where $\beta_1, \beta_2, r_1, r_2$ are technical configurations for a given impeller model (Bacharoudis et al., 2008). The polar angle $\varphi(r)$ is

$$\varphi(r) = \int \frac{1}{r\tan(\beta(r))}dr \quad (4)$$

with $\varphi(r)$ satisfying the differential equation

$$\varphi'(r) = \frac{1}{r\tan\beta(r)} \quad (5)$$

To simplify programming issues the function $Q(r)$ is determined such that, $\sin\beta(r) = Q(r)$ and is:

$$Q(r) = Arc\sin\frac{w(r)}{c(r)} + \frac{(r - r_1)}{r} \quad (6)$$

The next step is to define on $\gamma(r)$ curve, a point identified as:

$$\gamma(r_0) = [x(r_0), y(r_0)] \quad (7)$$

In this point, the vectors $T(r)$ and $N(r)$ can be created to form an orthogonal coordinate system (see Figure 2). Thereby, the curvature circle

Figure 2. The curve approximated by the circular interpolation approach with the predefined chord error

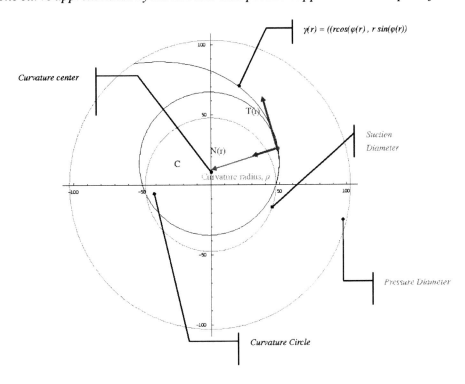

approaching $\gamma(r_0)$ curve, has the curvature center of:

$$\vec{c} = \vec{\gamma}(r_0) + \rho \vec{N}(r_0) \qquad (8)$$

where, $\rho = 1/k$ the curvature radius and k, the curvature.

The determination of the curvature circle can be done by calculating k term of $\vec{\gamma}(r)$ on the point $\vec{\gamma}(r_0)$. According to differential geometry it is determined that,

$$T(r) = \frac{\gamma'(t)}{u(r)} \qquad (9)$$

where $u(r) = |\gamma'(r)|$, and $\rho = 1/k$ as it is already been mentioned.

In order to calculate the curvature which is computationally costly, the procedure can be continued by calculating the fraction N/k. Utilizing and applying the chain rule we obtain,

$$\overline{T}'(r) = ku\overline{N} \mathbin{P} |\overline{T}'(r)| = ku| \qquad (10)$$

because $k(r) = \dfrac{dt}{ds} = \dfrac{dt}{dr}\dfrac{dr}{ds} = T'(r)\dfrac{1}{u(r)}$, according to "chain rule". Further on, Equation (10) is then written as,

$$u\overline{T}'(r) = ku^2 \overline{N} \qquad (11)$$

The calculus is carried on by dividing with "$|\overline{T}'(r)|^2$" so, Equation (11) becomes:

$$u\frac{\overline{T}'(r)}{|\overline{T}'(r)|^2} = \frac{ku^2 \overline{N}}{|\overline{T}'(r)|^2} \mathbin{P} u\frac{\overline{T}'(r)}{|\overline{T}'(r)|^2} = \frac{ku^2 \overline{N}}{k^2 u^2} \rightarrow \left[u\frac{\overline{T}'(r)}{|\overline{T}'(r)|^2} = \frac{N}{k} \right] \qquad (12)$$

in order to obtain the following equation of the curvature circle,

$$\vec{c} = \vec{\gamma}(r_0) + u(r_0) \frac{\overline{T}'(r_0)}{|\overline{T}'(r_0)|^2} \vec{N}(r_0) \qquad (13)$$

The programming procedure is executed by determining an initial point, that is to say the $\gamma(r_1)$ point. Then, the curvature center coordinates can be easily obtained through equation (14) by using a commercial technical computing software as follows:

$$\left\{ \frac{Q(r)(\sqrt{1-Q(r)^2}\, y(r) + rx(r)Q(r))}{Q(r)^2 + rQ'(r)Q(r) - 1}, \frac{Q(r)(-\sqrt{1-Q(r)^2}\, x(r) + ry(r)Q(r))}{Q(r)^2 + rQ'(r)Q(r) - 1} \right\} \qquad (14)$$

Equation (14) reveals that curvature center coordinates are linear combination of $x(r)$ and $y(r)$. In order to obtain all the possible combinations of $x(r)$ and $y(r)$ the Equation (13) in matrix form is created.

$$\begin{bmatrix} \dfrac{rQ(r)Q_i(r)}{Q(r)^2 + rQ_i(r)Q(r) - 1} & \dfrac{Q(r)\sqrt{1-Q(r)^2}}{Q(r)^2 + rQ_i(r)Q(r) - 1} \\ -\dfrac{Q(r)\sqrt{1-Q(r)^2}}{Q(r)^2 + rQ_i(r)Q(r) - 1} & \dfrac{rQ(r)Q_i(r)}{Q(r)^2 + rQ_i(r)Q(r) - 1} \end{bmatrix} \qquad (15)$$

Having presented the strategy that calculates and generates the curvature circles and having specified the initial point $r=r_1$ and angle $\varphi=0$, the algorithm needs to perform an error analysis to determine the next point for another curvature circle. By using Equation (13) and circle's equation $x^2 + y^2 = r^2$, an equation system is calculated providing two solutions of point coordinates whilst the solution that provides a 'next-in-order' point to the respective x, y coordinates is the one kept. Polar angle $\varphi(r)$ of that point is then calculated as a derivative of the radius $\left(\dfrac{d\varphi(r)}{dr}\right)$. Finally the chord error between two successive approximated points on the curve is determined as:

$$Err_{ch} = \left|\dfrac{d\varphi(r)}{dr} - \varphi_{,,}(r)\right| \quad (16)$$

The tolerance specified for the calculations is 10^{-2} mm). Hence, the calculated error should occur $Err_{ch} = \left|\dfrac{d\varphi(r)}{dr} - \varphi_{,,}(r)\right| \leq 10^{-2} mm$. The error is a second-grade polynomial form; hence the maximum error can be expressed as:

$$Err_{ch} = \dfrac{3d}{(r_2 - r_1) r_2 Q(r_2)} \quad (17)$$

where 'd' is the value of 10^{-2} mm.

The total error was found to be, $Err_{Ch}^{Tot} = 0.0000100441 = 10^{-5}$m. In order to find the displacement on the curvature circle until the chord error reaches its maximum value, the non-linear equation

$$\dfrac{d\varphi(r)}{dr} - Err_{Ch}^{Tot} \leq r_2 \quad (18)$$

is solved utilizing '*Secand*' method. Further on, the equation system is recalculated to indicate the next circle position on the curve. Finally the circular arc approximation which describes the curve was concluded after 120 iterations and hence; 120 points were obtained. Figure 3(a) depicts the curve's approximation under the specified error whilst Figure 3(b) shows the group of points after the 120 iterations. The calculus presented was integrated to a computer program. The workflow diagram illustrated in Figure 4 depicts the process steps of this method.

3.2.2 Automated CAD Based on Generated Data

Having obtained the curve's points it is easy to import them into a CAD environment so as to generate the spline from which the solid model of the impeller's fin will be built. This would involve programming for inputting data from files, calling respective CAD workbenches to utilize tools for sketching and featuring. The sequential order of automated operations is illustrated in the workflow diagram depicted in Figure 5. To run this mod-

Figure 3. Graphical representation of the calculated chord error

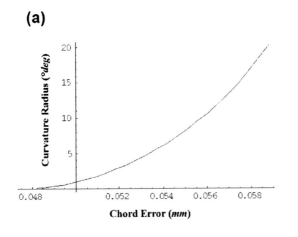

(a)

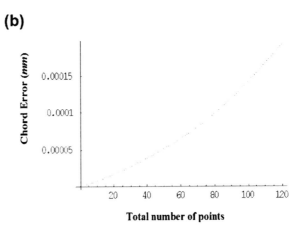

(b)

Figure 4. Flow chart of the circular approximation numerical programming method

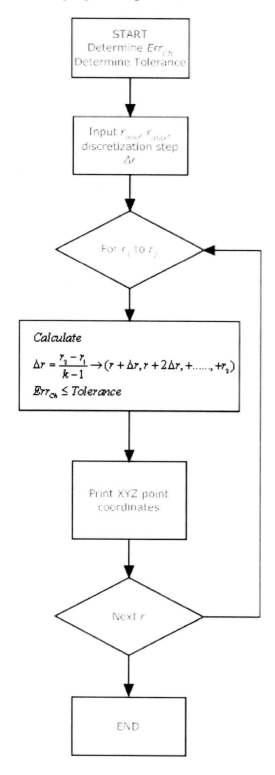

ule, a CAD system's database should be active as operations have been internally programmed.

The process steps followed to generate the program of which the workflow diagram is depicted in Figure 5 are enumerated below:

1. General consideration of the procedure to generate the shape
2. Use macro recording to get a sample code
3. Declarations for variables and sets of user inputs
4. Replace unnecessary code stings with new developed
5. Sample code modifications through Visual Basic statements
6. Develop new code to handle sketch and modeling workbench objects for design
7. Graphical user interface (GUI) preparation and proper coding

Figure 6a and Figure 6b illustrate parts of the code developed to declare proper variables for points and to execute loops of operations for the points and curve segments.

The GUI was formulated by adding a simple frame with its respective controls and is depicted in Figure 7.

The resulting outputs obtained from the CAD environment's host application are depicted in Figure 8. Figure 8a depicts the point locations regarding their coordinates. Based on special studies concerning centrifugal impellers respective configurations are adopted to further use feature tools for the modeling operation of the impeller blade (Figure 8b) and finally design the entire impeller model (Figure 8c).

Having verified the model produced through the employment of automated CAD strategies by programming, process planning is next to start considering manufacturing issues for production.

3.3 Automated CAPP

Despite the fast development and the continuous evolution of product design, analysis and manufacturing, an unlinked gap is appeared at the interface of computer-aided design (CAD) and computer-aided process planning (CAPP) interfaces. Various software systems have been developed to provide solutions to this problem by trying to generate reliable manufacturing instructions for shop-floors. To support the manufacturing trends and contribute to the research efforts for the realization of precise, reliable and efficient process plans, a set of programming functions are presented and implemented to a case study in the form of an object-oriented application with appropriate sequence of operations which may enable process planners to automatically produce instructions concerning machining processes.

3.3.1 Automation Entities for Process Planning

In general, process planning involves several entities to be determined such as stock selection, manufacturing resources, tooling and fixturing, etc. These entities usually demand an expert knowledge to various fields of manufacturing engineering. What really perplexes the problem of fully automated process planning approach is the fact that huge expertise should be incorporated to knowledge databases in the form of rules; increasing thus, processing time performed by computers. And yet, when it comes to the process planning of components with high specificity degree, knowledge rules cannot be efficiently implemented since parts fall into this category are one of a kind (i.e. aerospace parts).

It is thus; healthier to support process planning by programming special support functions which will be manipulated by human expertise. It is expected that such approach will be based on the proper handling of automation entities found to process planning. The automation entities to

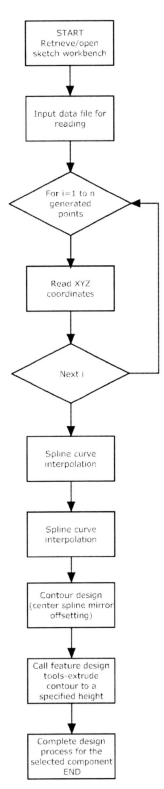

Figure 5. Workflow of automated operations performed in CAD environment

Figure 6. Code samples for CAD automation (a)Variable declarations/Loop for point importation; (b) Loop for number of curves and control points

```
hybridBodies1 As HybridBodies                                    (a)
hybridBodies1 = part1.HybridBodies

hybridBody1 As HybridBody
hybridBody1 = hybridBodies1.Add()
hybridBody1.Name = "Control_Points"


nC = 1 To numberOfCurve

Dim CurrSet As HybridBody
Set CurrSet = hybridBody1.HybridBodies.Add()
CurrSet.Name = "CurvePoints_" & NumStr(nC)

Dim pointX As Double
Dim pointY As Double
Dim pointZ As Double


For nP = 1 To numberOfPoint

    Dim hybridShapePointCoord1 As HybridShapePointCoord
    Set hybridShapePointCoord1 = hybridShapeFactory1.AddNewPointCoord(pointX, pointY, pointZ)

    CurrSet.AppendHybridShape hybridShapePointCoord1

    part1.InWorkObject = hybridShapePointCoord1

    part1.Update

Next


hybridShapePointCoord2 As HybridShapePointCoord
hybridShapePointCoord2 = hybridShapeFactory1.AddNewPointCoord(100#, 100#, 50#)
```

```
For nC = 1 To numberOfCurve                                      (b)
    For nP = 1 To numberOfControlPoint

Dim hybridShapeSpline1 As HybridShapeSpline
Set hybridShapeSpline1 = hybridShapeFactory1.AddNewSpline()

hybridShapeSpline1.SetSplineType 0

hybridShapeSpline1.SetClosing 0

Dim reference5 As Reference
Set reference5 = part1.CreateReferenceFromObject(hybridShapePointCoord1)

hybridShapeSpline1.AddPointWithConstraintExplicit reference5, Nothing, -1#, 1, Nothing, 0#

Dim reference6 As Reference
Set reference6 = part1.CreateReferenceFromObject(hybridShapePointCoord2)

hybridShapeSpline1.AddPointWithConstraintExplicit reference6, Nothing, -1#, 1, Nothing, 0#

hybridBody2.AppendHybridShape hybridShapeSpline1

part1.InWorkObject = hybridShapeSpline1

part1.Update

    Next nP
Next nC

Dim hybridShapeSpline2 As HybridShapeSpline
Set hybridShapeSpline2 = hybridShapeFactory1.AddNewSpline()
```

Software Development Tools to Automate CAD/CAM Systems

Figure 7. Graphical User Interface (GUI) for the automated CAD project

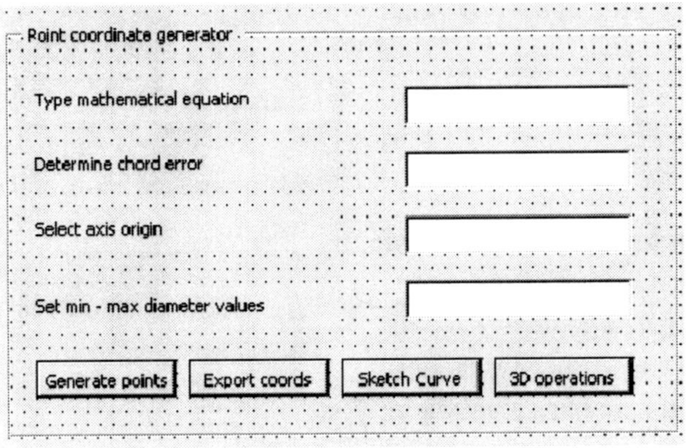

Figure 8. Resulting CAD outputs (a) Point importation relative to XYZ coordinates; (b) Impeller blade 3D model; (c) Impeller 3D CAD model

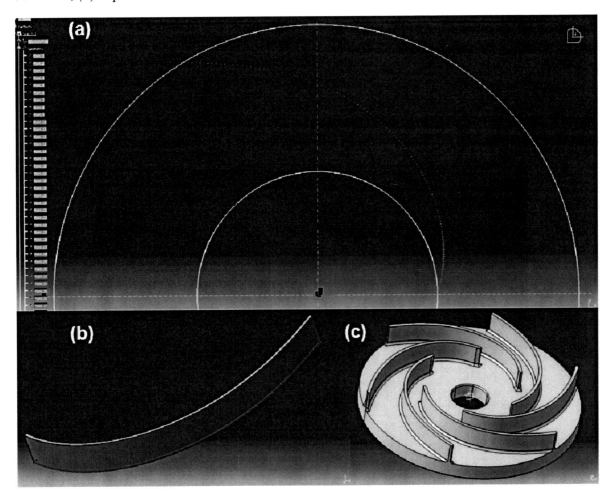

be presented in the following sub section are also the ones which integrate CAPP through programming techniques.

3.3.1.1 Stock Selection

The primary investigation a planner conducts is involved with the suitable selection of raw material form related to a final part's geometry and its nominal dimensions. The raw material is associated with the final part that is used as an input to the system (Figure 9a). Two cases are recognized for stock type determination. The former case deals with the specification of a simple boundary box which encloses the designed model regarding its maximum X, Y and Z dimensions Figure 9b). The latter case specifies a stock type which occurs by offsetting the designed model's features by a specific value thus; resulting to a stock type similar to the part's final geometry Figure 9c).

The type of the stock to be determined plays important role mainly to setup definition. Should the first case is implemented; common practices to prepare the setup and clamp it onto the machine tool table may be used. When it comes to the second case of stock preparation, more complex setup definitions have to be implemented. The importance of this function is the ability to evaluate available manufacturing resources mainly in terms of machine tools and propose the most suitable equipment for each case. That is, the parts' basic dimensions are taken into consideration so as to find the machine tool which provides the optimum workspace for machining operations. It is to mention that non-compatible machine tools in relation to a part's basic dimensions may increase idle movements hence result-

Figure 9. Stock types depending the part's geometry (a) Designed part of a NACA 0025 airfoil; (b) Stock created as bounding box; (c) Stock created by offsetting the part's basic dimensions

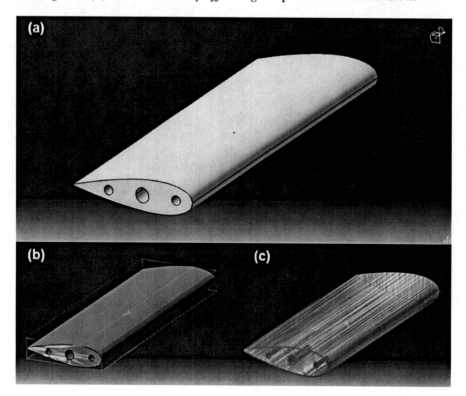

Software Development Tools to Automate CAD/CAM Systems

Figure 10. Part setup regarding the machine tool's table geometrical configurations (a) Typical manufacturing setup; (b) setup positioning on a square type machine tool table; (c) setup positioning on an orthogonal machine tool table type; (d) setup positioning on a semi circular machine tool table type; (e) setup positioning on a circular machine tool table type

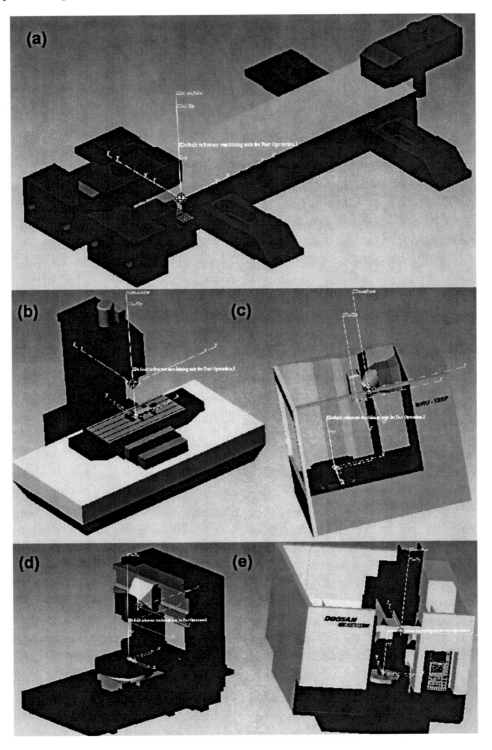

Figure 11. Association of end mill basic dimensions for machining (a) check on available cutting length; (b) check on available cutting diameter; (c) check on available cutting diameter and cutting length

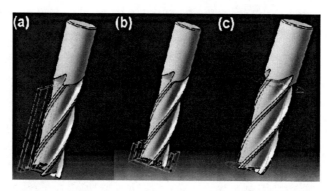

ing to longer non-cutting time (larger machine tool table) or yield tool accessibility constraints (smaller machine tool table). Through the geometrical type of parts, the function is capable of proposing the milling mode among 3 and 5 axis. This is achieved by developing code for the association of a specific CNC machine tool depending on the part's unique geometry. Common geometries of machine tools along with their associated tables found in industry are illustrated in Figure 10.

3.3.1.2 Cutting Tools Selection

Feature Recognition (FR); either manual or automated (Jones et al., 2006) is the main characteristic of this function. The programming module developed to support this planning stage, reads part documents and recognizes the machinable features of a part body. The code written is capable of recognizing all kinds of machinable features but special emphasis has been given to pockets and slots since these two machining feature categories are widely processed in aerospace industry. After proper retrieval of the features' basic dimensions (width, depth, etc), the module checks all tools enlisted to a database (*.xls or *.csv document) and associates a specific tool capable to machine a selected feature, or multiple features. Mainly referring to pockets, the depth is associated to a tool's cutting length (Figure 11a), whilst the width is associated to a tool's cutting diameter (Figure 11b and Figure 11c). Hence, these two basic dimensions are used to characterize the feasibility of a machining operation using a cutting tool and applied to features. At the end of the process, all tools proposed by the system are sequentially presented to the printed setup sheet.

3.3.1.3 Priority for Machinable Features

Machining and geometric interactions among features along with technological constraints for a part's machining operations generate precedence constraints. (Faheem et al., 1998). Precedence constraints, are of paramount importance when machining complex parts such the parts found in aerospace. When the sequence of machining operations for features of each setup is not well-prepared, overall manufacturing program may be not feasible at all ("Hard" constraints) or may result to longer cycle time ("Soft" constraints). Moreover, tool wear and tool/part deflections may also be faced if machining priority of features is not suitably considered. Definitions, classifications and illustrative examples can be found in the work presented by Li and Qiu (2006).

For this function the respective module developed, evaluates the features to be machined according a pre-programmed code that contributes to the process as a knowledge base system. The rules reflecting knowledge were developed through "Select Case" and "If Then" routines created by information mainly summoned from actual industrial strategies and shops' practices, general process planning techniques and summaries of process documents for manufacturing. For each constraint type, an indicative number has been assigned to reflect the yielded violation to a process. The machining priority number for a machinable feature is then determined by evaluating the indices (indicative numbers) of interaction types. Hence, a new ranking for the machinable features existed in a part's unique geometry is performed to finally define the order of machining operations to be executed.

3.3.1.4 Job Setup and Work Holding

Having defined the ranking in term of features machining priority, work holding comes next.

Figure 12. Part center of gravity related to job setup (a) Top positioning; (b) Vertically inclined positioning

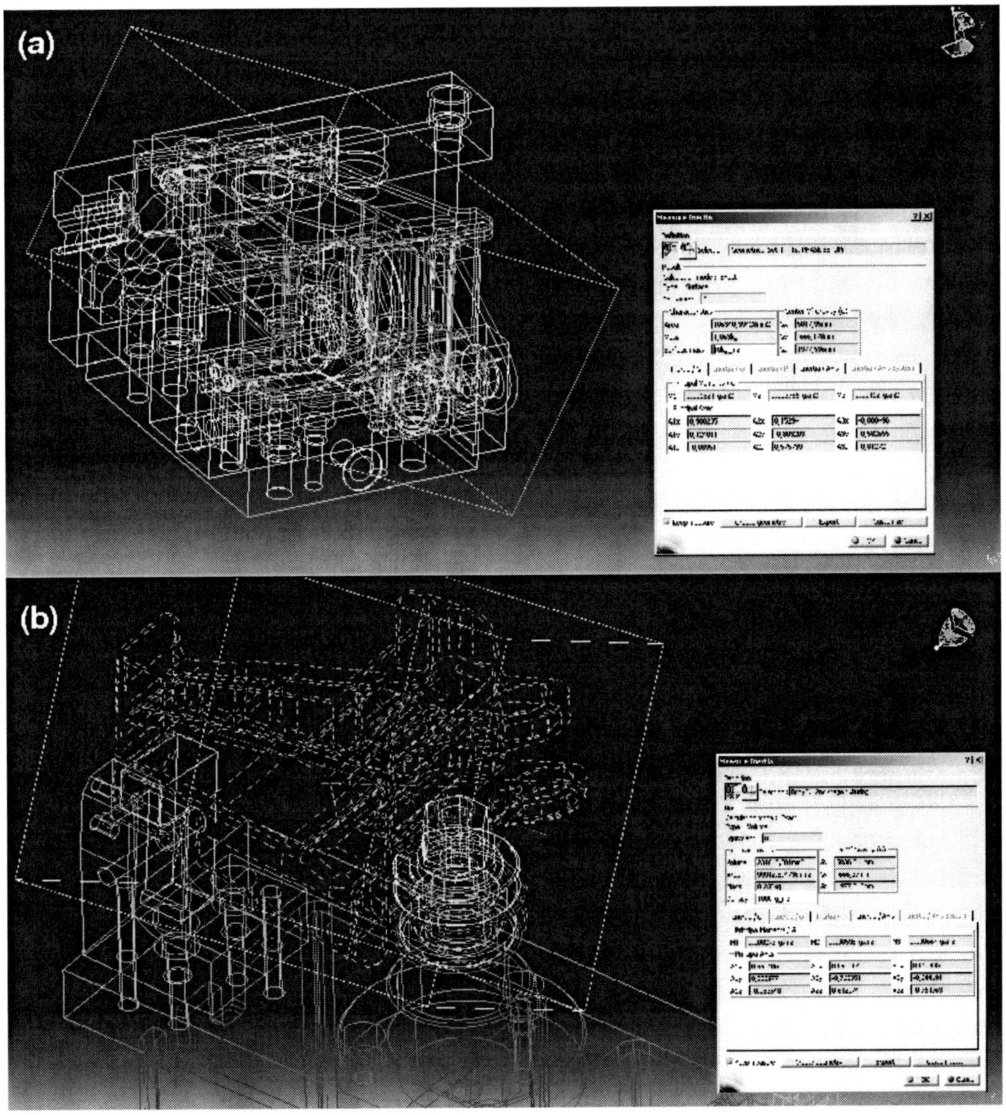

The function of part positioning deals with the entire assembly the planner prepares to support the stock onto the machine table for machining. Fixture design and stock support may vary from planner to planner, however; basic principles should always be considered (Stampfer, 2004). The module for this function programmatically retrieves the assembly document and analyzes it so as to check whether the setup conforms with certain technical specifications for fixture design or not. Should this check is not satisfactory the planner might consider to change a part of the overall support design concept, or re-design the assembly. Thus; the function has a consultative character.

The check on which the programmable function is based, lies heavily on the relationship between the stock's Center of Gravity position (CoG position) and the selected support base of the setup scenario. That should be the lowest possible in distance with reference to the machine tool's table to maintain stability and rigidity during machining operations. As an example of considering the setup definition with respect to the center of gravity position, Figure 12 illustrates two different setups for the same machining phase. In the first case (Figure 12a), the part's CoG position, is lower than the one computed for the second case (Figure 12b).

3.3.2 CAPP Integration through Programming

The CAPP support functions were developed within the CAD module of *Dassault Systèmes*® CATIA V5®. Internally hosted Visual Basic language was the programming development environment. To stress the validity and the practicality of the approach presented, the CAPP support functions were implemented to a typical aircraft part to generate its manufacturing process plan sheet. The workflow of the functions is described in the following subsections and depictions of user forms are explained.

3.3.2.1 Stock Selection Function

The programming module developed to select the raw material for a given part starts with the part document retrieval. Then, according to the part's geometry, nominal dimensions and the user's preferences, the work material is applied to the part model as a stock. Further on, NC machining resources are examined to trace the best relation among part's nominal dimensions and machine tool's table configurations. To do so, the code stays in a loop to evaluate all available NC machine tools. Once a machine tool is proposed, it's ID number stands by to be later printed to the process document. The workflow of this procedure is depicted in Figure 13a and its user interface in Figure 13b.

3.3.2.2 Cutting Tools Selection Function

Special tools and milling cutters are chosen regarding the machine tool to be used, the material's properties and available ranges of cutting conditions, and so on. More or less, computerized systems may provide proper tools regarding these entities; however additional information concerning the features to be machined is yet to be provided. The programming module developed for this operation, retrieves and applies manufacturing feature recognition (MFR) commands so as to ensure that proposed tools fit features' dimensions; hence, to provide feasible feature machining solutions. The code operates using two main loops; one to scan all of the machinable features a CAD model has and one to scan all tools existed in the database to suit a tool for each feature. In the end, a list of all tools about to be involved, are stored for printing to the setup sheet. The workflow of this procedure is depicted in Figure 14a and its user interface in Figure 14b.

3.3.2.3 Priority for Machinable Features Function

Having recognized manufacturing features, it is essential that their machining priority should be specified. Basically, priority for machinable

Software Development Tools to Automate CAD/CAM Systems

Figure 13. Stock selection function (a) workflow; (b) user interface

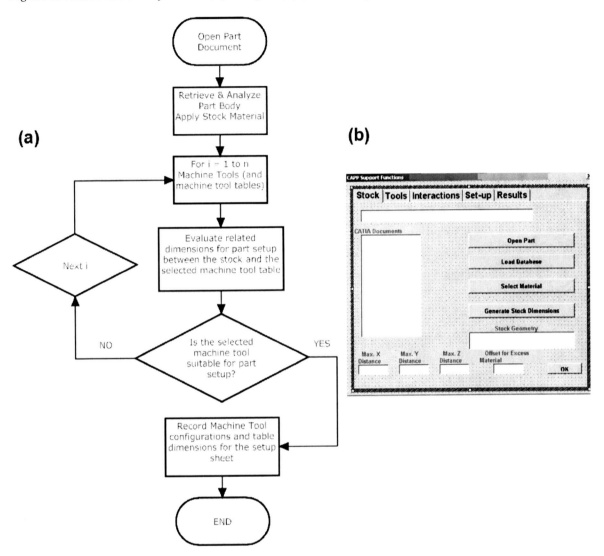

features aims at facilitating production through the minimization of pat setup number, idle time reduction, collision danger, etc. The part of code developed to perform this task, evaluates machining operations that will formulate the manufacturing program by assigning knowledge rules. These rules correspond to precedence constraints as it has already been mentioned. When the loop of evaluations is concluded; machining operations are ordered not only to generate the most convenient manufacturing program but also to prepare the part setup definition. The workflow of this function is depicted in Figure 15a; whereas, the user interface is illustrated in Figure 15b.

3.3.2.4 Job Setup and Work Holding Function

It is possible that a planner may prepare more than one setup definitions for the same operation in order to simulate and verify their resulting outputs using CAM software. As it has been mentioned, certain fixture design rules have been established in order to ensure proper work holding towards the achievement of a safe and productive manu-

Figure 14. Cutting tools selection function (a) workflow; (b) user interface

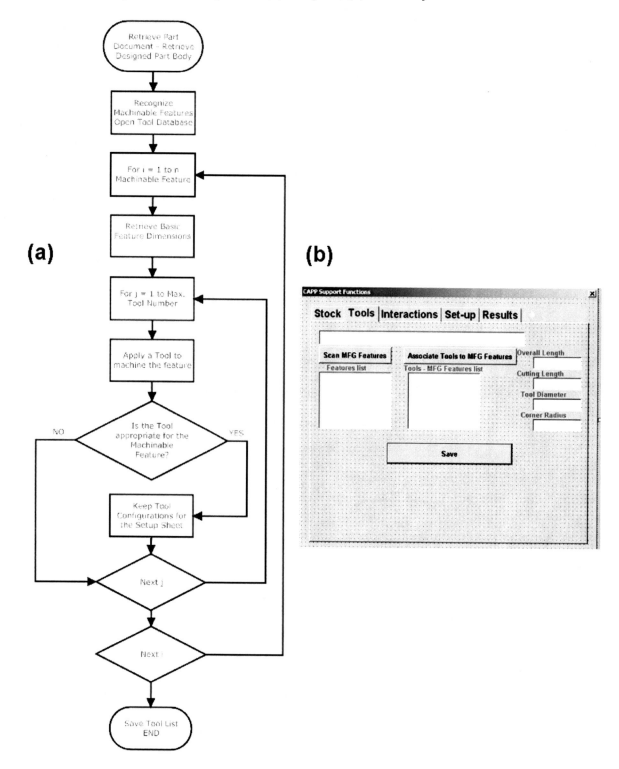

Software Development Tools to Automate CAD/CAM Systems

Figure 15. Priority for machinable features function (a) workflow; (b) user interface

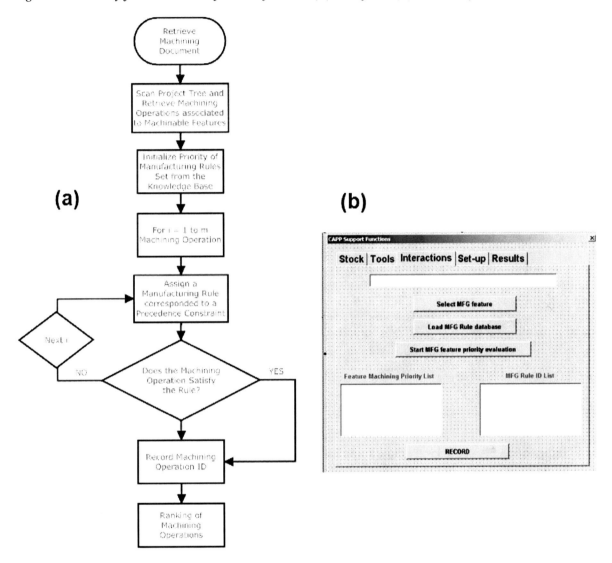

facturing process. Most important fixture design rules lie heavily on part's stability and rigidity when the latter is clamped on to the machine tool table. That is; its center of gravity position (CoG) with reference to the machine tool table should be the lowest possible to ground yielded cutting forces during the process. The code developed to undertake this issue evaluates each of the setups being proposed and then calculates the distance from the part's CoG position from its support base and/or the machine tool table. Thus; solutions are clustered to the best work holding scenario which will be the one to be adopted. The workflow depicted in Figure 16a represents the process to arrive at a satisfactory part setup and Figure 16b illustrates the user interface.

Results obtained from the operation of process planning support functions are stored in the setup sheet (process plan) which formulates the documentation for the shop floor. This setup sheet summons all information in a sequential order as the application executes the commands for the functions involved. The setup sheet contains:

Figure 16. Job setup and work holding function (a) workflow; (b) user interface

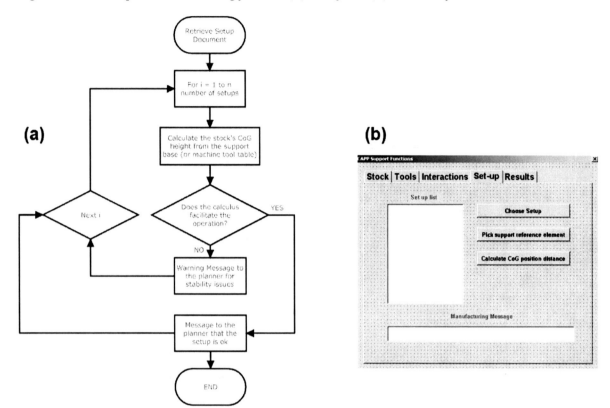

- Information about the kind of stock and its basic dimensions;
- Information about the machine tool to be used and its technological configurations;
- The cutting tools related to manufacturing features in a sequential order;
- The machining sequence of the features related to technological constraints and manufacturing rules; and,
- The setup, or the number of setups which are most convenient for the machining operation.

3.4 Automated CAM

Machining processes are specified using a special environment dedicated for this task. CAM software is such an environment. Specifications of machining processes involve machining parameter settings, cutting strategies selections, tool definitions etc. CAM software aims at calculating the tool path a cutting tool will follow to produce a part according the inputs of machining parameters. Machining simulation can be then done to check the feasibility of machining processes and obtain statistical data such as material removal rate (MRR), machining or cycle time, tool path length, etc.

This section presents programming modules built "in house" of a typical CAM system's platform to facilitate CNC programming and sculptured surface machining modeling stage, by automatically passing process parameter values to respective inputs for process evaluation purposes. The programming modules manage two major machining operations namely roughing and finishing (three and five axis).

3.4.1 Process Parameter Identifications for Roughing and Finishing

Machining parameters are adjusted regarding a machining phase's characterization. Roughing and finishing are two of the most important machining phases applied to manufacturing industries. While roughing aims at maintaining a high material removal rate and approximate final surface as close as possible, finishing undertakes to achieve a satisfied surface finish along with good geometrical – dimensional accuracy regarding design specifications. Prior to the programming development for automatic machining parameters specifications it is mandatory that roughing and finishing parameters are to be introduced. Note that process parameters are common for all commercially available CAM systems.

3.4.1.1 Process Parameters for Roughing

Process parameters for rough machining operations are as follows:

- **Axial Cutting Depth (Step Down):** The distance joining two successive passes in the Z-level. This parameter can be adjusted by determining the number of levels in Z direction considering the total depth of cut, or by calculating cutting passes through the tool height ratio, or by assigning specific arithmetic values for the cutting passes.
- **Radial Cutting Depth (Step Over):** The distance joining two successive passes in X-Y plane. The parameter can be adjusted by determining the total number of paths, specifying a cutting depth ratio related to the tool's diameter (%Ø), or by defining the maximum scallop height to be left between passes.
- **Feeds and Speeds:** Feed rate in (*mm/rev*) and Spindle speed in (*rpm*). The choice of spindle speed determines the cutting speed which is equivalent to the surface speed of the cutting tool. This not only depends on the spindle speed but also on the cutter diameter (the higher the spindle speed and the larger the cutter diameter, the higher the cutting speed). Feed is the movement of the milling cutter in the machining direction.
- **Cutting Tool:** Different types of cutting tools are existed regarding their dimensional and geometrical configurations. The tools are selected according the part's material properties and special features. In the particular case of roughing operations, flat-end mills are usually programmed to remove the material from the raw stock.

3.4.1.2 Process Parameters for Finishing

Process parameters related to finishing are distinguished regarding the machining technology adopted. Since finishing may be performed by adopting either 3; or 5 axis machining technology, process parameters are differently determined. The process parameters for finish machining depending on the technology adopted, are bulleted below:

- **3 axis finish machining** *Radial cutting depth (step over)* and *Feeds-speeds* are as described for roughing operation. *Cutting tool:* In 3 axis finishing, the tool axis is fixed; thus, possible degrees of freedom are restricted. Since these limitations are existed, tools having spherical geometries (ball end mills) are used to finish sculptured parts due to their advantage of moving over all surface regions and machining any point around the curvature.
- **5 axis finish machining** *Lead angle:* The angle between the tool's vertical axis and the machined surface. *Tilt angle:* The side angle between the tool's vertical axis and the machined surface. *Tool type:* As in 5 axis machining more freedom degrees are existed, all types of cutting tools can be im-

plemented. Hence; flat ended; ball ended and corner radius mills are used. *Radial cutting depth (step over)* and *Feeds-speeds* are as described for roughing operation.

3.4.2 Manufacturing Objectives for Quality Assessment

Towards their efforts of finding best settings among machining parameter values, NC programmers perform multiple machining simulations in the CAM environment in order to test various experimental scenarios and obtain important statistics referring to quality objectives such as machining time, idle time, material removal rate, etc. Such scenarios are formulated by trying several settings of process parameters and checking whether they suite industrial specifications or not.

Most common manufacturing objectives utilized to assess CNC machining operations in relation to their manufacturing phases (roughing and finishing) are enumerated and explained below:

- **Rough Machining Time:** The time needed to remove the material from a part. Its magnitude is automatically calculated on CNC machine controls and CAM systems after the tool path calculation. Through its calculus, it is easy to observe where productivity is decelerated and how it can be improved.
- **Remaining Volume:** The uncut material of a roughed or semi finished part that remains to be removed at the finishing stage. Lower remaining volume amount leads to quicker finishing operations, lower chip loads and cutting forces thus; drastically reducing overall machining time whilst improving part quality.
- **Finish Machining Time:** The time needed to reach a final part surface as close as possible to the ideally designed surface geometry.
- **Surface Deviation:** The maximum allowed deviation from the mean area between the finished and the nominal surface geometry computed as follows:

$$SD = \left| \frac{A_{FINISHED} - \overline{A}}{\overline{A}} \right| \qquad (19)$$

Where,

SD: The Surface deviation in (μm);
$A_{FINISHED}$: The Area of the finished model's surface;
$\overline{A}$: The mean area:

$$\overline{A} = \frac{A_{FINISHED} + A_{DESIGNED}}{2} \qquad (20)$$

3.4.3 Program Development for Automated CAM

Manual operations to set values for machining parameters and cutting conditions are both costly and time consuming. Therefore, CAM software customization needs to be done so as to automate the overall task and arrive at optimum scenarios, simulate their outputs and rapidly deliver manufacturing documentation to the shop floor. CAM software customization for such a task involves the automatic adjustment of machining parameters by taking into account their value ranges, technological constraints and quality specifications. As machining parameters should be set in such a way that quality characteristics are satisfied, loops of operations must be programmed in order to arrive at the optimum solution to the task. In addition technological characteristics and configurations should be specified to limit parameter selections to a feasible range for a process. Such entities play the role of constraints and selected elements should always be validated through them. Material properties, machining equipment capabilities (motor power, maximum feed rate, axes limits), quality objective value ranges (machining time, material removal rate, etc) are usually treated as constraints.

In order to achieve a proper conjugation among two software entities and ensure data exchange, additional code should be written for the functions to be performed. For a CAM environment such functions would be meant to automatically execute operations like process parameter settings, tool changes, machining simulations, importing/exporting statistical data and so on. Thereby, proper CAM software customization for automated tasks imposes the need of code development capable of performing the following operations:

1. Scanning of manufacturing documents to retrieve products, processes and resources;
2. Manipulation of objects and their properties to execute specific tasks for products, processes and resources;
3. Importation and extraction of computed data for evaluation through loops of operations.

The following macro workflow presented in Figure 17 illustrates the procedure executed to first scan a machining process document and access its parameters for the automatic changes to be performed. Having developed a code which workflow is as Figure 17 depicts, trial and error scenarios may be conducted by end users to perform simulations in the CAM environment so as to evaluate manufacturing outputs like the ones mentioned in sub section 3.4.2 depending on each set of process parameters.

Another important contribution of CAM automation can be the process of optimizing machining operations using Artificial Intelligent (AI) systems, i.e; Genetic or Evolutionary Algorithms (GAs-EAs). Under this prism, evaluations of several machining simulation scenarios are repetitively performed so as to assess each process parameter combination for the quality of its outputs, hence; optimizing an objective function referred to one; or more quality objectives subjected to optimization and depending on the problem is maximized or minimized. Should the case is a multi objective optimization then the objective function is formulated by determining a weighted sum of quality attributes along with their weight coefficients which specify the influential importance for each objective (Fountas et al., 2012).

Figure 18.a and Figure 18b illustrate the operations of manually obtaining the numerical results for a rough machined part's remaining volume in the form of scallops and a finished part's surface deviation among its machined surface and its ideally designed one.

It is obvious that manual operations for process simulations take enormous time to be executed since end users should determine each process parameter (spindle speed, feed rate, stepover, stepdown, etc) and then analyze resulting models to extract the outputs being of particular interest. These efforts become even more complex when more than one quality outputs are to be assessed. Should this is the case, automation of machining process simulations is almost inevitable. Figure 19a and Figure 19b depict the workflows of two macro commands developed to automate the overall procedure of collecting outputs resulted from machining process simulations so as to be further evaluated and finally trace optimal solutions. Specifically the workflow illustrated in Figure 19a refers to rough machining processes when aiming at automatically optimizing the combinatorial function specified among remaining volume and rough machining time, whilst, Figure 19a refers to finish machining processes when aiming at automatically optimizing the combinatorial function specified among surface deviation and finish machining time. Evaluations may be performed by a Genetic or an Evolutionary algorithm as it was previously mentioned.

3.5 Intelligent Post Processing

CNC machining of industrial products require lengthy G-codes due to the large number of tool

Figure 17. Macro programming workflow to automate process parameter settings and store results

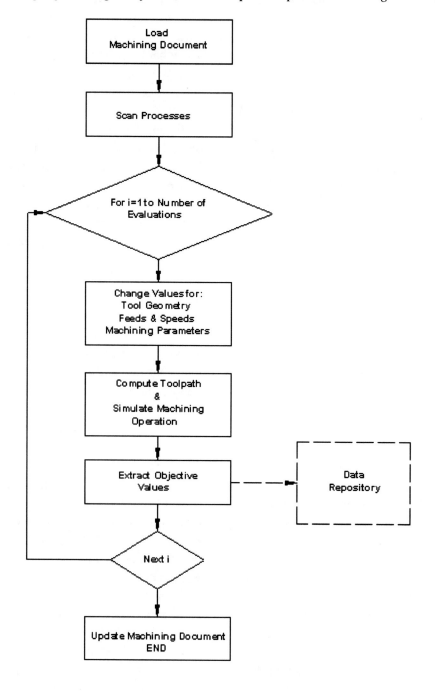

path points. This section will present a new post processor application able to generate standard ISO codes; based on the cutter location (CL) data which CAM systems generate for 3 axis milling operations. By employing the post processing application, resulting G code files occur shorter than those extracted from typical post processors found in most of commercial CAM systems, while at the same time they are produced quicker. The post processing application was developed with

Software Development Tools to Automate CAD/CAM Systems

Figure 18. Manual analysis for machined parts in a CAM environment (a) analysis to obtain remaining volume after a roughing operation; (b) analysis to obtain surface deviation after finishing operation

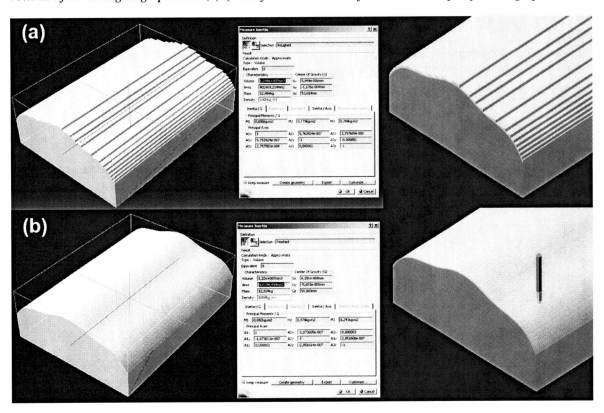

the use of Visual Basic for Applications and tested in the G code production mainly for sculptured parts. It was indicated that the new post processing application is capable of rapidly generating NC programs with both efficiency and precision. Verification experiments were conducted for the system's evaluation followed by their respective results.

3.5.1 Program Development for Automated Post Processing Software

A typical post processing application should suitably handle the variety of functions that facilitate the achievement of better quality and more efficient machining operations. Such functions are for example the warnings required by the initial movement of the tool before cutting, or tool compensation to allow users to control specific parts of the program. These functions cannot be ignored during cutting processes since they are considered essential for increasing both productivity and surface quality. The information entered in a block type should be specific and correctly defined. Moreover, the type of information should be represented by an area corresponded to a specified address (G, M, F, S, T, etc). It is of paramount importance to mention that a post processor that has been developed with precise specifications after installation can support any of the preset modes, being under user control (Krimpenis et al., 2011).

The application developed, utilizes CL data files generated from typical CAM systems. CL data files contain all the information needed for tool paths, cutting conditions, rapid traverses, canned cycles etc, for a specific work piece. During the CL data conversion, specific steps are followed in

Figure 19. Automated analysis for machined parts in a CAM environment (a) analysis to obtain remaining volume after a roughing operation; (b) analysis to obtain surface deviation after finishing operation

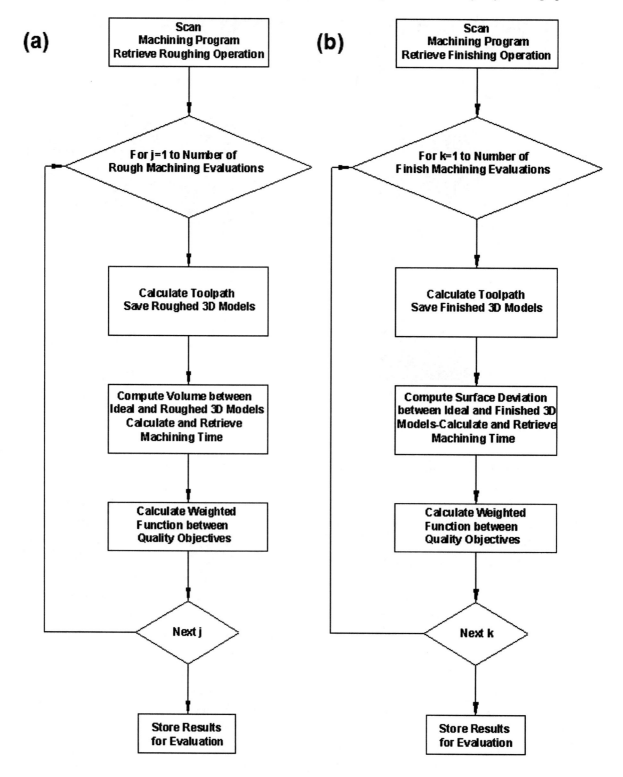

order to spot major information for the machining process while generating outputs ready to be executed in a CNC control unit. Each CL Data file corresponds to a given work piece being machined. The steps followed to produce G codes using the proposed post processor developed, are:

- Declaration of variables,
- Opening and reading CL-data files,
- Step by step processing of CL-data,
- Creation of matrices to save information processed,
- Generation of a two dimensional binary encoded matrix in which programming conditions handle "0" and "1" elements ("0" goes for blank characters; whilst "1" for non-blank characters),
- Scanning of binary matrix's elements – generation of new one-dimensional matrices word characters,
- Execution of a subroutine to create a two dimensional alphanumeric matrix in which identity matrices is addressed for the representation of words,
- Execution of a combinatorial subroutine module to scan tables so as to track down spindle speed, feed rate and maximum "z" level,
- Execution of a subroutine to scan elements of matrices to registrate x, y and z coordinates,
- Loop until all cutting conditions are evaluated and tool motions are assigned to each coordinate,
- Check to avoid reprinting of already processed coordinates,
- Creation of a "*.dat" file for output (The G-code is printed and saved).

The procedure followed by the application to generate the G-code, is executed via two command buttons. The first is the "Open" command button, which reads and processes the "CL-data" file for a given work-piece and extracts machining outputs. The second ("Write" command button) is responsible for the G-code generation, exploiting data obtained by the "Open" command button. The main purpose of dividing the post-processor module in two parts - corresponding to two programming "buttons" - was to facilitate its use even by inexperienced users. The program is created in such a way that it can easily be transformed into a single code, that runs without the need of user interaction.

The flowcharts describing the functions of these buttons ("open" and "write") are illustrated in Figure 20 and Figure 21 respectively.

3.5.2 Experimental Validation and Results

The proposed post-processing module was tested on three different sculptured parts that represent real manufacturing products of mechanical industries. The first part (Figure 22a) represents a typical hip joint mould applied in orthopedics; the second (Figure 22b) is a mould for the manufacturing of the front side of a mobile phone and the third (Figure 22c) is a mould part used for fluid mechanics studies.

Machining simulations were conducted in *Dassault Systèmes*® CATIA® and CL data were collected for all three parts. The presented post-processing application was then applied to translate the data for the G-code generation. The NC programs (G-codes) were verified by employing a specialized NC verification software. Despite the perplexity of the three parts used for this study, the presented approach performed very consistently by producing robust and efficient G-codes free of errors and generally small in block size considering their nature.

Simulations were performed on a typical personal computer with a 2Gb RAM and a 2.1Ghz processor. CL-Data extracted from CAM software were extremely large (files of 10.000 blocks approximately). It was verified that the proposed post processing application needed less than a minute

Figure 20. Operational workflow of "Open" button to read and process CL data

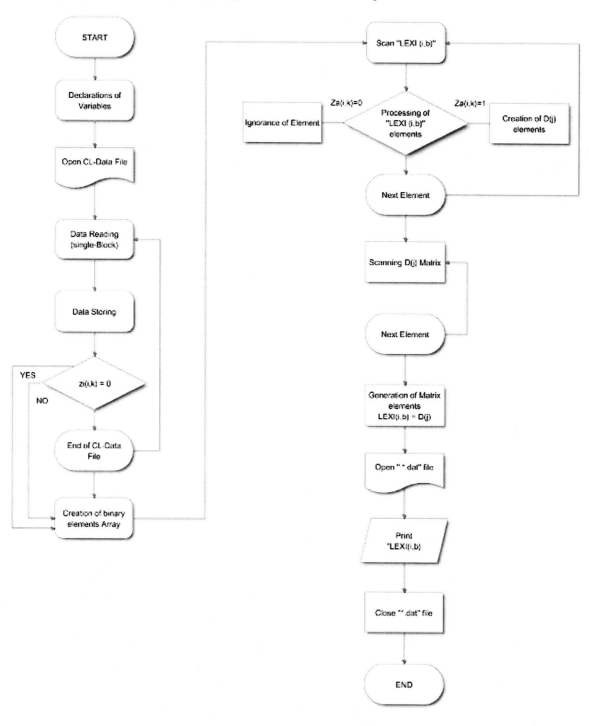

Figure 21. Operational workflow of "Write" button to print the ISO code

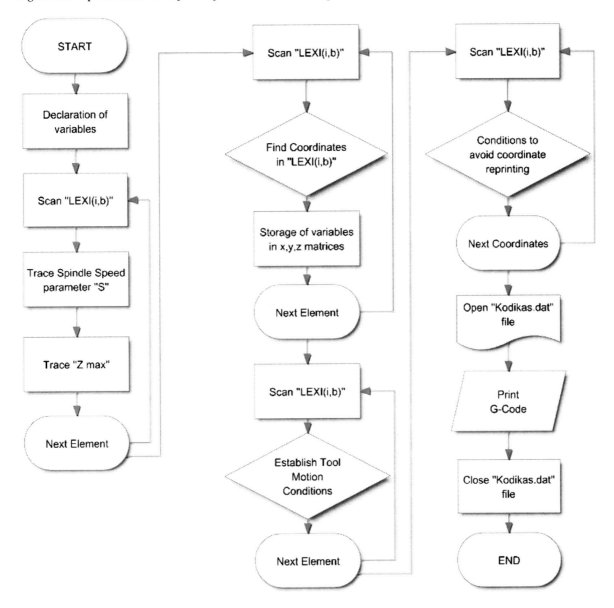

to read each of the CL data files, whilst the time spent to write the G code files was approximately 4 to 5 seconds for each model. In addition, the post processor is able to run on personal computers with typical configurations and quite low computational abilities. For the part depicted in Figure 22a machining simulations were performed using the G code created (Figure 23) whilst the statistical results obtained for all machining simulations for the parts of Figure 22, are illustrated in Table 1.

4. CONCLUSION

Modern manufacturing imposes the application of advanced manufacturing systems and software, suitably customized to meet industrial needs and

Figure 22. CAD models for intelligent post processing (a) Hip joint mould, (b) mobile phone mould (c) mould part from fluid mechanics

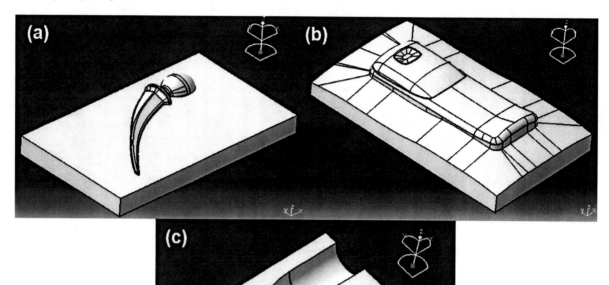

maintain competitiveness. Moving towards the direction of accelerating processes of product lifecycle management, programming and automation seem to be a trustworthy approaches to facilitate problem solving for important engineering areas. Under this aspect programming applications were developed either to automate repetitive procedures to engineering stages or to further expand software capabilities.

The stages investigated were those of computer aided design, computer aided process planning and computer aided manufacturing. For each of these phases programming techniques were implemented using different philosophy depending the task and the goal to be achieved. Yet, the main characteristic involved to all software modules studied was the methodology of retrieving, customizing and handling programming objectives available by engineering software interfaces through their open architecture. It was shown that such strategies only offer benefits since intelligence may be applied under proper computational thinking and problem mapping by end users. Hence; human expertise can still be present and manipulate operations whilst simultaneously eliminating shortcomings owned to large computational time and cost by automating repetitive tasks and/or achieve communication among two of more software platforms.

To solidify the above; representative examples were illustrated in the form of case studies. For computer aided design, it was shown that complex 3D CAD sculptured models may rapidly be created

Software Development Tools to Automate CAD/CAM Systems

Figure 23. Machining simulations performed using the G - code created for the hip joint mould part depicted in Figure 22a (a) Machining setup; (b) roughing operation; (c) finishing operation

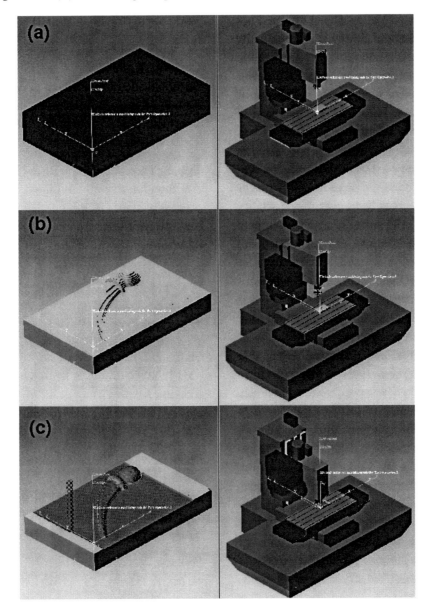

Table 1. Intelligent NC program creation results for the models processed

	Intelligent NC program creation results		
Component	Hip joint model	Mobile phone model	Fluid mechanics model
G-code creation time (sec)	4.94	3.99	3.18
Simulated Machining time (sec)	2114	1703	122

by utilizing their mathematical representations for curves and surfaces. For computer aided process planning a case study of an object oriented application was presented aiming at automatically generating documentation necessary for the shop floor. Support functions programmatically developed were presented along with workflows of procedures and user interfaces. For the case of computer aided manufacturing a case study was presented concerning the automatic specification of machining parameters when trial and error scenarios are to be conducted so as to satisfy one or more quality objectives. Moreover, the possibility of conjugating CAM software to an intelligent optimization algorithm was also mentioned and main aspects were discussed. On the basis of generating efficient and accurate CNC programs for complex industrial parts a case study was presented introducing a programming application for intelligent post processing.

Results obtained and discussed reveal that programming may be a powerful tool to perform almost whatever task when software is involved. According the manner of software development solutions to engineering problems may be both deterministic or stochastic depending on the application and its goal. Under this assumption, macro commands or scripts can be used to always provide exact solutions to problems or others may be developed to generate optimum solutions. This implies that both intelligent and conventional methods (at least automated) may be employed to reassure robustness and efficiency of solutions.

REFERENCES

Bacharoudis, E. C., Filios, A. E., Mentzos, M. D., & Margaris, D. P. (2008). Parametric study of a centrifugal pump impeller by varying the outlet blade angle. *The Open Mechanical Engineering Journal*, *2*, 75–83. doi:10.2174/1874155X00802010075

Chorin, A. J., & Marsden, J. E. (1979). *A mathematical introduction to fluid mechanics*. New York: Springer-Verlag. doi:10.1007/978-1-4684-0082-3

Ćuković, S., Devedžić, G., & Ghionea, I. (2010). Automatic determination of grinding tool profile for helical surfaces machining using CATIA/VB Interface. *U.P.B. Scientific Bulletin, Series D*, *72*(2), 85–96.

Deb, S., Parra-Castillo, J. R., & Ghosh, K. (2011). An integrated and intelligent computer-aided process planning methodology for machined rotationally symmetrical parts. *International Journal of Advanced Manufacturing Systems*, *13*(1), 1–26.

Faheem, W., Castano, J. F., Hayes, C. C., & Gaines, D. M. (1998). What is a manufacturing interaction? In *Proceedings of 1998 ASME Design Engineering Technical Conferences* (pp. 1-6), Atlanta, GA.

Farhan, U. H., Tolourei-Rad, M., & O'Brien, S. (2012). An automated approach for assembling modular fixtures using SolidWorks. *World Academy of Science. Engineering and Technology*, *72*, 394–397.

Fountas, N. A. (2008). *Advanced operations supporting process planning for aircraft structural parts in aerospace manufacturing* (MSc Thesis). Faculty of Science Engineering and Computing, Kingston University, London, UK.

Fountas, N. A., Krimpenis, A. A., Vaxevanidis, N. M., & Davim, J. P. (2012). Single and multi-objective optimization methodologies in CNC machining. In J. P. Davim (Ed.), *Statistical and computational techniques in manufacturing* (pp. 187–218). New York: Springer. doi:10.1007/978-3-642-25859-6_5

Harik, R. F., Derigent, W. J. E., & Ris, G. (2008). Computer aided process planning in aircraft manufacturing. *Computer-Aided Design & Applications*, *5*(6), 953–962. doi:10.3722/cadaps.2008.953-962

Ispas, C., Anania, F. D., Mohora, C., & Ivan, I. (2007). New methods for compensating the machining errors by CAD modeling of the machining surfaces on milling machines in coordinates. *Annals of the University of Petroşani. Mechanical Engineering (New York, N.Y.), 9*, 169–184.

Izamshah, R. A., Mo, J. P. T., & Ding, S. (2011). task automation for modeling deflection prediction on machining thin-wall part with Catia V5. *Advances in Mechanical Engineering, 1*(1), 8–14.

Jeba Singh, K. D., & Jebaraj, C. (2005). Feature-based design for process planning of machining processes with optimization using genetic algorithms. *International Journal of Production Research, 43*(18), 3855–3887. doi:10.1080/00207540500032160

Jones, T. J., Reidsema, C., & Smith, A. (2006). Automated feature recognition system for supporting conceptual engineering design. *International Journal of Knowledge-Based and Intelligent Engineering Systems, 10*(6), 477–492.

Krimpenis, A. (2008) *CNC rough milling optimization of complex sculptured surface parts using artificial intelligence algorithms* (PhD Thesis). National Technical University of Athens, Athens, Greece.

Krimpenis, A. A., Fountas, N. A., Skolias, J., Tzivelekis, C., & Vaxevanidis, N. M. (2011). Intelligent post-processor creation for sculptured surfaces in CAM software. In *Proceedings of the 4th International Conference on Manufacturing Engineering -ICMEN* (pp. 287-294), Thessaloniki, Greece.

Lamarche, B., & Rivest, L. (2007). Dynamic product modeling with inter-features associations: comparing customization and automation. *Computer-Aided Design & Applications, 4*(6), 877–886. doi:10.1080/16864360.2007.10738519

Li, W. D., & Qiu, Z. (2006). State-of-the-art technologies and methodologies for collaborative product development systems. *Computer Aided Design, 37*(9), 931–940. doi:10.1016/j.cad.2004.09.020

Reddy, B., & Brioso, R. G. (2011). *Automated and generic finite element analysis for industrial robot design* (MSc thesis). Linköping University, Department of Management and Engineering, Sweden.

Stampfer, M. (2004). Integrated setup and fixture planning system for gearbox casings. *International Journal of Advanced Manufacturing Technology, 26*(4), 310–318. doi:10.1007/s00170-003-1997-z

Wayzode, N. D., & Tupkar, A. B. (2013). Customization of Catia V5 for design of shaft coupling. In *Proceedings of the International Conference on Emerging Frontiers in Technology for Rural Area 2012*, (pp. 30-33).

Wayzode, N. D., & Wankhade, N. (2013). Design of flange coipling using CATSCript. *Indian Streams Research Journal, 2*(12), 1–7.

Wikström, J. (2011). *3D model of fuel tank for system simulation - A methodology for combining CAD models with simulation tools* (MSc thesis). Linköping University, Department of Management and Engineering, Sweden.

KEY TERMS AND DEFINITIONS

Artificial Intelligence (AI): Technology and a branch of computer science that studies and develops intelligent machines and software.

Artificial Neural Networks (ANNs): Models inspired by central nervous systems -in particular the brain- that are capable of machine learning and pattern recognition. They are usually presented as systems of interconnected "neurons" able to compute values from inputs by feeding information through the network.

Computer Aided Design (CAD): The technology concerned with the development of new products in the form of 3D models with the use of computers.

Computer Aided Manufacturing (CAM): The use of computer software to control machine tools and related resources to plan, simulate, and verify manufacturing operations for products.

Computer Aided Process Planning (CAPP): Concerned with determining the sequence of individual manufacturing operations needed to produce a given part or product with the aid of computers. The resulting operation sequence is documented on a form typically referred to as a route sheet containing a listing of the production operations and associated machine tools for a work part or assembly.

Cutter Location Data (CL-data): Files that contain the exact locations that a cutting tool will follow to machine a raw material, transforming it to a final product. CL-data files are, in fact, CAM software outputs to be further post processed according the control unit that a CNC machine tool implements.

Computer Numerical Control (CNC): The technology concerned with the control of machine tools through programming. CNC programs contain functions that specify machining parameters, determine cutting tool trajectories, and handle commands such as spindle turning, coolant supply, etc.

Evolutionary Algorithms (EAs): Heuristic search routines like GAs capable of providing optimum solutions to optimization problems using techniques inspired by natural evolution (genetic operators), such as selection, crossover, and mutation.

Genetic Algorithms (GAs): Heuristic search routines that mimic the process of natural selection. GAs are applied in many scientific areas to generate useful solutions to optimization problems.

Manufacturing Feature Recognition (MFR): A special property of computer aided design software capable of recognizing geometrical properties of designed features that constitute a 3D model.

Section 4
Smart Manufacturing Interconnection

Chapter 9
The Interaction between Design Research and Technological Research in Manufacturing Firm

Satoru Goto
Ritsumeikan University, Japan

Shuichi Ishida
Ritsumeikan University, Japan

Kiminori Gemba
Ritsumeikan University, Japan

Kazar Yaegashi
Ritsumeikan University, Japan

ABSTRACT

Design has significantly affected innovation and the discipline of design management focuses on meanings that it brings about a drastic change in life style of consumers. Although the relationship between design and technology is one of the important issues for the innovation of meanings, there were only a few studies which suggested the comprehensive model that includes design and technology. Verganti proposed the concept of design driven innovation, which regarded a design process of NPD as a research activity, and demonstrated a relationship between the technological research and the design research. In particular, he examined deeply the mechanism of the design research. In order to deepen his discussion, this chapter aims to suggest some propositions and a comprehensive model related to the interaction between the design research and the technological research. The authors utilize the concepts of exploration and exploitation for their framework. It shows that an augmenting of both researches may create effectively radical meanings or technologies and an integration of both researches may create radical meanings and technologies concurrently. In the case study of FPD industry, this chapter examines how some companies create competitive advantages by both researches and the commoditization of technology may cause the transition to the design research from the technological research as source of the competitive advantages. Additionally, the chapter suggests the strategic and organizational issues for conducting the design research and the technological research interactively in the discussion section.

DOI: 10.4018/978-1-4666-5836-3.ch009

INTRODUCTION

Design has recently been recognized as an important source that achieves competitive advantages. Many design-intensive companies have actually made a good profit (Hertenstein, Platt, & Veryzer, 2005; Gemser & Leenders, 2001; Walsh, 1995; Walsh, Roy, & Bruce, 1998). Many researchers in the design management discipline have demonstrated the efficacy of design (Bayazit, 2004; Borja, 2003; Bloch, 2011; Goodrich, 1995; Hertenstein & Platt, 1997). Additionally, the researchers in the marketing discipline have demonstrated the aesthetic value (Bloch, 1995; Sewall, 1978) and branding (Kreuzbauer & Malter, 2005). In the past, industrial designers have played a significant role in designing a package for finished products, and a product form for differentiation from competing products. As preferences of consumers become more multifaceted, to change emotional and symbolic values of products (meanings and languages) have been required for industrial designers (Dell'Era & Verganti, 2007). Therefore, design has become important not only for marketing, but also for NPD (Clark & Fujimoto, 1990; Oakley, 1986; Walsh, 1996).

The research in the marketing discipline focuses on what to create. On the other hand, the research in the innovation and engineering discipline focuses on how to produce (Rao & Patel, 2011; Andrade-Campos, 2011). A significant difference between the viewpoint of marketing and innovation exists in the perspective of technology. Therefore, it is important to describe the relationship between technology and design. Verganti (2011) proposed the concept of technological epiphanies. According to the framework of design driven innovation (Verganti, 2006), he showed how to link new meanings to technologies. In his model, the definition of design is "Design is making sense (of things)" (Krippendorff, 1989). He explained that a radical meaning was created to improve emotional and symbolic value of a product and defined the process to create radical meanings as the design research that was different from the traditional design activity such as styling. In addition, most of researchers define the irrational value as a meaning. Chiksentmihy and Rochberg-Halton (1981) studied the significance of material possessions in contemporary urban life, and of the ways people carve meaning out of their domestic environment. According to their definition, the things don't have the meaning unless consumers pay attention to them. Hirchman (1982) discussed an importance of symbolic innovations as well as technological innovations. He defined the symbolic innovations as communicative devices representative of different lifestyles. He mentioned that the technology innovations changed the tangible attributes of products and the symbolic innovations changed intangible attributes of ones. However, technologies are likely to change intangible attributes, so that to make clear what kind of the technologies change the intangible attributes and create the symbolic innovations. This chapter defines the research activity to create the radical meanings as a design research (Verganti, 2009) whereas a technology research aims at creating the radical technology.

In practical NPD, many companies cannot conduct the design research because they view design as styling in the final stage of NPD or as cause of cost increases. In order to measure the impact of product form or meaning numerically is very difficult, top managements or project managers tend to downplay the irrational values. Their understanding of the irrational values has an impact on product performance. For example, top managements of Dyson and Apple lead to commercial success of their product. James Dyson and Steve jobs directly seized the initiative with their products and insisted on the irrational values, such as product form and sensuousness. Organizational hierarchy could result in disturb of radical irrational values because to communicate them between the hierarchy is difficult for depending on sensuousness. Therefore, the individuals who are responsibility for the product specification tend to rely on the rational values, such as spec and cost.

The issue how a company creates radical meanings effectively without depending on the individual resource is important for practical operations. In academic, the issue with regard to the relationship between the design research and the technology research is remained although a lot of researchers recognized the importance of the radical meanings and technologies for establishing the corporate competitive advantage. The purpose of this paper is to make clarify its relationship and to propose some issue for future research.

In the following section, we first introduce the prior research with regard to meanings. In the second section, we propose our framework to describe the relationship between technologies and meanings and some propositions. The third section presents the case study in a FPD industry to validate some propositions. The forth section discusses the result of the case study and proposes some issues with respect to the organizational approach and the resource to create the meaning.

At the end of this session, we provide the definitions for the key terms: the definition of design means "Design is making sense (of things)" (Krippendorff, 1989), meaning is an irrational values such as emotional and symbolic ones, design research is an activity to aim at creating the radical meanings and technological research is an activity to aim at creating radical technologies.

LITERATURE REVIEW

A product form, which is an only way to contact with users, should play an important role in conveying the meanings (Townsend, Montoya, & Calantone, 2011). Even though functionalities of a product don't change, to improve its product form could offer a new value of a product to consumers. In addition, consumers may interpret functionalities from its product form (Hoegg & Alba, 2011). Creusen and Schoormans (2005) categorized the roles of visual designs as follows: (1) communication of aesthetic, (2) symbolic, (3) functional, (4) ergonomic information, (5) attention drawing, and (6) categorization. With these categories, they demonstrated the relationships with evaluations of consumers and their purchase behavior. With regard to aesthetics of a product, there were a lot of researches in the marketing discipline (Veryzer, 1993, 1995; Veryzer & Wesley, 1998). Additionally, a product form can activate human recognition along with sensory stimulus (Grossbart, Mittelstaedt, Curtis, & Rogers, 1975). Therefore, product forms have been regarded as a source of competitive advantage (Maidique & Zigger, 1985; Dahl, Chattopdhyay, & Gorn, 1999; Swift, 1997; Cruesen, 2011).

In addition, industrial designers utilize the user-centered (oriented) design method (Veryzer & Borja de Mozota, 2005; Vredenburg, Isensee, & Righi, 2002; Kelley, 2001; Lojacono & Zaccai, 2004; Kumar & Whitney, 2003; Redstrom, 2006; Oikonomou, Moulianitis, Lekkas, & Koutsabasis, 2011) and ethnography (Rosenthal & Capper, 2006). Their focuses are always placed on users, and they are able to extract user needs by visualization techniques such as prototypes and sketches. Thus, it is very important that designs collaborate with marketing (Mertes, 1965; Zhang & Kotabe, 2011; Rusut, Thompson, & Hamilton, 2006). These visualization techniques contribute to the effectiveness of NPD itself (Hayes, 1990). Prototypes and sketches enable engineers and marketers to discuss their opinion in early stage of NPD. In particular, the prototype performs quite well in discontinuous product development (Veryzer, 2005) because it is not easy for consumers to adopt its radical product concept (Cox & William, 1987). The same may be true among individuals who are in charge of NPD. Therefore, prototypes can continue to keep its product concept consistently from the upstream of product development to the downstream. In particular, cooperation between different departments such as R&D and marketing should be important (Johne & Snelson, 1998; Griffin & Hauser, 1996), and the consistency in the product concept is affected significantly by communication between departments (Bailetti & Litva, 1995). Additionally, designers play a role as a technology broker in NPD (Hargadon & Sutton, 1997), and contribute to making technical decisions (Cooper, Prendiville, & Jones, 1995).

Designs have an impact on all activities related to NPD (Dahl, 2011). Therefore, the focus of design management has transited from marketing to innovation (Luchs & Swan, 2011). Utterback et al. (2006) suggested Design Inspired Innovation that the uniqueness underlies the balance between technology, markets, and product meanings. They indicated that the emotional and symbolic values of the product, which is defined as product meanings, are the most important for consumers. With respect to the meanings, there are a lot of studies in the marketing and semantics discipline (Csikszentmihalyi & Rochberg-Halton, 1981; Butter & Krippendorff, 1984). Verganti (2009) also defined the creation of radical meanings as Design Driven Innovation. He divided meanings and technologies into two dimensions (Figure 1). He also proposed that a design research was the process to analyze a sociocultural and clearly distinguished it from traditional design activity such as styling and user-centered design. Since a lot of companies are able to obtain the technology externally (Chesborough, 2003; Pisano & Verganti, 2008), an important issue is how effectively technological innovation can be combined with innovation of meanings. Therefore, innovation management is important research issue in the design management and technology management discipline. However, some issues to solve still remain. For example, the strategic issue is whether a company should select technological research or design research. The organizational issues also exist as to what kind of resources a company need to both researches. To bring about solutions for such issues, first of all, comprehensive models that describe the relationship between design and technology is needed and more considerable amount of additional research should be done using them.

CONCEPTUAL FRAMEWORK

Exploration of Meanings

The purpose of the design research is to create radical meanings (Verganti, 2009). However, radical technologies are not necessity for radical meanings (Hirschman, 1982). This is represented by the case of the Walkman created by SONY.

The prototype of the Walkman was a small cassette recorder called Pressman. The technology that was adopted was so simple that other Japanese electronic manufacturers could have easily made such a device at that time. In those days, general devices for listening to music were cassette decks that had a recording function. Even consumers and other companies in the same industry were not able to forecast the need for a device that had only the function listening to music. SONY had the unique corporate culture that didn't imitate other products and it proposed a new life style for users (NIKKEI BIZTECH, 2004). SONY created a radical meaning of carrying around music outside. It also put their focus on advertising in order to strengthen this meaning. In addition to traditional news releases, it launched the advertising in magazines targeting young people. In the press conference held to release the Walkman, it invited the press to the park and showed them the actual scenes of young people actively using the Walkman. The proposal of consistent meaning from NPD to advertisings reinforced the meaning of the Walkman. Many similar products were released after the announcement of the Walkman; however, the Walkman could maintain its strong brand identity. In this case, SONY proposed the radical meaning to consumers on purpose, not by chance. This process had uncertainty because of the lack of consumer needs. Therefore, this process to create the radical meaning was regarded as exploration one (March, 1991). Additionally, the technology that SONY adopted was achievable by other companies. For this reason, this process to adopt the technology was exploitation one.

Figure 1. Innovation Strategies and the Positioning of Radical Design and Technological Epiphanies (Verganti, 2011)

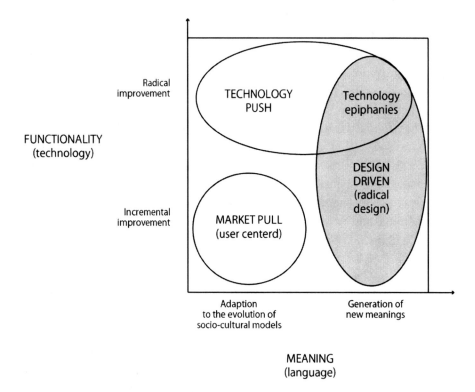

Table 1. Categories of the literatures about design

	Marketing				Innovation
Categories	Corporate performance	Product form	Role of design on NPD	user-centered design	Innovation
Authors	Gemser and Leenders, 2001	Bloch, 1995	Bailetti and Litva, 1995;	Kelley, 2001	Dell'Era and Verganti, 2007
	Hertenstein, Platt and Veryzer, 2005	Creusen and Schoormans, 2005	Clark and Fujimoto, 1990	Kumar and Whitney, 2003	Utterback et al., 2006
	Walsh, 1995	Dahl, Chattopdhyay and Gorn, 1999	Hayes, 1990	Lojacono and Zaccai, 2004	Verganti, 2009
	Walsh, Roy and Bruce, 1998	Grossbart, Mittelstaedt, Curtis, and Rogers, 1975	Luchs and Scott Swan, 2011	Redstrom, 2006	
		Hoegg and Alba, 2011	Mertes, 1985	Rosenthal and Capper, 2006	
		Kreuzbauer and Malter, 2005	Oakley, 1986	Veryzer and Borja de Mozota, 2005	
		Maidique and Zigger, 1985	Veryzer, 2005	Vredenburg, Isensee, and Righi, 2002	
		Sewall, 1978	Walsh, 1998		
		Swift, 1997	Zhang and Kotabe, 2011		
		Townsend, Montoya and Calantone, 2011			
		Veryzer and Wesley, 1998			
		Veryzer, 1993			
		Veryzer, 1995			

Actually, other companies also followed after the Walkman was launched. This fact suggested that the latest cutting-edge technology is not always the source of competitiveness. In fact, there are many cases that destructive technology dominates markets (Christensen, 2003). Exploration is believed to be an appropriate activity for creating radical technologies (Henrich & Greve, 2007). It can be said that exploration is also appropriate for creating radical meanings. In conclusion, we suggest that design research conducts exploration of meanings and the following proposition with regard to design research.

Proposition 1: *Design research can effectively conduct exploration of meanings by exploiting technology.*

Exploration of Technologies

Technological research generates innovation that leads to significant change in technological structures (Latour, 1987). However, radical technologies enhance product functionality but may not change product meaning. Verganti (2009) defined this concept as technological substitution. Development of digital cameras can be included as a case of technological substitution.

Development of a digital camera was started by CASIO in 1985. The original concept was to shoot photos electronically. Two years later, the electronic still camera was launched. This camera had a function to save images on a floppy desk as analog data and to play such data on a TV monitor. This product did not success commercially; however, it was sold again as a digital camera with the LCD monitor equipped in 1994. At that time, a lot of consumers used personal computers in their homes and they could connect his digital camera to computer and view their photos on their PC monitor. This function enabled the digital camera to be commercially successful.

Not only the existing film camera manufacturers but also a number of home appliance manufacturers participated in this market. The technology of lens, CCD and software had a significant impact on its performance. The integration of these technologies, which were implicit, was very important. Originally, Japanese companies had the competitive advantages of technology in the conventional film cameras market. In addition, each manufacturer continued to generate technology innovations because of intensifying competition in the domestic market (NIKKEI ELECTRONICS, 2005; NIKKEI ELECTRONICS, 2006; NIKKEI BUSINESS, 2007). This resulted in developing a wide variety of innovations such as increased resolution, large LCD monitors, and downsizing of the camera body and price reduction was also realized. These resulted in the advantages of Japanese companies that captured about an 80% of share of the global market.

When CASIO started developing digital cameras, their purpose was technology substitution. This case is indicated that the existing meaning was exploited and the radical technology was explored. After digital cameras launched, Japanese companies continued to conduct continuous the technological research without changing the meaning of the conventional camera. This technology innovation resulted in providing sophisticated products at low price. In conclusion, we suggest the following proposition.

Proposition 2: *Technological research can conduct effective exploration of technology by exploiting meanings.*

Design and Technological Research

Baldwin and Clark (2000) indicated the following modularized operators as design rules: splitting, substituting, augmenting, excluding, inverting, and porting. In accordance with these rules, design research and technological research could be regarded as a result of splitting a research activity. How can these two kinds of research be conducted in order to lead to competitive advantages?

Meanings are also imitated as well as technologies (Dell'Era & Verganti, 2007). After SONY launched the Walkman, many companies actually followed this meaning. On the other hand, companies that have developed their own technologies earlier tried to employ a variety of strategies in order to protect their technologies, such as acquisition of patents and putting their technologies into a black box. How can we deal with meaning imitation? Meanings are social and cultural reasons of consumers for using certain products (Verganti, 2009). In other words, meanings can be built when consumers actually use the product in a society. An American major retail company Target hired Michael Graves, who created powerful meanings in Alessi, and imitated the meaning that Alessi created. However, the product that Target imitated could not reach the sales of Alessi at all despite of selling the same product (Verganti, 2009). This case shows that to create the radical meaning makes the company brand strong.

However, meanings could become obsolete. In order to maintain the strength of the meaning, SONY provided new models of Walkman one after another with incremental innovations of technology. In the digital camera market, Japanese manufacturers maintain an 80% of the market share by continuing to provide new products with specifications that were higher than the existing products without changing the meanings of cameras. Therefore, the important thing is to select the technological research or design research to fit their market environment strategically. In addition, it also becomes important to create radical technologies after to create radical meanings and vice versa.

Consider a case of digital audio players. Apple should be regarded as the company that conducted exploration of meanings in the market of the digital audio player. Some precedent products already existed before Apple broke into this market. Those products utilized a new technology, but they had just the same product meanings as the Walkman. In other words, they might be developed as a result of technological substitution. The digital technology enabled users to take a lot of songs in comparison to the Walkman. However, the user interface was not good; for example, it took a lot of time to select their favorite songs because a user needed to push many buttons. Apple had already obtained the technology for a digital audio player from the external sources. This can be estimated as their exploitation of technology. However, Apple developed a product form and a user interface that allow users to operate the device instinctively. Additionally, the radical meaning was created under the Apple's corporate vision "digital hub" (Levy, 2006). The iPod incorporated all experiences around music, such as purchasing, managing, and enjoying music based on the personal computer. Its function changed Having-Level Goals of consumers (Ligas, 2000). Apple certainly conducted excellent development of UI technology; however, they acquired the basic technology from an external source.

This case suggests that separate utilization of the design research and the technological research should become a significant strategic issue for companies. If a company can realize these two kinds of research in-house, this company might predominate in competitive advantages. However, companies do not have to conduct all such things in-house. For example, if some companies are quite good at the design research, cooperation with other companies that do well at the technological research may achieve competitive advantages. If some companies have good proficiency in the technological research, they should cooperate with the external design research. Based on these suggestions, the following proposition can be provided.

Proposition 3: *Technological research is not to enough to create competitive advantages. Interactive augmenting of design research with technological research can achieve competitive advantages.*

Integration of Design and Technological Research

If a company creates a radical meaning and technology concurrently, it can achieve a competitive advantage. Consider the development of cyclone vacuums by Dyson.

The founder of Dyson, James Dyson, saw cyclone technology used in a powder processor for professional use. He discovered that this technology could be utilized in vacuum cleaners. At that time, the major mechanism of vacuum cleaners was to collect the absorbed dust into a paper bag. Vacuum cleaner manufactures and even though consumers took them for granted. However, Dyson produced a prototype of a radical vacuum cleaner without a paper bag. His propose was continually rejected by major manufacturers. Amidst such circumstances, he continued cultivating and enhancing his technical skills, and finally he succeeded in commercializing his proposal. He proposed the emotional value that was a pleasure to clean. (NIKKEI BIZTECH, 2004). The traditional vacuum cleaners hide the dust in sight of consumers because they regarded the dust as unclean. On the other hand, the vacuum cleaner that Dyson produced had the transparent body in order to show dust on purpose. He attempted to create the radical meaning and the radical technology by the distinguishing appearance and the cyclone technology.

The feature of product development of Dyson is that designers and engineers are not separated physically. Designers originally get absorbed in the socio-cultural context while engineers get absorbed in the technological context. Integration of these functions by identical individuals probably means that design research and technological research may be conducted concurrently. They were likely to create radical meanings for vacuum cleaners while exploiting the existing cyclone technology in their design research. In parallel, while exploiting the meanings of traditional vacuum cleaners, they probably conducted technological research by which they adopted cyclone technology for a small vacuum cleaner. In conclusion, they might conduct the design research and the technological research concurrently. In other words, both researches are integrated, which is opposite concept of modularization that is one of the concepts of design rule (Baldwin & Clark, 2000). Therefore, we suggest the following proposition.

Proposition 4: *New meanings and technologies can be realized concurrently by the integrating design research and the technological research.*

Design Research and Fuzzy Front-End

The fuzzy front-end of NPD is the time and activity prior to an organization's first screen of a new product idea (Moenaert, De Meyer, Souder, & Deschoolmeester, 1995) and the important phase in a product strategy (Khurana & Rosenthal, 1998; Zhang & Doll, 2001). The fuzzy have been used in the fuzzy system theory originally, which had been developed from Fuzzy Sets (Zadeh, 1965). The fuzzy system is capable of handling many complex situations such as some control systems with large uncertainties in process parameters and/or systems structures (Chen, 1996). The discipline of management applied the fuzzy system theory to the front end stage of NPD because of its uncertainly. The fuzzy front end premises that information flow in the early development of such innovations moves from the environment into the firm, facilitated by individuals playing three key roles at three decision-making interfaces: (1) the boundary spanner at the boundary interface, (2) the gatekeeper at the gatekeeping interface, and (3) what is identified in this paper as the "project broker" at the project interface. (Reid & Brentani, 2004). This phase consists of the following four steps: generation of ideas, screening in the initial stage, explorative evaluations, and concept evaluations. Meanings created as results of design researches could be the foundation of the product concept. At the front end, it is a significant task to correct user needs from the market. User-centered design is a very effective method in this phase (Veryzer & Borja de Mozota, 2005). However, this method could turn ineffective in the case of radical innovation (Veryzer, 1998). Therefore, design research creates meanings by utilizing different methods from user-centered design at the front end. Italian manufacturing industry created radical meanings in a dialogue with design discourse (Verganti, 2008). Apple had their unique concept "digital hub" although other competitors considered that the PC would be an obsolete device for consumer. Dyson also proposed the prototypes of vacuum cleaners with cyclone technology to many companies, but no one accepted his proposal. The common point in these cases is that none of consumers noticed the meanings when they created them. It is not always true that information from the market serves as help for the radical innovation (Lukas & Ferrell, 2000). In fact, these meanings were created as a result of ignoring the market (Verganti, 2009). Therefore, we suggest the following proposition.

Proposition 5: *Although traditional design activity is conducted after front-end phase, the design research is the pre-front-end phase, and they contribute to creating innovative concepts at the front end.*

The above-mentioned propositions can be summarized as shown in Figure 2. The design research and the technological research create meanings and technologies, and augment each other or are integrated. Based on this relationship, they create product concepts, and confirm the effectiveness of them in the fuzzy front-end phase. Development processes after both researches will lead to the traditional NPD ones.

CASE STUDY

Methodology

In order to the exploratory nature of this paper, a qualitative case study methodology was adapted. It is difficult to bring theoretical findings about our framework from existing theory. Case study is a valid methodology to explain the relationship between cause and effect under novel framework (Yin, 1994).

This paper focuses on the industry of the Flat Panel Display (FPD) as a case study, especially Plasma Display panel (PDP) in Japanese manufacture Panasonic Corporation. The reason why we adopt the FPD industry are (1) the market has been matured, (2) it is possible to compare it to CRT TV industry, (3) there are a lot of market players in the world, (4) Japanese companies contributed to technology improvement since the early days of FPD industry. Mature of market enable to trace a transition of design researches and technological researches. Figure 3 shows the diffusion rate of FPD industry has been over 80% and market in 2011. FPD was made replacement with CRT. Therefore, we can reveal how meanings of FPD took place of them of CRT. In this case study, semi-structured interviews and secondary date was collected. In the semi-structured interviews, we conducted 2 hours interviews on 2 top managements of Panasonic. We collected secondary data from 1998 to 2012. In this case, the facts that don't have any reference are collected from interviews.

Framework of Analysis

This study analyzes FPD case through our framework. A technological research exploits existing meanings and explores new technologies. Therefore, the premise to realize a technological research is to exist the meanings when a technological research is conducted. A design research exploits existing technologies and explores new meanings. Therefore, the premise to realize a

Figure 2. Framework that shows the relationship between design and technological research

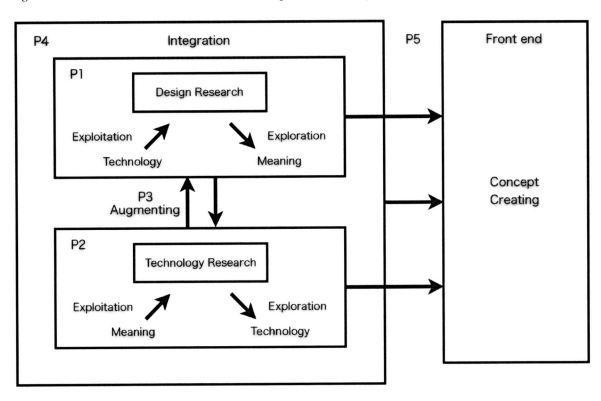

Figure 3. Diffusion rate of FPD(Nagano, Ishida, & Ikeda, 2011)

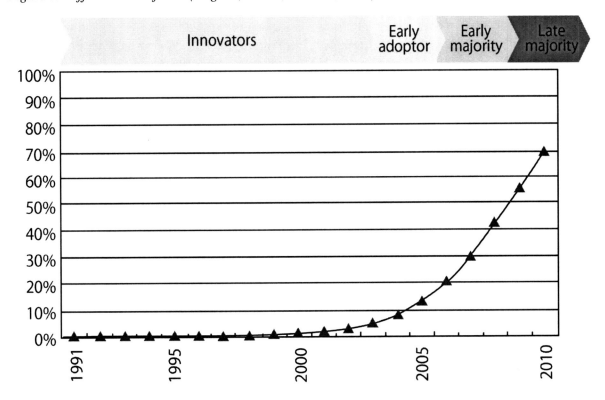

design research is to exist the technologies when a design research is conducted. If the meanings and technologies which new product has are novel, both researches may be realized concurrently. The purpose of both researches is not to realize consumer needs, but to give consumer suggestions of product concept (Verganti, 2009). Therefore, if output of both researches becomes the product concept, we guess that both researches conduct before front-end.

Foundation of PDP Technology

The original technology of the PDP was developed by the University of Illinois in the United States. Panasonic (Matsushita Electric corp.) began participating in the research of PDP in around 1973. The first PDP they launched was a monochrome PDP for automated teller machines (ATM). Afterwards, Panasonic started to develop high definition (HD) monochrome PDP for laptop computers and launched it in the market in 1985. In 1996 Panasonic, which adopted the DC system consistently, acquired Plasmaco Inc. of the United States that conducted the research on the AC-system color PDP and Panasonic made a shift in their mass-produced products from the DC system to the AC system.

After the Nagano Olympics in 1998, Panasonic launched AC-system color PDP monitors (NIKKEI BIZTEC, 2005). At that time, the product development of this PDP monitor and mass production was conducted in their laboratory. The monthly production volume of this PDP monitor was a few hundred. This product was sold as a monitor with a separate tuner and speakers. The engineering department that had conducted the research took the initiative in this product development, and there was no product planning department within the company. In 2001 the task force was organized to achieve high image quality and incorporated engineers who had developed other devices (NIKKEI BUSINESS, 2003). Panasonic established their first product planning department in the company around this time. This department, however, was positioned in the engineering department, working on adjustments between the marketing department and the engineering department. On the other hand, Panasonic Design Company, which was assigned to the product form, spent a year elaborating on the design of the VIERA series, sold in 2002 (NIKKEI DESIGN, 2004).

In 2003, experimental digital terrestrial broadcasting was started. Since the 1990's, Panasonic has invested a lot of management resources into development of digital technology. In Japan, there were only a few companies to internally manufacture LSI, which is the core digital TV technology, except for Panasonic. In 2003, the product planning department became independent from the engineering department. Additionally, the department of PDP development was integrated with the existing department for the CRT TV set. This integration kicked off a full-scale product development of PDP. The VIERA series was launched based on the design idea which was worked on by Panasonic Design during the previous year. This VIERA series was able to control a large screen by installing LSI that Panasonic had developed. In order to realize the concept that was the flat-screen for interior value, Panasonic solved the significant technical issue that was the speaker capacity. FPD could ensure only small capacity for their speakers compared to the traditional CRT TV; therefore, a new speaker technology was necessary. The technology called "Candy Speaker" brought the solution for this issue (NIKKEI DESIGN, 2004). After launching this VIERA series, Panasonic reached the third in the domestic market share of FPD.

Increased Competition of the FPD MARKET

The market for FPD larger than 40 inches in 2004 was dominated by the PDP. However, the LCD began to compete with the PDP due to the gradual increase in size (NIKKEI BUSINESS, 2006). The technical focus was placed on how high image quality could be achieved at low cost in order to differentiate the PDP from the LCD. In addition to image quality, reduction of electricity consumption by 35% was the sales appeal to consumers. In 2005, Panasonic's PDP had a 35% share of the market. In order to expand this market share furthermore for the next five years, the top management of Panasonic announced in public that they would try to enhance image quality and cost reduction (NIKKEI MONOZUKURI, 2006).

Market conditions changed around 2005 with the presence of Samsung Electronics Co., Ltd. of Korea that gradually expanded a their share of the market. After 1993, as a companywide goal, Samsung aimed to improve their industrial design (NIKKEI ELECTRONICS, 2007). In addition, Samsung made their in-house designers transfer to various places in the world in order to experience the local lifestyle. Therefore, the industrial design department has strong authority in NPD. As a result, Samsung wined the gold award in IDEA, which was one of the most prestigious design awards. Samsung was actually better than Japan in power-source technology to make the FPD thinner and other technologies that produced aesthetic. In particular, the Bordeaux TV that was launched in April of 2006 was designed not as a traditional TV, but as an interior product and became a huge commercial success on the global market. Bordeaux TV had speakers embedded in the bottom and not visible to consumers whereas other FPD generally had speakers embedded on both sides. This clearly made a distinctive difference in Samsung's FPD from other company's FPD (NIKKEI ELECTRONICS, 2007). On the other hand, in Panasonic, the technical engineering department took initiative in the styling of products until 2006. After this year, however, the product planning department had taken charge of the styling, but the engineering department still had total authority in NPD. For this reason, the product styling was always conducted during the final phase of product development.

Drastic Change of the Market Condition

Previously, Panasonic had its own FPD segmentation standard; 37-inch or larger FPD were PDP while 32-inch or smaller were LCD. In the summer of 2006, the global shipments of LCD, even though 40 inch or larger, exceeded those of PDP. Pioneer Corporation withdrew from self-manufacturing of PDP in 2008 (NIKKEI BUSINESS, 2008). Around this time, the price of TV dropped at an annual rate of 20% to 30%. In addition, Chinese manufacturers also expanded the share of their self-manufactured products. In such a market environment, Panasonic tried to expand the production of PDP by starting to operate a new factory in 2007 (NIKKEI BUSINESS, 2011). In NPD, Panasonic coped with their competitors by improving image quality and adding the functions based on the efforts made by the engineering department. Panasonic launched the 42-inch full HD PDP in 2007, two years after the launch of LCD. From around this time, each manufacturer started to focus on the technology to make FPD thinner. There was a shift in technical trends into making a thinner power-source system and power saving by LED back lights technology. Samsung was technically ahead of any other manufacturers in the technology of producing thinner TV modules. Samsung presented the thinnest LCD panel module, 10mm, in the FPD International 2007 (NIKKEI ELECTRONICS, 2007). On the other hand, Panasonic launched research and development of the 3D TV system.

Commoditization of the Technology

Taiwanese chip manufacturers started to sell single-chip image processing engines and this made it possible to produce a FPD of high image quality. Many of Japanese manufacturers that had produced FPD based on their vertical integration systems also began to entrust Original Design Manufacturing (ODM) to Taiwanese manufacturers. As a result, Japanese manufacturers could no longer maintain their technical competitiveness in image quality with their FPD including PDP. Given such a situation, Panasonic brought out the 3D PDP, ahead of FPD produced by other manufacturers (NIKKEI BUSINESS, 2010). However, other manufacturers also released 3D TVs immediately. Actually, Samsung's 3D FPD was the top-seller in the market of North America in this year. As shown by this fact, it became very difficult to achieve technical differentiations because overseas manufacturers rapidly caught up with Japanese manufacturers. This trend actually moved the focus of differentiation from the image quality to applications. As Apple began to participate in the TV business, the mainstream of product development transitioned into integration of the Internet (NIKKEI BUSINESS, 2008). Just like smartphones took the place of the existing cell phones, TV also began to be used as one of the digital devices. Samsung and LG Electronics featured the "Smart TV" at CES held in Las Vegas in January of 2011 (NIKKEI BUSINESS, 2011). In the same year, Panasonic presented a major policy change of NPD. The purpose of this policy change was to bring about change in its technology-driven product development. Panasonic sent out a questionnaire survey to consumers by showing its own PDP with various FPD such as LCD of other manufacturers and asked which TV has the highest image quality. Users' answers were very clear. It was not Panasonic's TV that has the highest image quality, but the other manufacturer. This survey result was just enough to change the thoughts of those engineers who had absolute confidence in their technology. This changed from their policy of making image quality the first priority to design and styling, and experiences users can enjoy through watching TV which appeals to users. At the same time, Panasonic changed the initiative of product development, from the engineering department to the product planning department. This shift resulted in producing the Smart VIERA, released in February 2012, featuring high aesthetic and integration with the Internet by connecting with smartphone. In addition, Panasonic provided consumers with enjoyment and experience through software, by developing Panasonic's original applications.

FINDINGS AND DISCUSSION OF CASE STUDY

Technological Research

Though Panasonic launched its first PDP in 1998, the research on PDP itself started in 1973. Technical exploration was conducted along with improvement in image quality in order to make PDP fit for practical use as TV. The first PDP has the tuner and speakers installed separately. In other words, this PDP was just a monitor. The research process of the PDP had the purpose of technology substitution (Verganti, 2009) of CRT. Therefore, Panasonic might exploit the meaning of CRT, and conducted the technological research. As a result, the engineering department could put their all efforts into improvement of image quality without any sociocultural analysis.

Samsung launched Bordeaux TV series in 2006. Samsung proposed its TV not as a device to watch a TV program but an interior because of high aesthetic value. In Panasonic, the responsibility for determining product forms was moved from the engineering department to the product planning department, which showed that the significance of product forms had been recognized. Around this time, each company began competing to develop producing thinner TV panels. In the technological research done by Panasonic, they might be likely to focus on the technical development of achieving thinner panels along with electric power saving and exploited the existing meaning of the TV as interior value. In conclusion, we examine the second proposition; technological research can conduct effective exploration of technology by exploiting meanings.

Design Research

The PDP that launched by Panasonic for the first time was a result of the technology substitution from CRT. The engineering department led to its NPD, and the analysis of sociocultural model was not conducted. It is unlikely that Panasonic intentionally tried to change the meaning of TV that the traditional CRT had. In 2002, technical development made it possible to integrate the tuner and speakers within FPD. For the new TV series, Panasonic design company began considering how PDP should be in the domestic life of consumers. This could indicate that they conducted analysis of sociocultural model in their research process. As a result, they presented a TV had a high aesthetic in comparison to a traditional CRT. Therefore, the process of creating this meaning may be considered as design research, and also as the exploration of the meaning. The most significant technical issue to improve aesthetic is the speakers and it was also solved by the existing technology that was called "Candy Speaker." This means that major technologies that compose PDP were actually available at that time. This fact may show that technology was exploited in the process of design research. By doing so, Panasonic might be able to concentrate on the exploration of the new meaning.

Integration with the Internet brought about the next change in the meaning. Just as the traditional cell phone changed into the smartphone, this integration changed the TV, not as just mere hardware, but into the TV as one of the digital devices on the Internet. Technically, FPD had already been digitized, and it was highly likely that there was no significant hardware-related issue in order to achieve this new meaning. However, Panasonic was not the one who created this meaning first, but the Korean manufacturer, so that it is impossible to investigate the creation process of this meaning in detail in this case. We cannot make a hasty conclusion, though, judging from then technical conditions, we estimate that the new meaning was explored by exploiting technology. Additionally, Panasonic released the Smart VIERA series one year after the announcement of the concept of the smart TV in 2011. This might be the result of exploiting the meaning produced by other companies. In conclusion, we examine the first proposition; design research can effectively conduct exploration of meanings by exploiting technology.

Interaction between Technological Research and Design Research

As FPD has entered into the early adopter stage of diffusion rate from the innovator stage, and furthermore the early majority stage, FPD has experienced several turning points of the technologies and meanings. In this section, turning point of meanings will be considered. It is the meaning of TV as an interior product. Samsung made this meaning of stronger after Panasonic had proposed this meaning to consumers in 2002. Samsung had industrial designers analyze sociocultural models by sending them to the local sites and having them experience local life. Therefore, with the dominant power of the industrial design department, Samsung's product development is led by industrial design. When the Bordeaux TV was developed, Samsung's technology related to aesthetic already had greater excellence than Panasonic. Samsung might intend to work on technological research in order to achieve the meaning as an interior design. Samsung was the first company that commercialized LED back light technology, which had a technically critical impact on the thinness of TV panels. This may show that technological research exploited the meaning. We guess that the sequence of the technological research and the design research should affect product sales significantly. Panasonic afterward put a focus on the product form as similar to image quality because Panasonic might exploit the meaning as an interior product. This fact shows that it is important, not to conduct both technological research and design research individually, but to combine them together. In addition, the key to success may be to arrange their sequence strategically. Based on the points described above, we examine the third proposition; competitive superiority can be obtained by augmenting both technological research and design research.

Consider the fourth proposition; the integration of technological research and design research. As earlier mentioned, we examine that technological research and design research are conducted in a mutually complementary manner. In this case, however, we could not confirm any fact where the technological research and the design research are integrated, and we estimate two factors here. The first factor is Panasonic's organizational factor. The second factor is about the recognition of the industrial design department. As for the first factor, there existed an organizational barrier that separates the engineering department and design department that conduct design research within Panasonic. If there is smooth communication even though such a barrier exists, integration can be achieved. However, there are other factors that block such communication. It was the Panasonic's company culture that engineers have predominance and advantages in NPD, as well as other Japanese companies (NIKKEI BUSINESS, 2009). Panasonic especially increased production volume by starting operation of new factories for cost reduction of 20% to 30% annually. Additionally, the top management clearly declared that cost reduction was an important issue in product development. Under such circumstances, Panasonic might provide a company environment that the design department faced difficulty in expressing their opinions. As for the second factor, designing may be generally recognized as styling in Japan. In NPD, industrial designers often take responsible only for the final phase of development, which is usually styling. Design research, therefore, can be integrated with technological research and they might not be conducted concurrently. In conclusion, we were unable to confirm the fourth proposition in this case.

Technological Research and Design Research before the Front End

Until the launch of VIERA in Panasonic, the diffusion rate of FPD including the PDP was 10% or under (Figure 3). Therefore, the technological research and the design research conducted for VIERA may be mainly driven not by the market, but by Panasonic. Research of technology and design may be positioned in the previous stage of the front end and the result of both researches probably became a strong concept.

As products become widespread, however, the needs of consumers have become easier to understand and Panasonic also has begun to pursue their needs. In particular, image quality, power consumption and cost were specifications that were easy to understand for consumers. For this reason, Panasonic probably made TV as an interior product so as to appeal to consumers at first, but they were likely to change their approach to make the easy-to-understand specifications appeal to consumers.

In 2011, Panasonic practically transferred the initiative of product development from the engineering department to the product planning department because products might be differentiated only by the software aspect. Panasonic had to make new experience realized by the TV. In other words, the purpose of this transfer may be to present a strong concept not by response to consumer needs, but by conducting design research in the stage prior to the front end.

In conclusion, we can discover the possibility that design research and technological research are conducted in the stage prior to the front end; however, no detailed process has been examined in this case.

Another Fact Discovered

In this section, another fact that was discovered except for fifth proposition is described. It is the possibility that the proportion between technological research and design research varies along with the diffusion of products. We found that the technological research might be conducted during the innovator stage. In 2002, when the diffusion rate transitioned to the early adopter stage, the product that reflected the output of the design research began to be launched. Yet during this time, the technological research remained proportionally high because there was still a great deal of room to improve technology about image quality, cost reduction and so on. When the diffusion rate reached the early majority stage, Samsung that had given priority to the design research gradually increased their market share. A lot of companies in FPD industry might come to exploit technologies gradually and to increase the proportion of the design research. After 2008 when the diffusion rate reached the late majority stage, Japanese manufacturers also started ODM for image processing chips with Taiwanese manufacturers, and this indicates that technologies differentiation became very difficult. It might become necessary to manage the design research in order to create new meanings. Most of the companies in FPD industry improve emotional and symbolic value by software as well as digital audio player industry. The iPod was able to gain a high share in its market. One of the keys to success should be to dramatically change the product context by the software, the iTunes, in addition to innovative user interface in the hardware (Linzmayer, 2004). Apple, however, obtain the major technologies from the outside the company. In other words, Apple could exploit them at that time. These facts may show that to set the priority of technological research and design research in the product life cycle could be the keys to success.

DISCUSSION AND ISSUES FOR FUTURE RESEARCH

Organizational Approach

It is not an easy task to conduct design research and technological research concurrently. There are probably only a few companies that have the resource like Dyson. Many companies separately organize a department that analyzes sociocultural models and the engineering department. In many cases, a sales and a marketing department, a production department and a quality control department may be independent. This could be one of the causes that make it difficult to maintain the concept consistency. How can such an organization achieve the product development system like that of Dyson?

Product development methods of Japanese manufacturers may be a key to success. These methods include the following parallel methods and approaches: cross-functional integration (Song & Parry, 1996, 1997), rugby approach (Imai, Nonaka, & Takeuchi, 1985), and concurrent engineering (Hartley, 1992). A lot of Japanese manufacturers reduce uncertainty of NPD by having close communication between each department from the early stage of NPD (Herstatt, Verworn, & Nagahira, 2004; Verworn, Herstatt, & Nagahira, 2008). To put it simply, each department interferes mutually in Japanese-style NPD. Most of the previous study about cross-functional integration, however, focused on a fuzzy front end. On the other hand, this paper assumes that design research and technological research are conducted in the stage prior to a front end. As indicated by Verganti (2009), the design research does not take any information from the market as well as the technological research (but lead users are still an important information source (Chesborough, 2003)). Information from the market does not always serve as help (Lukas & Ferrell, 2000). To create a product that can easily be predicted by consumers might result in a short-term product life cycle. Therefore, the Italian manufacturing industry always seeks not those designers that have become famous, but those up-and-coming designers in order to have them create radical meanings that do not exist in the market (Verganti, 2009). How can cross-functional process be carried out in the design research and the technological research?

In the case study of FPD, this paper could not confirm the fact that both technological research and design research are integrated. However, our interview survey has confirmed that there is a recent movement by Panasonic that aims for the integration of this research of technology and design. This movement promotes the idea that strong authority should be given to the product planning department that exists between the industrial design department and the technical engineering department. Previously, the industrial design department was responsible for analyzing the sociocultural model; however, it was transferred to the product planning department. In addition, the members of the product planning department have deep understanding of technology. This makes it easier to bring about coordination with the engineering department. Additionally, the product planning department can conduct the design research while they are observing the trends of the engineering department. Therefore, the product planning department may play role more than the Heavy Weight Product Manager (Kim & Fujimoto, 1990). Conversely, the industrial design department specializes in styling and they have only slight knowledge of the technology. Therefore, it could be more difficult to coordinate with the engineering department if the industrial design department took initiative in conducting design research. The product planning department understands the importance of product forms as well as technology. They also admitted that Panasonic lost their market share to Samsung because of their poor design sense including product forms.

As described above, the need to establish resources between each research should be recognized in order to make a cross-functional team during the stage of the design research and the technological research. It is important not to derive any conclusion from the research of either technological and design, but to just show a direction and leave room for each research to come about. In order to enable communication and knowledge transfer between departments, it becomes necessary to understand common languages between departments (Carlile, 2002). A broker between both researches be required the understanding two languages, design and technology.

Resources to Create Meanings

A design research starts not with market information, but with analysis on sociocultural models (Verganti, 2009). What kind of resources do conduct a sociocultural analysis? Panasonic, previously described, creates meanings based on their own product planning department. According to Verganti (2009), Italian manufacturing industry creates meanings in terms of dialogue with the design discourse and this design discourse is an external resource. What is common between these two cases is the importance of company vision. Italian manufacturing industry always maintains its own vision fresh in the dialogue with the design discourse (Verganti, 2009). Then, is it possible to maintain their vision always fresh for the company that relies on in-house resources just like Panasonic? This matter may be similar to the discussion of the open innovation (Chesborough, 2003). In addition, Samsung may adopt so-called semi-opening method by dispatching their employees to various places in the world and has them experience local life in order to create new meanings. However, this is a discussion that is devoted to design research and does not consider mutual interference with technological research. Technological research in itself may also be the resource for radical meanings. Actually, there were many cases of silent design (Gorb & Dums, 1987) that industrial designers do not get involved in the design process. This paper clearly distinguishes between the design research and the industrial design; however, the industrial design is one of the important elements for creating radical meanings (Steiner, 1995). Silent design actually implies that an analysis of sociocultural models is not conducted in NPD process of an excellent design product, but something else not the design research takes initiative in creating radical meanings. Additionally, radical technologies could create radical meanings that have not been achieved in the past (Geels, 2004). In the case of digital cameras, CASIO does not conduct any design research and their purpose was technological substation, but they actually created a new feature that users could confirm and enjoy what they have just shot on the scene. This feature made a context of life be different from the one of a traditional camera and might change the emotional and symbolic value of a camera. Therefore, we can presume that technological feature created a radical meaning without the analysis of the sociocultural model. In other words, technological research could become resources for radical meanings, thus it is important to integrate design research and technological research.

The company vision should closely be associated with the thoughts of the top management. Actually, it can be observed in Apple and Dyson that the top executives affect the creation of meanings. It has also been found that the thoughts of the top managements on design have a great deal of impact on the achievements of the company (Hart & Service, 1988). In the future, a lot of discussions should be done with regard to the resources for the design research, as similar to the organizational approach that conducts the design research and the technological research.

CONCLUSION

Implication and Conclusion

This paper suggests a framework that comprehends design and technology. By using the concept of exploration and exploitation, this framework indicates what kind of interaction there is between meaning and technology in the design research and the technological research and its interaction is described by the FPD case study. The design research may explore radical meanings and exploit existing technologies, and the technological research may explore radical technologies and exploit existing meanings. Interactive augmenting of the design research with the technological research may achieve competitive advantages Additionally, the FPD case study indicates the possibility that to set the priority of the technological research and the design research in the product life cycle could be the keys to success. Thus, this framework may supports the concept in Figure 4 which is suggested by Verganti (2009). Although he described the effectiveness of design research and to analyze its process in detail, the analysis with regard to interaction between the design research and the technological research was not enough. Moreover, theoretical analysis about the technological research has not been addressed in the design management disciplinary. Therefore, our framework might play a significant role in order to address the interdisciplinary research in the design management disciplinary and the technology management disciplinary.

We also suggest the possibility that competitive advantages are brought about by conducting the design research and the technological research concurrently. In order to achieve competitive advantages, a cross-functional approach should be necessary at each stage of the research. With regard to this point, we suggest that the Japanese-style NPD could be the key to success. Actually, a representative Japanese company, Panasonic, has just started applying this approach to the stage

Figure 4. Verganti's framework (2009)

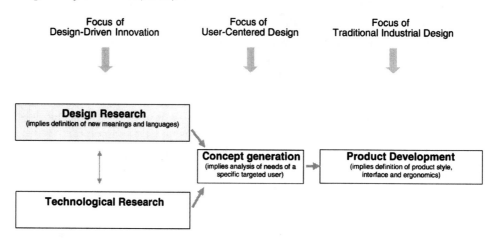

of the design research and the technological research. However, the issues with regard to resources of design research and a strategy for both researches are remained.

Limitations and Future Research

This paper suggests a few issues in the discussion part. They are based on a wide range of points of view; therefore, many studies should be addressed on this matter by the interdisciplinary research such as strategic theory and organizational theory (Noble, 2011). The following themes will be mainly discussed in strategic theory: (1) in what kind of environments design research and technological research can be conducted in the most effective way, (2) what kind of resources should be acquired (external or internal), and (3) relationship between the design research and the fuzzy front end. In organizational theory, the following will be considered: (1) what organization conducts design research (what rolls does the design department have?), (2) the relationship between organizations that conducts technological research and design research, and (3) how to incorporate the external resources internally.

This case study suggests that to set the priority of the design research and the technological research may affect the success of products significantly and importance of design research gradually increases in the product life cycle. However, we cannot examine individual processes of product development in detail, and what kind of technologies and meanings are existing or new. Therefore, we need to analyze other case studies of each individual product development in order to clarify its issue in a future research. In addition, meanings are created within the society of consumer; therefore, it is essential for us to conduct surveys of consumers in order to evaluate such meanings. However, the purpose of this paper is to suggest the comprehensive model of interaction between the design research and the technological research. We hope this stimulates further investigations in the field of technology management and design management.

REFERENCES

Andrade-Campos, A. (2011). Development of an optimization framework for parameter identification and shape optimization problems in engineering. *International Journal of Manufacturing, Materials, and Mechanical Engineering*, *1*(1), 57–79. doi:10.4018/ijmmme.2011010105

Bailetti, A. J., & Litva, P. F. (1995). Integrating customer requirements into product designs. *Journal of Product Innovation Management, 12*(1), 3–15. doi:10.1016/0737-6782(94)00021-7

Baldwin, C. Y., & Clark, K. B. (2000). *Design rules: The power of modularity.* Cambridge, MA: MIT Press.

Bayazit, N. (2004). Investigating design: A review of forty years of design research. *Design Issues, 20*(1), 16–29. doi:10.1162/074793604772933739

Bloch, P. H. (1995). Seeking the ideal form: Product design and consumer response. *Journal of Marketing, 59*(3), 16–29. doi:10.2307/1252116

Bloch, P. H. (2011). Product design and marketing: Reflections after fifteen years. *Journal of Product Innovation Management, 28*(3), 378–380. doi:10.1111/j.1540-5885.2011.00805.x

Borja de Mozota, B. (2003). *Design management.* New York: Allworth Press.

Butter, R., & Krippendorff, K. (1984). Product semantics—Exploring the symbolic qualities of form. *Journal of the Industrial Designers Society of America, 3,* 4–9.

Carlile, P. R. (2002). A pragmatic view of knowledge and boundaries: Boundary objects in new product development. *Organization Science, 13*(4), 442–455. doi:10.1287/orsc.13.4.442.2953

Chen, G. (1996). Conventional and fuzzy PID controller: An overview. *International Journal of Intelligent Control and Systems, 1,* 235–246. doi:10.1142/S0218796596000155

Chesborough, H. (2003). *Open innovation: The new imperative for creating and profiting from technological.* Boston, MA: Harvard Business School Press.

Christensen, C. M. (2003). *The innovator's dilemma: When new technologies cause great firms to fail.* Boston, MA: Harvard Business School Press.

Clark, K., & Fujimoto, T. (1990). The power of product integrity. *Harvard Business Review,* (November-December): 107–118. PMID:10107956

Cooper, R., Prendiville, A., & Jones, T. (1995). High technological NPD. *CoDesign, 3,* 14–22.

Cox, D., & William, L. (1987). Product novelty: Does it moderate the relationship between ad attitudes and brand attitudes? *Journal of Advertising, 16,* 39–44. doi:10.1080/00913367.1987.10673084

Creusen, H., & Schoormans, L. (2005). The different role of product appearance in consumer choice. *Journal of Product Innovation Management, 22*(1), 63–82. doi:10.1111/j.0737-6782.2005.00103.x

Creusen, M. E. H. (2011). Research opportunities related to consumer response to product design. *Journal of Product Innovation Management, 28*(3), 405–408. doi:10.1111/j.1540-5885.2011.00812.x

Csikszentmihalyi, M., & Rochberg-Halton, E. (1981). *The meaning of things: domestic symbols and the self.* Cambridge, UK: Cambridge University Press. doi:10.1017/CBO9781139167611

Dahl, D. W. (2011). Clarity in defining product design: inspiring research opportunities for the design process. *Journal of Product Innovation Management, 28*(3), 425–427. doi:10.1111/j.1540-5885.2011.00816.x

Dahl, D. W., Chattopdhyay, A., & Gorn, G. (1999). The use of visual mental imagery in new product design. *JMR, Journal of Marketing Research, 36*(1), 18–28. doi:10.2307/3151912

Dan Zhang, P. H., & Kotabe, M. (2011). Marketing–industrial design integration in new product development: The case of China. *Journal of Product Innovation Management, 28*(3), 360–373. doi:10.1111/j.1540-5885.2011.00803.x

Dell'Era, C., & Verganti, R. (2007). Strategies of innovation and imitation of product languages. *Journal of Product Innovation Management, 24,* 580–599. doi:10.1111/j.1540-5885.2007.00273.x

Geels, F. W. (2004). From sectoral systems of innovation to socio- technical systems. Insights about dynamics and change from sociology and institutional theory. *Research Policy, 33,* 897–920. doi:10.1016/j.respol.2004.01.015

Gemser, G., & Leenders, M. A. A. M. (2001). How integrating industrial design in the product development process impacts on company performance. *Journal of Product Innovation Management, 18*(1), 28–38. doi:10.1016/S0737-6782(00)00069-2

Goodrich, L. L. (1995). The design of the decade: Quantifying design impact over ten years. *Design Management Journal, 5*(2), 47.

Gorb, P., & Dums, A. (1987). Silent design. *Design Studies, 8,* 150–156. doi:10.1016/0142-694X(87)90037-8

Greve, H. R. (2007). Exploration and exploitation in product innovation. *Industrial and Corporate Change, 16*(5), 945–975. doi:10.1093/icc/dtm013

Griffin, A., & Hauser, J. R. (1996). Integrating R&D and marketing: A review and analysis of the literature. *Journal of Product Innovation Management, 13,* 191–213. doi:10.1111/1540-5885.1330191

Grossbart, S., Mittelstaedt, R. A., Curtis, W. W., & Rogers, R. D. (1975). Environmental sensitivity and shopping behaivor. *Journal of Business Research, 3*(4), 281–294. doi:10.1016/0148-2963(75)90010-7

Hargadon, A., & Sutton, R. I. (1997). Technological brokering and innovation in a product development firm. *Administrative Science Quarterly, 42*(4), 716–749. doi:10.2307/2393655

Hart, S., & Service, L. (1988). The effects of managerial attitude to design on company. *Journal of Marketing Management, 4*(2), 217–230. doi:10.1080/0267257X.1988.9964070

Hartley, R. (1992). *Concurrent engineering.* Combridge, USA: Productivity Press.

Hayes, R. (1990). Design: Putting class into 'world class'. *Design Management Journal, 1*(2), 8–14.

Herstatt, C., Verworn, B., & Nagahira, A. (2004). Reducing project related uncertainty in the "fuzzy front end" of innovation: A comparison of German and Japanese product innovation projects. *International Journal of Product Development, 1*(1), 43–65. doi:10.1504/IJPD.2004.004890

Hertenstein, J. H., & Platt, M. B. (1997). Developing a strategic design culture. *Design Management Journal, 2*(2), 10–19.

Hertenstein, J. H., Platt, M. B., & Veryzer, R. W. (2005). The impact of industrial design effectiveness on corporate financial performance. *Journal of Product Innovation Management, 22*(1), 3–21. doi:10.1111/j.0737-6782.2005.00100.x

Hirschman, E. C. (1982). Symbolizm and technology as sources for the generation of innovations. *Advances in Consumer Research. Association for Consumer Research (U. S.), 9*(1), 537–541.

Hoegg, J., & Alba, J. W. (2011). Seeing is believing (too much): The influence of product form on perceptions of functional performance. *Journal of Product Innovation Management, 28*(3), 346–359. doi:10.1111/j.1540-5885.2011.00802.x

Imai, K., Nonaka, I., & Takeuchi, H. (1985). *Managing new product development process: How Japanese learn and unlearn.* Boston, MA: Harvard Business School Press.

Johne, A. F., & Snelson, P. A. (1988). Success factors in product innovation: A selective review of the literature. *Journal of Product Innovation Management, 5*(2), 114–128. doi:10.1016/0737-6782(88)90003-3

Kelley, T. (2001). *The Art of innovation.* New York: Curreny.

Khurana, A., & Rosenthal, S. R. (1998). Towards holistic "front ends" in new product development. *Journal of Product Innovation Management, 15*(1), 57–74. doi:10.1016/S0737-6782(97)00066-0

Kreuzbauer, R., & Malter, A. J. (2005). Embodied cognition and new product design: changing product form to influence brand categorization. *Journal of Product Innovation Management, 22*(2), 165–176. doi:10.1111/j.0737-6782.2005.00112.x

Krippendorff, K. (1989). On the essential contexts of artifacts or on the proposition that "design is making sense (of things)". *Design Issues, 5*(2), 9–38. doi:10.2307/1511512

Kumar, V., & Whitney, P. (2003). Faster, deeper user research. *Design Management Journal, 14*(2), 50–55.

Latour, B. (1987). *Science in action: How to follow scientists and engineers through society.* Cambridge, MA: Harvard University Press.

Levy, S. (2006). *The perfect thing: How the iPod shuffles commerce, culture, and coolness.* New York: Simon & Schuster.

Ligas, M. (2000). People, products, and pursuits: Exploring the relationship between consumer goals and product meanings. *Psychology and Marketing, 17*(11), 983–1003. doi:10.1002/1520-6793(200011)17:11<983::AID-MAR4>3.0.CO;2-J

Linzmayer, O. W. (2004). *Apple confidential 2.0: The definitive history of the world's most colorful company.* San Francisco: No Starch Press.

Lojacono, G., & Zaccai, G. (2004). The evolution of the design-inspired enterprise. *Sloan Management Review, 45*, 75–79.

Luchs, M., & Scott Swan, K. (2011). The emergence of product design as a field of marketing inquiry. *Journal of Product Innovation Management, 28*(3), 327–345. doi:10.1111/j.1540-5885.2011.00801.x

Lukas, B. A., & Ferrell, O.bC. (2000). The effect of market orientation on product innovation. *Journal of the Academy of Marketing Science, 28*(2), 239–247. doi:10.1177/0092070300282005

Maidique, M. A., & Zigger, B. J. (1985). The new product learning cycle. *Research Policy, 14*(6), 299–313. doi:10.1016/0048-7333(85)90001-0

March, J. G. (1991). Exploration and exploitation in organizational learning. *Organization Science, 2*(1), 71–87. doi:10.1287/orsc.2.1.71

Mertes, J. (1965). Visual design and marketing manager. *California Management Review, 8*(2), 29–39. doi:10.2307/41165669

Moenaert, R. K., De Meyer, A., Souder, W. E., & Deschoolmeester, D. (1995). R&D/marketing, communications during the fuzzy front-end. *IEEE Transactions on Engineering Management, 42*(3), 243–258. doi:10.1109/17.403743

Nagano, H., Ishida, S., & Ikeda, J. (2011). A study about boundaries of firm in FPD industry. *Journal of Japan Association for Management Systems, 28*(1), 1–8.

NIKKEI BIZTECH. (2004, December 20). Study finds free care used more, pp. 20-25.

NIKKEI BIZTECH. (2004, August 15). Study finds free care used more, pp. 168-175.

NIKKEI BUSINESS. (2003, October 6). Study finds free care used more, pp. 28-33.

NIKKEI BUSINESS. (2006, January 16). Study finds free care used more, pp. 13.

NIKKEI BUSINESS. (2007, December 24/31). Study finds free care used more, pp. 46-49.

NIKKEI BUSINESS. (2008, January 28). Study finds free care used more, pp. 53-55.

NIKKEI BUSINESS. (2009, March 31). Study finds free care used more, pp. 7-9.

NIKKEI BUSINESS. (2010, July 5). Study finds free care used more, pp. 34-40.

NIKKEI BUSINESS. (2011, January 31). Study finds free care used more, pp. 22-24.

NIKKEI BUSINESS. (2011, October 31). Study finds free care used more, pp. 16.

NIKKEI DESIGN. (2004, October). Study finds free care used more, pp. 50-61.

NIKKEI ELECTRONICS. (2005, September 26). Study finds free care used more, pp. 112-119.

NIKKEI ELECTRONICS. (2006, October 9). Study finds free care used more, pp. 112-121.

NIKKEI ELECTRONICS. (2007, August 27). Study finds free care used more, pp. 168-170.

NIKKEI ELECTRONICS. (2007, November 19). Study finds free care used more, pp. 54-61.

NIKKEI MONOZUKURI. (2006, November). Study finds free care used more, pp. 54-55.

Noble, C. H. (2011). On elevating strategic design research. *Journal of Product Innovation Management, 28*, 389–393. doi:10.1111/j.1540-5885.2011.00808.x

Oakley, M. (1986). *Managing design. Product design and technological innovation*. University Press.

Oikonomou, D., Moulianitis, V., Lekkas, D., & Koutsabasis, P. (2011). DSS for health emergency response: A contextual, user-centred approach. *International Journal of User-Driven Healthcare, 1*(2), 39–56. doi:10.4018/IJUDH.2011040120110401 04

Pisano, G. P., & Verganti, R. (2008). Which kind of collaboration is right for you? The new leaders in innovation will be those who figure out the best way to leverage a network of outsiders. *Harvard Business Review, 84*(12), 78–86.

Rao, R. V., & Patel, B. K. (2011). Material selection using a novel multiple attribute decision making method. *International Journal of Manufacturing, Materials, and Mechanical Engineering, 1*(1), 43–56. doi:10.4018/ijmmme.2011010104

Redstrom, J. (2006). Towards user design? On the shift from object to user as the subject of design. *Design Studies, 27*(2), 123–139. doi:10.1016/j.destud.2005.06.001

Reid, S. E., & De Brentani, U. (2004). The fuzzy front end of new product development for discontinuous innovation: a theoretical model. *Journal of Product Innovation Management, 21*(3), 170–184. doi:10.1111/j.0737-6782.2004.00068.x

Rosenthal, S. R., & Capper, M. (2006). Ethnographies in the front end: designing for enhanced customer experiences. *Journal of Product Innovation Management, 23*(3), 215–237. doi:10.1111/j.1540-5885.2006.00195.x

Rusut, R. T., Thompson, D. V., & Hamilton, R. W. (2006). Defeating feature fatigue. *Harvard Business Review, 84*(2), 98–107. PMID:16485808

Sewall, M. A. (1978). Market segmentation based on consumer ratings of proposed product designs. *JMR, Journal of Marketing Research, 15*(4), 557–564. doi:10.2307/3150625

Song, X. M., & Parry, M. E. (1996). What separates Japanese new product winners from losers. *Journal of Product Innovation Management, 13*(5), 422–439. doi:10.1016/0737-6782(96)00055-0

Song, X. M., & Parry, M. E. (1997). The determinants of Japanese new product successes. *JMR, Journal of Marketing Research, 34*(1), 64–76. doi:10.2307/3152065

Song, X. M., & Parry, M. E. (1997). A cross-national comparative study of new product development processes: Japan and the United States. *Journal of Marketing, 61*(2), 1–18. doi:10.2307/1251827

Steiner, C. J. (1995). A Philosophy for innovation: The role of unconventional individuals in innovations success. *Journal of Product Innovation Management, 12*(5), 431–440. doi:10.1016/0737-6782(95)00058-5

Swift, P. W. (1997). Science drives creativity: A methodology for quantifying perceptions. *Design Management Journal, 8*(2), 51–57.

Townsend, J. D., Montoya, M. M., & Calantone, R. J. (2011). Form and function: A matter of perspective. *Journal of Product Innovation Management, 28*(3), 327–345. doi:10.1111/j.1540-5885.2011.00804.x

Utterback, J. M., Vedin Bengt-Arne, A. E., Ekman, S., Sanderson, S., Tether, B., & Verganti, R. (2006). *Design-Inspired Innovation*. New York: World Scientific.

Verganti, R. (2006). Innovating through design. *Harvard Business Review, 84*(12), 114–122.

Verganti, R. (2008). Design, meanings, and radical innovation: A metamodel and a research agenda. *Journal of Product Innovation Management, 25*(3), 436–456. doi:10.1111/j.1540-5885.2008.00313.x

Verganti, R. (2009). *Design-driven innovation: Changing the rules of competition by radically innovating what things mean*. Boston, MA: Harvard Business Press.

Verganti, R. (2011). Radical design and technological epiphanies: A new focus for research on design management. *Journal of Product Innovation Management, 28*(3), 384–388. doi:10.1111/j.1540-5885.2011.00807.x

Verganti, R. (2011). Designing breakthrough products. *Harvard Business Review, 89*(10), 115–120.

Verworn, B., Herstatt, C., & Nagahira, A. (2008). The fuzzy front end of Japanese new product development projects: impact on success and differences between incremental and radical projects. *R & D Management, 38*(1), 1–19. doi:10.1111/j.1467-9310.2007.00492.x

Veryzer, R. W. (1993). Aesthetic response and the influence of design principles on product preferences. *Advances in Consumer Research. Association for Consumer Research (U. S.), 20*(1), 224–228.

Veryzer, R. W. (1995). The place of product design and aesthetics. *Advances in Consumer Research. Association for Consumer Research (U. S.), 22*(1), 641–645.

Veryzer, R. W. (1998). Discontinuous innovation and the new product development process. *Journal of Product Innovation Management, 15*(4), 304–321. doi:10.1016/S0737-6782(97)00105-7

Veryzer, R. W. (2005). The roles of marketing and industrial design in discontinuous new product development. *Journal of Product Innovation Management, 22*(1), 22–41. doi:10.1111/j.0737-6782.2005.00101.x

Veryzer, R. W., & Borja de Mozota, B. (2005). The impact of user-oriented design on new product development. *Journal of Product Innovation Management, 22*(1), 128–143. doi:10.1111/j.0737-6782.2005.00110.x

Veryzer, R. W., & Hutchinson, W. (1998). The influence of utility and prototypicality on aesthetic responses to new product designs. *The Journal of Consumer Research, 24*(4), 224–228. doi:10.1086/209516

Vredenburg, K., Isensee, S., & Righi, C. (2002). *User-centered design: An integrated approach*. Upper Saddle River, NJ: Prentice Hall.

Whalsh, V. (1995). The evaluation of design. *International Journal of Technological Management, 10*(4/5/6), 489-509.

Whalsh, V. (1996). Design, innovation and the boundaries of the firm. *Research Policy, 25*(4), 509–529. doi:10.1016/0048-7333(95)00847-0

Whalsh, V., Roy, R., & Bruce, M. (1998). Competitive by design. *Journal of Marketing Management, 4*(2), 201–216. doi:10.1080/0267257X.1988.9964069

Yin, R. (1994). *Case study research: Design and METHODS* (2nd ed.). Thousand Oaks, CA: Sage Publications Inc.

Zadeh, L. A. (1965). Fuzzy sets. *Information and Control, 8*(3), 338–353. doi:10.1016/S0019-9958(65)90241-X

Zhang, Q., & Doll, W. J. (2001). The fuzzy front end and success of new product development: A causal model. *European Journal of Innovation Management, 4*(2), 95–112. doi:10.1108/14601060110390602

ADDITIONAL READING

Armstrong, L. (1991). It Started with Egg. *Business Week, (December 2)*, 142-146.

Blichfeldt, B. S. (2005). On the development of brand and line extensions. *Brand Management, 12*(3), 177–190. doi:10.1057/palgrave.bm.2540214

Bollen, K. A. (1989). *Structural Equations with Latent Variables*. New York: John Wiley & Sons.

Crilly, N., Moultrie, J., & Clarkson, P. J. (2004). Seeing things: Consumer response to the visual domain in product design. *Design Studies, 25*(6), 547–577. doi:10.1016/j.destud.2004.03.001

Crozier, W. R. (1994). *Manufactured pleasures: Psychological response to design*. Manchester, UK: Manchester University Press.

Dahlin, K. B., & Behrens, D. M. (2005). When Is an Invention Really Radical? Defining and Measuring Technological Radicalness. *Research Policy, 34*, 717–737. doi:10.1016/j.respol.2005.03.009

Ernst, H., Hoyer, W. D., & Rübsaamen, C. (2010). Sales, Marketing, and Research-and-Development Cooperation Across New Product Development Stages. *Journal of Marketing, 74*, 80–92. doi:10.1509/jmkg.74.5.80

Fournier, S. (1991). A Meaning-Based Framework for the Study of Consumer-Object Relations. *Advances in Consumer Research. Association for Consumer Research (U. S.), 18*(1), 736–742.

Friedmann, R. (1986). Psychological Meaning of Products: Identification and Marketing Applications. *Psychology and Marketing, 3*(1), 1–15. doi:10.1002/mar.4220030102

Goode, M. R., Dahl, D. W., & Moreau, C. P. (2013). Innovation Aesthetics: The Relationship between Category Cues, Categorization Certainty, and Newness Perceptions. *Journal of Product Innovation Management, 30*(2), 192–208. doi:10.1111/j.1540-5885.2012.00995.x

Goto, S. (2012). Exploration and Exploitation of Meanings: The Interaction between Design Research and Technological Research. *The R&D Management Conference 2012*.

Johansson, J. K., & Hirano, M. (1995). Brand Reality: The Japanese Perspective. *Journal of Marketing Management, 15*, 93–105. doi:10.1362/026725799784870478

Kang, I., Lee, G. C., Park, C., & Shin, M. M. (2012). Tailored and targeted communication strategies for encouraging voluntary adoption of non-preferred public policy. *Technological Forecasting and Social Change, 80*(1), 24–37. doi:10.1016/j.techfore.2012.08.001

Klein, R. E. III, & Kernan, J. B. (1988). Measuring the Meaning of Consumption objects: An Empirical Investigation. *Advances in Consumer Research. Association for Consumer Research (U. S.), 15*(1), 498–504.

Klein, R. E. III, & Kernan, J. B. (1991). Contextual Influence on the Meanings Ascribed to Ordinary Consumption. *The Journal of Consumer Research, 18*(3), 311–324. doi:10.1086/209262

Levy, S. J. (1959). Symbols for sale. *Harvard Business Review, 37*(4), 117–124.

McCracken, G. (1986). Culture and Consumption: A Theoretical Account of the Structure and Movement of the Cultural Meaning of Consumer Goods. *The Journal of Consumer Research, 13*(1), 71–84. doi:10.1086/209048

Mono, R. (1997). *Design for product understanding*. Stockholm, Sweden: Liber.

Norman, D. A. (2004). *Emotional design: why we love (or hate) everyday things*. New York, NY: Basic Books.

O'Connor, G. C., & Veryzer, R. W. (2001). The nature of market visioning for technology-based radical innovation. *Journal of Product Innovation Management, 18*, 231–246. doi:10.1016/S0737-6782(01)00092-3

Paul, P. J., & Olson, J. (1987). *Consumer Behavior, Marketing Strategy Perspective*. Homewood, IL: Richard D, Irwin, Inc.

Ranscombe, C., Hicks, B., Mullineux, G., & Singh, B. (2012). Visually decomposing vehicle images: Exploring the influence of different aesthetic features on consumer perception of brand. *Design Studies, 33*(4), 319–341. doi:10.1016/j.destud.2011.06.006

Rubera, G., & Droge, C. (2013). Technology versus Design Innovation's Effects on Sales and Tobin's Q: The Moderating Role of Branding Strategy. *Journal of Product Innovation Management, 30*(3), 448–464. doi:10.1111/jpim.12012

Shannon, C. E. (n.d.). A mathematical theory of communication. *The Bell System Technical Journal, 27*, 379–423. doi:10.1002/j.1538-7305.1948.tb01338.x

Shibata, T. (2012). Unveiling the successful process of technological transition: A case study of Matsushita Electric. *R & D Management, 42*(4), 358–376.

Solomon, M. R. (1983). The Role of Products as Social Stimuli: A Symbolic Interactionism Perspective. *The Journal of Consumer Research, 10*(3), 319–329. doi:10.1086/208971

Tsai, S. (2005). Utility, cultural symbolism and emotion: A comprehensive model of brand purchase value. *International Journal of Research in Marketing, 22*, 277–291. doi:10.1016/j.ijresmar.2004.11.002

KEY TERMS AND DEFINITIONS

Design Research: An activity to aim at creating radical meanings.

Design: Making sense (of things).

Exploitation: To use existing ideas or methodologies to address some challenge.

Exploration: To seek and find novel ideas or methodologies in order to address some challenge.

Fuzzy Front End: The fuzzy front-end of NPD is the time and activity prior to an organization's first screen of a new product idea and the important phase in a product strategy.

Meaning: Irrational values such as emotional and symbolic ones.

Technological Research: An activity to aim at creating radical technologies.

Chapter 10
The Role of Brand Loyalty on CRM Performance:
An Innovative Framework for Smart Manufacturing

Kijpokin Kasemsap
Suan Sunandha Rajabhat University, Thailand

ABSTRACT

This chapter introduces the framework and causal model of customer value, customer satisfaction, brand loyalty, and customer relationship management performance in terms of the innovative manufacturing and marketing solutions. It argues that dimensions of customer value, customer satisfaction, and brand loyalty have mediated positive effect on customer relationship management performance. Furthermore, brand loyalty positively mediates the relationships between customer value and customer relationship management performance and between customer satisfaction and customer relationship management performance. Customer value is positively correlated with customer satisfaction. Understanding the theoretical learning is beneficial for organizations aiming to increase customer relationship management performance and achieve business goals.

INTRODUCTION

Customer value and customer satisfaction are becoming the important factors for successful business competition for either manufacturers or service providers (Bolton & Drew, 1991; Buzzel & Gale, 1987; Chang & Chen, 1998; Parasuraman, 1997; Parasuraman et al., 1988, 1991; Phillips et al., 1983; Wang et al., 2004; Zeithaml et al., 1996). Delivering superior customer value is a matter of an ongoing concern in building and sustaining competitive advantage by driving customer relationship management (CRM) performance (Wang et al., 2004). In addition, customer value is a strategic weapon in attracting and retaining customers and it is one of the most significant factors in the success of both manufacturing businesses and service providers (Parasuraman, 1997; Woodruff, 1997). Firms should reorient their operations toward the creation and delivery of superior customer value if they need to improve customer relationship

DOI: 10.4018/978-1-4666-5836-3.ch010

management performance (Jensen, 2001; Slater, 1997). The growing body of knowledge about customer value is rather fragmented; the different points of view are advocated with no widely accepted way of pulling views together and the related empirical study of customer value is very limited (Jensen, 2001; Slater, 1997).

Furthermore, relevant studies have not yet yielded any unambiguous interpretations of the key dimensions of customer value (Lapierre, 2000; McDougall & Levesque, 2000). Little is known about the relative importance of each dimension of customer value in improving different dimensions of customer relationship management performance. Although there is a significant body of knowledge about the concept of customer value and its relationships with service quality and customer satisfaction, there are the relatively little empirical research on the customer value (Lapierre, 2000). Only a few studies have focused on how superior customer value is constituted in the perspective of customers and how a more reliable measurement scale for such an important construct might be developed (Sheth et al., 1991; Sweeney & Soutar, 2001). Furthermore, Sweeney & Soutar (2001) considered the price as the only sacrifice of customers in measuring customer value. Many other types of sacrifices (i.e., opportunity cost, psychological cost, and maintenance and learning cost) can exert determining influences on the perception of customer value besides price (Slater, 1997; Woodruff, 1997). In addition, market power is growing, but the market infrastructure is not well developed and the application of customer value knowledge and customer relationship management is rather limited (Peng & Health, 1996). By drawing on a growing body of literature on customer value, customer relationship management and other findings, the research measures the customer value in terms of get (benefit) and give (sacrifice) components (Slater, 1997; Woodruff, 1997) in contrast to those researchers who have argued that perceived value consists only of benefits (Hamel & Prahalad, 1994), thus adopting a broader concept of sacrifices rather than the simple price assessment suggested by Sweeney and Soutar (2001); and several key dimensions apart from price and quality are identified conceptually and tested empirically.

The evaluation of customer relationship management performance is discussed on the basis of models of customer behavior, customer equity, and customer asset (Blattberg et al., 2001; Rust et al., 2000). Brand as a basic of the competitive game, must be carefully defined, created and managed as branding enables a producer to obtain the benefits of offering products with superior quality, thus providing an opportunity to transfer this identifiable relationship to other products or services (Motameni & Shahrokhi, 1998). In addition, the strong brand names lead competitive advantages (Lee & Back, 2010), increase organizational cash flow and accelerate liquidity (Miller & Muir, 2004), and provide premium price, profitability and more loyalty for customers (Madden et al., 2006). Revitalized brand images might help attract such customers who might remain loyal over a significantly longer period (Guo et al., 2011). Management teams need to promote customer orientation in their organization in order to implement customer relationship practices (Cai, 2009). Furthermore, organizational customer orientation is a prerequisite for the successful implementation of customer relationship practices. Consequently, customer orientation emphasizes creating superior customer value (Homburg et al., 2002; Slater & Narver, 1995). Research objective was to construct an innovative framework and a causal model of customer value, customer satisfaction, brand loyalty, and customer relationship management performance in terms of the innovative manufacturing and marketing solutions of pulp and paper company employees in Thailand.

BACKGROUND

The details of constructs such as customer value, customer satisfaction, brand loyalty, and customer relationship management performance for smart manufacturing related to this chapter are shown as follows.

Customer Value

Driven by demanding customers, keen competition, and technological change, many firms have sought to deliver superior customer value (Butz & Goodstein, 1996; Woodruff, 1997). Zeithaml (1988) reviewed the relevant literature and reported four common features of customer value (i.e., price, trade-off between costs and benefits, trade-off between perceived product quality and price, and overall assessment of subjective worth). Furthermore, Zeithaml (1988) considered customer value as the customer's overall assessment of the utility of a product based on the perception of what is received and what is given. Dodds et al. (1991) stated that buyers' perceptions of customer value represent a trade-off between the quality or benefits that they receive in the product and the sacrifice that they perceive in paying the price. Gale (1994) considered customer value as the market perceiving quality to be adjusted for relative product price. Butz and Goodstein (1996) defined customer value as the emotional bond established between a customer and a producer after the customer has used a salient product or service produced by the supplier. Woodruff (1997) defined customer value as a customer-perceived preference for, and evaluation of, product attributes, attribute performances, and consequences in terms of the customer's goals and purposes in situations based on customer perspectives on value derived from empirical research into how customers really think about value.

In addition, Woodruff (1997) stated that customer value is derived from the perception, preference, and evaluation of customers, and that any consideration of customer value should take account of these factors. The concept of customer value is often perceived as a multifaceted word (Mentzer et al., 1997; Parasuraman, 1997). Although various theoretical concepts of customer value coexist, and at least partly overlap each other (i.e., perceived customer value, utility, worth, benefits, and perceived quality) (Parasuraman, 1997; Teas & Agarwal, 2000), Monroe (2005) indicated that customer value is the buyers' mental trade-off between the quality or benefits they perceive in the product related to the sacrifice that they perceive by paying the price. Thus, perceived customer value is affected by both intrinsic and extrinsic dimensions of product quality and perceived price (Akshay & Monroe, 1988; Monroe, 2005). In addition, Holbrook (1996) defined customer value as an interactive preference experience. The interaction of customer value will happen between a consumer and a product (Holbrook, 1996). Furthermore, Holbrook (1996) constructed a typology of customer value in which the value construct is divided into extrinsic and intrinsic customer values. An extrinsic (or other-oriented) customer value refers to a means-ends relationship, which might include efficiency, excellence, status, and esteem (Holbrook, 1996). On the other hand, an intrinsic (or self-oriented) customer value consists of play, aesthetics, ethics and spirituality (Holbrook, 1996).

Information of customer value is a necessary element of accurate pricing and other marketing decisions (Munnukka & Farvi, 2012). Furthermore, companies aiming to improve the customer value of their high-tech products should especially focus on price satisfaction, perceived excellence of the product, and visual appeal (Munnukka & Farvi, 2012). According to Sweeney and Soutar (2001), firms enquire about other factors that might constitute the perceived benefits and sacrifices, and the managerial implications of these factors, if they need to understand the perception and evaluation process of customers and reconfigure their resources and activities accordingly. For example,

Kotler (1997) indicated that customer value can be understood in terms of product value, service value, employee value, and image value. However, this approach is largely derived from the standpoint of a firm or at least not totally customer based (Kotler, 1997). The broad theoretical framework developed by Sheth et al. (1991) is different in that they suggested five dimensions of customer value from the customer's perspective (i.e., social, emotional, functional, epistemic, and conditional perspective) as providing the best foundation for extending the customer value construct. However, it is worth noting that not all these dimensions have equal significance at any time, although they are related in some sense (Sheth et al., 1991).

The concept of customer value is applied to many settings in the management, strategy, finance, information systems, and marketing literatures (Wikstrom & Normann, 1994). Furthermore, the concept of customer value is multifaceted and complicated by numerous interpretations and biases. For instance, the derivations of various customer value dimensions are evident from the aspect of strategic value (Katz, 1993), the economic contribution that customer value makes to profit maximization (Banker & Kauffman, 1991), value as manifest in continuous quality improvement initiatives (Richardson & Gartner, 1999), intrinsic value created through scientific approaches to organization (Meredith et al., 1994) and the customer value in organization (Blattberg & Deighton, 1996; Engel et al., 1990), and the principle of customer perceived value (Zeithaml, 1988). Furthermore, creation of superior customer value is a key element for organizational success (Higgins, 1998; Kordupleski & Laitamaki, 1997; Milgrom & Roberts, 1995; Porter, 1996; Woodruff, 1997; Wyner, 1996). Likewise, building superior customer value is a major goal for market-driven firms (Day, 1990; Narver & Slater, 1990). The meaning of customer value is a level of return in the product benefits for a customer's payment in a purchase exchange (Normann & Ramirez, 1993). The number of customers can be increased by delivering more customer value than the competition (Day & Wensley, 1988; Gale, 1994; Porter, 1985; Woodruff & Gardial, 1996).

Companies realize commercial success through the satisfaction of customers by ensuring they perceive the customer value that they expect (Zeithaml et al., 1990). Creation of customer value through closer and more special relationships leads to customer satisfaction, trust, affective commitment and loyalty (Bakanauskas & Jakutis, 2010; Bick, 2009; Cailleux et al., 2009). Building and maintaining long-term relationships with customers should encompass a value co-creation process of experiences (Smith & Colgate, 2007). Smith and Colgate (2007) proposed a customer-value framework on the basis of a thorough literature study and analysis. In addition, Smith and Colgate (2007) designed a customer value framework with the practical implication for marketers by proposing categories of customer value that can differentiate marketing offerings. The customer value framework identifies four types of value, including functional/instrumental, experiential/hedonic, symbolic/expressive and cost/sacrifice value from five value sources (i.e., information, products, interactions, environment, and environment) (Smith & Colgate, 2007). Most brand managers aim to build a positive brand relationship with their customers to prevent customers from searching for alternative brands and engaging in cross-shopping (Choo et al., 2012).

Customer value can be better understood in terms of the four key dimensions of emotional value, social value, functional value, and perceived sacrifice (Wang et al., 2004). Emotional value refers to utility derived from the affective aspects that a product or service generates; social value refers to the social utility derived from the product or service; functional value refers to the utility derived from the perceived quality and expected performance of a product or service, and perceived sacrifice refers to the loss derived from the product or service due to the increment of its perceived short-term and long-term costs (Wang et al., 2004).

Furthermore, a strong long-term brand relationship with customers is the ultimate goal of customer relationship management programs (Choo et al., 2012). Many firms are transforming their focus within the firm for improvement in terms of quality management, downsizing, business process reengineering, and agile manufacturing process to pursue superior customer value delivery (Band, 1991; Butz & Goodstein, 1996; Day, 1990; Gale, 1994; Naumann, 1995, Woodruff, 1997). Any factors, influencing benefits that customer can get, will cause different evaluation of customer value even though different customers may provide different opinions over time, for example, product-related factors (i.e., product quality and product customization), service-related factors (i.e., responsiveness, flexibility, reliability, and technical competencies), and relationship-related factors (i.e., image, time, effort, energy, and solidarity) are all customer-value drivers (Bolton & Drew, 1991; Lapierre, 2000; Ravald & Gronroos, 1996; Zeithaml, 1988).

Furthermore, customer value has a relationship not only with what customers can get, but also with what they have to give up, that is the customer perceived sacrifice. For example, Lapierre (2000) identified the key drivers of customer perceived value and defined the customer perceived sacrifice as one of the two key factors (the other is benefit). Customer perceived sacrifice refers to what is given up or sacrificed to acquire a product or service (Heskett et al., 1997; Zeithaml, 1988). However, not only price is considered as the element of customer perceived sacrifice, but also other non-monetary factors are believed to be closely related to customer perceived sacrifice. Furthermore, Carothers and Adams (1991) stated that many customers count time rather than dollar cost as their most previous asset. Therefore, it is obvious that there are two broad types of customer perceived sacrifice: monetary costs and non-monetary costs (De Ruyter et al., 1997). Monetary costs can be assessed by a direct measure of the dollar price of the service or product (De Ruyter et al., 1997). Non-monetary costs can be defined as the time, effort, energy, distance, and conflict invested by customers to obtain the products or services or to establish a relationship with a supplier (De Ruyter et al., 1997). In addition, Flint et al. (1997) defined customer value as the customers' perception of what they want to be happened in a specific kind of use situation, by offering a product or service in order to accomplish a desired purpose or goal. This definition of customer value implies that customer value is created by products and services when the benefits help customers to achieve their goals in various situations (Flint et al., 1997). Furthermore, Plakoyiannaki and Saren (2006) suggested that the process of customer value is concerned with transforming the firm's understanding from customers to offerings that deliver customer value.

Customer Satisfaction

Customer satisfaction is defined as an affective state with positive feelings (Cronin et al., 2000) resulting from an evaluation of the overall consumption experiences. Customer satisfaction is the consumer's fulfillment response (Andreassen, 2000). Torres and Kline (2006) defined customer satisfaction as the individual's perception of the performance of the product or service in relation to his or her expectations. Moreover, customer satisfaction is an attitude change obtaining from the consumption experience (McCollough et al., 2000; Oliver, 1981). Furthermore, customer satisfaction is a judgment that a product or service feature, or the product of service itself, provided a pleasurable level of consumption-related fulfillment, including levels of under- or over-fulfillment (Oliver, 1997; Tronvoll, 2011). There are at least two different conceptualizations of customer satisfaction: transaction-specific customer satisfaction and cumulative customer satisfaction (Andreassen, 2000; Boulding et al., 1993; Johnson et al., 1995). From a transaction-specific perspective, customer satisfaction is viewed as a post-choice evaluative judgment of a specific purchase occasion (Hunt, 1977; Oliver, 1977, 1980, 1981, 1993). Past studies have also used a more transaction-specific measure of service satisfaction (Andreassen, 2000; Boulding et al., 1993).

Cumulative customer satisfaction is an overall evaluation based on the total purchase and consumption experiences with a product or service over time which is more useful than transaction-specific customer satisfaction in predicting a customer's subsequent behavior and a company's past, present and future performance (Anderson et al., 1994; Fornell, 1992; Johnson & Fornell, 1991). In addition, cumulative customer satisfaction motivates a firm's investment in customer satisfaction (Anderson et al., 1994; Fornell, 1992; Johnson & Fornell, 1991). Customer satisfaction is also influenced by customer value, and the two concepts (i.e., brand loyalty and customer satisfaction) exert their effects on customer behavior-based customer relationship management performance simultaneously (Fornell et al., 1996). In addition, customer satisfaction has a positive impact on service recovery, leading to a high-level of customer loyalty through positive repurchase intention (Komunda & Osarenkhoe, 2012). Higher customer satisfaction has been proposed to be related to higher customer loyalty (Morrisson & Huppertz, 2010; Sousa & Voss, 2009). Customer satisfaction is a complex construct and is defined in various ways (Barsky, 1995; Besterfield, 1994; Fecikova, 2004; Kanji & Moura, 2002). Customer satisfaction is a major goal of business organizations, since it is considered to affect customer retention and companies' market share (Hansemark & Albinsson, 2004). Zeithaml (1988) developed a body of literature on the antecedents and consequences of this type of customer satisfaction at the individual level.

Customer satisfaction is a consequence of customer-perceived value (Fornell et al., 1996; Hallowell, 1996). Customer satisfaction is usually perceived to be a key indicator of a firm's market share and profitability, and an important indicator of a firm's overall financial health (Fornell et al., 1996). A satisfied customer will show a strong tendency to be loyal and repeat the purchase of the goods or services, and increase a firm's market share and profits, which signifies its significance to the successful competition in customer-centered era (Fornell et al., 1996). Customer satisfaction is increasingly considered as a baseline standard of performance and a possible standard of excellence for any business organization (Mihelis et al., 2001). Companies with a bigger share of loyal customers gain from increasing repurchase rates, increasing cross-buying potential, higher price willingness, positive recommendation behavior and less switching tendency (Rust et al., 2000). Organizational commitment, process-driven approach and reliability of customer relationship management effectiveness are found to positively affect customer satisfaction (Padmavathy et al., 2012). In addition, it is necessary to identify customer needs and expectations and ensure that they are met to improve customer satisfaction (Chalmeta, 2006).

Brand Loyalty

Brand loyalty reflects the desirability of product functionality and services (Kim et al., 2001). Assael (1992) defined brand loyalty as a favorable attitude toward a brand, thus resulting in consistent purchase of the brand over time supported by Keller (1993). Oliver (1997, 1999) defined brand loyalty as a deeply held commitment to re-buy or re-patronize a preferred brand consistently in the future, thereby causing repetitive purchasing of same brands or same products, despite situational influences, despite situational influences and marketing efforts having the potential to cause switching behavior. The positive relationship is found among customer satisfaction, customer loyalty and firm performance (Anderson et al., 1994; Fornell, 1992; Hallowell, 1996; Reichfeld & Sasser, 1990; Rust et al., 1995; Storbacka et al., 1994; Zeithaml et al., 1990). Brand loyalty has received the most attention by academicians and practitioners (Malai & Speece, 2005; Tsao & Chen,

2005). For that reason, there are several definitions and measures of brand loyalty; some focus on the attitudinal dimension and others focus on the behavioral aspect of brand loyalty (Gee et al., 2008; Oliver, 1997). The concept of brand loyalty is extended to encompass both attitudinal loyalty and behavioral loyalty (Jacoby & Kyner, 1973). Attitudinal loyalty is assumed to be more stable than behavioral loyalty and represents consumers' commitment or preferences when considering unique values associated with a brand (Chaudhuri & Holbrook, 2001).

In addition, Gounaris and Stathakopoulos (2004) stated that attitudinal loyalty should lead to an increase in behavioral loyalty. Attitudinal loyalty reflects customer's predispositions to continue relating to the brand or company (Oliver, 1999) often involving customer commitment (Bloemer & Kasper, 1994; Morgan & Hunt, 1994; Solomon, 1992) and involving trust and commitment after an initial experience of customer satisfaction (Vasquez-Parraga & Alonso, 2000; Zamora et al., 2004). The cost of recruiting new customers is very high due to advertising, personal selling, establishing new accounts, and customer training (Mittal & Lassar, 1998). Brand commitment leads to permanent purchase of the brand during usage (Lee et al., 2009). The high cost involved in the acquisition of new customers makes the early stages of a new customer relationship unprofitable (Reichheld & Sasser, 1990). The decreased cost does such a customer relationship become more profitable (Reichheld & Sasser, 1990). The individual who consistently buys the same brand is said to be loyal to this brand (Kuehn, 1962). According to Zeithaml et al. (1996), loyal customers tend to build and strengthen the relationship with a firm and behave differently from non-loyal customers. By influencing directly both purchase and non-purchase behaviors of customers, loyal customers contribute to the financial performance of a firm (Zeithaml et al., 1996). For example, loyal customers emphasize a close relationship with a firm with lower price elasticity (Zeithaml et al., 1996).

Furthermore, loyal customers pass on the favorable word-of-mouth referral (Brown et al., 1987). Brand loyalty brings about other marketing advantages such as word-of-mouth marketing and competitiveness (Hawkins et al., 2006). Thus, the perception of factors influencing customers' brand loyalty is becoming more and more important to the marketers (Hawkins et al., 2006). Loyal customers put greater emphasis on social and emotional value, and studies have found that superior customer value creation and delivery can help firms to build close emotional links with targeted customers (Butz & Goodstein, 1996). Howard and Sheth (1969) stated that highly involved consumers might perceive greater product importance and commit to their choice. Iwasaki and Havitz (1998) suggested a three-stage model to present the relationship between involvement and behavioral loyalty. In addition, the psychological progress includes the formation of a high level of involvement in an activity, then the development of a psychological commitment to a brand and the maintenance of strong attitudes toward resisting changes in brand preference (Iwasaki & Havitz, 1998).

Customer Relationship Management Performance

Customer relationship management (CRM) is part of an organizational strategy to identify customers, attract their customer satisfaction and turn them into the repeat customer (Pamsari et al., 2013). Customer relationship management is a unified system for program development and the timing and control in the previous activities and after the sale of the organization, and its aim is to attract and maintain long-term relationship and evaluation with customers (Mohebbi et al., 2012). Consequently, customer relationship management is a combination of people, processes and technology that seeks to understand a company's customers (Chen & Popovich, 2003). Customer relationship management is an integrated approach to manage customer relationships by focusing on customer retention and developing relationships (Chen & Popovich, 2003). The purpose of customer relationship management is to increase the acquisition and retention of profitable customers by selectively initiating, building and maintaining appropriate relationships with customers (Payne & Frow, 2006). Customer relationship management can be defined as the management approach that involves identifying, attracting, developing and maintaining successful customer relationships over time in order to increase the retention of profitable customers (Bradshaw & Brash, 2001; Massey et al., 2001). In order to develop successful relationships with customers, firms should focus on the economically valuable customers while keeping away and eliminating the economically invaluable aspects (Verhoef & Donkers, 2001).

Customer relationship management enables firms to deploy such strategies by managing individual customer relationships with the support of customer databases and mass customization technologies (Verhoef & Donkers, 2001). In addition, Ciborra and Failla (2000) conceptualized customer relationship management beyond a front office contact management system. Levine (2000) stated that customer relationship management is the utilization of customer-related information or knowledge to deliver relevant products or services to customers. Relationship marketing emphasizes that customer retention affects company profitability in that it is more efficient to maintain an existing relationship with a customer than create a new one (Payne et al., 1999; Reichheld, 1996). Newell (2000) explored strategic methods for maintaining or improving customer retention. Customer relationship management normally involves business process change and the introduction of new information technology; an effective leadership of customer relationship management is important (Galbreath & Rogers, 1999). Managers in firms are influential in the authorization and control of expenditure, the setting and monitoring of performance and the empowerment and motivation of key personnel related to customer relationship management (Pinto & Slevin, 1987). As firms learn how to manage customer relationship management effectively over time, they develop a one-to-one relationship with customers, thereby reducing cost efficiency and increasing profit efficiency (Krasnikov et al., 2009). Furthermore, the accurate customer data is essential to successful customer relationship management performance (Abbott et al., 2001) and, consequently, technology plays an important role in customer relationship management in adding to firm intelligence (Boyle, 2004).

MAIN FOCUS OF THE CHAPTER

This chapter focuses on constructing an innovative framework and a causal model of customer value, customer satisfaction, brand loyalty, and customer relationship management performance for smart manufacturing of pulp and paper company employees in Thailand.

The Role of CRM in Smart Manufacturing

Interest in customer relationship management (CRM) began to grow in 1990s (Ling & Yen, 2001; Xu et al., 2002). The changing environment forces the companies to change their focus from customer acquisition to customer retention (Sheth, 2002) by building relationships with customers and adding more value to goods and services (Lindgreen & Wynstra, 2005). In the late 1990s, customer relationship management is formed as a popular business term, which holds the same roots with relationship marketing and enhances the paradigm with the emerging information technologies. Zablah et al. (2004) stated that there are five perspectives of CRM (i.e., strategy, process, capability, philosophy and technology). According to its complex structure, the positive financial return rates of CRM-related projects vary from ten to 30 percent (Brewton & Schiemann, 2003; Richards & Jones, 2008). Furthermore, the low level of success rates, the unsuccessful CRM implementations may lead to decrease in customer satisfaction (CS) and customer loyalty (CL) (Richards & Jones, 2008). As a result of the low success level of CRM, measurement models are proposed by both academicians and professionals. The mentioned frameworks can be classified in two groups: partial measurement model and holistic measurement model. In addition, the processes of a company are defined as one of the main components of CRM (Chen & Popovich, 2003; Mendoza et al., 2007).

CRM processes are largely investigated by academicians (Bueren et al., 2005; Kohli et al., 2001; Leigh & Tanner, 2004; Parvatiyar & Sheth, 2001; Payne & Frow, 2005; Zablah et al., 2004). In addition, Zablah et al. (2004) analyzed these processes in two levels as the macro level and micro level. The macro-level process approaches CRM process as company-wide process. Payne and Frow (2005) defined these macro-level processes as the strategy development, value creation, channel integration, knowledge management and performance evaluation process. The micro-process approach focuses on customer interaction management in order to maintain long-term profitable customer relationships (Kohli et al., 2001). Academicians and professionals proposed frameworks in order to measure CRM. The proposed methods are categorized in ten groups according to their application of CRM and the CRM measurement processes as follows:

1. Indirect measurement models. CRM is measured by brand building or customer equity terms and methods (Kellen, 2002; Richards & Jones, 2008). The components of brand equity, brand loyalty, brand awareness, perceived quality and brand associations are related to CRM activities (Aaker, 1991). In addition, customer equity is divided into three groups; value equity, brand equity and retention equity and researchers defined the relations between each component and CRM (Richards & Jones, 2008).

2. Measurement of customer facing operations. According to the data collected from customers facing activities within the company, the operational CRM systems automate the customer-related processes that why companies evaluate process metrics in order to measure the success of CRM (Kellen, 2002).

3. Critical success factors scoring (CSFS). Proposed by Mendoza et al. (2007), this tool uses critical success factors as the basis for scoring. There are 13 critical success factors, and over 50 metrics defined in the mode (Mendoza et al., 2007). Regarding the variety in the nature of the factors and the metrics, the model also provides the options for each metric (Mendoza et al., 2007).

4. Behavioral dimensions of CRM effectiveness. Jain et al. (2003) identified the need for a tool focusing on behavioral dimensions for measuring the CRM effectiveness. In addition, Jain et al. (2003) stated that behavioral dimensions of CRM effectiveness are defined in ten factors (i.e., attitude, understanding expectations, quality perceptions, reliability, communication, customization, recognition, keeping promises, satisfaction audit and retention).

5. CRM scale. Proposed by Sin et al. (2005), the CRM scale conceptualizes CRM in four dimensions (i.e., key customer focus, CRM organization, knowledge management and technology based CRM). Key customer focus is the dimension that emphasizes the customer-centric marketing, personalization and communication between the company and the customers (Sin et al., 2005). CRM organization focuses on organizational structure, commitment of resource, and human resource management (Agarwal et al., 2004). Knowledge management highlights the creation, transfer and application of the

knowledge in a company (Sin et al., 2005). The last perspective, technology-based CRM stresses the companies' technology availability (Sin et al., 2005).

6. Relationship quality (RQ). The quality of relationship between the customer and the vendor has been in the interest area of researchers (Crosby et al., 1990; Dorsch et al., 1998; Kumar et al., 1995; Lagace et al., 1991). The aim of RQ measurement models is to define and measure the issues that compose customers' perception about the relationship. The models are based on statistical studies and utilize a survey as a measurement tool. Roberts et al. (2003) identified the dimensions of RQ in consumer services as; trust in partner's honesty, trust in partner's benevolence, affective commitment, satisfaction and affective conflict.

7. Customer measurement assessment tool (CMAT). CMAT is developed by QCi consultancy as a customer assessment tool that considers a questioner composed of 260 questions. The questions and the scoring of the tool are based on the best practices of each category (Starkey et al., 2002). The tool defines nine crucial areas of customer management assessment; information technology, people and organization, process, analysis and planning, the proposition, customer management activity, measuring the effect, customer experience and competitor (Woodcock et al., 2003).

8. Customer management process (CMP). CMP is a part of balanced scorecard (BSC) (Kaplan & Norton, 1992) is one of these processes that focus on customer management. The four sub-processes of this category is defined as; customer selection, customer acquisition, customer retention and customer growth (Kaplan & Norton, 1992). For each sub processes, companies define objectives and metrics that are used for measurement. CMP measurement should be identified for each company (Kaplan & Norton, 1992).

9. Relationship management assessment tool (RMAT). Proposed by Lindgreen et al. (2006), this tool aims to help managers to make self-assessments about the stages of relationships. The elements analyzed in RMAT are; customer strategy, customer-interaction strategy, brand strategy, value-creation strategy, culture, people, organization, information technology, relationship management process and finally, knowledge management and learning (Lindgreen et al., 2006).

10. CRM scorecards (CRM-SC). Balanced scorecard (BSC) is originally developed in order to provide an outline for organizational performance measurement. CRM-SC is the CRM performance measurement models that are based on BSC (Kim et al., 2003; Kim & Kim, 2009). In the CRM-SC, the main dimensions are defined, the elements in each dimension are identified and a map that specifies the relationship between the dimensions and the elements are maintained (Kim et al., 2003; Kim & Kim, 2009).

Many firms have invested heavily in customer relationship management systems (Smith & Chang, 2010). Firms which pay more attention to a customer-centric approach can benefit significantly from the implementation of CRM systems (Smith & Chang, 2010). In addition, a fundamental change in CRM thinking is needed to shift the focus of CRM from empowering firms to empowering customers (Saarijarvi et al., 2013). Refining and giving customer data back to customers may represent a future mechanism through which firms develop their customer relationship management to a whole new level (Saarijarvi et al., 2013). An analytical CRM system should be developed to support customer knowledge acquisition (Xu & Walton, 2005). Managing a successful CRM implementation requires an integrated approach to technology, process, and people (Chen & Popovich, 2003). CRM has evolved from advances in information technology and organizational changes in customer-centric processes (Chen & Popovich, 2003). Companies implementing the CRM will reap the rewards in customer loyalty and long run profitability (Chen & Popovich, 2003). The key ways to build a strong competitive position of organization are through customer relationship management (CRM) and product/service quality (Zineldin, 2006). CRM adoption has a significant positive effect on both customer satisfaction and organizational performance in business-to-business (B2B) settings (Ata & Toker, 2012). In addition, CRM adoption is also found to significantly affect organizational marketing performance (Ata & Toker, 2012). At the strategic level, management identifies key customers and customer groups to be targeted as part of the firm's business mission (Lambert, 2010). The CRM process provides the structure for how the relationships with customers will be developed and maintained (Lambert, 2010).

CRM Organization in Smart Manufacturing

CRM means the fundamental changes in the way that firms are organized (Ryals & Knox, 2001) and business processes are conducted (Hoffman & Kashmeri, 2000). Firms should pay the heightened attention to the organizational challenges inherent in any CRM initiative (Agarwal et al., 2004). In the past, various studies have examined CRM implementation in different kinds of industries such as hotels (Lo et al., 2010), retailing (Minami & Dawson, 2008), financial services (Dimitriadis, 2010), tourism (Ozgener & Iraz, 2006), transport services (Cheng et al., 2008), business markets (Gummesson, 2004) and public services (Pan et al., 2006). Chan (2005) proposed a similar conceptualization of CRM that integrated business processes, organizational structures, analytical structures and technological representation to present a unified view of a customer. In addition, Kim et al. (2003) put forwarded a conceptual model of CRM effectiveness (CRME) that consists of four customer-centric perspectives namely, customer knowledge, interaction, value, and satisfaction. The key considerations to organize the whole firm around CRM include organizational structure, organization-wide commitment of resources, and human resources management (Agarwal et al., 2004): (1) Organizational structure. CRM requires that the entire organizational work toward the common goal of forging and nurturing strong customer relationships. In addition, the customer relationships include the establishment of process teams, customer-focused teams (Sheth & Sisodia, 2002), cross-discipline segment teams, and cross-functional teams (Ryals & Knox, 2001). All these structural designs demand strong inter-functional coordination and inter-functional integration (Sheth et al., 2000); (2) Organization-wide commitment of resources. The organization-wide commitment of resources should follow after crafting the design of organizational structure and integrating properly those involved components. Furthermore, sales and marketing resources, technical expertise, as well as resources promoting service excellence should be in place (Nykamp, 2001). The success of customer acquisition, development, retention, and reactivation hinges on the company's commitment of time and resources through key customer needs (Nykamp, 2001); and (3) Human resources management. The strategy, people, technology, and processes are important to CRM, but these are the individual employees who are the building blocks of customer relationships (Horne, 2003; McGovern & Panaro, 2004; Ryals & Knox, 2001). According to Krauss (2002), the hardest part of becoming CRM-oriented aspect is not the technology, it is the people. In addition, internal marketing brings in employees, the utmost importance of service-mindedness, and customer orientation (Gronroos, 1990).

In a business-to-business environment, customer relationship management is the business process that provides the structure for how relationships with customers are developed and maintained. In addition, customer relationship management is being viewed as strategic (Lambert, 2004; Payne & Frow, 2005; Zablah et al., 2004), process-oriented (Lambert, 2004; Payne & Frow, 2005; Zablah et al., 2004), cross-functional (Lambert, 2004; Payne & Frow, 2005), value-creating for buyer and seller (Lambert, 2004; Boulding et al., 2005; Payne & Frow, 2005), and a means of achieving superior financial performance (Bohling et al., 2006; Boulding et al., 2005; Lambert, 2004; Payne & Frow, 2005). The content of a CRM strategy consists of six mutually dependent criteria (Donaldson & O'Toole, 2002) as follows:

1. Emphasis on quality. Poor service is the dominant reason for losing business. The core product alone is no longer enough, and service quality is stressed as the key for business to be successful.
2. Measuring customer satisfaction but managing customer service. This implies the understanding of various benefits that the prospect expects to manage the gap between expectations and performance after the purchase process.
3. Investing in people. Internal relationships are as important as external relationships. To implement relationship, orientation can come from the understanding of the set objectives and the standardized meeting.
4. Maintaining dialogue with customers. Building long-term relationships is a key factor in CRM. Companies that listen and adapt to customers' preferences have a higher propensity to retain them and make them loyal.
5. Setting realistic targets and assessing performance. Organizations must have an understanding of customer perceptions of the various elements in the offering and the important elements to each individual customer.
6. Relationship-based interfaces. This view means being in touch with both internal and external customers in a responsive and flexible manner. In practice, there is a gap between what firms do, what they should do, and what the most desirable is to do. The means of communication should be adapted to the needs of the individual customer.

Relationships among Variables

Research studies indicated that retaining current customers is much less expensive than attempting to attract new ones (Bitran & Mondschein, 1997; Chattopadhyay, 2001; Massey et al., 2001; Stone et al., 1996). Customer satisfaction can lead to brand loyalty, repurchase intention and repeat sales (Oliver, 1999; Parasuraman & Grewal, 2000; Swan & Oliver, 1989), in short to customer retention. Customer retention seems to be related to profitability (Oliver, 1999). Customer satisfaction is related to the perceived performance and expectations (Anderson, 1973; Hunt, 1977). If performance matches expectations or exceeds expectations, the customer is satisfied or highly satisfied respectively (Anderson, 1973; Hunt, 1977). If performance falls short of expectations, the customer is dissatisfied (Anderson, 1973; Hunt, 1977). Moreover, customer satisfaction is at the core of the marketing concept, which is the guiding force for most of the leading companies (Kotler, 2000; McCarthy & Perreault, 1987; Webster, 1988). Furthermore, emphasis of customer satisfaction is given in obtaining additional customers and encouraging brand switching from competitors, that is, in offensive strategies (Fornell & Wernerfelt, 1987). Customer satisfaction and retaining has labeled as defensive strategy (Fornell, 1992). The goal of defensive strategy is the minimization of customer turnover (maximization of customer retention) through the protection of products and markets from competitive brands and generally from competitive inroads (Fornell, 1992).

Because the fundamental objective of customer relationship management is to ensure steady streams of revenue and maximization of customer lifetime value or customer equity, customer behaviors that might bring revenue streams become strategically significant (Grant & Schlesinger, 1995). For example, researchers have tried to understand relationship in terms of customer retention, intensity, or usage level of services or products over time, cross-buying or add-on purchase, and word-of-mouth (Bettencourt, 1997; Blattbert et al., 2001), which usually implies a fundamental increment of customer lifetime value or customer equity. According to Mazumdar (1993), customers are becoming more value-oriented and are not simply influenced by high quality or lower price. Rather, customers tend to make a reasonable trade-off between the perceived benefits and perceived sacrifices in the process of obtaining and consuming products or services (Mazumdar, 1993). Such customer behaviors are influenced by factors such as customer satisfaction and brand loyalty and that customer value has significant influence on customer satisfaction and brand loyalty (Mittal & Kamakura, 2001; Szymanski & Henard, 2001). Customer value not only has a direct influence on customer behavior-based customer relationship management performance, but also has the indirect effects on customer behavior-based customer relationship management performance (Mittal & Kamakura, 2001; Szymanski & Henard, 2001). Furthermore, customer satisfaction is positively related to brand loyalty (Flint et al., 1997). Likewise, the definition of high customer satisfaction and brand loyalty means that fewer customers will defect, and the long-term effects on firm performance can be significant (Mittal & Kamakura, 2001; Szymanski & Henard, 2001).

In addition, Reichheld and Teal (1996) stated that a five percent increase in customer retention can have a 30-95 percent effect on customers' net present value (NPV) and a similar effect on corporate profits. Furthermore, customer satisfaction is used as a predictor of future consumer purchases (Kasper, 1988; Newman & Werbel, 1973). Satisfied customers have a higher likelihood of repeating purchases in time (Zeithaml et al., 1996), of recommending that others try the source of customer satisfaction (Reynolds & Arnold, 2000; Reynolds & Beatty, 1999), and of becoming less receptive to the competitor's offerings (Fitzell, 1998). More specifically, customer satisfaction is found to be a necessary precursor of customer loyalty (Fitzell, 1998; Fornell, 1992; Reynolds & Beatty, 1999; Sivadas & Baker-Prewitt, 2000; Zeithaml et al., 1996). Whereas customer satisfaction and customer loyalty are recognized as strongly related by most studies (Anderson & Sullivan, 1993; Fornell, 1992; Rust & Zahorik, 1993; Taylor & Baker, 1994), some researchers stated that the relationship is interchangeable (Hallowell, 1996; Oliver, 1999), and some researchers stated that the relationship is unidirectional, that is, progressing from satisfaction to loyalty only (Strauss & Neuhaus, 1997). Customer satisfaction is one of the most influential variables affecting brand loyalty (Eskafi et al., 2013). Satisfied customers tend to be the loyal customers with other variables (Rowley, 2005) or without the mediation of other variables (Coyne, 1989; Fornell, 1992; Oliva et al., 1992).

Customer value contributes to an improvement in customer satisfaction (Bojanic, 1996; Fornell et al., 1996). Customer value should be enhanced by raising expectations and should be positively related to customer satisfaction (Fornell et al., 1996). It is obvious that customer satisfaction depends on customer value to some extent (Caruana et al., 2000; De Ruyter et al., 1997; Howard & Sheth, 1969; Kotler & Levy, 1969; Rust & Oliver, 1994). Furthermore, customer satisfaction is the result of a customer's perception of the customer value received (Athanassopoulos, 2000, Fornell et al., 1996). Therefore, customer value is viewed as another key driver of customer satisfaction (Athanassopoulos, 2000, Fornell et al., 1996). Consequently, customer value is a key element in achieving long-term success for trade organizations (Khalifa, 2004). Organizations need to develop structural and systematic processes of acquiring and evaluating market information, thereby understanding both the expressed and latent demands of customers (Hartline et al., 2,000; Slater & Narver, 1998). The information obtained should be shared broadly throughout the organization, thus customer value would be created in a far more coordinated and focused manner (Slater & Narver, 1995).

Furthermore, such customer relationship practices allow organizations to understand both the expressed and latent needs of their customers, and create customer value by sharing such information throughout the organization and enabling coordinated and focused actions to serve customer needs (Slater & Narver, 1995). Customer relationship practices allow organizations to collect customer feedback related to their products and services, and modify their offerings based on such feedback (Forza & Filippini, 1998; Slater & Narver, 1998). In addition, customer relationship practices can also enhance a company's production performance by enhancing product design efficiency (Reed et al., 1996). Thus, quality problems can be reduced by including customer requirements in the new product and service design stage (Kaynak, 2003). Customer relationship management efforts enable firms to collect customer information across various interactions and customize the offers to suit individual tastes and preferences (Mithas et al., 2005). This view of customer relationship management enhances perception of perceived quality and affects customer satisfaction (Mithas et al., 2005). By understanding customer needs, companies can eliminate excess components from their products (Reed et al., 1996). The improved product design can allow companies to enhance their product quality, reduce costs of defective production, and improve productivity (Reed et al., 1996). Furthermore, Samson and Terziovski (1999) stated that customer focus practices are positively related to various operational performance criteria, including production performance.

Material and Methods

Data for this study were collected out of 593 operational employees from 11,274 operational employees working in the 25 pulp and paper companies in Thailand by using Yamane's formula (Yamane, 1970) for a 96% confidence level with a 4% margin of error by the proportional random sampling method. A five-point Likert scale ranging from 1 (strongly disagree) to 5 (strongly agree) was used with all of the constructs. Data were analyzed with descriptive statistics using SPSS (version 20) and assessed with confirmatory factor analysis (CFA) to confirm the heterogeneity of all constructs and path analysis (Joreskog & Sorborn, 1993) to detect the cause-effect relationships among various dimensions of main constructs of the study using LISREL (version 8.8) on a structured questionnaire containing standard scales of customer value, customer satisfaction, brand loyalty, and customer relationship management performance in terms of the innovative manufacturing and marketing solutions, besides some demographic details like age, education, and tenure with the organization. Customer value was measured using the questionnaire developed by Sheth et al. (1991) and Sweeney and Soutar (2001). Customer satisfaction was measured using the questionnaire developed by Wang and Lo (2002). Brand loyalty was measured using the questionnaire developed by Keller (1993) and Assael (1992). Customer relationship management performance was measured by the questionnaire developed by Bettencourt (1997), Blattberg et al. (2001), and Reichheld and Teal (1996). This chapter has a research question: what are a framework and a causal model of customer value, customer satisfaction, brand loyalty, and customer relationship management performance for smart manufacturing?

SOLUTIONS AND RECOMMENDATIONS

An innovative framework and a causal model are constructed shown in Figure 1. Research findings indicated that dimensions of customer value, customer satisfaction, and brand loyalty have mediated positive effect on customer relationship management performance in terms of the innovative manufacturing and marketing solutions. Brand loyalty positively mediates the relationships between customer value and customer relationship management performance and between customer satisfaction and customer relationship management performance. Furthermore, customer value is positively correlated with customer satisfaction. Regarding the innovative framework and causal model, there are lots of researchers studying the relationships of customer value, customer satisfaction, brand loyalty, and customer relationship management performance in a wide variety of fields. The innovative framework was positively compatible with the following research findings. Customer value, customer satisfaction, and brand loyalty are positively linked to customer relationship management performance. It is important that the other organizations implementing large-scale manufacturing reformations need to pay great attention to customer value, customer satisfaction, brand loyalty, and customer relationship management performance in order to effectively achieve business success. When firms adopt brand loyalty activities to promote customer relationship management performance, the role of customer value and customer satisfaction becomes more evident and critical. Therefore, the strength of

Figure 1. Innovative Framework and Causal Model (Key: CV = Customer Value, EV = Emotional Value, SV = Social Value, FV = Functional Value, PS = Perceived Sacrifice, CS = Customer Satisfaction, BL = Brand Loyalty, CP = Customer Relationship Management Performance)

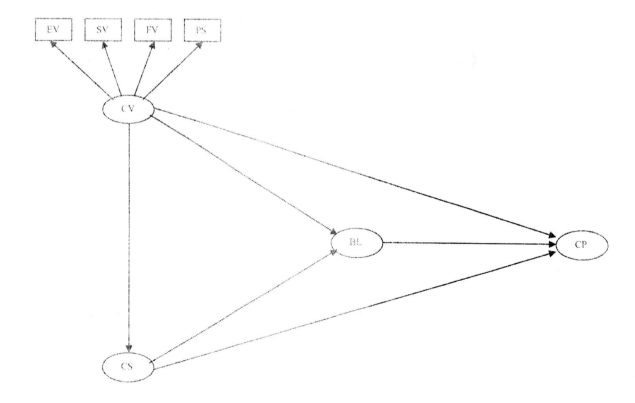

brand loyalty affecting customer relationship management performance will be influenced by what kinds of customer value and customer satisfaction that firms adopt. If firms possess a higher degree of customer value and customer satisfaction, the degrees of brand loyalty would be more enhanced. These findings show that customer value and customer satisfaction can promote a higher degree of brand loyalty within firms. In addition, the present results are also quite instructive in helping to explain the effect of customer value, customer satisfaction, and brand loyalty on customer relationship management performance. Furthermore, the managers in firms should carefully design and nurture appropriate organizational contexts to facilitate brand loyalty and customer relationship management performance. It is important that the other organizations implementing large-scale manufacturing reformations need to pay great attention to customer value, customer satisfaction, brand loyalty, and customer relationship management performance in terms of the innovative manufacturing and marketing solutions in order to effectively achieve business success.

FUTURE RESEARCH DIRECTIONS

The author indicates some limitations in this study and suggests possible directions for future research. This study is based on self-report data that may have the possibility of common method variance. Future research is suggested to benefit from using objective measures for customer relationship management performance that can be independently verified. The low return rate of the survey is still noted as a potential limitation in this study. Future research can benefit from a larger sample to bring more statistical power and a higher degree of representation. This study was done by empirically investigating pulp and paper companies in Thailand. Potential cultural limitation should be noted and it is suggested that future research be done in different cultural contexts to generalize or modify the concepts. Furthermore, this study mainly concerns the effects of customer value, customer satisfaction, and brand loyalty on customer relationship management performance. Other variables (i.e., organizational justice, organizational citizenship behavior, organizational learning, leader-member exchange, job involvement, and employee engagement) may potentially affect customer relationship management performance in terms of the innovative manufacturing and marketing solutions as well. Future research may work on examining their impacts on customer relationship management performance for smart manufacturing.

CONCLUSION

The purpose of this study was to construct an innovative framework and a causal model of customer value, customer satisfaction, brand loyalty, and customer relationship management performance in terms of the innovative manufacturing and marketing solutions for pulp and paper company employees in Thailand. The findings showed that the customer value, customer satisfaction, and brand loyalty have the strengths to mediate positive effect on customer relationship management performance. In relation to the innovative framework and causal model, this result was the extent to which brand loyalty positively mediates the relationships between customer value and customer relationship management performance and between customer satisfaction and customer relationship management performance. Furthermore, customer value is positively correlated with customer satisfaction. Whenever the organizations want to promote their customer relationship management performance, they would better adopt the suitable brand loyalty based on the conditions of their organizational structure, environment, and context, thus aligning their customer value and customer satisfaction. This functional fit will enhance not only brand loyalty but also customer relationship management performance. Given the need for brand loyalty and customer relationship management performance as a solution to the complex challenges, firms need to be aware of the implications of customer value and customer satisfaction that may affect brand loyalty and customer relationship management performance. Furthermore, firms should recognize the importance of brand loyalty and need to put more efforts in building up the brand loyalty mechanisms to promote their customer relationship management performance for smart manufacturing.

REFERENCES

Abbott, J., Stone, M., & Buttle, F. (2001). Customer relationship management in practice – A qualitative study. *Journal of Database Marketing*, *9*(1), 24–34. doi:10.1057/palgrave.jdm.3240055

Agarwal, A., Harding, D. P., & Schumacher, J. R. (2004). Organizing for CRM. *The McKinsey Quarterly*, *3*, 80–91.

Akshay, R. R., & Monroe, K. B. (1988). The moderating effect of prior knowledge on cue utilization in product evaluations. *The Journal of Consumer Research*, *15*(2), 253–264. doi:10.1086/209162

Anderson, E., & Sullivan, M. (1993). The antecedents and consequences of customer satisfaction for firms. *Marketing Science*, *12*(2), 125–143. doi:10.1287/mksc.12.2.125

Anderson, E. W., Fornell, C., & Lehman, D. R. (1994). Customer satisfaction, market share, and profitability: Findings from Sweden. *Journal of Marketing*, *58*(3), 53–66. doi:10.2307/1252310

Anderson, R. E. (1973). Consumer dissatisfaction: The effect of disconfirmed expectancy on perceived product performance. *JMR, Journal of Marketing Research*, *10*(1), 38–44. doi:10.2307/3149407

Andreassen, T. W. (2000). Antecedents to satisfaction with service recovery. *European Journal of Marketing*, *34*(1-2), 156–175. doi:10.1108/03090560010306269

Assael, H. (1992). *Consumer behavior and marketing action*. Boston, MA: PWS-Kent.

Ata, U. Z., & Toker, A. (2012). The effect of customer relationship management adoption in business-to-business markets. *Journal of Business and Industrial Marketing*, *27*(6), 497–507. doi:10.1108/08858621211251497

Athanassipoulos, A. D. (2000). Customer satisfaction: Cues to support market segmentation and explain switching behavior. *Journal of Business Research*, *47*(3), 191–207. doi:10.1016/S0148-2963(98)00060-5

Bakanauskas, A., & Jakutis, A. (2010). Customer value: Determination in undefined environment. *Management of Organizations: Systematic Research*, *53*, 7–18.

Band, W. A. (1991). *Creating value for customers*. New York, NY: John Wiley & Sons.

Banker, R. D., & Kauffman, R. J. (1991). *Quantifying the business value of information technology: An illustration of the business value linkage framework*. New York, NY: Stern School of Business, New York University.

Barsky, J. (1995). *World-class customer satisfaction*. Burr Ridge, IL: Irwin Professional.

Besterfield, D. H. (1994). *Quality control*. Englewood Cliffs, NJ: Prentice-Hall.

Bettencourt, L. A. (1997). Customer voluntary performance: Customers as partners in service delivery. *Journal of Retailing*, *73*(3), 383–406. doi:10.1016/S0022-4359(97)90024-5

Bick, G. (2009). Increasing shareholder value through building customer and brand equity. *Journal of Marketing Management, 25*(1), 117–141. doi:10.1362/026725709X410061

Bitran, G. R., & Mondschein, S. V. (1997). A comparative analysis of decision making procedures in the catalog sales industry. *European Management Journal, 15*(2), 105–116. doi:10.1016/S0263-2373(96)00080-1

Blattberg, R. C., & Deighton, J. (1996). Manage marketing by the customer equity test. *Harvard Business Review, 74*(4), 136–144. PMID:10158473

Blattberg, R. C., Getz, G., & Thomas, J. S. (2001). *Customer equity: Building and managing relationships as valuable assets*. Boston, MA: Harvard Business School Press.

Bloemer, J. M. M., & Kasper, H. D. P. (1994). The impact of satisfaction on brand loyalty: Urging on classifying satisfaction and brand loyalty. *Journal of Consumer Satisfaction. Dissatisfaction and Complaining Behavior, 7*, 152–160.

Bohling, T., Bowman, D., LaValle, S., Mittal, V., Narayandas, D., Ramani, G., & Varadarajan, R. (2006). CRM implementation: Effectiveness, issues and insights. *Journal of Service Research, 9*(2), 184–194. doi:10.1177/1094670506293573

Bojanic, D. C. (1996). Consumer perceptions of price, value and satisfaction in the hotel industry: An exploratory study. *Journal of Hospitality & Leisure Marketing, 4*(1), 5–22. doi:10.1300/J150v04n01_02

Bolton, R., & Drew, J. H. (1991). A multistage model of customers' assessment of service quality and value. *The Journal of Consumer Research, 17*(4), 375–384. doi:10.1086/208564

Boulding, W., Kalra, A., Staelin, R., & Zeithaml, V. A. (1993). A dynamic process model of service quality: From expectation. *JMR, Journal of Marketing Research, 30*(1), 7–27. doi:10.2307/3172510

Boulding, W., Staelin, R., Ehret, M., & Johnston, W. J. (2005). A customer relationship management roadmap: What is known, potential pitfalls, and where to go. *Journal of Marketing, 69*(4), 155–166. doi:10.1509/jmkg.2005.69.4.155

Boyle, M. J. (2004). Using CRM software effectively. *The CPA Journal, 74*(7), 17.

Bradshaw, D., & Brash, C. (2001). Managing customer relationships in the e-business world: How to personalize computer relationships for increased profitability. *International Journal of Retail and Distribution Management, 29*(12), 520–529. doi:10.1108/09590550110696969

Brewton, J., & Schiemann, W. (2003). Measurement: The missing ingredient in today's CRM strategy. *Coastal Management, 17*(1), 5–14.

Brown, J., Johnson, M., & Reingen, P. H. (1987). Social ties and word-of-mouth referral behavior. *The Journal of Consumer Research, 14*(3), 350–362. doi:10.1086/209118

Bueren, A., Schierholz, R., Kolbe, L. M., & Brenner, W. (2005). Improving performance of customer-processes with knowledge management. *Business Process Management Journal, 11*(5), 573–588. doi:10.1108/14637150510619894

Butz, H. E., & Goodstein, L. D. (1996). Measuring customer value: Gaining the strategic advantage. *Organizational Dynamics, 24*(3), 63–77. doi:10.1016/S0090-2616(96)90006-6

Buzzel, R. D., & Gale, B. T. (1987). *The PIMS principles: Linking strategy to performance*. New York, NY: Free Press.

Cai, S. (2009). The importance of customer focus for organizational performance: A study of Chinese companies. *International Journal of Quality & Reliability Management, 26*(4), 369–379. doi:10.1108/02656710910950351

Cailleux, H., Mignot, C., & Kapferer, J. (2009). Is CRM for luxury brands? *Journal of Brand Management, 16*(5-6), 406–412. doi:10.1057/bm.2008.50

Carothers, G. H., & Adams, M. (1991). Competitive advantage through customer value: The role of value-based strategies. In M. Stahl, & G. M. Bounds (Eds.), *Competing globally through customer value* (pp. 32–66). Westport, CT: Quorum Books.

Caruana, A., Money, A. H., & Berthon, P. R. (2000). Service quality and satisfaction: The moderating role of value. *European Journal of Marketing, 34*(11-12), 1338–1352. doi:10.1108/03090560010764432

Chalmeta, R. (2006). Methodology for customer relationship management. *Journal of Systems and Software, 79*(7), 1015–1024. doi:10.1016/j.jss.2005.10.018

Chan, J. O. (2005). Toward a unified view of customer relationship management. *Journal of American Academy of Business, 6*(1), 32–38.

Chang, T. Z., & Chen, S. J. (1988). Market orientation, service quality and business profitability: A conceptual model and empirical evidence. *Journal of Services Marketing, 12*(4), 246–264. doi:10.1108/08876049810226937

Chattopadhyay, S. P. (2001). Relationship marketing in an enterprise resource planning environment. *Marketing Intelligence & Planning, 19*(2), 136–139. doi:10.1108/02634500110385444

Chaudhuri, A., & Holbrook, M. B. (2001). The chain of effects from brand trust and brand affect to brand performance: The role of brand loyalty. *Journal of Marketing, 65*(2), 81–93. doi:10.1509/jmkg.65.2.81.18255

Chen, I. J., & Popovich, K. (2003). Understanding customer relationship management (CRM): People, process and technology. *Business Process Management Journal, 9*(5), 672–688. doi:10.1108/14637150310496758

Cheng, J. H., Chen, F. Y., & Chang, Y. H. (2008). Airline relationship quality: An examination of Taiwanese passengers. *Tourism Management, 29*(3), 487–499. doi:10.1016/j.tourman.2007.05.015

Choo, H. J., Moon, H., Kim, H., & Yoon, N. (2012). Luxury customer value. *Journal of Fashion Marketing and Management, 16*(1), 81–101. doi:10.1108/13612021211203041

Ciborra, C., & Failla, A. (2000). Infrastructure as a process: The case of CRM in IBM. In C. Ciborra (Ed.), *From control to drift: The dynamics of corporate information infrastructures* (pp. 105–124). Oxford, UK: Oxford University Press.

Coyne, K. (1989). Beyond service fads – Meaningful strategies for the real world. *Sloan Management Review, 30*(4), 69–76.

Cronin, J. J., Brady, M. K., & Hult, G. T. M. (2000). Assessing the effects of quality, value, and customer satisfaction on consumer behavioral intentions in service environments. *Journal of Retailing, 76*(2), 193–218. doi:10.1016/S0022-4359(00)00028-2

Crosby, L. A., Evans, K. R., & Cowles, D. (1990). Relationship quality in services selling: An interpersonal influence perspective. *Journal of Marketing, 54*(3), 68–81. doi:10.2307/1251817

Day, G. S. (1990). *Market driven strategy: Processes for creating value*. New York, NY: Free Press.

Day, G. S., & Wensley, R. (1988). Assessing advantage: A framework for diagnosing competitive advantage. *Journal of Marketing, 52*(2), 1–20. doi:10.2307/1251261

De Ruyter, K., Bloemer, J., & Pascal, P. (1997). Merging service quality and service satisfaction: An empirical test of an integrative model. *Journal of Economic Psychology, 18*(4), 187–206. doi:10.1016/S0167-4870(97)00014-7

Dimitriadis, S. (2010). Testing perceived relational benefits as satisfaction and behavioral outcomes drivers. *International Journal of Bank Marketing, 28*(4), 207–213. doi:10.1108/02652321011054981

Dodds, W. B., Monroe, K. B., & Grewal, D. J. (1991). Effects of price, brand, and store information on buyers' product evaluations. *JMR, Journal of Marketing Research, 28*(3), 307–319. doi:10.2307/3172866

Donaldson, B., & O' Toole, T. (2002). *Strategic marketing relationship*. Chichester, UK: John Wiley & Sons.

Dorsch, M. J., Swanson, S. R., & Kelley, S. W. (1998). The role of relationship quality in the satisfaction of vendors as perceived by customers. *Journal of the Academy of Marketing Science, 26*(2), 128–142. doi:10.1177/0092070398262004

Engel, J. F., Blackwell, R. D., & Miniard, P. W. (1990). *Consumer behavior*. Chicago, IL: Dryden Press.

Eskafi, M., Hosseini, S., & Yazd, A. M. (2013). The value of telecom subscribers and customer relationship management. *Business Process Management Journal, 19*(4), 737–748. doi:10.1108/BPMJ-Feb-2012-0016

Fecikova, I. (2004). An index method for measurement of customer satisfaction. *The TQM Magazine, 16*(1), 57–66. doi:10.1108/09544780410511498

Fitzell, P. (1998). *The explosive growth of private labels in North America*. New York, NY: Global Books.

Flint, D. J., Woodruff, R. B., & Gardial, S. F. (1997). Customer value change in industrial marketing relationships. *Industrial Marketing Management, 26*(2), 163–175. doi:10.1016/S0019-8501(96)00112-5

Fornell, C. (1992). A national customer satisfaction barometer: The Swedish experience. *Journal of Marketing, 56*(1), 6–21. doi:10.2307/1252129

Fornell, C., Johnson, M. D., Anderson, E. W., Cha, J., & Bryant, B. E. (1996). The American customer satisfaction index: Nature, purpose, and findings. *Journal of Marketing, 60*(4), 7–18. doi:10.2307/1251898

Fornell, C., & Wernerfelt, B. (1987). Defensive marketing strategy by customer complaint management: A theoretical analysis. *JMR, Journal of Marketing Research, 24*(4), 337–346. doi:10.2307/3151381

Forza, C., & Filippini, R. (1998). TQM impact on quality conformance and customer satisfaction: A causal model. *International Journal of Production Economics, 55*(1), 1–20. doi:10.1016/S0925-5273(98)00007-3

Galbreath, J., & Rogers, T. (1999). Customer relationship leadership. *The TQM Magazine, 11*(3), 161–171. doi:10.1108/09544789910262734

Gale, B. T. (1994). *Managing customer value*. New York, NY: Free Press.

Gee, R., Coates, G., & Nicholson, M. (2008). Understanding and profitably managing customer loyalty. *Marketing Intelligence & Planning, 25*(4), 359–374. doi:10.1108/02634500810879278

Gounaris, S. P., & Stathakopoulos, V. (2004). Antecedents and consequences of brand loyalty: An empirical study. *Journal of Brand Management, 11*(4), 283–306. doi:10.1057/palgrave.bm.2540174

Grant, A. W. H., & Schlesinger, L. A. (1995). Realize your customer's full profit potential. *Harvard Business Review, 73*(5), 59–62.

Gronroos, C. (1990). *Service management and marketing: Managing the moments of truth in service competition*. Lexington, MA: Lexington Books.

Gummesson, E. (2004). Return on relationships (ROR), The value of relationship marketing and CRM in business-to-business contexts. *Journal of Business and Industrial Marketing, 19*(2), 136–148. doi:10.1108/08858620410524016

Guo, X., Hao, A. W., & Shang, X. (2011). Consumer perceptions of brand functions: An empirical study in China. *Journal of Consumer Marketing, 28*(4), 169–279. doi:10.1108/07363761111143169

Hallowell, R. (1996). The relationships of customer satisfaction, customer loyalty, and profitability: An empirical study. *International Journal of Service Industry Management, 7*(4), 27–42. doi:10.1108/09564239610129931

Hamel, G., & Prahalad, C. K. (1994). *Computing for the future*. Boston, MA: Harvard Business School Press.

Hansemark, O. C., & Albinsson, M. (2004). Customer satisfaction and retention: The experience of individual employees. *Managing Service Quality, 14*(1), 40–57. doi:10.1108/09604520410513668

Hartline, M. D., Maxham, J. G., & McKee, D. O. (2000). Corridors of influence in the dissemination of customer-oriented strategy to customer contact service employees. *Journal of Marketing, 64*(2), 35–50. doi:10.1509/jmkg.64.2.35.18001

Hawkins, D., Best, R., & Kenneth, C. (2006). *Consumer behavior*. Tehran, Iran: Sargol.

Heskett, J. L., Sasser, W. E., & Schlesinger, L. A. (1997). *The service profit chain: How leading companies link profit and growth to loyalty, satisfaction and value*. New York, NY: Free Press.

Higgins, K. T. (1998). The value of customer value analysis. *Marketing Research, 10*(4), 39–44.

Hoffman, T., & Kashmeri, S. (2000). Coddling the customer. *Computerworld, 34*(50), 58–60.

Holbrook, M. B. (1996). Customer value – A framework for analysis and research. *Advances in Consumer Research. Association for Consumer Research (U. S.), 23*(1), 138–142.

Homburg, C., Hoyer, W. D., & Fassnacht, M. (2002). Service orientation of a retailer's business strategy: Dimensions, antecedents, and performance outcomes. *Journal of Marketing, 66*(4), 86–101. doi:10.1509/jmkg.66.4.86.18511

Horne, S. (2003). Needed: A cultural change. *Target Marketing, 26*(8), 53–56.

Howard, J. A., & Sheth, J. N. (1969). *The theory of buyer behavior*. New York, NY: John Wiley & Sons.

Hunt, H. K. (1977). CS/D - overview and future research directions. In H. K. Hunt (Ed.), *Conceptualization and measurement of CS/D* (pp. 455–488). Cambridge, MA: Marketing Science Institute.

Iwasaki, Y., & Havitz, M. E. (1998). A path-analytic model of the relationship between involvement, psychological commitment, and loyalty. *Journal of Leisure Research, 30*(2), 337–347.

Jacoby, J., & Kyner, D. B. (1973). Brand loyalty vs. repeat purchasing behavior. *JMR, Journal of Marketing Research, 10*(1), 1–9. doi:10.2307/3149402

Jain, R., Jain, S., & Dhar, U. (2003). Measuring customer relationship management. *Journal of Service Research, 2*(2), 97–109.

Jensen, H. R. (2001). Antecedents and consequences of consumer value assessments: Implications for marketing strategy and future research. *Journal of Retailing and Consumer Services, 8*(6), 299–310. doi:10.1016/S0969-6989(00)00036-9

Johnson, M. D., Anderson, E. W., & Fornell, C. (1995). Rational and adaptive performance expectations in a customer satisfaction framework. *The Journal of Consumer Research, 21*(4), 128–140. doi:10.1086/209428

Johnson, M. D., & Fornell, C. (1991). A framework for comparing customer satisfaction across individuals and product categories. *Journal of Economic Psychology, 12*(2), 267–286. doi:10.1016/0167-4870(91)90016-M

Joreskog, K. G., & Sorbom, D. (1993). *LISREL 8: User's reference guide*. Chicago, IL: Scientific Software International.

Kanji, G., & Moura, P. (2002). Kanji's business scorecard. *Total Quality Management, 13*(1), 13–27. doi:10.1080/09544120120098537

Kaplan, R. S., & Norton, D. P. (1992). The balanced scorecard – Measures that drive performance. *Harvard Business Review, 70*(1), 71–79. PMID:10119714

Kasper, H. (1988). On problem perception, dissatisfaction and brand loyalty. *Journal of Economic Psychology, 9*(3), 387–397. doi:10.1016/0167-4870(88)90042-6

Katz, A. (1993). Measuring technology's business value. *Information Systems Management, 10*(1), 33–39. doi:10.1080/10580539308906910

Kaynak, H. (2003). The relationship between total quality management practices and their effects on firm performance. *Journal of Operations Management, 21*(4), 405–435. doi:10.1016/S0272-6963(03)00004-4

Kellen, V. (2002). *CRM measurement frameworks* (Working paper). DePaul University, Chicago, IL.

Keller, K. L. (1993). Conceptualizing, measuring and managing customer-based brand equity. *Journal of Marketing, 57*(1), 1–22. doi:10.2307/1252054

Khalifa, A. S. (2004). Customer value: A review of recent literature and an integrative configuration. *Management Decision, 42*(5), 645–666. doi:10.1108/00251740410538497

Kim, C. K., Han, D., & Park, S. B. (2001). The effect of brand personality and brand identification on brand loyalty: Applying the theory of social identification. *The Japanese Psychological Research, 43*(4), 195–206. doi:10.1111/1468-5884.00177

Kim, H. S., & Kim, Y. G. (2009). A CRM performance measurement framework: Its development process and application. *Industrial Marketing Management, 38*(4), 477–489. doi:10.1016/j.indmarman.2008.04.008

Kim, J., Suh, E., & Hwang, H. (2003). A model for evaluating the effectiveness of CRM using balanced scorecard. *Journal of Interactive Marketing, 17*(2), 5–19. doi:10.1002/dir.10051

Kohli, R., Piontek, F., Ellington, T., VanOsdol, T., Shepard, M., & Brazel, G. (2001). Managing customer relationships through e-business decision support applications: A case of hospital-physician collaboration. *Decision Support Systems, 32*(2), 171–187. doi:10.1016/S0167-9236(01)00109-9

Komunda, M., & Osarenkhoe, A. (2012). Remedy or cure for service failure? Effects of service recovery on customer satisfaction and loyalty. *Business Process Management Journal, 18*(1), 82–103. doi:10.1108/14637151211215028

Kordupleski, R. T., & Laitamaki, J. (1997). Building and deploying profitable growth strategies based on the waterfall of customer value added. *European Management Journal, 15*(2), 158–166. doi:10.1016/S0263-2373(96)00085-0

Kotler, P. (1997). *Marketing management: Analysis, planning, implementation, and control.* Upper Saddle River, NJ: Prentice-Hall.

Kotler, P. (2000). *Marketing management: The millennium edition.* Upper Saddle River, NJ: Prentice-Hall.

Kotler, P., & Levy, S. J. (1969). Broadening the concept of marketing. *Journal of Marketing, 33*(1), 10–15. doi:10.2307/1248740 PMID:12309673

Krasnikov, A., Jayachandran, S., & Kumar, V. (2009). The impact of customer relationship management implementation on cost and profit efficiencies: Evidence from the US commercial banking industry. *Journal of Marketing, 73*(6), 61–76. doi:10.1509/jmkg.73.6.61

Krauss, M. (2002). At many firms, technology obscures CRM. *Marketing News, 36*(6), 5.

Kuehn, A. (1962). Consumer brand choice as a learning process. *Journal of Advertising Research, 2*(4), 10–17.

Kumar, N., Scheer, L. K., & Steenkamp, J. E. M. (1995). The effects of supplier fairness on vulnerable resellers. *JMR, Journal of Marketing Research, 32*(1), 54–65. doi:10.2307/3152110

Lagace, R. R., Dahlstrom, R., & Gassenheimer, J. B. (1991). The relevance of ethical salesperson behavior on relationship quality: The pharmaceutical industry. *Journal of Personal Selling & Sales Management, 11*(4), 39–47.

Lambert, D. M. (2004). *Supply chain management: Processes, partnerships, performance.* Sarasota, FL: Supply Chain Management Institute.

Lambert, D. M. (2010). Customer relationship management as a business process. *Journal of Business and Industrial Marketing, 25*(1), 4–17. doi:10.1108/08858621011009119

Lapierre, J. (2000). Customer-perceived value in industrial contexts. *Journal of Business and Industrial Marketing, 15*(2-3), 122–140. doi:10.1108/08858620010316831

Lee, J. S., & Back, K. J. (2010). Reexamination of attendee-based brand equity. *Tourism Management, 31*(3), 395–401. doi:10.1016/j.tourman.2009.04.006

Lee, Y. K., Back, K. J., & Kim, J. Y. (2009). Family restaurant brand personality and its impact on consumer's emotion, satisfaction, and band loyalty. *Journal of Hospitality & Tourism Research (Washington, D.C.), 33*(3), 305–328. doi:10.1177/1096348009338511

Leigh, T. W., & Tanner, J. F. (2004). Introduction: JPSSM special issue on customer relationship management. *Journal of Personal Selling & Sales Management, 24*(4), 259–262.

Levine, S. (2000). The rise of CRM. *America's Network, 104*(6), 34.

Lindgreen, A., Palmer, R., Vanhamme, J., & Wouters, J. (2006). A relationship management assessment tool: Questioning, identifying, and prioritizing critical aspects of customer relationships. *Industrial Marketing Management, 35*(1), 57–71. doi:10.1016/j.indmarman.2005.08.008

Lindgreen, A., & Wynstra, F. (2005). Value in business markets: What do we know? Where are we going? *Industrial Marketing Management, 34*(7), 732–748. doi:10.1016/j.indmarman.2005.01.001

Ling, R., & Yen, D. C. (2001). Customer relationship management: An analysis framework and implementation strategies. *Journal of Computer Information Systems, 41*(3), 82–97.

Lo, A. S., Stalcup, L. D., & Lee, A. (2010). Customer relationship management for hotels in Hong Kong. *International Journal of Contemporary Hospitality Management, 22*(2), 139–159. doi:10.1108/09596111011018151

Madden, T. J., Fehle, F., & Fournier, S. (2006). Brands matter: An empirical demonstration of the creation of shareholder value through branding. *Journal of the Academy of Marketing Science, 34*(2), 224–235. doi:10.1177/0092070305283356

Malai, V., & Speece, M. (2005). Cultural impact on the relationship among perceived service quality, brand name value, and customer loyalty. *Journal of International Consumer Marketing, 17*(4), 7–39. doi:10.1300/J046v17n04_02

Massey, A. P., Montoya-Weiss, M., & Holcom, K. (2001). Re-engineering the customer relationship: Leveraging knowledge assets at IBM. *Decision Support Systems, 32*(2), 155–170. doi:10.1016/S0167-9236(01)00108-7

Mazumdar, T. (1993). A value-based orientation to new product planning. *Journal of Consumer Marketing, 10*(1), 28–41. doi:10.1108/07363769310026557

McCarthy, E. J., & Perreault, W. D. (1987). *Basic marketing: A managerial approach*. Homewood, IL: Irwin.

McCollough, M. A., Berry, L. L., & Yadav, M. S. (2000). An empirical investigation of customer satisfaction after service failure and recovery. *Journal of Service Research, 3*(2), 121–137. doi:10.1177/109467050032002

McDougall, H. G., & Levesque, T. (2000). Customer satisfaction with services: Putting perceived value into equation. *Journal of Services Marketing, 14*(5), 392–410. doi:10.1108/08876040010340937

McGovern, T., & Panaro, J. (2004). The human side of customer relationship management. *Benefits Quarterly, 20*(3), 26–33.

Mendoza, L. E., Marius, A., Perez, M., & Griman, A. C. (2007). Critical success factors for a customer relationship management strategy. *Information and Software Technology, 49*(8), 913–945. doi:10.1016/j.infsof.2006.10.003

Mentzer, J. T., Rutner, S. M., & Matsuno, K. (1997). Application of the means-end value hierarchy model to understanding logistics service value. *International Journal of Physical Distribution & Logistics Management, 27*(9-10), 630–643. doi:10.1108/09600039710188693

Meredith, J. R., McCutcheon, D. M., & Hartley, J. (1994). Enhancing competitiveness through the new market value equation. *International Journal of Operations & Production Management, 14*(11), 7–22. doi:10.1108/01443579410068611

Mihelis, G., Grigoroudis, E., Siskos, Y., Politis, Y., & Malandrakis, Y. (2001). Customer satisfaction measurement in the private bank sector. *European Journal of Operational Research, 130*(2), 347–360. doi:10.1016/S0377-2217(00)00036-9

Milgrom, P., & Roberts, J. (1995). Complementarities and fit: Strategy, structure and organizational change in manufacturing. *Journal of Accounting and Economics, 19*(2-3), 179–208. doi:10.1016/0165-4101(94)00382-F

Miller, J., & Muir, D. (2004). *The business of brands*. Chichester, UK: John Wiley & Sons.

Minami, C., & Dawson, J. (2008). The CRM process in retail and service sector firms in Japan: loyalty development and financial return. *Journal of Retailing and Consumer Services, 15*(5), 375–385. doi:10.1016/j.jretconser.2007.09.001

Mithas, S., Krishnan, M. S., & Fornell, C. (2005). Why do customer relationship management applications affect customer satisfaction? *Journal of Marketing, 69*(4), 201–209. doi:10.1509/jmkg.2005.69.4.201

Mittal, B., & Lassar, W. M. (1998). Why do consumers switch? The dynamics of satisfaction versus loyalty. *Journal of Services Marketing, 12*(3), 177–194. doi:10.1108/08876049810219502

Mittal, V., & Kamakura, W. A. (2001). Satisfaction, repurchase intent, and repurchase behavior: Investigating the moderating effect of customer characteristics. *JMR, Journal of Marketing Research*, *38*(1), 131–142. doi:10.1509/jmkr.38.1.131.18832

Mohebbi, N., Hoseini, M. A., & Esfidani, M. R. (2012). Identification & prioritization of the affecting factors on CRM implementation in edible oil industry. *Journal of Basic and Applied Scientific Research*, *2*(5), 4993–5001.

Monroe, K. (2005). *Pricing: Making profitable decisions*. New York, NY: McGraw-Hill.

Morgan, R. M., & Hunt, S. D. (1994). The commitment - trust theory of relationship marketing. *Journal of Marketing*, *58*(3), 20–38. doi:10.2307/1252308

Morrisson, O., & Huppertz, J. W. (2010). External equity, loyalty program membership, and service recovery. *Journal of Services Marketing*, *24*(3), 244–254. doi:10.1108/08876041011040640

Motameni, R., & Shahrokhi, M. (1998). Brand equity valuation: A global perspective. *Journal of Product and Brand Management*, *7*(4), 275–290. doi:10.1108/10610429810229799

Munnukka, F., & Farvi, P. (2012). The price-category effect and the formation of customer value of high-tech products. *Journal of Consumer Marketing*, *29*(4), 293–301. doi:10.1108/07363761211237362

Narver, J. C., & Slater, S. F. (1990). The effect of a market orientation on business profitability. *Journal of Marketing*, *54*(4), 20–35. doi:10.2307/1251757

Naumann, E. (1995). *Creating customer value*. Cincinnati, OH: Thompson Executive Press.

Newell, F. (2000). *Loyalty.com: Customer relationship management in the new era of Internet marketing*. New York, NY: McGraw-Hill.

Newman, J. W., & Werbel, R. A. (1973). Multivariate analysis of brand loyalty for major household appliances. *JMR, Journal of Marketing Research*, *10*(4), 404–409. doi:10.2307/3149388

Normann, R., & Ramirez, R. (1993). From value chain to value constellation: Designing interactive strategy. *Harvard Business Review*, *71*(4), 65–77. PMID:10127040

Nykamp, M. (2001). *The customer differential: The complete guide to implementing customer relationship management*. New York, NY: AMACOM.

Oliva, T., Oliver, R. L., & McMillan, I. (1992). A catastrophe model for developing service satisfaction strategies. *Journal of Marketing*, *56*(3), 83–95. doi:10.2307/1252298

Oliver, R. L. (1977). Effects of expectations and disconfirmation on post exposure product evaluation. *The Journal of Applied Psychology*, *62*(4), 480–486. doi:10.1037/0021-9010.62.4.480

Oliver, R. L. (1980). A cognitive model of the antecedents and consequences of satisfaction decisions. *JMR, Journal of Marketing Research*, *17*(4), 460–469. doi:10.2307/3150499

Oliver, R. L. (1981). Measurement and evaluation of satisfaction process in retail settings. *Journal of Retailing*, *57*(1), 25–48.

Oliver, R. L. (1993). A conceptual model of service quality and service satisfaction: Compatible goals, different concepts. In T. A. Swartz, D. E. Bowen, & S. W. Brown (Eds.), *Advances in marketing and management* (pp. 65–85). Greenwich, CT: JAI Press.

Oliver, R. L. (1997). *Satisfaction – A behavioral perspective on the consumer*. New York, NY: McGraw-Hill.

Oliver, R. L. (1999). Whence customer loyalty? *Journal of Marketing*, *63*(4), 33–44. doi:10.2307/1252099

Ozgener, S., & Iraz, R. (2006). Customer relationship management in small-medium enterprises: The case of Turkish tourism industry. *Tourism Management, 27*(6), 1356–1363. doi:10.1016/j.tourman.2005.06.011

Padmavathy, C., Balaji, M. S., & Sivakumar, V. J. (2012). Measuring effectiveness of customer relationship management in Indian retail banks. *International Journal of Bank Marketing, 30*(4), 246–266. doi:10.1108/02652321211236888

Pamsari, M. B., Dehban, M., & Lulemani, H. K. (2013). Assessment of the key success factors of customer relationship management. *Universal Journal of Management and Social Sciences, 3*(4), 23–29.

Pan, S. L., Tan, C. W., & Lim, E. T. K. (2006). Customer relationship management (CRM) in e-government: A relational perspective. *Decision Support Systems, 42*(1), 237–250. doi:10.1016/j.dss.2004.12.001

Parasuraman, A. (1997). Reflections on gaining competitive advantage through customer value. *Journal of the Academy of Marketing Science, 25*(2), 154–161. doi:10.1007/BF02894351

Parasuraman, A., Berry, L. L., & Zeithaml, V. A. (1991). Perceived service quality as a customer-focused performance measure: An empirical examination of organizational barriers using and extended service quality model. *Human Resource Management, 30*(3), 335–364. doi:10.1002/hrm.3930300304

Parasuraman, A., & Grewal, D. (2000). The impact of technology on the quality-value-loyalty chain: A research agenda. *Journal of the Academy of Marketing Science, 28*(1), 168–175. doi:10.1177/0092070300281015

Parasuraman, A., Zeithaml, V. A., & Berry, L. L. (1988). SERVQUAL: A multiple-item scale for measuring consumer perceptions of service. *Journal of Retailing, 64*(1), 12–40.

Parvatiyar, A., & Sheth, J. N. (2001). Customer relationship management: Emerging practice, process, and discipline. *Journal of Economic & Social Research, 3*(2), 1–34.

Payne, A., Christopher, M., Clark, M., & Peck, H. (1999). *Relationship marketing for competitive advantage.* Oxford, UK: Butterworth Heinemann.

Payne, A., & Frow, P. (2005). A strategic framework for customer relationship management. *Journal of Marketing, 69*(4), 652–671. doi:10.1509/jmkg.2005.69.4.167

Payne, A., & Frow, P. (2006). Customer relationship management: From strategy to implementation. *Journal of Marketing Management, 22*(1-2), 135–168. doi:10.1362/0267257067776022272

Peng, M. W., & Health, P. S. (1996). The growth of firm in planned economies in transition: Institutions, organizations, and strategic choice. *Academy of Management Review, 21*(2), 492–528.

Phillips, L. W., Chang, D. R., & Buzzel, R. D. (1983). Product quality, cost position and business performance: A test of some key hypotheses. *Journal of Marketing, 47*(2), 26–43. doi:10.2307/1251491

Pinto, J. K., & Slevin, D. P. (1987). Critical factors in successful project implementation. *IEEE Transactions on Engineering Management, 34*(1), 22–27. doi:10.1109/TEM.1987.6498856

Plakoyiannaki, E., & Saren, M. (2006). Time and the customer relationship management process: Conceptual and methodological insights. *Journal of Business and Industrial Marketing, 21*(4), 218–230. doi:10.1108/08858620610672588

Porter, M. E. (1985). *Competitive advantage.* New York, NY: Free Press.

Porter, M. E. (1996). What is strategy? *Harvard Business Review, 74*(6), 61–78. PMID:10158474

Ravald, A., & Gronroos, C. (1996). The value concept and relationship marketing. *European Journal of Marketing*, *30*(2), 19–30. doi:10.1108/03090569610106626

Reed, R., Lemak, D. J., & Montgomery, J. C. (1996). Beyond process: TQM content and firm performance. *Academy of Management Review*, *21*(1), 173–202.

Reichheld, F. F. (1996). *The loyalty effect*. Boston, MA: Harvard Business School Press.

Reichheld, F. F., & Sasser, W. E. (1990). Zero defections: Quality comes to services. *Harvard Business Review*, *68*(5), 105–111. PMID:10107082

Reichheld, F. F., & Teal, T. (1996). *The loyalty effect: The hidden force behind growth, profits, and lasting value*. Boston, MA: Harvard Business School Press.

Reynolds, K., & Arnold, M. (2000). Customer loyalty to the salesperson and the store: Examining relationship customers in an upscale retail context. *Journal of Personal Selling & Sales Management*, *20*(2), 89–97.

Reynolds, K., & Beatty, S. (1999). Customer benefits and company consequences of customer-salesperson relationships in retailing. *Journal of Retailing*, *75*(1), 11–32. doi:10.1016/S0022-4359(99)80002-5

Richards, K. A., & Jones, E. (2008). Customer relationship management: Finding value drivers. *Industrial Marketing Management*, *37*(2), 120–130. doi:10.1016/j.indmarman.2006.08.005

Richardson, M. L., & Gartner, W. H. (1999). Contemporary organizational strategies for enhancing value in healthcare. *International Journal of Health Care Quality Assurance*, *12*(5), 183–189. doi:10.1108/09526869910280339

Roberts, K., Varki, S., & Brodie, R. (2003). Measuring the quality of relationships in customer services: An empirical study. *European Journal of Marketing*, *37*(1-2), 169–196. doi:10.1108/03090560310454037

Rowley, J. (2005). The four Cs of customer loyalty. *Marketing Intelligence & Planning*, *23*(6), 574–581. doi:10.1108/02634500510624138

Rust, R. T., Danaher, P., & Varki, S. (2000). Using service quality data for competitive marketing decisions. *International Journal of Service Industry Management*, *11*(5), 438–469. doi:10.1108/09564230010360173

Rust, R. T., & Oliver, R. L. (1994). Service quality: Insights and managerial implications from the frontier. In R. T. Rust, & R. L. Oliver (Eds.), *Service quality: New directions in theory and practice* (pp. 1–19). London, UK: Sage. doi:10.4135/9781452229102.n1

Rust, R. T., & Zahorik, A. J. (1993). Customer satisfaction, customer retention and market share. *Journal of Retailing*, *69*(2), 145–156. doi:10.1016/0022-4359(93)90003-2

Rust, R. T., Zahorik, A. J., & Keiningham, T. L. (1995). Return on quality (ROQ), Making service quality financially accountable. *Journal of Marketing*, *59*(2), 58–70. doi:10.2307/1252073

Rust, R. T., Zeithaml, V. A., & Lemon, K. N. (2000). *Driving customer equity: How customer lifetime value is reshaping corporate strategy*. New York, NY: Free Press.

Ryals, L., & Knox, S. (2001). Cross-functional issues in the implementation of relationship marketing through customer relationship management. *European Management Journal*, *19*(5), 534–542. doi:10.1016/S0263-2373(01)00067-6

Saarijarvi, H., Karjaluoto, H., & Kuusela, H. (2013). Extending customer relationship management: From empowering firms to empowering customers. *Journal of Systems and Information Technology*, *15*(2), 140–158. doi:10.1108/13287261311328877

Samson, D., & Terziovski, M. (1999). The relationship between total quality management practices and operational performance. *Journal of Operations Management*, *17*(4), 393–409. doi:10.1016/S0272-6963(98)00046-1

Sheth, J. N. (2002). The future of relationship marketing. *Journal of Services Marketing*, *16*(7), 590–592. doi:10.1108/08876040210447324

Sheth, J. N., Newman, B. I., & Gross, B. L. (1991). *Consumption values and market choice*. Cincinnati, OH: South Western.

Sheth, J. N., & Sisodia, R. S. (2002). Marketing productivity: Issues and analysis. *Journal of Business Research*, *55*(5), 349–362. doi:10.1016/S0148-2963(00)00164-8

Sheth, J. N., Sisodia, R. S., & Sharma, A. (2000). The antecedents and consequences of customer-centric marketing. *Journal of the Academy of Marketing Science*, *28*(1), 55–66. doi:10.1177/0092070300281006

Sin, L. Y. M., Tse, A. C. B., & Yim, F. H. K. (2005). CRM: Conceptualization and scale development. *European Journal of Marketing*, *39*(11-12), 1264–1290. doi:10.1108/03090560510623253

Sivadas, E., & Baker-Prewitt, J. (2000). An examination of the relationship between service quality, customer satisfaction, and store loyalty. *International Journal of Retail & Distribution Management*, *28*(2), 73–82. doi:10.1108/09590550010315223

Slater, S. F. (1997). Developing a customer value-based theory of the firm. *Journal of the Academy of Marketing Science*, *25*(2), 162–167. doi:10.1007/BF02894352

Slater, S. F., & Narver, J. C. (1995). Market orientation and the learning organization. *Journal of Marketing*, *59*(3), 63–74. doi:10.2307/1252120

Slater, S. F., & Narver, J. C. (1998). Customer-led and market-oriented: Let's not confuse the two. *Strategic Management Journal*, *19*(10), 1001–1006. doi:10.1002/(SICI)1097-0266(199810)19:10<1001::AID-SMJ996>3.0.CO;2-4

Smith, J. B., & Colgate, M. (2007). Customer value creation: A practical framework. *Journal of Marketing Theory and Practice*, *15*(1), 7–23. doi:10.2753/MTP1069-6679150101

Smith, M., & Chang, C. (2010). Improving customer outcomes through the implementation of customer relationship management: Evidence from Taiwan. *Asian Review of Accounting*, *18*(3), 260–285. doi:10.1108/13217341011089658

Solomon, M. R. (1992). *Consumer behavior: Buying, having and being*. Needham Heights, MA: Allyn and Bacon.

Sousa, R., & Voss, C. A. (2009). The effects of service failures and recovery on customer loyalty in e-services: An empirical investigation. *International Journal of Operations & Production Management*, *29*(8), 834–864. doi:10.1108/01443570910977715

Starkey, M. W., Williams, D., & Stone, M. (2002). The state of customer management performance in Malaysia. *Marketing Intelligence & Planning*, *20*(6), 378–385. doi:10.1108/02634500210445437

Stone, M., Woodcock, N., & Wilson, M. (1996). Managing the change from marketing planning to customer relationship management. *Long Range Planning*, *29*(5), 675–683. doi:10.1016/0024-6301(96)00061-1

Storbacka, K., Strandvik, T., & Gronroos, C. (1994). Managing customer relationships for profit: The dynamics of relationship quality. *International Journal of Service Industry Management*, *5*(8), 21–38. doi:10.1108/09564239410074358

Strauss, B., & Neuhaus, P. (1997). The qualitative satisfaction model. *International Journal of Service Industries Management*, *8*(3), 236–249. doi:10.1108/09564239710185424

Swan, J. E., & Oliver, R. L. (1989). Postpurchase communications by consumers. *Journal of Retailing*, *65*(4), 516–533.

Sweeney, J. C., & Soutar, G. N. (2001). Consumer-perceived value: The development of a multiple-item scale. *Journal of Retailing*, *77*(2), 203–220. doi:10.1016/S0022-4359(01)00041-0

Szymanski, D. M., & Henard, D. H. (2001). Customer satisfaction: A meta-analysis of the empirical evidence. *Journal of the Academy of Marketing Science*, *29*(1), 16–35. doi:10.1177/0092070301291002

Taylor, S., & Baker, T. (1994). An assessment of the relationship between service quality and customer satisfaction in the formation of consumers' purchase intentions. *Journal of Retailing*, *70*(2), 163–178. doi:10.1016/0022-4359(94)90013-2

Teas, K. R., & Agarwal, S. (2000). The effects of extrinsic product cues on consumers' perceptions of quality, sacrifice, and value. *Journal of the Academy of Marketing Science*, *28*(2), 278–290. doi:10.1177/0092070300282008

Torres, E. N., & Kline, S. (2006). From satisfaction to delight: A model for the hotel industry. *International Journal of Contemporary Hospitality Management*, *18*(4), 290–301. doi:10.1108/09596110610665302

Tronvoll, B. (2011). Negative emotions and their effect on customer complaint behavior. *Journal of Service Management*, *22*(1), 111–134. doi:10.1108/09564231111106947

Tsao, H. Y., & Chen, L. W. (2005). Exploring brand loyalty from the perspective of brand switching costs. *International Journal of Management*, *22*(3), 436–441.

Vasquez-Parraga, A. Z., & Alonso, S. (2000). Antecedents of customer loyalty for strategic intent. In J. P. Workman, & W. D. Perreault (Eds.), *Marketing theory and applications* (pp. 82–83). Chicago, IL: American Marketing Association.

Verhoef, P. C., & Donkers, B. (2001). Predicting customer potential value: An application in the insurance industry. *Decision Support Systems*, *32*(2), 189–199. doi:10.1016/S0167-9236(01)00110-5

Wang, Y., Lo, H. P., Chi, R., & Yang, Y. (2004). An integrated framework for customer value and customer-relationship-management performance: A customer-based perspective from China. *Managing Service Quality*, *14*(2-3), 169–182. doi:10.1108/09604520410528590

Wang, Y. G., & Lo, H. P. (2002). Service quality, customer satisfaction, customer value and behavior intentions: Evidence from China's telecommunication industry. *Info – The Journal of Policy. Regulation and Strategy for Telecommunications*, *4*(6), 50–60. doi:10.1108/14636690210453406

Webster, F. E. (1988). Determining the characteristics of the socially conscious consumer. *Business Horizons*, *31*(3), 29–39. doi:10.1016/0007-6813(88)90006-7

Wikstrom, S., & Normann, R. (1994). *Knowledge and value: A new perspective on corporate transformation*. London, UK: Routledge.

Woodcock, N., Stone, M., & Foss, B. (2003). *The customer management scorecard: Managing CRM for profit*. London, UK: Kogan Page.

Woodruff, R. B. (1997). Customer value: The next source for competitive advantage. *Journal of the Academy of Marketing Science*, *25*(2), 139–153. doi:10.1007/BF02894350

Woodruff, R. B., & Gardial, S. (1996). *Know your customer: New approaches to understanding customer value and satisfaction*. Oxford, UK: Blackwell.

Wyner, G. A. (1996). Customer valuation: Linking behavior and economics. *Marketing Research*, *8*(2), 36–38.

Xu, M., & Walton, J. (2005). Gaining customer knowledge through analytical CRM. *Industrial Management & Data Systems*, *105*(7), 955–971. doi:10.1108/02635570510616139

Xu, Y., Yen, D. C., Lin, B., & Chou, D. C. (2002). Adopting customer relationship management technology. *Industrial Management & Data Systems*, *102*(8-9), 442–452. doi:10.1108/02635570210445871

Yamane, T. (1970). *Statistics – An introductory analysis*. Tokyo, Japan: John Weatherhill.

Zablah, A. R., Bellenger, D. N., & Johnston, W. J. (2004). An evaluation of divergent perspectives on customer relationship management: Towards a common understanding of an emerging phenomenon. *Industrial Marketing Management*, *33*(6), 475–489. doi:10.1016/j.indmarman.2004.01.006

Zamora, J., Vasquez-Parraga, A. Z., Morales, F., & Cisternas, C. (2004). Formation process of guest loyalty: Theory and empirical test. *Estudios y Perspectivas en Turismo*, *13*(3-4), 197–221.

Zeithaml, V. A. (1988). Consumer perceptions of price, quality and value: A means-end model and synthesis of evidence. *Journal of Marketing*, *52*(3), 2–22. doi:10.2307/1251446

Zeithaml, V. A., Berry, L., & Parasuraman, A. (1996). The behavioral consequences of service quality. *Journal of Marketing*, *60*(2), 31–46. doi:10.2307/1251929

Zeithaml, V. A., Parasuraman, A., & Berry, L. (1990). *Delivering quality service: Balancing customer perceptions and expectations*. New York, NY: Free Press.

Zineldin, M. (2006). The royalty of loyalty: CRM, quality and retention. *Journal of Consumer Marketing*, *23*(7), 430–437. doi:10.1108/07363760610712975

ADDITIONAL READINGS

Abbasi, M. R., & Torkamani, M. (2010). Theoretical models of customer relationship management (CRM). *Business Survey*, *41*, 19–35.

Agrawal, M. L. (2003). Customer relationship management (CRM) & corporate renaissance. *Journal of Service Research*, *3*(2), 149–171.

Alam, A., Arshad, M. U., & Shabbir, S. A. (2012). Brand credibility, customer loyalty and the role of religious orientation. *Asia Pacific Journal of Marketing and Logistics*, *24*(4), 583–598. doi:10.1108/13555851211259034

Baek, T. H., Kim, J., & Yu, J. H. (2010). The differential roles of brand credibility and brand prestige in consumer brand choice. *Psychology and Marketing*, *27*(7), 662–678. doi:10.1002/mar.20350

Becker, J. U., Greve, G., & Albers, S. (2009). The impact of technological and organizational implementation of CRM on customer acquisition, maintenance, and retention. *International Journal of Research in Marketing*, *26*(3), 207–215. doi:10.1016/j.ijresmar.2009.03.006

Buttle, F. (2004). *Customer relationship management: Concept and tools*. Burlington, MA: Elsevier.

Choi, J., Seol, H., Lee, S., Cho, H., & Park, Y. (2008). Customer satisfaction factors of mobile commerce in Korea. *Internet Research*, *18*(3), 313–335. doi:10.1108/10662240810883335

Corteau, A. M., & Li, P. (2003). Critical success factors of CRM technological initiatives. *Canadian Journal of Administrative Sciences*, *20*(1), 21–34. doi:10.1111/j.1936-4490.2003.tb00303.x

Eid, R. (2007). Towards a successful CRM implementation in banks: An integrated model. *The Service Industries Journal*, *27*(8), 51–90. doi:10.1080/02642060701673703

Gounaris, S. P., Tzempelikos, N. A., & Chatzipanagiotou, K. (2007). The relationships of customer- perceived value, satisfaction, loyalty and behavioral intentions. *Journal of Relationship Marketing*, *6*(1), 63–87. doi:10.1300/J366v06n01_05

Hennig-Thurau, T. (2004). Customer orientation of service employees: Its impact on customer satisfaction, commitment and retention. *International Journal of Service Industry Management*, *15*(5), 460–478. doi:10.1108/09564230410564939

Hoest, V., & Knie-Andersen, M. (2004). Modeling customer satisfaction in mortgage credit companies. *International Journal of Bank Marketing*, *22*(1), 26–42. doi:10.1108/02652320410514915

Ingenbleek, P. (2007). Value-informed pricing in its organizational context: Literature review, conceptual framework, and directions for future research. *Journal of Product and Brand Management*, *16*(7), 441–458. doi:10.1108/10610420710834904

Jae, W. K., Johio, C., Qualls, W., & Keyessok, H. (2008). It takes a marketplace community to raise brand commitment: The role of online communities. *Journal of Marketing Management*, *24*(2-4), 409–431.

Jamali, R., Moshabaki, A., Aramoon, H., & Alimohammadi, A. (2013). Customer relationship management in electronic environment. *The Electronic Library*, *31*(1), 119–130. doi:10.1108/02640471311299173

Jean, R. B., Sinkovics, R. R., & Kim, D. (2010). Drivers and performance outcomes of relationship learning for suppliers in cross-border customer–supplier relationships: The role of communication culture. *Journal of International Marketing*, *18*(1), 63–85. doi:10.1509/jimk.18.1.63

Kanagal, N. (2009). Role of relationship marketing in competitive marketing strategy. *Journal of Management and Marketing Research*, *2*(1), 1–17.

Kassim, N., & Abdullah, N. A. (2010). The effect of perceived service quality dimensions on customer satisfaction, trust, and loyalty in e-commerce settings: A cross cultural analysis. *Asia Pacific Journal of Marketing and Logistics*, *22*(3), 351–371. doi:10.1108/13555851011062269

Keller, K. L., & Lehman, D. R. (2006). Brand and branding: Research findings and future priorities. *Marketing Science*, *25*(6), 740–759. doi:10.1287/mksc.1050.0153

Kemp, E., & Bui, M. (2011). Healthy brands: Establishing brand credibility, commitment and connection among consumers. *Journal of Consumer Marketing*, *28*(6), 429–437. doi:10.1108/07363761111165949

Keramati, A., Zavareh, J. T., Ellioon, A., Moshki, H., & Sajjadiani, S. (2008). The adoption of customer relationship management: An empirical study on Iranian firms. *International Journal of Electronic CRM*, *1*(4), 30–37.

Ko, E., Kim, S. H., Kim, M., & Woo, J. Y. (2008). Organizational characteristics and the CRM adoption process. *Journal of Business Research*, *61*(1), 65–74. doi:10.1016/j.jbusres.2006.05.011

Lee, K., Joshi, K., & Bae, M. (2009). A cross-national comparison of the determinants of customer satisfaction with online stores. *Journal of Global Information Technology Management, 12*(4), 25–51.

Leone, R. P., Rao, V. R., Keller, K. L., & Luo, A. M. (2006). Linking brand equity to customer equity. *Journal of Service Research, 9*(2), 125–138. doi:10.1177/1094670506293563

Liao, N. N. H., & Wu, T. C. H. (2009). The pivotal role of trust in customer loyalty, empirical research on the system integration market in Taiwan. *The Business Review, Cambridge, 12*(2), 277–384.

Lin, C. H., & Peng, C. H. (2005). The cultural dimension of technology readiness on customer value chain in technology-based service encounters. *Journal of American Academy of Business, 7*(1), 176–181.

Lin, R., Chen, R., & Chui, K. K. (2010). Customer relationship management and innovation capability: An empirical study. *Industrial Management & Data Systems, 110*(1), 111–133. doi:10.1108/02635571011008434

Lindgreen, A., Michael, A., Palmer, R., & Van, H. T. (2009). High-tech innovative products: Identifying and meeting business customers' value needs. *Journal of Business and Industrial Marketing, 24*(3-4), 182–197. doi:10.1108/08858620910939732

Mainela, T., & Ulkuniemi, P. (2013). Personal interaction and customer relationship management in project business. *Journal of Business and Industrial Marketing, 28*(2), 103–110. doi:10.1108/08858621311295245

Mosad, Z. (2006). The royalty of loyalty: CRM, quality and retention. *Journal of Consumer Marketing, 23*(7), 430–437. doi:10.1108/07363760610712975

Orth, U., & Marchi, R. D. (2007). Understanding the relationships between functional, symbolic, and experiential brand beliefs, product experiential attributes, and product schema: Advertising-trial interactions revisited. *Journal of Marketing Theory and Practice, 15*(3), 219–233. doi:10.2753/MTP1069-6679150303

Osarenkhoe, A. (2009). The business culture of a firm applying a customer-intimate philosophy: A conceptual framework. *International Journal of Business and Systems Research, 3*(3), 257–278. doi:10.1504/IJBSR.2009.026183

Osarenkhoe, A., & Bennani, A. Z. (2007). An exploratory study of implementation of customer relationship projects. *Business Process Management Journal, 13*(1), 139–164. doi:10.1108/14637150710721177

Payne, A., Storbacka, K., Frow, P., & Knox, S. (2009). Co-creating brands: Diagnosing and designing the relationship experience. *Journal of Business Research, 62*(3), 379–389. doi:10.1016/j.jbusres.2008.05.013

Raman, P., Wittmann, C. M., & Rauseo, N. A. (2006). Leveraging CRM for sales: The role of organizational capabilities in successful CRM implementation. *Journal of Personal Selling & Sales Management, 26*(1), 39–53. doi:10.2753/PSS0885-3134260104

Reinartz, W. J., Krafft, M., & Hoyer, W. D. (2004). The customer relationship management process: Its measurement and Impact on performance. *JMR, Journal of Marketing Research, 41*(3), 293–305. doi:10.1509/jmkr.41.3.293.35991

Sichtmann, C. (2007). An analysis of antecedents and consequences of trust in a corporate brand. *European Journal of Marketing, 41*(9-10), 999–1015. doi:10.1108/03090560710773318

Swait, J., & Erdem, T. (2007). Brand effects on choice and choice set formation under uncertainty. *Marketing Science, 26*(5), 679–697. doi:10.1287/mksc.1060.0260

Ting, D. H. (2006). Further probing of higher order in satisfaction construct: The case of banking institutions in Malaysia. *International Journal of Bank Marketing, 24*(2), 98–111. doi:10.1108/02652320610649914

Varela-Neira, C., Vazquez-Casielles, R., & Iglesias, V. (2010). Explaining customer satisfaction with complaint handling. *International Journal of Bank Marketing, 28*(2), 88–112. doi:10.1108/02652321011018305

Veloutsou, C., & Moutinho, L. (2009). Brand relationship through brand reputation and brand tribalism. *Journal of Business Research, 62*(3), 314–322. doi:10.1016/j.jbusres.2008.05.010

Yim, F. H., Anderson, R. E., & Swaminathan, S. (2004). Customer relationship management its dimensions and effect on customer outcomes. *Journal of Personal Selling & Sales Management, 24*(4), 263–278.

Zablah, A. R., Bellenger, D. N., & Johnston, W. J. (2004). An evaluation of divergent perspectives on customer relationship management: Towards a common understanding of an emerging phenomenon. *Industrial Marketing Management, 33*(6), 475–489. doi:10.1016/j.indmarman.2004.01.006

Zhang, J., & Bloemer, J. M. (2008). The impact of value congruence on consumer-service brand relationships. *Journal of Service Research, 11*(2), 161–178. doi:10.1177/1094670508322561

Zikmund, W. G., McLeod, R., & Gilbert, F. W. (2003). *Customer relationship management: Integrating marketing strategy and information technology*. Hoboken, NJ: John Wiley & Sons.

Zineldin, M. (2005). Quality and customer relationship management (CRM) as competitive strategy in the Swedish banking industry. *The TQM Magazine, 17*(4), 329–344. doi:10.1108/09544780310487749

KEY TERMS AND DEFINITIONS

Brand: A symbol that identifies a product and differentiates it from its competitors.

Brand Loyalty: The extent of the faithfulness of consumers to a particular brand, expressed through their repeat purchases, irrespective of the marketing pressure generated by the competing brands.

Brand Management: The process of maintaining and improving a brand, thus involving a number of important aspects such as cost, customer satisfaction, in-store presentation, and competition.

Customer: The person that consumes products and has the ability to choose between different products and suppliers.

Customer Relationship Management: A management philosophy according to which the company's goal can be best achieved through identification and satisfaction of the customers' stated and unstated requirement.

Customer Satisfaction: The degree of satisfaction provided by the goods or services of a company as measured by the number of repeat customers.

Customer Value: The difference between what a customer gets from a product and what a customer has to give in order to get it.

Framework: Broad overview, outline, or skeleton of interlinked items which support a particular approach to a specific objective of the study.

Manufacturing: The process of converting raw materials and components into finished goods that meet a customer's expectations.

Chapter 11
Smart, Innovative and Intelligent Technologies Used in Drug Designing

S. Deshpande
Data Consulting, New Delhi, India

S. K. Basu
University of Lethbridge, Canada

X. Li
Industrial Crop Research Institute, Yunan Academy of Agricultural Sciences, China

X. Chen
Institute of Food Crops, Yunan Academy of Agricultural Sciences, China

ABSTRACT

Smart and intelligent computational methods are essential nowadays for designing, manufacturing and optimizing new drugs. New and innovative computational tools and algorithms are consistently developed and applied for the development of novel therapeutic compounds in many research projects. Rapid developments in the architecture of computers have also provided complex calculations to be performed in a smart, intelligent and timely manner for desired quality outputs. Research groups worldwide are developing drug discovery platforms and innovative tools following smart manufacturing ideas using highly advanced biophysical, statistical and mathematical methods for accelerated discovery and analysis of smaller molecules. This chapter discusses novel innovative applications in drug discovery involving use of structure-based drug design which utilizes geometrical knowledge of the three-dimensional protein structures. It discusses statistical and physics based methods such as quantum mechanics and classical molecular dynamics which can also play a major role in improving the performance and in prediction of computational drug discovery. Lastly, the authors provide insights on recent developments in cloud computing with significant increase in smart and intelligent computational power thus allowing larger data sets to be analyzed simultaneously on multi processor cloud systems. Future directions for the research are outlined.

DOI: 10.4018/978-1-4666-5836-3.ch011

INTRODUCTION

Rapid and steady growth of low-cost computer power in discovering and designing new drugs has become a central topic in modern biology and medical chemistry. These includes methods and techniques ranging from high-throughput screening of compounds in databases to protein inhibition through molecular dynamic and thermodynamic approaches using distributed and cloud computing. Computational drug design approaches can be divided into structure and ligand-based approaches (Reddy et al., 2007). Former approach focuses on available X-ray crystallographic protein structure, NMR-based structure or homology model obtained from known template from protein databank. Once a rigid structure of the target is available, ligand-based virtual screening identifies novel ligands by three-dimensional (3D) similarity searching or by pharmacophore protein matching (Reddy et al., 2007). Ligand based virtual screening detects compounds that are similar to the active known compounds. Next, the target compounds are identified through virtual screening following which the docking algorithms are applied to position the available compounds in the binding sites derived from several biochemical studies (Reddy et al., 2007). These compounds are then ranked according to their steric and electrostatic interactions between the receptor and ligand (Reddy et al., 2007). On the other hand, ligand-based approach is useful in absence of experimental 3D structure. Under such circumstances, known ligands are explored to understand the basic structural and chemical properties correlating with specific pharmacological properties. Design of new ligands is performed in several steps mostly by manual inspection and qualitative interpretation of ligand-binding site interactions. Ligand-based approaches use Quantitative Structure Activity Relationship (QSAR) method and pharmacophore modeling. The QSAR is a computational methodology for quantifying correlation between structures of chemical compounds and chemical/biological process (Acharya et al., 2011).

The QSAR model is applied to optimize active compounds to increase their relevant biological activities to the maximum possible level. The success of this model is closely related to the selection of appropriate molecular descriptors and its ability to generate mathematical relationships between the molecular descriptor and biological activity (Acharya et al., 2011). Some of the major statistical methods used to select molecular features are multivariate linear regression analysis (MLA), Principle Component Analysis (PCA) and Partial Least Square Analysis (PLS), all of which will be discussed in this chapter review (Acharya et al., 2011). Structure-based approaches also involve virtual screening of compound libraries which can be manifested using molecular docking approach and Molecular Dynamics (MD) simulations (Okimoto et al., 2009; Durrant & McCammon, 2011). Molecular docking allows prediction of different complex protein-ligand conformations and estimation of binding affinity using mathematical algorithms for faster conformational search of ligands. MD simulations allow the estimation of the effect of explicit and implicit water molecules around the binding site of ligand and calculation of binding free energy. The techniques involved are thermodynamic integration (TI), free energy perturbation (FEP) and molecular mechanics/Poisson-Boltzmann surface area (MM/PBSA) (Okimoto et al., 2009; Durrant & McCammon, 2011).

One of the robust methodologies currently being developed and extensively used is cloud computing in computational drug discovery. Cloud computing is a term which involves services provided over the Internet. This idea has been implemented from distributed/grid computing that integrates several individual computers to generate a super computer system of superior computational abilities (Shudong et al., 2004). Cloud computing utilizes computing resources allocated on-demand over the Internet. Use of massive computational power can thus be applied for virtual screening of principal compounds and undergoing complex calculations within a very short time span.

In this chapter we will discuss all the computational methods developed and currently used in drug discovery and design. We will focus on the ability of these smart, intelligent and innovative methods and approaches for identifying novel drug candidates and latest advances in intelligent computing power and their interconnection and smart manufacturing applications in computational biochemistry.

Computational Drug Design Strategies

Application of computational tools in drug discovery has grown steadily over the past several years and there have been significant improvements over the methods. A number of computational techniques are used to resolve complex chemical and biological problems including statistical analysis of large datasets, structure-based drug designs, computational genomics, molecular simulations and pharmacophore based molecular library designing. Computational chemistry tools are regularly employed to resolve detailed atomic structures of different ligand-receptor complexes; and for evaluating small-molecule drug characters. Statistical and information sciences are being increasingly used nowadays to generate and manage chemical and biological activity databases. With the aid of computers, new techniques have been developed for automated derivation of drug derivatives. Libraries of compound derivatives can be assembled by targeted structure-based combinatorial chemistry. Improved methods have developed for rapid generation and prediction of potential drug binding to proteins. In addition, refinement of drug candidates can also be performed in most effective manner. Therefore, use of computers has become an important asset for development of new therapeutics for most common and complex diseases.

Structure-Based Drug Designing

SBDD is one of the powerful approaches in drug discovery as it combines both combinatorial chemistry and high-throughput screening for synthesis of potential drug candidates. First step in drug development is to find small molecule which can bind to target proteins and alter its biological function by increasing or inhibiting its activity (Wilson & Muftuoglu, 2012). Development of small molecule drug candidate against a given disease requires proper understanding of metabolic pathways. Therefore, first step in drug designing is identification of potential biological target from the given metabolic pathway. Therefore, identification and development of potential ligands specific to the biological target is the second step in Structure-Based Drug Design (SBDD) (Reddy et al., 2007). There are many automated screening tools and software that allow screening, testing and selection of large number of compounds from online Web databases which is based on their ability for interaction with biological targets. Some of the tools such as PyRx developed by The Scripps Research Institute allow screening libraries of compounds against potential drug targets (http://mgltools.scripps.edu/documentation/links/pyrx-virtual-screening-tool). The VSDocker is an important tool allowing parallel high-throughput virtual screening on windows based computer cluster systems (Prakhov, Chernorudskiy, & Gainullin, 2010).

The biological target has pocket where the ligand binds and alters the activity of the target. The pocket comprises of different types of possible hydrogen bond donors and acceptors and hydrophobic characteristics (Reddy et al., 2007). Compounds (or fragments of compounds) from the available databases are then positioned over a selected region or in other words, the active site of the structure. These are next scored and ranked depending on the steric and electrostatic interactions with the target site. Top selected compounds are next screened using biochemical assays (Anderson, 2003).

Molecular Docking Approach from Virtual Screening of Lead Compounds

Structure-based computational methodology involves computational screening of compounds which can be performed by molecular docking. It is often useful when the 3D structure of the biological target is available. Molecular docking involves knowledge of putative binding sites where the ligand can bind and initiate a biological response. The knowledge can either be acquired from biochemical studies or obtained from co-crystallizing the target protein with a ligand. Once the target binding sites are known, correct binding modes (binding pose) of a ligand molecule is conducted by running number of trials by searching different translational as well as rotational degrees of freedom. The ligand-receptor interaction energy is calculated for each binding pose and a '**score**' is assigned for each pose. Score is calculated mainly by three scoring function techniques: force-field, knowledge and empirical based scoring functions (Kroemer, 2007; Cheng et al., 2010).

Force field based functions employ a force-field to compute non-covalent ligand-receptor interactions in an implicit solvent. Generalized Born (GB/SA) or Poisson-Boltzmann (PB/SA) methods are used for computing the interactions. Knowledge-based scoring functions are generated from the inter-atomic distances available in the protein-ligand complex structures databases. Frequency distributions are computed from the distances and the distribution is converted to potential of mean force or knowledge-based potentials. Empirical based method is based on the principle that free energy change upon ligand binding (ΔG_{bind}) can be decomposed into the summation of the internal energy (ΔG_{int}), the solvation energy (ΔG_{solv}), the conformational energy (ΔG_{conf}) and the free energy of motion (ΔG_{motion}).

Virtual Screening

Small-molecule compound library is prepared. Table 1 illustrates a list of databases which can be used for library preparation. Both the biological target and the compound library require preprocessing such as assigning proper tautumeric, stereoisomeric, and protonation states (Cheng et al., 2010). Each compound in the library is docked to the biological target molecule using a docking program or tool which models the ligand-receptor interactions for achieving optimal complimentarity between the active site of target and the compound. Scoring function evaluates the fitness between the target and compound and calculates a score for each binding pose. Binding poses are ranked based on their scores. Lee et al. (2010) performed virtual screening for determining vascular endothelial growth factor receptor (VEGFR)-2 kinase inhibitors using pharmacophore modeling and docking studies.

Table 1. Number of compounds (entries) in the commonly screened databases (Note: All are approximate figures)

Databases	Number of compounds
PubChem	30 million
ChEMBL	1 million
NCI (National Chemical Institute)	250,000
ChemSpider	26 million
ACX	140,000
ASINEX	140,000
ChemBridge	900,000
ChemDiv	1.2 million
ChemNavigator	55 million

Recent Advances in Docking

4D Docking

Giovanni et al. utilized a novel procedure of ligand docking by exploiting the receptor flexibility (Bottegoni et al., 2009). In four dimensional (4D) docking procedure, an ensemble of receptor conformers are generated. Generated structures are converted to conformational stack and matching structure with complete sequence in Uniprot database is selected as reference structure. All the other structures are superimposed to the reference structure by an iterative weighted process. This procedure of making a conformational stack with one reference molecule is called 4D grid. Ligands are placed in the 4D plane and allowed to move from one 4D plane to another allowing transfer between the different receptor conformations. Receptor-ligand poses with lowest energy conformation then undergoes all atom scoring procedures. This approach was tested on 99 therapeutically relevant proteins and 300 diverse ligands. The approach reproduced correct ligand-binding geometry matching the success rate of the traditional process.

Drug Repositioning by Structure-Based Virtual Screening and Molecular Docking

Drug repositioning is a strategy which involves generation of new or additional value from drug by targeting diseases. In this approach, candidate compounds are screened against the biological targets for higher potency and selectivity against the targets. Combination of structure-based virtual screening and drug repositioning represents highly efficient approach for development of novel medicines. For drug repositioning, specialized virtual libraries are available through databases which contain information about existing drugs and compounds which have made through later stages of clinical trials and commercialization (Ma et al., 2013). When potential compounds are identified by virtual screening, molecular docking can be performed using "single target" approach which identifies potential interactions between the drug candidates and biological target. Molecular docking involves three main steps: (1) generation of molecular model of the biological target and pre-treatment of ligands, (2) conformational sampling of the ligands on the biological target binding site, (3) Assignment of score for the poses of the ligand-receptor complexes. Docking can be performed by various algorithms. Some of the commonly used algorithms are Monte Carlo (MC) or Genetic Algorithms (GA) (Hart & Read, 1992; Morris et al., 1998). Drug repositioning has been applied in several applications. A well-known example is its use in discovery of sildenafil for the treatment of erectile dysfunction in human males (Ashburn & Thor, 2004). Sildenafil was originally known for its anti-anginal activity for inhibition of phosphodiesterase family of proteins. By performing drug repositioning, male erectile dysfunction was treated in phase I clinical trials.

Drug Docking via Systems Theorem

Computational docking of drug candidates is performed by various docking approaches in which the flexible drug compound is docked to the rigid receptor and various receptor-drug conformational poses are generated. Score is calculated for each pose of the complex. However the docking algorithms considers receptor structures as rigid which does not provide accurate docking scores. Flexible protein and ligand motions are required for efficient docking at the docking site. The Lyapunov approach provides focused stability during protein folding for drug docking calculations (Sung & Chen, 2012). Applying the Lyapunov equation and the molecular dynamics approaches, lowest energy protein structure can be determined. Ligand-receptor docking procedure consists of 3 major steps, namely, (1) creation of docking box around the receptor binding site, (2) placement of ligands in the docking box, (3) generate various poses and score the poses based on various interacting energy terms. The Lyapunov function provides dynamic stability in both linear and non-linear systems which are based on Hess's law by which free energy of binding of ligand and receptor is calculated in solvent system. Lyapunov stability function illustrates relationship between minimum energy and modeling. Reaction coordinates are generated and stability basins of the energy landscape for system of interest are characterized using biomolecular simulations. This approach is based on stability theory which decomposes the molecular systems into subsystems. Then it constructs a suitable vector Lyapunov function. Authors have tried this approach on Cyclooxygenase-2 protein by docking with Flurbiprofen. However, practical application of the approach remains unexplored.

Molecular Dynamics Approach

Macromolecular structures (target biological receptors) play a vital role in drug discovery. For novel drug discovery prediction, virtual screening of the compounds is performed using drug database. This is followed by selection of potential compounds and docking to the receptor active site. However, the docking algorithms were unable to take into consideration the role of protein flexibility in ligand binding involving complex calculations essential for describing quantum-mechanical motions and chemical interactions of large molecular systems. MD simulations were able to remove this problem by applying approximations based on Newtonian physics for simulating atomic motions [4]. By undertaking such approach, computational complexity can be further reduced. First, a crystal structure or homology model of the receptor is taken and forces acting on each atom are calculated from the equation:

$$E_{Total} = \sum_{bonds} K_r (r - r_{eq})^2 \\ + \sum_{angles} K_\theta (\theta - \theta_{eq})^2 \\ + \sum_{dihedrals} \frac{V_n}{2}[1 + \cos(n\varphi - \gamma)] \\ + \sum_{i<j} \left[\frac{A_{ij}}{R_{ij}^{12}} - \frac{B_{ij}}{R_{ij}^6} + \frac{q_i q_j}{\epsilon R_{ij}} \right]$$

(Durrant, & McCammon, 2011)

Forces are calculated from contributions from bonded and non-bonded atom types. Bonded atom contribution is composed of chemical bonds and atomic angles from virtual springs and dihedral angles generated from rotations around the bonds. However, the non-bonded contribution is estimated from Van der Waals interactions that are modeled applying the Lennard-Jones potential and the electrostatic interactions based on the Coulomb's law (Leonnard-Jones, 1924). For reproducing the behavior of the target molecule in motion, the energy terms are derived from experimental and quantum-mechanical data. To perform MD simulations, force fields from CHARMM (Brooks et al., 1983), GROMACS (Christensen et al., 2009) and AMBER (Wang et al., 2009) are generally used. Once the forces on each are calculated, each atom is moved to different position according to Newton's laws of motion. Simulation is advanced by a femtosecond and is repeated till the simulation time specified by the user, usually 40-50 nanoseconds. Results from MD simulations explain various conformational changes of protein and ligands. Experimental studies involving protein and ligand have shown good agreement between the experimental and computational measurements using MD simulations (Markwick et al., 2009; Showalter, & Brüschweiler, 2007).

MD simulations provide structural insights into the identification of drug binding sites of the β-adrenergic receptors. MD simulations were performed for 0.5 microseconds which produced distinct conformational changes. MD simulations revealed binding pockets at both the solvent and lipid-exposed areas which serves for undertaking various structure-based approaches for designing new compounds (Ivetac & McCammon, 2010). Using β-adrenergic receptor protein, MD simulations were performed for testing the association of several ligands such as alprenolol and propranolol (Dror et al., 2011). The long MD simulations performed for ~20 microseconds revealed the entire process of association of drug molecules to the GPCR binding site. The process of drug binding explained the energetic barriers required for drug binding and unbinding kinetics.

Statistical Mechanics and Thermodynamic Approaches

The MD calculations utilize virtual screening of ligands for a receptor protein by screening drug databases for available compounds/libraries. However, docking fails to perform in certain cases which limit its use in drug designing. Methods like MM/PBSA and MM/GBSA are applied for calculating binding free energies using molecular mechanics and implicit solvation models. Several thermodynamic approaches have been identified recently which can be used for calculating binding free energy of ligands. Pathway method employs a technique which transforms the system from one state (complex) to another state (unbound) (Shirts, Mobley, & Chodera, 2007). Using FEP method, the system is transformed from one thermodynamic state to another. The two states (A and B) can be characterized by potential energy functions where state A corresponds to unbound ligand state and state B corresponds to complex state. The free energy difference between the two states is calculated by:

$$U_{AB} = U_A - U_B$$
(Shirts, Mobley, & Chodera, 2007)

Where U_{AB} is the free energy difference between the two states. By solving the above equation, we obtain the final *free energy perturbation (FEP) formula*:

$$U_{AB} = -kT \ln \left\langle e^{-\beta(U_B - U_A)} \right\rangle$$
(Shirts, Mobley, & Chodera, 2007)

Where β is the Boltzmann constant and free energy difference between states A and B needs to be calculated by natural log of average Boltzmann factor of potential energy difference between states A and B over equilibrium configurations from the state A. Such a transformation between two end states is termed as *alchemical free energy calculation*. A recent application of the approach was demonstrated the use of FEP coupled with MD and MC used for developing and optimizing lead compounds for HIV-1 protease and non-nucleoside reverse transcriptase inhibitors (Acevedo et al., 2012).

Another widely used approach for binding energy calculation are the MM/PBSA or MM/GBSA methods (Kaukonen et al., 2008). Free energy calculations using MM/PBSA or MM/GBSA methodology are conducted on collective assemblage of structures that are sampled during MD simulations using implicit solvent. By the application of energy minimized structure of receptor; the ligand binding free energies can be estimated. Implicit solvation approximations speed up the calculations, reduce overall completion time and utilize lower computational power in comparison to conventional MD simulation. The free energy is calculated applying PB implicit solvent technique based on SASA using thermodynamic variables based on experimental conditions for regulating specific parameters (for *e.g.* ionic strength, external and internal dielectric constants) (Suenaga et al., 2012; Miller et al., 2012). Free energies calculated using GB also uses implicit solvent method based on SASA. GB method controls other variables such as salt concentration. The biding free energies between the ligand and receptor are calculated as:

$$\Delta G_{bind} = \Delta H - T\Delta S \approx \Delta E_{MM} + \Delta G_{sol} - T\Delta S$$
(Hou et al., 2012)

Where,

$$\Delta E_{MM} = \Delta E_{internal} + \Delta E_{electrostatic} + \Delta E_{vdw}$$
(Hou et al., 2012)

And,

$$\Delta G_{sol} = \Delta G_{\frac{PB}{GB}} + \Delta G_{SA}$$
(Hou et al., 2012)

Where ΔG_{sol}, ΔE_{MM} and $-T\Delta S$ are solvation energy, change in gas phase molecular mechanical energy and conformational entropy upon binding of ligand. The ΔE_{MM} is calculated as sum of internal energy, electrostatic solvation energy and Van der Waals energy. The ΔG_{sol} is calculated as contribution of polar (PB/GB) and non-polar (surface area) components (Hou et al., 2012).

PBSA and GBSA methods are integrated in MD simulation software suites such as GROMACS (Brooks et al., 1983), CHARMM (Christensen et al., 2009) and AMBER (Wang et al., 2009). Both the methods have been successfully applied for drug binding to the receptor where 59 ligands were tested for binding to 6 different proteins (Hou et al., 2012). The sensitivity of these approaches was evaluated by applying MD simulations for a timescale of 400-4800 picoseconds. The simulation results demonstrated that the length of the simulations did not impact the binding energies seriously. However the solute dielectric constant impacted protein-ligand binding more significantly. The results clearly showed that PBSA and GBSA are important tools in drug designing that can carefully estimate energies and rank inhibitors depending on interactions within the targeted biomolecules.

Ligand-Based Drug Design

Ligand-based drug designing is an indirect method for generating of pharmacologically active compounds interacting with target biological compounds. The first step in drug development will be proper identification of the target molecule associated with a disease present in the biochemical pathway. In the absence of experimental structure or homology model, ligand molecules are analyzed to comprehend the structural and physico-chemical characteristics of the ligand that correlates with targeted pharmacological activity of the ligand.

Ligand-based drug design is generally performed using QSAR and pharmacophore modeling. QSAR is an approach for correlating structures of chemical compounds and biological activities. The first step of QSAR involves identification of the active ligands. The ligands are screened from the drug databases such as PubChem, ChEMBL, NCI, etc by a process called "virtual screening." Next suitable molecular descriptors are identified for the ligand. The molecular descriptors can be characterized into 1D, 2D, 3D or 4D that comprises of the description of shape, volume, density, dipole moment, solvation energy, water accessible surface area, potential energy, torsion energy, electronic energy, electrostatic energy, Van der Waals energy, etc. describing the general characteristics of the ligand.

In the following step a mathematical expression is written for correlating the molecular descriptors and the biological activity. Finally, the QSAR model is constructed and validated for its statistical and predictive abilities (Acharya et al., 2011). The application of QSAR using HypoGen/Discovery Studio softwares for constructing pharmacophore models with hydrophobic and hydrogen bond acceptor features for cathepsin D inhibitors for prevention of various degenerative diseases such as Alzheimer's disease and neuronal ceroid lipofuscinosis (Sakkiah, Thangapandian, & Lee, 2012). Fischer's randomization method was used for identifying correlation between experimental and predicted activity. They obtained 49 compounds as potent cathepsin D inhibitors based on consensus scoring values (Sakkiah, Thangapandian & Lee, 2012).

QSAR approach can be performed using three major statistical methods, namely: MLR, PCA and PLS. The MLR is performed using stepwise regression for finding the best model. MLR has also been used in developing Web-based software (PreADME) which performs *in silico* ADMET (Absorption, Distribution, Metabolism and Toxicity) predictions which is dependent on relationship between molecular descriptors and activity profiles of the compounds (Khan, Mahmud, & Ingebrigt, 2009). One such study using MLR analysis was performed on thermolysin protein responsible for several bacterial infections such as cholera, gastritis and peptic ulcer (Jin et al., 1996; Winfield, Inniss, & Smith, 2009). Thermolysin inhibitors were constructed applying ADMET parameters using QSAR models derived from MLR analysis. This was performed using a Web-based tool called PreADME (http://www.bmdrc.org/04_product/01_preadme.asp). The MLR has also been used successfully in prediction of lead compounds in the development of anti-proliferative agents (Winfield, Inniss, & Smith, 2009). Celecoxib, an FDA approved drug used for treatment of colon cancer was used for development of pyrazole-based molecules as celecoxib was found to cause cardiovascular and cerebral vascular diseases.

There are various steps required for QSAR analysis using MLR. In the *first step*, database of molecules is built and the initial conformations of molecules are energy minimized using a force field. Then molecules are aligned using an alignment tool. In the *second step*, relationship between anti-proliferative activity and structure of compounds is calculated which can include several 2D and 3D molecular descriptors such as *pKa, protein binding, volume of distribution, etc.* From a subset of descriptors, those compounds are selected which shows significant association with anti-cancer activity performed using Contingency coefficient, Cramer's V, uncertainty coefficient and correlation coefficient (Winfield, Inniss, & Smith, 2009). In the *third step*, selected compounds are subjected to conformational analysis and alignment. Alignment generates an overlap score and the compounds with highest overlap score are selected. Compounds which best fits the pharmacophore are separated into training and test sets. In the *fourth step*, multiple linear regression equation is determined using statistical software such as SAS for anti-cancer activity prediction. Then, stepwise regression is performed in which descriptors are added or removed based on p-values of F-statistics. Those descriptors which satisfy the minimum significance criteria are considered as 'predictors'. PCA on the other hand generate data from multiple redundant variables into smaller uncorrelated variables (Wold, 1987). This method is used to find collinearities and multicollinearities in multiple and multivariate QSAR. This has been used previously for finding trimethoprim derivatives (Mager, 1982). PLS is an improvised technique which uses combination of MLR and PCA techniques. An improved method of PLS called the non-linear PLS (NPLS) overcomes the inability of PLS for obtaining nonlinear model between principle component of dependent and independent variables (2; Tang & Li, 2008).

QSAR

2D-QSAR

In the QSAR method, several hydrophobic, electronic and steric features are correlated with the biological activities represented by (Acharya et al., 2011):

$$\log\left(\frac{1}{c}\right) = k_1\pi - k_2\pi^2 + k_3\sigma + k_4 E_s + k_s$$

(Acharya et al., 2011)

C is the concentration of compound needed for biological activity, π is partition coefficient, σ is Hammett electronic substituent constant and E_s is the steric substituent constant. Different chemical groups of the compound mediate specific biological activity. When the chemical group is substituted, the specific biological activity of the substituted compound is evaluated by the summation of parent molecule activity and individual substituent contribution. The molecular descriptors used (for correlation with the biological activity) represent fragments of the parent molecule. The advantages of using fragment-based descriptors include computational ease, diversity of available substituents and their convenient mathematical implementations (Acharya et al., 2011).

3D-QSAR

3D-QSAR uses 3D properties of ligands for correlating and predicting the biological activity. This can be performed using several statistical approaches such as PLS and MLR described above. Several 3D-QSAR techniques such as CoMFA and CoMSA methods have been developed which utilizes the data from the compounds stored in drug databases and derive correlation between the activity of a ligand and its 3D features.

CoMFA

CoMFA (Comparative Molecular Field Analysis) is a technique for developing correlation between 3D properties of ligands and biological activities. Several optimized conformations of a ligand are generated. The optimized conformations of a ligand are placed in 3D grid and a probe atom is used for calculation of steric, electrostatic, potential and Van der Waals energies (Kubinyi, 2008) A cutoff value for the energies is specified and the conformations which exceed the cutoff values are rejected. Probe atom not only helps in determining the energies but it also helps in determining the activity of the ligand using PLS method (Acharya et al., 2011). Result of the analysis is represented by contour maps which show favorable and unfavorable steric regions around the molecules. The CoMFA has been successfully applied in the development of novel N1-amino-acid substituted 2,4,5-triphenyl imidazoline derivatives for inhibition of p53-MDM2 binding which is responsible for malignant tumor growth in humans (Hu et al., 2012).

CoMSA

CoMSA (Comparative Molecular Surface Analysis) is a method similar to CoMFA (Acharya et al., 2011) and calculates similarity indices by comparison between individual ligand molecule with a common probe having charge, radius, hydrophobicity and hydrogen bond properties similar to (Reddy et al., 2007).. The similarity indices describe receptor-ligand binding interaction energy. CoMSA has also been successfully used for designing molecules for synthesis of new HIV-1 integrase inhibitors which uses Kohonen Self-Organizing Maps (SOM) which allows projection of 3D molecular data into 2D maps (Niedbala et al., 2006).

Cloud Computing Advances in Drug Discovery

Cloud computing is a technology for delivering computational resources using Web technology. Molecular docking and MD simulations require extensive computational power and time for performing high throughput analysis. Cloud computing technology reduces the huge investment in computer power and time by providing same resources over the Internet. The end user can utilize such resources on-demand which means that user can demand amount of computational power for performing high throughput analysis. Such technology has been recently incorporated by InhibOx (http://www.inhibox.com). It has developed next generation drug discovery platform and delivers cloud-based drug development services. Virtual screening of active site of biological receptor protein has been performed by molecular docking tools such as AutoDock (Morris et al., 2009). It applies distributed/grid computing technology that integrates several individual computers generating a super computer with huge computational power (Shudong et al., 2004). Similar research has been performed by cloud computing using Internet version of AutoDock. Using 100 million compound databases of 3D candidate compounds with pre-calculated shape, charge and stereochemistry can be used to guide the process of virtual screening. Using such massively computer intensive and rigorous approach, cloud computing makes the whole approach flexible.

CONCLUSION

In this review, we discussed various computational approaches for drug design from virtual screening of ligands from drug databases to computation of ligand binding free energies using statistical and thermodynamic methods. We clearly outlined both the structure and ligand-based approaches requiring known target 3D structure or clear description of 3D features of the ligand for optimal binding to the receptor for QSAR approach, respectively. Ligand binding to the receptor active site can be performed using long MD simulations, however it requires the extensive processing and graphical power which states current limitation of the approach for larger macromolecules. However, by employing certain molecular mechanics approaches such as PBSA or GBSA can speed up the computation otherwise done for classical all-atom MD approach. Development of cloud-based technology can also help drug discovery projects with higher computational power delivered at a reduced cost.

FUTURE DIRECTIONS

Advances in computational techniques and resources have led to the development of smart and innovative methods in drug designing. However, current research and developments are targeted mostly towards a single receptor protein responsible for causing disease. This approach however creates a targeted drug but neglects the potential effects caused due to various interactions with other proteins and biomolecules which creates a plethora of potential side effects. Current approaches outlined can only be used for designing derivatives of known compounds by various statistical and thermodynamic approaches which completely fail to consider several design parameters aimed at targeting causal biomolecule at targeted site. Certain design parameters and smarter algorithms which can consider the toxicological and other side effects of the compounds during pharmacophore modeling can be manufactured which can specifically target the protein of interest. Also, development of nanoparticles can also enhance the specificity of the ligand by allowing the nanoparticle to carry the drug concentration to the specific target site. Such smarter approaches and innovative thinking can help in the development of novel therapeutic compounds for combating target diseases such as cancer.

REFERENCES

Acevedo, O., Ambrose, Z., Flaherty, P. T., Aamer, H., Jain, P., & Sambasivarao, S. V. (2012). Identification of HIV inhibitors guided by free energy perturbation calculations. *Current Pharmaceutical Design*, *18*(9), 199–216. doi:10.2174/138161212799436421 PMID:22316150

Acharya, C., Coop, A., Polli, J. E., & Mackerell, A. D. (2011). Recent advances in ligand-based drug design: relevance and utility of the conformationally sampled pharmacophore approach. *Curr Comp-Aid Drug Design, 7*(1), 10–22. doi:10.2174/157340911793743547 PMID:20807187

Anderson, A. C. (2003). The process of structure-based drug design. *Chemistry & Biology, 10*, 787–797. doi:10.1016/j.chembiol.2003.09.002 PMID:14522049

Ashburn, T. T., & Thor, K. B. (2004). Drug repositioning: Identifying and developing new uses for existing drugs. *Nature Reviews. Drug Discovery, 3*(8), 673–683. doi:10.1038/nrd1468 PMID:15286734

Bottegoni, G., Kufareva, I., Totrov, M., & Abagyan, R. (2009). Four-dimensional docking: A fast and accurate account of discrete receptor flexibility in ligand docking. *Journal of Medicinal Chemistry, 52*(2), 397–406. doi:10.1021/jm8009958 PMID:19090659

Brooks, B. R., Bruccoleri, R. E., Olafson, B. D., States, D. J., Swaminathan, S., & Karplus, M. (1983). CHARMM: A program for macromolecular energy, minimization, and dynamics calculations. *Journal of Computational Chemistry, 4*, 187–217. doi:10.1002/jcc.540040211

Cheng, T., Li, Q., Zhou, Z., Wang, Y., & Bryant, S. H. (2012). Structure-based virtual screening for drug discovery: A problem-centric review. *The AAPS Journal, 14*(1), 133–141. doi:10.1208/s12248-012-9322-0 PMID:22281989

Christen, M., Hünenberger, P. H., Bakowies, D., Baron, R., Bürgi, R., & Geerke, D. P. et al. (2005). The GROMOS software for biomolecular simulation: GROMOS05. *Journal of Computational Chemistry, 26*(16), 1719–1751. doi:10.1002/jcc.20303 PMID:16211540

Dror, R. O., Pan, A. C., Arlow, D. H., Borhani, D. W., Maragakis, P., & Shan, Y. et al. (2011). Pathway and mechanism of drug binding to G-protein-coupled receptors. *Proceedings of the National Academy of Sciences of the United States of America, 108*(32), 13118–13123. doi:10.1073/pnas.1104614108 PMID:21778406

Durrant, J. D., & McCammon, J. A. (2011). Molecular dynamics simulations and drug discovery. *BMC Biology, 9*, 71. doi:10.1186/1741-7007-9-71 PMID:22035460

Hart, T. N., & Read, R. J. (1992). A multiple-start Monte Carlo docking method. *Proteins, 13*(3), 206–222. doi:10.1002/prot.340130304 PMID:1603810

Hou, T., Wang, J., Li, Y., & Wang, W. (2011). Assessing the performance of the MM/PBSA and MM/GBSA methods. 1. The accuracy of binding free energy calculations based on molecular dynamics simulations. *Journal of Chemical Information and Modeling, 51*(1), 69–82. doi:10.1021/ci100275a PMID:21117705

Hu, C., Dou, X., Wu, Y., Zhang, L., & Hu, Y. (2012). Design, synthesis and CoMFA studies of N1-amino acid substituted 2,4,5-triphenyl imidazoline derivatives as p53-MDM2 binding inhibitors. *Bioorganic & Medicinal Chemistry, 20*(4), 1417–1424. doi:10.1016/j.bmc.2012.01.003 PMID:22273545

Ivetac, A., & McCammon, J. A. (2010). Mapping the druggable allosteric space of G-protein coupled receptors: a fragment-based molecular dynamics approach. *Chemical Biology & Drug Design, 76*(3), 201–217. PMID:20626410

Jin, F., Matsushita, O., Katayama, S., Jin, S., Matsushita, C., Minami, J., & Okabe, A. (1996). Purification, characterization, and primary structure of Clostridium perfringens lambda-toxin, a thermolysin-like metalloprotease. *Infection and Immunity, 64*(1), 230–237. PMID:8557345

Kaukonen, M., Söderhjelm, P., Heimdal, J., & Ryde, U. (2008). QM/MM-PBSA method to estimate free energies for reactions in proteins. *The Journal of Physical Chemistry B, 112*(39), 12537–12548. doi:10.1021/jp802648k PMID:18781715

Khan Mahmud, T. H., & Ingebrigt, S. (2009). Multivariate linear regression models based on ADME descriptors and predictions of ADMET profile for structurally diverse thermolysin inhibitors. *Letters in Drug Design & Discovery, 6*(6), 428–436. doi:10.2174/157018009789057607

Kroemer, R. T. (2007). Structure-based drug design: Docking and scoring. *Current Protein & Peptide Science, 8*(4), 312–328. doi:10.2174/138920307781369382 PMID:17696866

Kubinyi, H. (2008). Comparative molecular field analysis (CoMFA). In J. Gasteiger (Ed.), *Handbook of Chemoinformatics: From Data to Knowledge in 4 Volumes*. Germany: Wiley-VCH.

Lee, K., Jeong, K. W., Lee, Y., Song, J. Y., Kim, M. S., Lee, G. S., & Kim, Y. (2010). Pharmacophore modeling and virtual screening studies for new VEGFR-2 kinase inhibitors. *European Journal of Medicinal Chemistry, 45*(11), 5420–5427. doi:10.1016/j.ejmech.2010.09.002 PMID:20869793

Lennard-Jones, J. E. (1924). On the determination of molecular fields: I: from the variation of the viscosity of a gas with temperature. *Proc Royal Soc A, 106*(738), 441–462. doi:10.1098/rspa.1924.0081

Ma, D. L., Chan, D. S., & Leung, C. H. (2013). Drug repositioning by structure-based virtual screening. [Epub ahead of print]. *Chemical Society Reviews*. doi:10.1039/c2cs35357a PMID:23288298

Mager, P. P. (1982). Theoretical approaches to drug design and biological activity: critical comments to the use of mathematical methods applied to univariate and multivariate quantitative structure-activity relationships (QSAR). *Medicinal Research Reviews, 2*(1), 93–121. doi:10.1002/med.2610020106 PMID:7109783

Markwick, R. L., Cervantes, C. F., Abel, B. L., Komives, E. A., Blackledge, M., & McCammon, J. A. (2010). Enhanced conformational space sampling improves the prediction of chemical shifts in proteins. *Journal of the American Chemical Society, 132*(4), 1220–1221. doi:10.1021/ja9093692 PMID:20063881

Miller, B. R., McGee, T. D., Swails, J. M., Homeyer, N., Gohlke, H., & Roitberg, A. E. (2012). MMPBSA.py: an efficient program for end-state free energy calculations. *Journal of Chemical Theory and Computation, 8*(9), 3314–3321. doi:10.1021/ct300418h

Morris, G. M., Goodsell, D. S., Halliday, R. S., Huey, R., Hart, W. E., Belew, R. K., & Olson, A. J. (1998). Automated docking using a Lamarckian genetic algorithm and an empirical binding free energy function. *Journal of Computational Chemistry, 19*, 1639–1662. doi:10.1002/(SICI)1096-987X(19981115)19:14<1639::AID-JCC10>3.0.CO;2-B

Morris, G. M., Huey, R., Lindstrom, W., Sanner, M. F., Belew, R. K., Goodsell, D. S., & Olson, A. J. (2009). AutoDock4 and AutoDockTools4: Automated docking with selective receptor flexibility. *Journal of Computational Chemistry, 30*(16), 2785–279. doi:10.1002/jcc.21256 PMID:19399780

Niedbala, H., Polanski, J., Gieleciak, R., Musiol, R., Tabak, D., & Podeszwa, B. et al. (2006). Comparative molecular surface analysis (CoMSA) for virtual combinatorial library screening of styrylquinoline HIV-1 blocking agents. *Combinatorial Chemistry & High Throughput Screening, 9*(10), 753–770. doi:10.2174/138620706779026042 PMID:17168681

Okimoto, N., Futatsugi, N., Fuji, H., Suenaga, A., Morimoto, G., & Yanai, R. et al. (2009). High-performance drug discovery: computational screening by combining docking and molecular dynamics simulations. *PLoS Computational Biology, 5*(10), e1000528. doi:10.1371/journal.pcbi.1000528 PMID:19816553

Prakhov, N. D., Chernorudskiy, A. L., & Gainullin, M. R. (2010). VSDocker: A tool for parallel high-throughput virtual screening using AutoDock on Windows-based computer clusters. *Bioinformatics (Oxford, England), 26*(10), 1374–1375. doi:10.1093/bioinformatics/btq149 PMID:20378556

Reddy, A. S., Pati, S. P., Kumar, P. P., Pradeep, H. N., & Sastry, G. N. (2007). Virtual screening in drug discovery- a computational perspective. *Current Protein & Peptide Science, 8*(4), 329–351. doi:10.2174/138920307781369427 PMID:17696867

Sakkiah, S., Thangapandian, S., & Lee, K. W. (2012). Ligand-based virtual screening and molecular docking studies to identify the critical chemical features of potent cathepsin D inhibitors. *Chemical Biology & Drug Design, 80*(1), 64–79. doi:10.1111/j.1747-0285.2012.01339.x PMID:22269155

Shirts, M., Mobley, D., & Chodera, J. (2007). Alchemical free energy calculations: Ready for prime time? *Annual Reports in Comput Chem, 3*, 41–59. doi:10.1016/S1574-1400(07)03004-6

Showalter, S. A., & Brüschweiler, R. (2007). Validation of molecular dynamics simulations of biomolecules using NMR spin relaxation as benchmarks: Application to the AMBER99SB force field. *Journal of Chemical Theory and Computation, 3*(3), 961–975. doi:10.1021/ct7000045

Shudong, C. S., Wenju, Z., Fanyuan, M., Jianhua, S., & Minglu, L. (2004). The design of a grid computing system for drug discovery and design. *Lecture Notes in Computer Science, 3251*, 799–802. doi:10.1007/978-3-540-30208-7_108

Suenaga, A., Okimoto, N., Hirano, Y., & Fukui, K. (2012). An efficient computational method for calculating ligand binding affinities. *PLoS ONE, 7*(8), e42846. doi:10.1371/journal.pone.0042846 PMID:22916168

Sung, W. T., & Chen, J. H. (2012). Enhancing Molecular Docking Efficiency for Computer-Aided Drug Design via Systems Theorem. *Int J Comp Consumer Control, 1*(1), 54–61.

Tang, K., & Li, T. (2003). Comparison of different partial least-squares methods in quantitative structure–activity relationships. *Analytica Chimica Acta, 476*(1), 85–92. doi:10.1016/S0003-2670(02)01257-6

Wang, J., Wolf, R. M., Caldwell, J. W., Kollman, P. A., & Case, D. A. (2004). Development and testing of a general amber force field. *Journal of Computational Chemistry, 25*(9), 1157–1174. doi:10.1002/jcc.20035 PMID:15116359

Wilson, G. M., & Muftuoglu, Y. (2012). Computational strategies in cancer drug discovery. In R. Mohan (Ed.), *Advances in cancer management* (pp. 237–254).

Winfield, L. L., Inniss, T. R., & Smith, D. M. (2009). Structure activity relationship of antiproliferative agents using multiple linear regression. *Chemical Biology & Drug Design, 74*(3), 309–316. doi:10.1111/j.1747-0285.2009.00863.x PMID:19703034

Wold, S. (1987). Principal component analysis. *Chemometrics and Intelligent Laboratory Systems*, 2, 37–52. doi:10.1016/0169-7439(87)80084-9

KEY TERMS AND DEFINITIONS

Cloud Computing: A term that involves services provided over the Internet. This idea has been implemented from distributed/grid computing that integrates several individual computers to generate a super computer system of superior computational abilities utilizing computing resources allocated on-demand over the Internet.

Computational Drug Design: Drug designs using highly specialized and sophisticated software.

Drug: Chemical substance(s) that can have either medicinal, hallucinogenic or performance improving properties administered as an external/foreign agent to either humans or animals.

Ligands: Small molecules forming complexes with other complex biomolecules such as nucleic acids and proteins to perform highly specialized biochemical reactions and/or transformations.

Molecular Docking: Molecular docking involves knowledge of putative binding sites where the ligand can bind and initiate a biological response.

Molecular Dynamics: Specialized software used for simulating physical movements of atoms and molecules.

Pharmacophore: Molecular characteristics of ligands for identifying specific target biomolecuels.

APPENDIX: ABBREVIATIONS USED IN THE CHAPTER

3D: Three-Dimensional;

4D: Four-Dimensional;

ADMET: Absorption, Distribution, Metabolism and Toxicity;

CADD: Computer Aided Drug Design;

CoMFA: Comparative Molecular Field Analysis;

CoMSA: Comparative Molecular Surface Analysis;

FEP: Free Energy Perturbation;

GB: Generalized-Born;

GBSA: Generalized-Born Surface Area;

HTS: High-throughput Screening;

MC: Monte Carlo;

GA: Genetic Algorithm;

MD: Molecular Dynamics;

MLA: Multivariate Linear Regression Analysis;

MM/GBSA: Generalized-Born Surface Area;

MM/PBSA: Molecular Mechanics/Poisson-Boltzmann Surface Area;

MM: Molecular Mechanics;

NMR: Nuclear Magnetic Resonance;

NPLS: Non-Linear PLS;

PB: Poisson-Boltzmann;

PBSA: Poisson-Boltzmann Surface Area;

PCA: Principle Component Analysis;

PLS: Partial Least Square Analysis;

QM: Quantum Mechanics;

QSAR: Quantitative Structure Activity Relationship;

SASA: Solvent Accessible Surface Area;

SBDD: Structure-Based Drug Design;

SOM: Self-Organizing Maps;

TI: Thermodynamic Integration.

Section 5
Smart Manufacturing Sustainability

Chapter 12
Fair Share of Supply Chain Responsibility for Low Carbon Manufacturing

Yu Mei Wong
The University of Hong Kong, Hong Kong

ABSTRACT

Large amounts of carbon emissions and pollution are generated during the manufacturing process for consumer goods. Low carbon manufacturing has been increasingly enquired or requested by stakeholders. However, international trade blurs the responsibility for carbon emissions reduction and raises the questions of responsibility allocation among producers and consumers. Scholars have been examining the nexus of producer versus consumer responsibility among supply chains. Recently, there have been discussions on the share of producer and consumer responsibility. Both producer and consumer responsibility approaches have intrinsic shortcomings and are ineffective in curbing the rise of carbon emissions in supply chains. Shared responsibility based on the equity principle attempts to address these issues. This chapter relates a case study of carbon impact on China's export and economy with scenarios which show that the benefits of carbon reduction by producers can trickle down along the supply chain and motivate the sharing responsibility under certain circumstances. The share of producer and consumer responsibility for low carbon manufacturing can be enabled when embodied carbon emissions in goods and services are priced and such accurate information is available. A mechanism engaging the global participation is recommended. The author calls for further research on the system pricing embodied carbon emission, the universal standard to calculate the embodied carbon emissions and to disclose the information, and the way to secure global cooperation and participation.

DOI: 10.4018/978-1-4666-5836-3.ch012

INTRODUCTION

Carbon dioxide is emitted at all phases of manufacturing process along the supply chain, from raw materials extraction, assembly, to distribution of the finished goods to final consumers. With escalating and alarming levels of global carbon emissions, the concept of low carbon manufacturing has been proposed. International trade and fragmented production complicate the responsibility for low carbon manufacturing and have an increasing influence on world carbon emission. Under the globalization of production and interconnectedness of economies, producers and consumers among the supply chain are often geographically separated. In one of the current key efforts – the Kyoto Protocol under the United Nations Framework Convention on Climate Change (UNFCCC), the responsibility for carbon emissions is allocated according to geographical area of the carbon emitters, which also follows producer responsibility approach. Only Annex B parties are legally bounded to emissions reduction targets against 1990 levels (UNFCCC, 2013a). Comparative advantage results in the relocation of labour and pollution intensive industries from countries with stringent environmental policies to countries without environmental policy. Carbon leakage and free riding are resulted. The responsibility of reducing carbon emission is then shifted to developing countries, while developed countries can still enjoy the goods and services through imports from developing countries. The increasing global carbon emission shows that carbon leakage undermines the efforts in achieving these mitigation targets and highlights the failure of the Kyoto Protocol.

The failure instigates many studies on consumer responsibility and quantitatively evaluating carbon emissions embodied in trade at global level, or regional level or bilateral trades in a bid to address carbon leakage. Under the consumer responsibility principle, consumers shoulder all the carbon emissions emitted in the manufacturing processes. However, it suffers some drawbacks like trespassing on the jurisdictional limits of national power (Cadarso et al., 2012).

Producers and consumers constitute the supply chain and both enjoy the benefits from trade. Responsibility for low carbon manufacturing should not be allocated to single party. Therefore, the intermediate approach - sharing the responsibility among producer and consumers, is proposed and studied by scholars. Its main advantage is to commit all agents including producers, industries, final consumers and/or countries in the supply chain to play a part to reduce the carbon emissions (Cadarso et al., 2012). Nevertheless, the shared responsibility was only examined in application of carbon accounting. So far, there is no mechanism or policy to enable the share of responsibility among supply chain. This chapter is one of the efforts in finding out the mechanism enabling the share of responsibility.

To examine the possibility of sharing responsibility for low carbon manufacturing among the supply chain, China – the largest exporter is studied in this chapter. Over the last thirty years, China has been experiencing rapid economic growth and enjoying the benefits brought by export. In 2010, China became the world's largest exporter, and the second largest importer (OECD, 2013). Since the launch of the 'Open Door Policy', China has been encouraging and supporting foreign trades and investments, which turned the country into the world's factory at the cost of extensive use of energy and environmental degradation. Within a very short period of time, from 2002 to 2007, the carbon emissions in China have almost doubled. In 2010, China was responsible for 17.5% of global energy demand (IEA, 2012a) and 23.8% of world carbon dioxide emission and became the largest carbon emitter in the world (IEA, 2012b). Manufacturing is believed to be one of the driving forces and certain portion of the carbon emission in China is embodied in goods produced for export and consumed by other countries. Meanwhile, China's economy partly relies on export. Direct restriction of carbon emissions of manufacturing industries for export would bring detrimental impact on economy and would discourage the exporting countries to reduce the emission. Sole responsibility shouldered by producer would also limit the degree of reduction. The detailed analysis of the carbon impact on China's export and economy aims to find out policy criteria that can facilitate the share of producer and consumer responsibility among the supply chain.

The objective of this chapter is to examine how the share of producer and consumer responsibility for low carbon manufacturing can address the current failure of carbon abatement measures in supply chain, and to give recommendations on how the mechanism is formed. This chapter begins with the failure of the Kyoto Protocol in combating climate change due to trade, and reviews the concept of producer responsibility, consumer responsibility and the share of producer and consumer responsibility. Then, after briefly examining the key international carbon abatement practices and discussing their strength and weakness, the chapter illustrates the embodied carbon emissions in China in 2009 and analyzes the carbon impact on China's export and economy. The positive influence on other sectors within the country and those in other countries by one sector reducing direct carbon emissions is demonstrated in scenario. The chapter then builds on the review and analysis in the previous sections to identify the criteria required for the mechanism to share the producer and consumer responsibility. Hence, concrete measures to build the mechanism and future research for the share of responsibility among supply chain are suggested. The chapter concludes by reiterating key points and emphasizing future research areas.

BACKGROUND

Trade, the Kyoto Protocol and Climate Change

The Kyoto Protocol came into force in 2005, it defines the operational details on how to stabilize greenhouse gas emissions based on the principle of the United Nations Framework Convention on Climate Change (UNFCCC) (2013a). It differentiated the obligation and set different binding emission reduction targets for Annex B parties and these targets added up to an average five percent emission reduction compared to 1990 levels over the first commitment period 2008 to 2012. The national emission target limits the further pollution and internalizes the global public good – carbon emission, which was unpriced externality. Annex B parties must implement domestic policies and measures to reach the reduction targets. They were allowed to seek optimal ways to achieve their targets and adjust their climate change strategies according to their technical and economic circumstances.

Despite the implementation of the Kyoto Protocol, the world carbon emissions increased by 44.4% in 2010 when compared with the carbon emissions level in 1990 (IEA, 2012b). Figure 1 shows the increasing trend of global carbon emissions by fuel.

There is much debate about the failure of the Kyoto Protocol and part of the failure was attributed to international trade. With globalization and increasing link of world economies, producers and consumers are geographic separated in international trade. Under the current Kyoto Protocol, only domestically produced carbon emissions are accounted for and legally binding emission reduction targets were just set in Annex B parties. Carbon abatement in large open economies not only alters the domestic production and consumption pattern but also influence international prices via changes in export and import. These mitigation measures increase the

Figure 1. Global carbon emissions by fuel from 1971 to 2010 (IEA, 2012b)

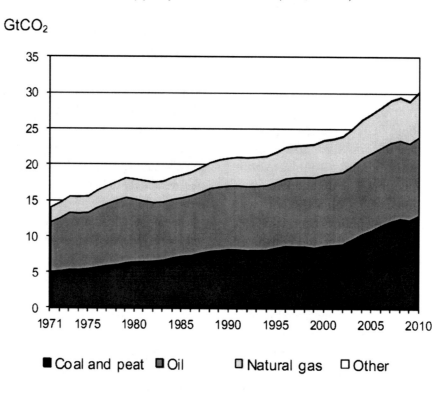

cost of production in Annex B countries and hence, it gives rise to competitiveness issues. The countries undertaking carbon abatement measures not only have higher cost in production of energy intensive goods but also give competitors in countries that do not have such measures or poor environmental legislation and enforcement a competitive edge and unfair advantage. Countries that originally import these goods from abating countries suffer from higher prices to the extent that they cannot but substitute the same goods from non-abating countries. Gradually, the producers in Annex B countries lose their market share and to prevent this loss, these companies would shift the carbon and energy intensive operations to non-abating countries. The exports of energy intensive goods and services in the abating countries are reduced and the imports of the same goods and services from non- abating countries increases.

Responsibility for Carbon Emissions in Context of Supply Chain

The complex interconnected web of supply chain consists of a succession of economically bound suppliers and consumers (Lenzen & Murry, 2010). Supply chain starts with a primary supplier, and then a number of intermediate suppliers and ends with a final consumer (Lenzen & Murry, 2010). Except for the final consumer, all actors may emit carbon dioxide during the production of goods and services along the supply chain. This leads to the question of who should be responsible for the emissions associated with the production of goods consumed in other countries through import and export. The essence of the question is the allocation of responsibility for carbon emissions. The concept of producer responsibility, consumer responsibility and the share of producer and consumer responsibility are discussed below.

Producer Responsibility

Producer responsibility is underpinned by the well known polluter pays principle, which means that the party directly producing pollution is responsible for paying for the damage done to the natural environment (Zhou & Kojima, 2010). This principle has some advantages. First, the producer has the best knowledge, capacity and jurisdiction to incorporate environmental considerations into the design, production of the goods and operation of the manufacturing plant and to control carbon abatement (Zhou & Kojima, 2010). Second, it is convenient for the government to regulate and monitor the producer as a business entity and keep track of their performance. Third, allocating emissions responsibility to the producer can create a strong and direct deterrent to polluters to reduce the emissions. The Kyoto Protocol is based on producer responsibility by nation and a country is responsible for the emissions within its territory when goods, services and energy are produced, independent of whether they are consumed within the country or not.

However, it has some drawbacks as well that can affect the degree of commitment by countries to achieve their final targets. The major concern is the "pollution heaven hypothesis," which states that the weaker environmental regulations in developing countries may be compounding the general shift of manufacturing in developed world and causing developing countries to specialize in the most pollution intensive manufacturing sectors (Cole, 2004). As a result of globalization, the growth of international trade boosts the transference of pollution through trade flow, since it is incorporated into exports and imports (Antweiler, 1996; Muradian et al., 2002). To meet national emission reduction targets, countries implement different measures to control emissions, which trigger the relocation of the production processes to other countries and then importing the goods. For instance, U.K.'s emissions had been decreasing over time but when including imports into the U.K., especially from China, overall emissions caused by the consumption of U.K. residents had actually increased (Lenzen & Murray, 2010; Wiedmann et al., 2008). Another argument involves the equity principle (Zhou & Kojima, 2010). Consumers, particularly those who reside in different regions than the producers, enjoy the goods and services and benefit from improvement in living standards and should play a part to reduce emissions. Moreover, full producer responsibility leaves little incentive to the consumer to conserve the environment while the supply chain is partially consumption driven (Zhou & Kojima, 2010).

Consumer Responsibility

As its name suggested, consumer responsibility means that the consumer is responsible for carbon emissions from all production processes of energy, good and services regardless of whether commodities are produced within the countries or imported. It is opposite to the producer responsibility and all emissions associated with the goods goes to consumer's account. Many literatures based on consumer responsibility like carbon emissions embodied in international trade and carbon footprint have been instigated (Dan, 2000; Guo et al., 2010; Kondo et al., 1998; Li & Hewitt, 2008; Munksgaard et al., 2005; Muñoz & Steininger, 2010; Peters & Hertwich, 2008; Yan & Yang, 2010).

This principle has certain advantages over the producer responsibility principle. It allows industries and consumers to exercise upstream responsibility, to guide economic development and to limit ecological footprint when inputs and imports of final goods are demanded. Also, it boosts the transfer of cleaner technology and share of knowledge to supplier countries because consuming countries also have to make an effort to reduce emissions within the import and final goods (Cadarso et al., 2012). Besides, it appears to be more beneficial and fairer to non Annex B countries, thus encourages more of their participation and supports (Zhou & Kojima, 2010).

However, there are several intrinsic shortcomings in its application. It is hard to quantify consumer responsibility since the carbon emissions emitted are by-product in the upstream production, which is difficult to trace back. Though emission accounting methods based on consumer responsibility like life cycle assessment are invented, they are data intensive and require more complex calculation. Massive harmonized data on international trade is essential and currently it is usually not consistently available and results in uncertainty (Lenzen et al., 2004; Lenzen, 2007; Zhou & Kojima, 2010). Consumer responsibility only works when consumers exert effective upstream pressure on suppliers or producers and urge for improvement. To achieve that, high environmental awareness among consumers and reliable environmental information are vital. Consumers don't have direct influence on the production processes or practices like advancing technologies and fuel used in other jurisdictions. It is beyond their power to control over emissions in other countries or regions. To adopt consumer responsibility effectively, international cooperation, negotiation and policies are needed (Zhou & Kojima, 2010). Moreover, allocating emission responsibility to consumers implies a weaker producer commitment to create cleaner and more efficient production processes (Bastianoni et al., 2004). It would become severe when the producer is monopoly of the sector.

Share of Producer and Consumer Responsibility

If the responsibility is deemed to be related to the benefits, both consumers and producers should have certain responsibility. Producers enjoy job creation, income and producer surplus while consumers enjoy the good and services and better quality of life (Mózner, 2013). Both have a part to play in causing carbon emissions (Lenzen et al., 2007). Therefore, neither full producer nor consumer responsibility is ideal but the shared responsibility should be the better solution. It allows the engagement of industries, countries, end consumers, all the agents involved in the network of global product chains in reduction of emissions arising in supply chain. This appears to be fairer because participants in different countries are also responsible for the impact. It has higher effectiveness due to concerted and coordinated effort to address the environmental impacts generated. Innovation to improve the environment can occur through interactions of parties, and buyer-suppliers relationship can stimulate environmental change in supply chain (Cadarso et al., 2012; Hall, 2000). The opportunities for environmental improvements are much greater through the cooperation of the supply chain than individual consumers or producers. Moreover, the share of producer and consumer responsibility can help ease the implementation of the Kyoto protocol and global climate change agreement for developing countries (Mózner, 2013; Ferng, 2003). In 2012 United Nations Climate Change Conference in Doha, it laid down the universal climate change agreement plan covering all countries to be adopted in 2015 and to be enforced in 2020 (UNFCCC, 2013b). Shared responsibility can reduce the burden on developing countries for emissions associated with exports to developed countries and fulfil their commitment in the future. The drawback of trespassing on the jurisdictional limits of national power when applying consumer responsibility can be overlooked in sharing the responsibility because in a global product chain, agents from different countries are integrated and firms can put pressure across borders easier.

So far, studies mostly focus on applying shared responsibility in the allocation of emission accounting and the responsibility follows emissions allocated. There are two main streams of allocating responsibility. The first one is to allocate the emissions by countries with reference to international trade flow (Mózner, 2013). Parameter were established to allow responsibility for emissions associated with trade to be distributed in an intermediate way between the producing and consuming countries. Ferng (2003) developed the framework based on benefit principle and ecological deficit and the sum of certain portion of emissions under consumer benefit principle and certain portion under producer benefit principle gives the national responsible emissions. Peters and Hertwich (2008) proposed balance of emissions embodied in trade, which is the difference between emissions embodied in export and import. The second stream allocates the responsibility for total emissions among various economic agents, suppliers and final consumers. Lenze et al. (2007) developed the 50%-50% emission allocation framework between supplier and consumer. Based on the value added created by the intermediate producer, 50% of the ecological footprint is then passed to its immediate consumer.

Theoretically, share of producer and consumer responsibility is more suitable in combating carbon emissions associated in the global supply chain when comparing with full producer responsibility or consumer responsibility. However, current literatures reviewing the share of producer and consumer responsibility do not study the mechanisms to facilitate the burden sharing. Deep discussion on how to define fairness is lacking. 50%-50% emission allocation proposed by scholars between supplier and consumers may not be equivalent to fairness. Share of producer and consumer responsibility is not necessarily linked to emissions accounting methods. More importantly, no matter what accounting principle is implemented, the abatement cost during in the production can be shared among different agents along the supply chain.

OVERVIEW OF CARBON ABATEMENT PRACTICES

Measures and polices for climate change mitigation and adaption intersect with supply chain in a number of ways. The following reviews current key carbon abatement practices. The overview shall form part of the basis for discussion on the mechanism to share producer and consumer responsibility in later section.

Emissions Trading Scheme

Emissions trading scheme is a market-based flexibility mechanism. Three elements are indispensable: 1) fixing a cap on the overall emissions of all the parties, 2) expressing the cap of the individual into allowed emissions or allowances and 3) creating a market for the allowed emissions to be auctioned or traded at the price determined by the market among the participating parties (Olhoff et al., 2009). The price of the allowed emissions should theoretically reflect the marginal cost of emission reductions and encourage the emitters to reach their targets (Olhoff et al., 2009).

The international emissions trading scheme for greenhouse gases was established under Article 17 of the Kyoto Protocol (UNFCCC, 2013g). The emission reduction targets of Annex B parties are expressed as the levels of allowed emissions and under emissions trading, the countries can sell their spare emission units to countries that exceed their targets (UNFCCC, 2013g). Emissions trading scheme could be both regional and domestic, and either mandatory or voluntary. The European Union Emission Trading Scheme is the prime example of regional and mandatory one.

The way the scheme allocates emissions has the implications on responsibility and the cost. All emissions trading schemes target direct emitters and so, allocate the allowed emission accordingly. Therefore, the costs are distributed among producers and then passed onto consumers in the form of higher prices. Ideally, it can limit emissions exceeding the overall emissions cap and achieve high environmental effectiveness and flexibility (Olhoff et al., 2009). However, so far, the schemes have limited scope and thus, limited ability to reduce emissions (Olhoff et al., 2009). Also, the assessment of the results is still at early stages since the schemes have only been operating for several years (Olhoff et al., 2009).

Clean Development Mechanism

Clean development mechanism (CDM) is an "offset mechanism" under the Article 12 of UNFCCC's Kyoto Protocol. Under the CDM, developing countries can earn certified emission reduction (CER) credits through emission reduction projects, each equivalent to one tonne of carbon emission (UNFCCC, 2013c). Industrialized countries can buy and use the CER credits to partially meet their emission reduction targets (UNFCCC, 2013c). Therefore, the implementation costs of reducing carbon emission shifts from Annex B countries to non Annex B countries (Rosendahl & Strand, 2009). In that sense, CDM maintains the overall greenhouse gases to neutral while reducing costs (Rosendahl & Strand, 2009). Up to 30th April 2013, there are 2294 CDM project activities that have issued CERs credits (UNFCCC, 2013d), which was equivalent to around 1307 million metric tonnes of carbon emissions. More than 60% of CERs credits were issued to China and India came the second, accounting for around 14% of CERs credits issued (UNFCCC, 2013e). CDM is not dedicated to mitigate the carbon emissions in the supply chain but to facilitate carbon reduction through the interaction between Annex B and non Annex B countries. There is slight indirect positive impact on the supply chain. Non Annex B countries can lower its carbon emissions in CDM projects through foreign investment and technology, particularly the projects for cleaner energy. However, its volume is small and the effects are limited. In 2010, the world emissions totals 30276.1 million metric tonnes. And, CDM projects were criticized for its likelihood of carbon leakage when aiming at reducing the use of fossil fuel (Rosendahl & St rand, 2009).

Joint Implementation

Similar to CDM, joint implementation is a market-based mechanism defined in Article 6 of the Kyoto Protocol (UNFCCC, 2013a). However, it is limited to Annex B parties. An Annex B country with an emission reduction commitment can earn emission reduction units (ERUs) from emission reduction or emission removal project in another Annex B country (UNFCCC, 2013a). Similar to CER, each ERU is equivalent to one tonne of carbon emission and it can be counted to meet its Kyoto target (UNFCCC, 2013a). It offers Annex B parties another flexible and cost efficient mean of fulfilling their Kyoto commitments. For parties, lower carbon abatement cost and selling ERUs reduces the carbon emission through foreign investment and technology transfer. When comparing with non Annex B parties, the abate cost of Annex B parties is higher and there is less room for abatement. Therefore, up to 2013, 796018576 ERUs were issued which is equivalent to around 796 million metric tonnes of carbon emission (UNFCCC, 2013f).

Carbon Tax

Carbon tax is a tax on carbon emissions. Tax is imposed based on the carbon content of fuels that generate carbon emissions during combustion or the carbon emissions directly the entity produce during the operation exported (Olhoff et al., 2009). For the former approach, the tax would be applied at a specific rate per tonne of different fuels based on their respective carbon content (Mani et al., 2008b). For the later approach, the tax is fixed per tonne of direct carbon emissions. In some countries, energy tax, which is based on the energy content of different energy sources, is adopted instead. Whether it is energy or carbon tax, it is normally applied within the jurisdiction of the same economy (Mani et al., 2008b). Carbon tax and its rates vary from country to country based on the category and quantity of fossil fuel, energy consumption and the type of greenhouse gas emission, and also varies in different sectors and geographical location. Carbon tax is one of the most widely adopted economic instruments, especially in Europe. The countries employing carbon tax also introduce some relief measures such as special tax reductions, exemption, exclusions to compensate the negative effect on competitiveness brought by the tax and so, lessen the economic impact (Mani et al., 2008b). In Australia, carbon-pricing mechanism was introduced by Clean Energy Legislation, which started on 1 July, 2012 in support of the transition to low carbon economy (Clean Energy Regulator, 2013). In this mechanism, polluters, which emits more than 25000 tonnes of carbon emissions per year, has to pay a price for their carbon emission (Clean Energy Regulator, 2013). The carbon price is fixed from 2012 to 2014 but will be controlled by the market starting from 1 July, 2015 (Clean Energy Regulator, 2013). Theoretically, the price should be set to ensure that emitters pay the full carbon cost of their operations, discourage the individuals and businesses from the use of carbon intensive goods or services and encourage investments in low carbon alternatives (Olhoff et al., 2009). But, in reality, the tax is normally not welcomed by the business and it is set to simply influence taxpayers' behaviour so as to achieve given environmental objectives (Olhoff et al., 2009).

Carbon tax directly internalizes the carbon cost generated in production and consumption. It can create economic incentive for the private sector to reduce carbon emissions. Additional benefit of carbon tax is the tax revenue, which can be included in the government's general budget or can finance specific programme, particularly environmental ones (Olhoff et al., 2009). Sometimes, carbon tax may be ineffective because of extensive tax exemptions and relatively inelastic demand in some sectors (Olhoff et al., 2009). Another concern is the competitiveness issue, which may drive the production process out of the original country and import similar products from the countries without carbon tax.

Border Tax Adjustment

Border tax adjustment is also called carbon border taxes (Gros & Egenhofer, 2010), embodied carbon tariff (Böhringer et al., 2011) or border carbon adjustment (Ghosh et al., 2012). It can occur in two ways: a tax is imposed on the imports of energy intensive goods which is proportional to its embodied carbon emissions and corresponding domestic products (Böhringer et al., 2011; Dissou & Eyland, 2011; Ghosh et al., 2012) and/or the domestic taxes are refunded when the products are exported (Olhoff et al., 2009). It works in complement to the domestic implementation of carbon tax or emission trading scheme (Olhoff et al., 2009). It arises from concerns about carbon leakage and reduced competitiveness, particularly in carbon intensive industries. Carbon abatement entails some costs and lowers the competitiveness in terms of price. In countries with stringent environmental policies, the carbon intensive industries lose the competitiveness in both international and domestic markets to counterparts without abatement (Böhringer et al., 2011; Dissou & Eyland, 2011). The relocation of the industry from countries with stringent environmental policy to those with weaker environmental policies results in the carbon leakage (Böhringer et al., 2011; Dissou & Eyland, 2011). Therefore, its objective is to remove the comparative advantage caused by different environmental policies, to provide a fair playing field, and to minimize carbon leakage (Böhringer et al., 2011; Dissou & Eyland, 2011; Ghosh et al., 2012). Border tax adjustment has attracted many developed countries' support and the EU and the United States are both considering its implementation (Dissou & Eyland, 2011; Godard, 2007). Nevertheless, no countries so far have implemented this measure yet (Dissou & Eyland, 2011; Ghosh et al., 2012). Information deficiencies on the production technology and the variation of production processes in different product, company and countries lead to the difficulty in assessing the product specific emissions (Olhoff et al., 2009). Also, the fixed border tax adjustment may be different from the fluctuating carbon price in emission trading scheme or carbon tax. Another challenge is the legality of this measure within World Trade Organization (WTO) (Dissou & Eyland, 2011). It doesn't violate WTO rules only if the members can justify either the need "to protect human, animal or plant life" or the relevance to "the conservation of exhaustible natural resources under articles XX (b) and (g) of General Agreement of Tariffs and Trade (Mani et al., 2008d).

Border tax adjustment may appear to be a tempting policy for developed countries seeking to restore their competitiveness, to reduce their domestic compliance costs and eliminate carbon leakage from their unilateral environmental policy initiatives (Ghosh et al., 2012). However, from developing countries perspective, the taxation of embodied carbon in their production process doesn't help them reduce the carbon emission. The taxation goes to the pocket of importing countries. This measure just discourages imports from developing countries.

Funding

Government funding or subsidy is the pool of financial reserve to enhance the deployment and utilization of climate friendly technologies and renewable energy (Olhoff et al., 2009). Financial incentive, funding or subsidy can facilitate the innovation, development and deployment of low carbon goods and technologies by compensating for their additional costs (Olhoff et al., 2009). Usually, there are three areas of focus: 1) increasing use of renewable energy and cleaner energy source, 2) development and deployment of low carbon goods and technologies and 3) deployment of carbon sequestration technologies (Olhoff et al., 2009). It can be in the form of one off research grant, one off subsidy on capital cost and energy feed-in tariff, etc (Olhoff et al., 2009). Very often, it is implemented in parallel with carbon tax and border tax adjustment. The tax revenue goes to the funding.

Carbon Labelling

Carbon labelling is the practice of publicly communicating greenhouse gas emissions associated with the life cycle of products or services via labels (Upham et al, 2011). Another form of carbon labelling communicates the emission reductions (Upham et al, 2011). The main aim is to enable the consumer to make informed decisions and put environmental considerations into purchasing decisions (Tan et al., 2012; Upham et al., 2011). The second form of carbon labelling further promotes the impressions that a strengthened and pro-active environmental commitment is undertaken by the producer (Upham et al, 2011). Also, carbon labelling which shows the life cycle carbon emission can offset carbon leakage (Cohen & Vandenbergh, 2012). Enhancing products with carbon footprint may support more sustainable consumption and hence, some policies and declarations like UN Agenda 21 (United Nations, 1992), the EU Sustainable Consumption and Production Action Plan (Commission of the European Communities, 2008), the UK Sustainable Development Strategy Report (Government of UK, 2005) highlighted the importance of carbon labelling (Upham et al, 2011). Voluntary carbon labelling schemes is implemented in countries like the United Kingdom, the Netherlands, Switzerland, and Japan (Cohen & Vandenbergh, 2012; Tan et al., 2012). The study for the EU Directorate-General for the Environment showed that 72% of EU citizens support carbon labelling scheme and reckon it should be mandatory in the future (The Gallup Organization, 2009; Upham et al, 2011).

Nevertheless, the usefulness and the effectiveness of carbon labelling have been frequently questioned. Tesco withdrawing the Carbon Trust Label in 2012 is the prime example. It claims it was necessary to reevaluate what works for its customers (Hubbbard, 2012). One of the challenges is the methodology to give reliable carbon labelling. The production processes of the products are highly complex and vary. Different carbon emitting activities interact to give different embodied carbon emissions of the product. The data requirements for detail carbon calculation could be daunting (Brenton et al., 2009). Doing the measurement along the supply chain could be costly and impractical but oversimplification and generalization also cause problems (Brenton et al., 2009). It has to balance accuracy with time and operational cost. The four key elements of the methodology are the use of primary versus secondary data, the use of emissions factors, the treatment of land use change and setting the system boundary (Brenton et al., 2009).

Now, there is no universal standard for calculating carbon footprint and carbon labelling. There are some similar protocols for standardized carbon footprint methodologies like Greenhouse Gas Protocol (Greenhouse Gas Protocol, 2011), PAS 2050:2011 (British Standards Institution, 2011), emerging ISO 14067 (ISO, 2012). Without a universal carbon labelling methodology, consumers may be confused by different standards and cannot make relevant purchase decision (Cohen & Vandenbergh, 2012). Another concern is the lack of impact assessment of the evolving standards (Brenton et al., 2009). Many standards and measurement methodologies were proposed. However, the costs of compliance remain unknown and so, there is no effort to identify ways to lower cost and raise the time efficiency (Brenton et al., 2009). Cost effective standards are important during economic and environmental consideration and it is easier to be adopted by market players.

Moreover, carbon labelling can only have significant impacts when it is widely applied (Upham et al., 2011). Current coverage of products with carbon labels is limited to voluntary scheme. Consumers cannot compare the products with labels with ones without label. Carbon labelling entails some additional cost on measurement, calculation and verification of the goods and makes producers unwilling apply carbon labelling.

Summary

This section has reviewed the key carbon abatement practices. All these measures involve the cooperation of multi-agents but the geographical scope and purpose of the measures vary. Carbon labelling, carbon tax and border tax adjustment are directly related to the supply chain. There is no perfect mechanism to address climate change. Each measure has its own purposes, merits and shortcomings. No single instrument is superior to all others in different settings. And, very often, multiple sets of instruments are adopted at the same time to complement each others. Goulder and Parry (2008) mentioned that the employment of instruments depend on cost effectiveness, distributional equity, minimization of risk in the presence of uncertainty and political feasibility. Therefore, countries adopt their own set of carbon abatement measures. And, for certain, currently there is no mechanism to facilitate the share of producer and consumer responsibility. Some barriers hindering the shared responsibility can be observed from the review of the measures – lack of standardization life cycle carbon emission information, carbon leakage issues, reduced competitiveness arising from individual countries internalizing carbon cost.

CARBON IMPACT ANALYSIS FOR CHINA'S EXPORT AND ECONOMY WITH SCENARIO

Methodology of Calculating of Embodied Carbon in this Chapter

This section aims to find out the embodied carbon emission in exports with respect to its economy by sector in China and analyze how carbon emissions reductions of one sector can positively influence other sectors in China and in other countries.

Multiregional Input-Output (MRIO) models have been widely adopted as the basic framework to examine the embodied environmental impacts in trade. Domestic input-output matrices are combined with import matrices from multiple regions to form a comprehensive and harmonized MRIO model (Sato, 2012). It can track the flow of import and export in and out of the specific regions involving all trading partners and hence, the embodied carbon emission can be calculated (Sato, 2012). Following the methods of Böhringer et al. (2011) and Qi et al. (2012), the MRIO model is adopted to calculate the embodied carbon content in production by sector using the formula below. Direct carbon emissions from fuel inputs and indirect carbon emissions associated with domestic non-fuel intermediate demand and foreign imports are captured. The total embodied carbon content of production by sector i in China is equal to the embodied carbon content per dollar of production, $Ay_{i,Chn}$ multiplied by the value of production, $y_{i,Chn}$. This product can be broken down into 3 components: direct emissions from burning of fuels inputs for the production process of sector $i - Ed_{i,Chn}$, indirect emissions associated with domestic intermediate for the production of sector $i - Eid_{i,Chn}$, indirect emissions associated with foreign import for the production of sector i in China $- Eim_{i,Chn}$. It is expressed in the Equation (1) below:

$$Ay_{i,Chn} \times y_{i,Chn} = Ed_{i,Chn} + Eid_{i,Chn} + Eim_{i,Chn} \quad (1)$$

Direct emissions associated with fuel consumption in production in sector i $Ed_{i,Chn}$ in China are directly taken from WIOD. Indirect emissions associated with domestic intermediate for the production of sector i $Eid_{i,Chn}$ are calculated by multiplying the embodied carbon emissions per dollars in sector j $Ay_{j,Chn}$ for intermediate demand of China per dollars and value flow from sector j to sector i in China $y_{j,i,Chn}$, as described by Equation (2).

$$Eid_{i,Chn} = \sum Ay_{j,Chn} \times y_{j,i,Chn} \quad (2)$$

Indirect emissions associated with non fuel import from sector j of region s for the production of sector i in China $Eim_{i,Chn}$ are calculated by multiplying embodied carbon emissions per dollars in sector j for intermediate demand of China per dollars $Ay_{j,s}$ and value flow from sector j to sector i $y_{j,i,s,Chn}$, as described by Equation (3).

$$Eim_{i,Chn} = \sum Ay_{j,s} \times y_{j,i,s,Chn} \quad (3)$$

It should be noted that transportations including the air, water and land transportation are separate sectors. Therefore, the emissions from transportations are calculated as one of the domestic intermediates.

Equations (1)-(3) represent a system of simultaneous equations, where the lifecycle per dollar carbon content of each good and services (Ay) are endogenous and other variables are exogenous. Values for exogenous variables are sourced from Word Input Output Database (WIOD). The simultaneous equation model is solved iteratively, after assigning initial values for (Ay). Embodied carbon emissions in export can be calculated by multiplying embodied carbon content per dollar of production, $Ay_{i,Chn}$ with China's export value.

Data Source and Accounting

Three sets of datasets are used in the analysis: world input output table, carbon emissions by sectors by country, sectoral composition of trade in value added. World input output table and carbon emissions data are extracted from the WIOD while sectoral composition of Trade in Value Added is extracted from OECD. By the definition of OECD, Trade in Value Added describes the value by country and industry added in producing goods and services for export (OECD, 2012a). Production fragmentation is common in current landscape of trading. Trade in Value Added is the better than export value as a parameter to measure the benefits of the exporting countries and is used in this analysis. It is further divided into domestic value added and foreign value added. Domestic value added means the direct contribution made by an industry in particular country in producing goods or services for export while foreign value added comes from imports of other countries that are embodied in export (OECD, 2012a). The domestic value added is the key measurement of the welfare China derives from its export sector (Chen et al., 2012).

Because of data availability, year 2009 is selected for analysis. For reasons of clarity and consistency, the data is aggregated into 18 sectors and 15 regions. The sectoral and regional aggregations are shown in the Table 1 and Table 2 respectively.

Table 1. Sectoral aggregation of MRIO table

Sector	Abbreviation
1. Agriculture, hunting, forestry and fishing	Agriculture
2. Mining and quarrying	Mining
3. Food products, beverages and tobacco	Food
4. Textiles, textile products; leather and footwear	Apparel
5. Wood, products of wood and cork; pulp, paper, paper products, printing and publishing	Paper
6. Coke, refined petroleum and nuclear fuel; chemicals and chemical products; rubber and plastics; Non-metallic mineral products	Chemical
7. Basic metals and fabricated metal products	Metal
8. Machinery and equipment, nec	Machinery
9. Electrical and optical equipment	Electrical
10. Transport equipment	Trans. Equip.
11. Manufacturing nec; recycling	Recycling
12. Electricity, gas and water supply	Electricity
13. Construction	Construction
14. Sale, Maintenance and Repair of Motor Vehicles and Motorcycles; Retail Sale of Fuel ; Wholesale Trade and Commission Trade; Retail Trade, Except of Motor Vehicles and Motorcycles; Repair of Household Goods; Hotels and Restaurants	Sales
15. Inland transport; water transport; air transport; other supporting auxiliary transport activities, activities of travel agencies; post and telecommunication	Transport
16. Financial intermediation	Financial
17. Real Estate Activities; Renting of M&Eq and Other Business Activities;	Business
18. Other services	Others

RESULTS

Total Embodied Carbon Emission by Sectors

In 2009, total embodied carbon emissions by sectors sums up to 13252.68 million metric tonnes in China. Direct carbon emissions is 6213.551 million metric tonnes while indirect carbon emissions due to domestic intermediate is 6533.469 million metric tonnes, accounting for 46.89% and 49.30% of the total carbon embodied emissions by sectors respectively while embodied carbon emissions by imports accounts for a relative small amount – 3.82% (Table 3). The relative significant amount of indirect carbon emissions due to domestic intermediate indicates the strong flow of carbon emissions among domestic sectors. It further reinforces production fragmentation in China.

Electricity sector clearly dominates all other sectors (Figure 2), accounting for 26.2% of the total embodied carbon emissions in China. Its direct carbon emissions accounts for more than 95% of the sectoral embodied carbon emissions. Without much demand for domestic intermediate or imports, embodied carbon emissions due to these components for electricity sector are relatively low. Comparing with other sectors, chemical and metal sectors are relatively carbon intensive (Figure 2).

The percentage of embodied carbon in import, direct carbon emissions and indirect carbon emissions due to domestic intermediates in each sector are shown in Table 3. Only electricity, transport, chemical and metal sectors have relative high share of direct carbon emissions, exceeding 40% of the respective sectoral total embodied carbon emissions. For the other sectors, the indirect carbon emissions due to domestic intermediate takes up large part of the respective sectoral total embodied carbon emissions.

Table 2. Regional aggregations in MRIO table

Regions		
1. Australia	12. Euro zone	13. Non Euro zone
2. Canada	• Austria	• Bulgaria
3. China	• Belgium	• Czech Republic
4. France	• Cyprus	• Denmark
5. Germany	• Estonia	• Hungary
6. Japan	• Finland	• Latvia
7. Korean	• Greece	• Lithuania
8. Mexico	• Ireland	• Poland
9. Taiwan	• Italy	• Romania
10. United Kingdom	• Luxembourg	• Sweden
11. USA	• Malta	14. BIIRT
	• Netherlands	• Brazil
	• Portugal	• India
	• Slovakia	• Indonesia
	• Slovenia	• Russia
	• Spain	• Turkey
		15. Rest of the world

Embodied Carbon Emissions and Economy in Export

Total embodied carbon emissions in export in 2009 is 1145 million metric tonnes, which is 8.6% of the total embodied carbon emissions in China. The calculations revealed that embodied carbon emissions by sectors for export (Figure 3) has different patterns than total embodied carbon emissions (Figure 2). Electrical sector has the highest embodied carbon emissions in export among sectors with 299 million metric tonnes, which is 26.1% of the total embodied carbon emissions in export. Chemical sector comes the second highest, followed by metal sector and then apparel sector. They are 19.2%, 10.3% and 9.0% of the total embodied carbon emissions respectively. Electricity sector has the highest sectoral embodied carbon emissions but relatively low embodied carbon emissions in export. It implies that most of the embodied carbon emissions in electricity sector is for domestic consumption and domestic intermediates.

According to the WIOD data, the imports and exports of China in 2009 are 1951553 and 1427486 million US$ respectively. However, embodied

Table 3. Percentages of embodied carbon emission by import, domestic direct emission and domestic intermediate in every sector

Sectors	CO2 embodied in import	Direct CO2	Indirect CO2 due to domestic intermediate
Agriculture	4.30%	38.04%	57.66%
Mining	2.30%	34.21%	63.49%
Food	5.90%	22.55%	71.55%
Apparel	6.04%	17.36%	76.60%
Paper	5.87%	27.05%	67.08%
Chemical	5.83%	49.44%	44.73%
Metal	6.09%	42.32%	51.58%
Machinery	5.48%	8.63%	85.89%
Electrical	7.35%	2.91%	89.74%
Trans. Equip.	5.65%	9.59%	84.76%
Recycling	4.74%	11.70%	83.56%
Electricity	0.35%	95.73%	3.93%
Construction	3.12%	6.17%	90.71%
Sales	2.66%	11.52%	85.83%
Transport	2.85%	53.81%	43.33%
Financial	3.93%	5.23%	90.84%
Business	5.94%	13.54%	80.52%
Others	5.11%	19.48%	75.41%
TOTAL	3.82%	46.89%	49.30%

carbon emissions in imports is 505.66 million metric tonnes while embodied carbon emissions for export is 1145 million metric tonnes. It shows that embodied carbon emissions in import per import value is much lower than embodied carbon emissions in export per export value (Figure 4). There are several possible reasons. First, imports from other countries are more carbon efficient and less carbon emissions is embodied. Second, the carbon cost of import is more internalized than exports through tax or other environmental policies. Third, China is the world factory and carbon intensive processes are concentrated there. Imports from other countries are further processed for exports and the carbon costs of these carbon intensive processes are not included. Also, carbon content of the energy source in China is higher than the importing countries. Some ideas about the difference between the embodied carbon emissions per value of imports and exports are being given in this chapter and further investigation is required to find out the reasons.

China is well recognized as a producer for manufacturing, processing and supplying goods and services to other countries. Therefore, the pattern of embodied carbon emissions in export and economy by export are worth investigation. Figure 5 plots sectoral embodied carbon emissions in export against economy by export in China in 2009. Economy by export is the sum of direct domestic trade in value added, domestic intermediate and re-imported domestic value added. The graph shows the variation between sectoral embodied carbon emissions in export and economy

Figure 2. Total embodied carbon emissions by sector in China in 2009

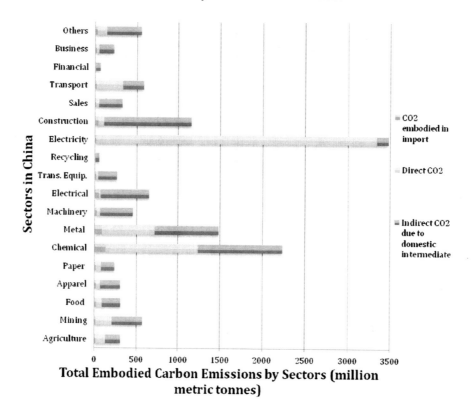

by export. Cluster analysis is carried out to investigate their relationship. 5 clusters were distinguished from the graph. They are electrical, chemical, apparel, cluster 1 and cluster 2. Electrical sector is the most embodied carbon intensive in export among all sectors and its embodied carbon emissions in export is 26.1% of the total. At the same time, its economy by sector is the largest, with 29.6% of the total economy by export. It shows that its embodied carbon emissions has strong influence on its export and economy. Electrical sector includes electrical and optical equipment in this study and most of its embodied carbon emissions come from indirect carbon emissions due to domestic intermediates. These products are at the later stages of the supply chain and closer to end consumers. Chemical sector is the second most embodied carbon intensive and its embodied carbon emission in export is 19.2% of the total embodied carbon emissions in export.

However, its economy by export is only 8.1% of the total economy by export. The manufacturing processes of products under the chemical sector are carbon intensive like refinery. Direct carbon emissions and indirect carbon emission due to domestic intermediates account for 49.44% and 44.73% of its total embodied carbon emission respectively. Contrary to the chemical sector, apparel sector's embodied carbon emission in export is 103.2 million metric tonnes which is only 9.01% while its economy by export is 16.13% of the total economy by export. Textile, textile product, leather and footwear are included in this sector and normally they are sold to end user without further processing.

Cluster 1 includes transport, metal and machinery. Their embodied carbon emission in export ranges from 84.6 million metric tonnes to 118 million metric tonnes, which is similar to apparel sector. However, their economies by sector

Figure 3. Embodied carbon emissions in export by sectors in China in 2009

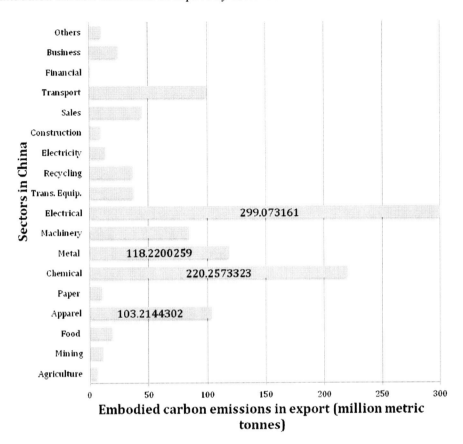

are lower than chemical and apparel sectors and accounts for 5.29% to 6.85% of the total economy by export. Both transport and metal sectors have relative high share of direct carbon emissions in the total embodied carbon emission while in machinery section, over 80% of its sectoral total embodied carbon emission come from indirect emission due to domestic intermediates. Cluster 2 includes 12 sectors which are the services, light industries or sectors with limited exports. Each sector in this cluster has relative low embodied carbon emission in export and economy by export. Their embodied carbon emission in export is less than 40 million metric tonnes and accounts for less than 3.5% of the total. Their total economy by exports represent 27.65% of the total. Services sectors like financial, business, sales sectors do not have high direct carbon emissions and the sectoral embodied carbon emissions are low too. Increasing the share and diversity of service industries is one of the strategies under the 12th Five Year Plan to meet the national carbon emission target (China Briefing, 2012) and at the same time maintain export economy. Electricity sector has the largest embodied carbon emissions among all the sectors but its embodied carbon emissions in export is low. This shows that this primary carbon intensive sector is not the primary sources of China's embodied carbon in exports and its impact on export economy is low.

Scenario Analysis

In order to examine the relationship between producers and consumers among supply chain and the ripple effect of carbon emissions reduc-

Figure 4. Embodied carbon emissions in export and import to respective value by sectors in China in 2009

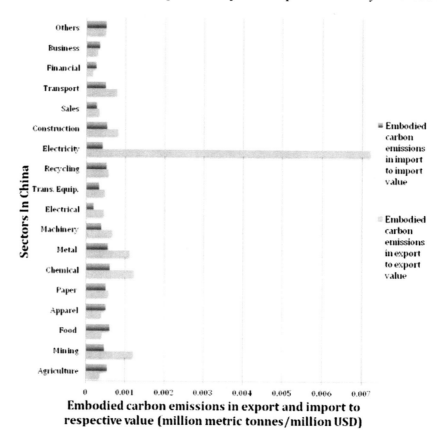

tions taken by one producer on the other sectors, China's embodied carbon emissions in 2009 above is used for the reference scenario. Electricity sector in China has the highest direct carbon emissions and supplies electricity to all other sectors. Therefore, in this scenario, electricity sector is assumed to reduce its direct carbon emissions by 30% but the cost of the carbon abatement is not included. The result of the reduction of direct carbon emissions in electricity sector was shown in Figure 6. Although different sectors vary in original embodied carbon emission, all sectors experience a decrease in embodied carbon emissions in export, ranging from 10% to 14%. In descending order, the embodied carbon emissions in export of electrical sector is reduced by 39.6 million metric tonnes, which is 13.24% while that of cluster 1 is reduced by 32.7 million metric tonnes, which is 10.81%. The embodied carbon emission in export of cluster 2 is decreased by 30.8 million metric tonnes, which is 13.96% while that of chemical decreases by 22.15 million metric tonnes, which 10.06%. The least reduction is in apparel sector and it is 13.14 million metric tonnes, which is 12.73%. Therefore, the benefits of carbon emission reduction of the electricity sector can be enjoyed by other domestic sectors and the embodied carbon emission in export can be reduced through it. Importing countries can enjoy the reduction of embodied carbon emission in imports when importing good and services from China. The benefits of carbon reduction measures taken by one sector are trickled down along the supply chain. Low carbon manufacturing adopted would potentially result in the new relationships among the supply chain, economy by export and

Figure 5. Embodied carbon emissions in export and economy by export in China in 2009

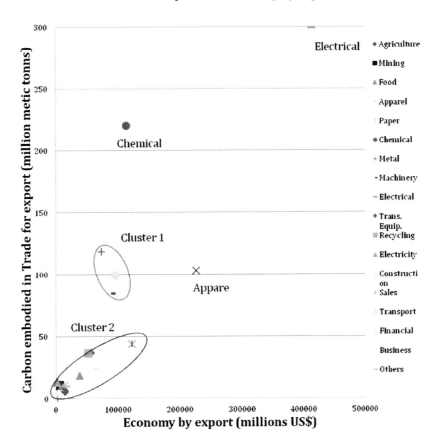

different pattern of reduction of sectoral embodied carbon emissions. Carbon reduction measures in the supply chain could be the reduction in direct carbon emissions of the sector, or choosing the upstream suppliers with lower embodied carbon emissions in imports or domestic intermediate.

Although different parties in supply chain can experience the reduction of embodied carbon emissions when one party reducing its direct carbon emissions, this information has to be linked to the benefits of the producers and consumers. Otherwise, no one will be motivated to take up the carbon reduction actions. The core concern comes from the capital cost of carbon reduction measures and how the cost and responsibility of the carbon abatement can be shared among producers and consumers. If the cost of carbon abatement taken by the producer is directly reflected on the price without additional measures, the competitiveness issue mentioned in earlier sections arises and carbon leakage may be resulted. Driven by the cost, consumers will opt for upstream suppliers with lower costs and no carbon abatement measures taken. Therefore, embodied carbon emissions in goods and services should be leveraged appropriately with supporting measures so that the benefits of carbon information can be associated with the cost. If the embodied carbon emissions in goods and services is priced, the reduction in the embodied carbon emission and lower price enjoyed by the consumer can be compared with the cost of carbon abatement measure taken by the producer. Producer must reflect the cost of carbon abatement measure into the goods and services. If the reduction in price due to embodied carbon emissions is equal to the

Figure 6. Embodied carbon emissions in export against economy by export in scenario

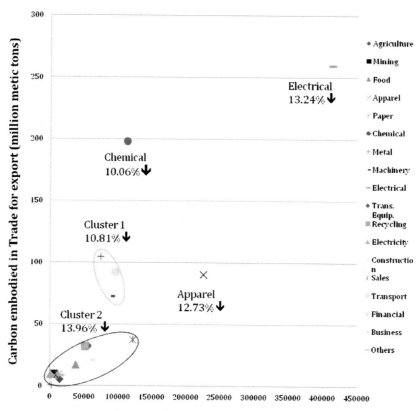

cost of taking the carbon abatement, the market is just neutral to the carbon abatement. If the reduction in price due to embodied carbon emissions is higher than the cost of taking the carbon abatement, the carbon abatement can definitely be welcomed by the both suppliers and consumers. However, in the case that the reduction in price due to embodied carbon emissions are less than the cost of carbon abatement, additional and complementary measures may have to be considered.

SOLUTION AND RECOMMENDATIONS

Low carbon manufacturing requires the responsibility sharing between producers and consumers instead of being shouldered by a single party. The previous section extensively examined how motivation of responsibility sharing for low carbon manufacturing could be created in the supply chain and offered three useful insights. The first insight is the propagating benefits of carbon emissions reduction taken by one producer. In the scenario, when one sector reduces its carbon emissions, the embodied carbon emissions for export also decreases. Second, sharing responsibility is possible when embodied carbon emissions are priced. The concept relates the benefits of embodied carbon emissions to the cost of the production so that consumer can directly make purchase decisions based on the price and associated environmental information. Third, the embodied carbon emissions in products should be counted to avoid carbon leakage. Carbon leakage is the main factor leading to the failure of the Kyoto protocol and UNFCCC. Regardless of where the product is produced, embodied carbon emissions accumulate along the supply chain and thus responsibility follows at each stage of the supply chain.

From the above analysis, to enable the share of producer and consumer responsibility among supply chain, the mechanism is proposed based on the criteria below.

Targeting Embodied Carbon Emission in Goods and Services

When embodied carbon emissions of goods and services are considered, the propagating benefits of carbon reduction taken by the producer can be demonstrated. Unlike direct carbon emissions, embodied carbon emissions in goods and services give the total carbon emissions of upstream suppliers and enhance connection among producers and consumers in the global supply chain. Targeting embodied carbon emissions implies that both producers and consumers are involved and enables the effort sharing for reducing the carbon emissions during the manufacturing processes. The environmental performance and direct carbon emissions of the producers are directly suggested by the embodied carbon emissions in goods and services they produced. Downstream consumers can trace embodied carbon emissions along the supply chain and easily identify the suppliers contributing relative high portion of embodied carbon emissions. So, it also increases the pressure for the carbon intensive emitters to reduce their carbon emissions. When producers conduct carbon abatement actions to reduce their direct carbon emissions, embodied carbon emissions in the goods and services consumed by the consumers can be reduced. The ripple effect can be visualized in the supply chain.

Availability of Embodied Carbon Emission Information

Information on embodied carbon emission has to be available to play a significant role in the mechanism. It is important for producers to know their own carbon emissions during the manufacturing process and for consumers to know about the embodied carbon emissions. As such, consumer can make informed choices, balancing economic preferences with environmental performance. On the other hand, producers can identify their positions and the differences among counterparts and limit their carbon emissions when faced with comparative disadvantage. Also, producers can exercise influence on upstream suppliers and choose raw materials with lower embodied carbon emission to achieve the lowest embodied carbon emission in finished goods and services. As described before, only with embodied carbon emissions information can the carbon relationship be shown and the potential benefits among supply chain be demonstrated.

Global Coverage

Nowadays, because of globalization and international trade, the end product in one country may involve the production and processes in another countries and material input from different part of the world. As a result, the scope of the mechanism should be global. That means estimating embodied carbon emission in goods and services and the availability of embodied carbon emissions should be applied to products in all countries. This is required to cease carbon leakage since it originates from some countries without environmental policy.

FUTURE RESEARCH DIRECTIONS

Recommendation for Sharing Responsibility Mechanism and Future Research

The above outlines three criteria required for the responsibility sharing mechanism for low carbon manufacturing. Although it appears to be simple, numerous implementation difficulties and operational challenges have to be overcome when transforming the theory to actual practice, particularly barriers highlighted in the review of current key carbon abatement measures. Directions for future research are suggested in order to truly enable the share of producer and consumer responsibility among supply chain.

Pricing the Embodied Carbon Emissions

As mentioned above, targeting and controlling embodied carbon emissions are essential in the mechanisms. But, not every carbon abatement measure discussed in previous sections is suitable for reducing the embodied carbon emissions. Three markets based flexibility mechanisms under the Kyoto Protocol – emissions trading scheme, CDM and joint investment are designed as the complementary measures for Annex B countries to meet their obligatory carbon emissions reduction target. These measures focus on swapping country-to-country benefits and target heavy polluting industries. Therefore, their concepts behind are not applicable to the sharing of responsibility in the supply chain context which involves the Web of connected producers and consumers. Moreover, they work effectively only under the cap on the emission but it may not be applicable for embodied carbon emissions. It is difficult to draw a reasonable and justifiable limit on embodied carbon emissions in goods and services. Voluntary carbon labelling incorporates environmental consideration into the purchase decisions of the consumers. Its effectiveness is low because it highly depends on the environmental awareness of the consumers and its limited application due to voluntary nature cannot let consumers compare among same category of products which some of them may not have carbon labels.

The second insight presented before suggested the possibility to share the cost when embodied carbon emissions are priced. When one producer supports low carbon manufacturing and reduces its direct carbon emissions, the embodied carbon emissions in goods and services produced are also reduced. The benefits are distributed to downstream consumers who enjoy the reduction of embodied carbon emissions. To reflect the benefits and enable the sharing the responsibility mechanism, embodied carbon emissions in goods and services should be priced. Consumers can choose a product with lower embodied carbon emission and subsequently lower associated carbon cost. It creates an incentive for the upstream producer to reduce its embodied carbon emissions and, at the same time, the cost of the carbon abatement measures can be shared with the consumers. For instance, a producer sells a product at the price of 100USD, with 10 USD for its embodied carbon emissions. If the carbon abatement measure is taken, the price of the product will be increased to 105 USD but the price for the embodied carbon emissions is reduced to 5 USD (Figure 7). The increase in price of the product can be compensated by the reduction of price for embodied carbon emissions. If the price for the embodied carbon emissions is reduced to 3 USD, the total price a consumer has to pay will be lowered (Figure 8). With comparative advantage, producer has the incentive to reduce embodied carbon emission in the product.

From the above discussion, tax on embodied carbon emissions is the most suitable among different measures. Tax is the fiscal measure to discourage behaviour. Taxation system includes the determination of tax rate, the use of tax revenue, whom to tax, administrative and opera-

Figure 7. Increased price with carbon cost reflected offset by the reduction of price for embodied carbon emissions

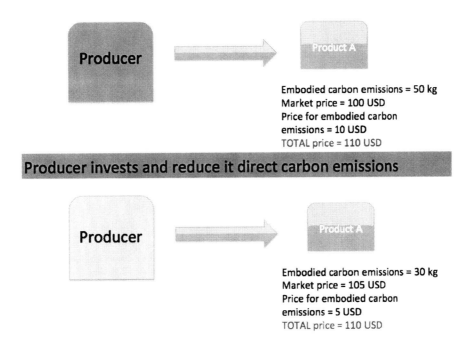

Figure 8. Price of the product can be lowered due to the reduction of price for embodied carbon emissions

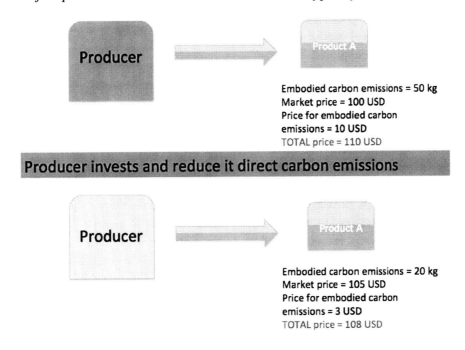

Figure 9. Price of the product become higher even with reduction of price for embodied carbon emissions

Producer → Product A
Embodied carbon emissions = 50 kg
Market price = 100 USD
Price for embodied carbon emissions = 10 USD
TOTAL price = 110 USD

Producer invests and reduce it direct carbon emissions

Producer → Product A
Embodied carbon emissions = 20 kg
Market price = 120 USD
Price for embodied carbon emissions = 5 USD
TOTAL price = 125 USD

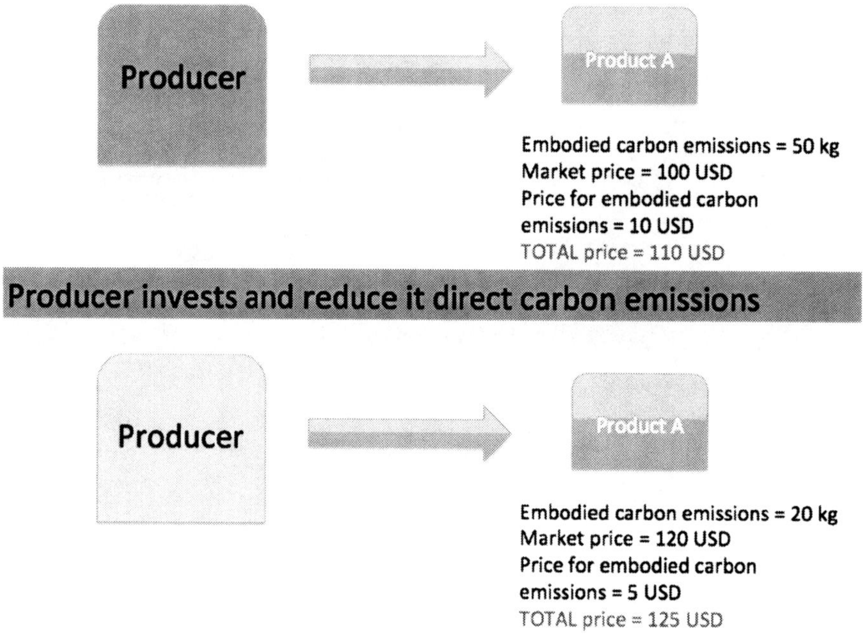

tional details. However, several difficulties are highlighted below and further study is required. There is no global taxation system and existing taxation systems are only enforced by the government within its jurisdiction. Therefore, the implementation the taxation system across different jurisdictions has to be further explored. Tax rates have to be decided carefully to encourage the reduction of embodied carbon emission but not to give negative impacts to the supply chain. Particular attention should be given to the imports. Rate of tax for different regions have to be studied. Those who oppose border tax adjustment regard it as a mean for the countries to protect their domestic market. The mechanism for responsibility sharing should not be exploited for such purpose. Also, how should the tax revenue be collected? Should it be used to create a fund that would be used to ease the transition of energy intensive industries and to invest in new technologies? The administration and operation details of such taxation system should be explored as well.

Supporting Funding or Subsidy

There are circumstances when the overall benefits enjoyed by consumers are less than the cost of carbon abatement and so, both producers and consumers do not have the motivation to reduce carbon emissions in the supply chain. Extend the example mentioned in the section above. If carbon abatement measures are taken, the price of the product has to be increased to 120 USD but the tax for embodied carbon emissions is reduced to 5 USD only (Figure 9). The total price consumers have to pay is higher than the original price. Because of comparative disadvantages, consumers prefer the product made by the producers without taking carbon abatement. Therefore, both producers and consumers don't want to make a move. As a result, the mechanism should be supported with external resources like subsidy or funding to avoid competitiveness issues and enable the responsibility sharing. If the taxation system is adopted, the tax revenue collected can be used as subsidy or funding.

Set Up Universal Standard for Embodied Carbon Emission Calculation Methodology and Information Disclosure

Regarding the second criteria – availability of embodied carbon emissions information, it is essential to set up universal standard to calculate and disclose the embodied carbon emission in goods and services. Currently, there are many methods and standards to quantify embodied carbon emissions in products. The scale, scope, assumption, definition and formula to calculate carbon emissions vary and lead to wide range of results. Therefore, a universal method to quantify carbon emissions is required and it is the basis for the pricing system. Standardized method can serve as guideline to information collection and database management for each national government. For instance, when using life cycle assessment, the emission factors of fuels used have to be found out. When using MRIO, the harmonized national input output table is required and database has to be maintained. However, it is not an easy task. Different products may involve different manufacturing processes and so, the methodology to calculate embodied carbon emission varies. Specific quantification of embodied carbon emission involves extensive data collection and calculation. Appropriate assumptions have to be made. Operation cost and efficiency is always the concern of embodied carbon emission calculation. Therefore, effort is required on assessing the impacts of the global standard and how to lower the operation cost and reach the optimal efficiency through adopting strategies or modifying the standard, etc.

Embodied carbon emission of goods and services should also be available and displayed. A standardized way to display information is required to ensure consistency and the information presented should be easy accessible and not misleading.

Securing Global Cooperation and Participation

Global coverage is an important criterion. Embodied carbon emissions in the supply chain is a global environmental problem as carbon emission of one party can affect the welfare of all other parties in the world. Global cooperation and participation is indispensable to ensure compliance of the mechanism in different countries. The most common way is through multilateral environmental agreements. Several international environmental agreements were designed earlier and their experience can serve as reference. For instance, the Kyoto Protocol is implemented under the auspices of the United Nations and representatives of all the nations gather regularly to curb global carbon emissions. Multilateral environmental agreement aims to gather as many interested parties into the agreement as possible to legitimize the outcome, secure proper implementation, compliance and evaluation and to establish and maintain long term cooperation. However, history showed that the introduction, implementation and operation of multilateral environmental agreements are extremely difficult in view of balancing the interest of different parties, establishment of the central body, settlement of disputes, etc. Therefore, multilateral environmental agreement for the responsibility sharing mechanism has to be further explored.

CONCLUSION

Escalating carbon dioxide emissions in the world is believed to be partly contributed by the manufacturing of consumer goods. Carbon dioxide is generated in all phases of the manufacturing process along the supply chain from raw materials extraction, assembly, to distribution of the finished goods to final consumers. Low carbon manufacturing is advocated among stakeholders yet it is not widely adopted. The responsibility for low carbon manufacturing is the crucial factor.

Traditionally, carbon responsibility was allocated to immediate producer. Yet, carbon leakage and the consumption-driven economy lead to the failure of curbing carbon emission during production. The consumer responsibility approach is also ineffective in reducing carbon emissions. Due to jurisdiction limits, it is hard for consumers to exert direct influence. It also weakens the producer responsibility. Therefore, the scholars have studied another responsibility allocation approach – the share of producer and consumer responsibility.

Nevertheless, at present, there is no mechanism to facilitate the share of producer and consumer responsibility from the review of current key carbon abatement measures. To bring about the share of producer and consumer responsibility, carbon impact on China's export and its economy was studied. China is the world factory and supplier of goods and services to consumers in other countries. In this scenario, benefits of carbon abatement practice by one sector can be shared among other sectors. The reduction of embodied carbon emissions in exports can benefit the importing countries. Motivation, which is the key of the share of producer and consumer responsibility, is demonstrated in this scenario.

Responsibility sharing between producers and consumers is not only feasible but also makes intuitive sense as it follows the natural flow of the market. Supply chain involves exchanging benefits between producers and consumers and such exchange is driven to be fair by itself. Therefore, to enable the fair share supply chain responsibility for low carbon manufacturing, the carbon consideration should be included into the supply chain context.

Carbon impact on China's export and its economy with scenario gives three useful insights in setting the criteria for the mechanism to share supply chain responsibility in low carbon manufacturing. Mechanism incorporated with the criteria is proposed. The proposed mechanism only gives a general framework and has to be refined by further research on 1. how the system prices embodied carbon emission, 2. the universal standard to calculate the embodied carbon emissions and disclose the information and 3. the system to secure global cooperation and participation.

REFERENCES

Antweiler, W. (1996). The pollution terms of trade. *Economic Systems Research, 8*, 361–365. doi:10.1080/09535319600000027

Bastianoni, S., Pulselli, F. M., & Tiezzi, E. (2004). The problem of assigning responsibility for greenhouse gas emissions. *Ecological Economics, 49*, 253–257. doi:10.1016/j.ecolecon.2004.01.018

Böhringer, C., Carbone, J. C., & Rutherford, T. F. (2011). *Embodied carbon tariffs* (National Bureau of Economic Research Working Paper No. 17376).

Brenton, P., Edwards-Jones, G., & Jensen, M. F. (2009). Carbon labelling and low-income country exports: A review of the development issues. *Development Policy Review, 27*, 243–267. doi:10.1111/j.1467-7679.2009.00445.x

British Standard Institution. (2011). [—*Specification for the assessment of the life cycle greenhouse gas emissions of goods and services*. London: British Standards Institution.]. *PAS, 2050*, 2011.

Cadarso, M.-A., López, L.-A., Gómez, N., & Tobarra, M.-A. (2012). International trade and shared environmental responsibility by sector. An application to the Spanish economy. *Ecological Economics, 83*, 221–235. doi:10.1016/j.ecolecon.2012.05.009

Chen, X., Cheng, L. K., Fung, K. C., Lau, L. J., Sung, Y.-W., & Zhu, K. et al. (2012). Domestic value added and employment generated by Chinese exports: A quantitative estimation. *China Economic Review, 23*, 850–864. doi:10.1016/j.chieco.2012.04.003

China Briefing. (2012). *China releases 12th five-year plan for energy saving and emission reduction*. Retrieved May 18, 2013, from http://www.china-briefing.com/news/2012/05/04/china-releases-12th-five-year-plan-for-foreign-trade-development.html

Clean Energy Regulator. (2013). *About the carbon pricing mechanism*. Australian Government, Australia. Retrieved May 18, 2013, from http://www.cleanenergyregulator.gov.au/Carbon-Pricing-Mechanism/About-the-Mechanism/Pages/default.aspx

Cohen, M. A., & Vandenbergh, M. P. (2012). The potential role of carbon labelling in a green economy. *Energy Economics, 34*, 553–563. doi:10.1016/j.eneco.2012.08.032

Cole, M. A. (2004). Trade, the pollution haven hypothesis and the environmental Kuznets curve: Examining the linkages. *Ecological Economics, 48*, 71–81. doi:10.1016/j.ecolecon.2003.09.007

Commission of the European Communities. (2008). *Communication from the Commission to the European Parliament, the Council, the European Economic and Social Committee and the Committee of the Regions on the Sustainable Consumption and Production and Sustainable Industrial Policy action plan*. Brussels, Belgium: Author.

Dan, W. (2010). Carbon emissions embodied in China's international trade based on input-output method. In *Proceedings of the 8th International Conference on Innovation & Management*, (pp. 81-85).

Dissou, Y., & Eyland, T. (2011). Carbon control policies, competitiveness, and border tax adjustment. *Energy Economics, 33*, 556–564. doi:10.1016/j.eneco.2011.01.003

Ferng, J.-J. (2003). Allocating the responsibility of CO2 over-emissions from the perspectives of benefit principle and ecological deficit. *Ecological Economics, 46*, 121–141. doi:10.1016/S0921-8009(03)00104-6

Ghosh, M., Luo, D., Siddiqui, M. S., & Zhu, Y. (2012). Border tax adjustments in the climate policy context: CO2 versus broad-based GHG emission targeting. *Energy Economics, 34*, 5154–5167. doi:10.1016/j.eneco.2012.09.005

Godard, O. (2007). *Unilateral European post-Kyoto climate policy and economic adjustment at EU borders* (EDF—Ecole Polytechnique, Cahier no. DDX 07–15).

Goulder, L. H., & Parry, I. W. H. (2008). *Instrument choice in environmental policy* (Discussion Paper DP 08-07). Washington, DC: Resources for the Future.

Government of UK. (2005). *Securing the future: Delivering UK sustainable development strategy*. United Kingdom: Government of UK.

Greenhouse Gas Protocol. (2011). *Product life cycle accounting and reporting standard*. Washington, DC: World Resources Institute and World Business Council for Sustainable Development.

Gros, D., & Egenhofer, C. (2010). *Global welfare implications for carbon border taxes climate change and trade taxing carbon at the border?* Brussels, Belgium: Centre for European Policy Studies.

Guo, J., Zou, L.-L., & Wei, Y.-M. (2010). Impact of inter-sectoral trade on national and global CO2 emissions: An empirical analysis of China and US. *Energy Policy, 38*, 1389–1397. doi:10.1016/j.enpol.2009.11.020

Hall, J. (2000). Environmental supply chain dynamics. *Journal of Cleaner Production, 8*, 455–471. doi:10.1016/S0959-6526(00)00013-5

Hubbard, B. (2012). Is there a future for carbon footprint labeling in the UK? *Ecologist*. Retrieved May 19, 2013, from http://www.theecologist.org/News/news_analysis/1231410/is_there_a_future_for_carbon_footprint_labelling_in_the_uk.html

International Energy Agency (IEA). (2012a). *2012 key world energy statistics*. Paris: IEA.

International Energy Agency (IEA). (2012b). CO2 emissions from fuel combustion. In *CO2 Highlights 2012*. Paris: IEA.

International Organization for Standardization (ISO). (2012). *ISO/TS 14067 Greenhouse gases -- Carbon footprint of products -- Requirements and Guidelines for Quantification and Communication*. ISO.

Kondo, Y., Moriguchi, Y., & Shimizu, H. (1998). CO2 emissions in Japan: Influences of imports and exports. *Applied Energy*, *59*, 163–174. doi:10.1016/S0306-2619(98)00011-7

Lenzen, M. (2007). Aggregation (in-)variance of shared responsibility: A case study of Australia. *Ecological Economics*, *64*, 19–24. doi:10.1016/j.ecolecon.2007.06.025

Lenzen, M., & Murray, J. (2010). Conceptualizing environmental responsibility. *Ecological Economics*, *70*, 261–270. doi:10.1016/j.ecolecon.2010.04.005

Lenzen, M., Murray, J., Sack, F., & Wiedmann, T. (2007). Shared producer and consumer responsibility — Theory and practice. *Ecological Economics*, *61*, 27–43. doi:10.1016/j.ecolecon.2006.05.018

Lenzen, M., Pade, L. L., & Munksgaard, J. (2004). co2 mulitpliers in multi-region input-output models. *Economic Systems Research*, *16*, 391–412. doi:10.1080/0953531042000304272

Li, Y., & Hewitt, C. N. (2008). The effect of trade between China and the UK on national and global carbon dioxide emissions. *Energy Policy*, *36*, 1907–1914. doi:10.1016/j.enpol.2008.02.005

Mani, M., Govindarajalu, C., Kee, H. L., Kishore, S., Tatsui, E., & Seki, C. et al. (2008). Climate change policies and international trade: Challenges and opportunities. In *International trade and climate change: Economic, legal, and institutional perspective*. Washington, DC: The World Bank.

Mózner, Z. F. V. (2013). A consumption-based approach to carbon emission accounting - Sectoral differences and environmental benefits. *Journal of Cleaner Production*, *42*, 83–95. doi:10.1016/j.jclepro.2012.10.014

Munksgaard, J., Wier, M., Lenzen, M., & Dey, C. (2005). using input-output analysis to measure the environmental pressure of consumption at different spatial level. *Journal of Industrial Ecology*, *9*, 169–185. doi:10.1162/1088198054084699

Muñoz, P., & Steininger, K. W. (2010). Austria's CO2 responsibility and the carbon content of its international trade. *Ecological Economics*, *69*, 2003–2019. doi:10.1016/j.ecolecon.2010.05.017

Muradian, R., O'Connor, M., & Martinez-Alier, J. (2002). Embodied pollution in trade: Estimating the environmental load displacement of industrialised countries. *Ecological Economics*, *41*, 51–67. doi:10.1016/S0921-8009(01)00281-6

Olhoff, A., Simmons, B., Abaza, H., Tamiotti, L., Teh, R., & Kulaçoğlu, V. (2009). *Trade and climate change: A report by the United Nations Environment Programme and the World Trade Organization*. Geneva, Switzerland: WTO and UNEP.

Organization of Economic Cooperation and Development (OECD). (2012a). *Measuring trade in value added: An OECD-WTO joint initiative*. Industry and globalisation. OECD. Retrieved February 14, 2013, from http://www.oecd.org/industry/industryandglobalisation/measuring-tradeinvalue-addedanoecd-wtojointinitiative.htm

Organization of Economic Cooperation and Development (OECD). (2013). *China and OECD - Trade*. Retrieved February 23, 2013, from http://www.oecd.org/tad/chinaandoecd-trade.htm

Peters, G. P., & Hertwich, E. G. (2008). CO2 embodied in international trade with implications for global climate policy. *Environmental Science & Technology*, *42*, 1401–1407. doi:10.1021/es072023k PMID:18441780

Qi, T., Winchester, N., Karplus, V. J., & Zhang, X. (2012). *CO2 emissions embodied in China's trade and reduction policy assessment global trade analysis project (GTAP), United States*. Retrieved February 14, 2013, from https://www.gtap.agecon.purdue.edu/resources/download/5991.pdf

Rosendahl, K. E., & Strand, J. (2009). *Simple model frameworks for explaining inefficiency of the clean development mechanism*. Washington, DC: The World Bank. doi:10.1596/1813-9450-4931

Sato, M. (2012). *Embodied carbon in trade: A survey of the empirical literature* (Centre for Climate Change Economics and Policy Working Paper No. 89, Grantham Research Institute on Climate Change and the Environment Working Paper No. 77). United Kingdom: Centre for Climate Change Economics and Policy and Grantham Research Institute on Climate Change and the Environment.

Tan, M. Q. B., Tan, R. B. H., & Khoo, H. H. (2012). Prospects of carbon labelling - A life cycle point of view. *Journal of Cleaner Production*, 1–13.

The Gallup Organisation. (2009). *Europeans' attitudes towards the issue of sustainable consumption and production: Analytical report*. The Gallup Organisation.

United Nation Framework Convention on Climate Change (UNFCCC). (2013a). *Kyoto Protocol*. Retrieved April 21, 2013, from http://unfccc.int/kyoto_protocol/items/2830.php

United Nation Framework Convention on Climate Change (UNFCCC). (2013b). *The Doha Climate Gateway*. Retrieved May 18, 2013, from http://unfccc.int/key_steps/doha_climate_gateway/items/7389.php

United Nation Framework Convention on Climate Change (UNFCCC). (2013c). *What is the CDM*. Retrieved May 18, 2013, from http://cdm.unfccc.int/about/index.html

United Nation Framework Convention on Climate Change (UNFCCC). (2013d). *Project Activities CDM Insights*. Retrieved May 18, 2013, from http://cdm.unfccc.int/Statistics/Public/CDMinsights/index.html

United Nation Framework Convention on Climate Change (UNFCCC). (2013e). *Distribution of CERs issued by host party*. Retrieved May 18, 2013, from http://cdm.unfccc.int/Statistics/Public/files/201304/cers_iss_byHost.pdf

United Nation Framework Convention on Climate Change (UNFCCC). (2013f). *Emission reduction units (ERUs) issued*. Retrieved May 18, 2013, from http://ji.unfccc.int/statistics/2013/ERU_Issuance.pdf

United Nation Framework Convention on Climate Change (UNFCCC). (2013g). *International emissions trading*. Retrieved May 18, 2013, from http://unfccc.int/kyoto_protocol/mechanisms/emissions_trading/items/2731.php

United Nations. (1992). *Agenda 21*. United Nations Conference on Environment & Development, Rio de Janerio, Brazil.

Upham, P., Dendler, L., & Bleda, M. (2011). Carbon labelling of grocery products: Public perceptions and potential emissions reductions. *Journal of Cleaner Production*, *19*, 348–355. doi:10.1016/j.jclepro.2010.05.014

Wiedmann, T., Wood, R., Lenzen, M., Minx, J., Guan, D., & Barrett, J. (2008). *Development of an embedded carbon emissions indicator – Producing a time series of input-output tables and embedded carbon dioxide emissions for the UK by using a MRIO data optimisation system.* London: Stockholm Environment Institute at the University of York and Centre for Integrated Sustainability Analysis at the University of Sydney.

Yan, Y., & Yang, L. (2010). China's foreign trade and climate change: A case study of CO2 emissions. *Energy Policy, 38*, 350–356. doi:10.1016/j.enpol.2009.09.025

Zhou, X., & Kojima, S. (2010). *Carbon emissions embodied in international trade: An assessment from the Asian perspective.* Japan: Institute for Global Environmental Strategies.

KEY TERMS AND DEFINITIONS

Carbon Leakage: A situation in which, despite some countries taking carbon abatement measures to limit their carbon emissions at the national level, the global carbon emissions are not reduced because carbon intensive industries are simply relocated to countries that do not take carbon abatement measures.

Consumer Responsibility: The consumer is responsible for carbon emissions from all production processes of good and services purchased regardless of where commodities are produced.

Economy by Export: In this chapter, it is defined as the sum of direct domestic trade in value added, domestic intermediate, and re-imported domestic value added.

Embodied Carbon Emissions in Export: All the carbon emissions generated during the production of goods for export to other countries.

Embodied Carbon Emissions in Goods and Services: All the carbon emissions generated during the production of goods and services.

Multiregional Input Output Model: The model where countries and world regions are distinguished and trade flows of the industries within and among the regions are demonstrated.

Producer Responsibility: The producer is responsible for the direct carbon emissions emitted during its operations.

Trade in Value Added: Developed by OECD, the value added in producing goods and services by country and industry for export.

Chapter 13
Antecedents of Green Manufacturing Practices:
A Journey towards Manufacturing Sustainability

Rameshwar Dubey
Symbiosis Institute of Operations Management, India

Surajit Bag
Tega Industries Limited, India

ABSTRACT

The purpose of this chapter is to identify green supply chain practices and study their impact on firm performance. In this study, the authors have adopted a two-pronged strategy. First, they reviewed extant literature published in academic journals and reports published by reputed agencies. They identified key variables through literature review and developed an instrument to measure the impact of GSCM practices on firm performance. The authors pretested this instrument using five experts drawn from industry having expertise in GSCM implementation and two academicians who have published their articles in reputed journals in the field of GSCM and sustainable manufacturing practice. After finalizing the instrument, the study then randomly targeted 175 companies from CII Institute of Manufacturing database and obtained response from 54 which represent 30.85% response rate. The authors also performed non-response bias test to ensure that non-response bias is not a major issue. They further performed PLSR analysis to test our hypotheses. The results of the study are very encouraging and provide further motivation to explore other constructs which are important for successful implementation of GSCM practices.

1. INTRODUCTION

The paradigm shift of policies towards green economy is forcing the companies to consider the green initiatives seriously. There is a need for the companies especially manufacturing companies to take a proactive approach rather than reactive approach in this aspect. According to UNEP (2011) report, global manufacturing industry consumes 35% of the total electricity consumed worldwide and responsible for 20% of the world's Co_2 emissions, which is detrimental to lives on the earth. While the above arguments indicate, that this the high time for empirical research on green manufacturing and implementation of green manufacturing framework, especially for countries like India which is becoming one of the global manufacturing hubs next to China. Before we delve into "green manufacturing" discussion, it is important to understand the evolution of green manufacturing.

1.1 Background

1.1.1. Evolution of Green Manufacturing

The field of Manufacturing and Operations has seen radical changes through the years. It started with the "The Industrial Revolution" in the late 1700s, has been through several phases but real challenge emerged in front of manufacturing was its sustainability. The issues that have forced developed economies like USA, Canada, Great Britain and European countries to shift their manufacturing base to China and India was cost and other was to put check on greenhouse gas emissions. However the approach of these developed economies was questionable. Rather than shifting location of manufacturing hubs is never going to solve problems but there is need for more sustainable solution which can optimize between cost and environment. The search of amicable solution has led to the growth of green manufacturing term. Green Manufacturing is defined as elimination of wastages and redefining existing process to minimize the carbon emissions during each process without increasing cost and affecting production targets. However except few literatures, green manufacturing has been used in context of green supply chain practices in recent years (Sarkis, 2003; Zhu, 2005; Srivastava, 2007; Vachon, 2007; Zhu et al., 2008; Simpson & Sampson, 2008; Darnell et al., 2008; Shukla, 2009; Yung et al., 2011; Bhateja et al., 2012; Luthra et al., 2012).

The green manufacturing addresses repeated processes (redundancy), frequent troubles related to man and machine and high cost due to traditional methods of producing goods. Today speed and cost are not only criteria to evaluate manufacturing performance but other factors such as types of materials used in manufacturing, generation of waste, effluents and their treatment, product life cycle and finally, treatment of the product after its useful life are all important considerations. Today, we have two manufacturing systems "Lean" and "Green." While lean focuses on removal of non-value added activities; green emphasizes on reducing waste having negative impact on environment. Green manufacturing is the need of the hour. Balancing the development with the earth's capacity to supply natural resources and process wastes will lead to environmentally sustainable manufacturing.

Operations management involves managing the processes which convert inputs into outputs of higher value. Green operations management involves integrating the Principles of Environment management to the Field of Operations Management. Green Operations Management embraces many concepts such as Green Building, Green Manufacturing, Green Supply chain, Reverse Logistics, and Green innovation. Hence, Green Manufacturing is an integral part of Green Operations Management. In its Sustainability Initiative, the Commerce Department defines Green manufacturing as "the creation of manufactured products that use processes

Antecedents of Green Manufacturing Practices

that minimize negative environmental impacts, conserve energy and natural resources, are safe for employees, communities, and consumers and are economically sound. Consumers have become more environmentally conscious and that reflects on the companies from which they buy .World Commission on Economic Development defines sustainability as a practice "to meet the needs of the present without compromising the ability of future generations to meet their own needs" (WCED, 1987, p. 43). Sustainable development encompasses the simultaneous adoption of Environmental, Economic and Social principles. This is called "Triple Bottom Line" approach. According to the Traditional View on Green Manufacturing, increased focus on the Environmental Initiatives would lead to Cost burden on the companies and as a result, companies would lose competitiveness in the market. Many Companies are being forced to take Green Initiatives in order comply with the various regulations of the Government. There is a need for the companies to see Green Management beyond the regulatory compliance as an opportunity to gain technological and market leadership. If you are a more responsible one, you're probably going to put the irresponsible ones out of business. Companies compete on Cost, Quality and Speed and various approaches were followed such as Just-In-Time, Six Sigma, Total Quality Management, and like. Although, all these approaches are still essential for the companies, a new Environmental dimension has emerged on which firms compete to differentiate themselves in the market. For example, Cummins supplies ReCon components where in old components are re-conditioned back to the Cummins standards. As a green or sustainable manufacturer you usually have three basic "levers" you can adjust to optimize the production of a product or component—process technology, energy source and material. The areas of focus in the green manufacturing practices are Energy consumption, Co2 emissions, water consumption and waste reduction. Apart from these, it also focuses on the facilities which achieve optimal environmental performance. For example, Sky lights in the high ceiling, windows lets in the short wave radiation from the sun essentially, Sun light and blocks long wave radiation, Heat. Usage of LED bulbs, rain water harvesting are some other features of green buildings. Green manufacturing encompasses many topics such as zero land-fill, Byproduct synergies, ELV (End-of Life-Vehicle). As a result of growing environmental awareness, corporate sustainability has been changing within the realm of international regulations. This so-called "green" issue has become a global vision. To mitigate environmental degradation, countries around the world have begun pushing for environmental protection and sustainable development. There are three ways companies compete: cost, speed and quality. As manufacturing progresses over time, different attributes become important. In the 1960s and 1970s, quality was the focal point. In the 1980s, speed was the most important factor of manufacturing. A decade later, the main issue was cost, all the while keeping a high level of quality and speed. Throughout history different methods have been used to achieve each of these goals. Many of these methods used can be considered a fad. A fad is a practice or interest followed for a time with exaggerated zeal. Concepts such as Deming management, just-in-time or even Six Sigma could all be considered fads depending on how they are implemented and sustained in an organization. Quality, speed and cost are all important in manufacturing today, but now it is so competitive that other standards have surfaced. One very important standard in particular has become a main concern and is known as green engineering (also known as environmentally conscious manufacturing). Green engineering originated in the early 1970s when it was realized that there were high levels of consumption and waste of products. Green engineering is the systems-level approach to product and process design where environmental attributes are treated as primary objectives or opportunities rather than simple constraints. It emphasizes the legitimacy of environmental objectives as consistent with the overall requirements of product quality and economy. As manufacturing sectors and the human population grow, more waste is produced, more energy is used and more products are needed. Lean manufacturing is the business model and collection of tactical methods that emphasize eliminating non-value added activities (waste) while delivering quality products at lowest cost with greater efficiency. In conjunction, six goals of green engineering are:

1. Select low environmental impact materials.
2. Avoid toxic or hazardous materials.
3. Choose cleaner production processes.
4. Maximize energy and water efficiencies.
5. Design for waste minimization.
6. Design for recyclability and reuse of material.

With these concepts in mind, lean manufacturing is a link to green engineering. The need for environmentally conscious manufacturing is becoming increasingly important due to these 7 points:

1. Environmental problems occur because people want and need products.
2. Consumer numbers will not reduce, and their habits will not change.
3. In modern society, mass production is the norm.
4. All products are designed.
5. All designs are manufactured.
6. There is an environmental impact for all product designs.
7. There is an environmental impact for all manufacturing processes.

Manufacturing has thus seen phases from Craft production, industrial revolution, assembly lines to lean and finally the present one i.e. the Green manufacturing.

1.1.2 Research Questions

There are two main research questions which will drive the present research are:

RQ1: What are the antecedents of "Green Manufacturing"?
RQ2: How these antecedents impact market performance?

1.1.3 Outline of this Chapter

The chapter is organized into the following sections. A brief summary of each section is given below:

Section 1 presents the introduction of the study. The overview of the study is briefly discussed along with the need for study.
Section 2 describes the concept of GSCM, GSCM dimensions and measures of overall firm performance based on the literature available from reputed journals such as IJPE, Journal of Cleaner Production, Energy Policy, IJPR, Production & Planning, World Journal of Science, Technology and Sustainable Development, Production Planning & Control, Benchmarking: An International Journal, International Journal of Operations and Production Management and other related journals.
Section 3 presents the methodological perspectives of this research. The strategies adopted in this research are discussed in greater detail. The issues of and how to conduct the empirical study are described in this chapter. A theoretical model showing the relationship between green procurement drivers and firm performance is developed. Finally, this chapter describes how to operationalize these theoretical constructs.
Section 4 presents data analysis and hypotheses testing.

Section 5 presents a brief summary of the research and the main conclusions with respect to the new knowledge derived from this research. The limitations of the research and issues requiring further study are also addressed.

2. LITERATURE REVIEW

In this section we have focused on state of art. We have conducted an in depth literature review on published research articles published in reputed journals, published reports of national & international agencies, edited books and expert opinions. We have presented some of reviews which we have derived from 116 papers between 2001 to 2013 in Table 1.

From above table researcher can draw conclusions:

- GSCM practices are source of competitive advantage;
- GSCM practices are source of sustainability;
- The companies practicing GSCM, enjoy high market share and superior business performance;
- GSCM is the necessity. The companies who will fail to implement will sooner or later will become obsolete;

The GSCM practices has helped firm to achieve superior performance is supported by literatures (e.g., Pauli, 1997; Farish, 2009; Franchetti et al., 2009; Deif, 2011; Murovec et al., 2012; Prajogo et al., 2012; Pereira-Moliner et al., 2012; Dues et al., 2013; Gavronski et al., 2013). The literature defines performance as combination of environmental performance (e.g., reduction in CO_2, SO_2, NO_x.., Converting waste into more useful product, reuse or recycling after use, increase in product quality, increase in customer satisfaction) and business performance (e.g., increase in market share, increase in profitability, increase in ROI, increase in ROA, improvement in Inventory Turnover, etc.).

However in present study we would like to confine our present research to environmental performance. We have identified literatures which have supported Green Supply Chain Management Practices and its positive impact on Environmental Performance (i.e., Zhu et al., 2005; Zhu., 2007; Zhu et al., 2008; Dues et al., 2011; Olugu & Wong, 2011; Bhateja et al., 2012; Seman et al., 2012; Gangele et al., 2012; Whitelock, 2012; Kim et al., 2012).

3. RESEARCH MODEL AND HYPOTHESES DEVELOPMENT

3.1 Proposed Theoretical Framework

On the basis of the preceding discussion and the synthesis of the existing literature, conceptual framework is proposed as shown in Figure 1. The two main components that constitute the conceptual framework include the drivers for Green Procurement implementation which will be measured in terms of firm performance variables.

Green supply chain management has its roots both in environmental management and supply chain management literature. Developed countries are those countries where we find high level development based on certain characteristics-consisting of economic, industrialization and human development index. Countries such as Japan, South Korea, Australia, Germany, Portugal, Italy, Sweden and Canada fall in this category and GCSM has been so far practiced majorly in these countries. In developing countries especially in Asian region (India, China, Philippines, Indonesia, Malaysia, Thailand, and Singapore) GCSM practices are in a very nascent stage. Most firms in developing countries adopt GCSM to reduce the environmental impact of various business activities rather than a proactive attitude to reduce the source of wastage or pollution.

Table 1. Research in green supply chain management

Author(s), Year	Research Objective	Research findings
Kumar and Chandrakar (2012)	To study the correlation of two major factors, i.e. organizational learning and management support in GCSM adoption in Indian manufacturing industries	Significant positive relationships exists between organizational learning and management support with respect to GCSM adoption
Luthra et al. (2012)	To identify the important factors to implement GCSM relevant to Indian manufacturing industry	1. Innovative green practices, Awareness level of customers 2. Supplier Motivation 3. Technology advancement and organization adoption, Organization encouragement, Quality of human resources 4. IT enablement 5. Top management commitment 6. Government support policies 7. International Environmental agreements
Bhateja et al. (2012)	To identify the critical factors related to evaluation of GCSM performance measurement in Indian manufacturing industry	A. Green Purchasing 1. Substitute for hazardous material 2. Improved quality of raw material 3. Minimal usage of raw material 4. Supplier Development 5. Reduced Resource B. Green Manufacturing 1. Process Design 2. Product Design 3. Higher Efficiency 4. Employee Satisfaction
Pandya et al. (2012)	To explore external factors affecting GCSM and understand the relationship between the GCSM practices and environmental performance and operational performance as well as financial performance in the context of Indian pharmaceutical industry	**Pressures/Drivers:** Environmental regulations, suppliers, consumers and community stakeholders GCSM practices can enhance the environmental, operational and financial performance of firms
Bhateja et al. (2011)	To study the activities of the supply chain process of various Indian manufacturing industries and evaluate their degree of greenness for the purpose of measuring performance.	51% of manufacturing industries feels that lack of awareness of environmental issues is the biggest issue facing manufacturing sectors. 36% of manufacturing industries have plans to implement GCSM initiatives within 2 years. 40% of companies use electronic processes to create efficiencies in procurement. 32% of companies are having active discussions regarding collaborating to reduce impact on environment. 64% of companies are not using e-tools extensively to support their supply chain operations. The biggest perceived barrier to adopting GCSM is that it is not cost effective.
Duarte et al. (2011)	To develop a conceptual model incorporating lean and green supply chain into performance measurement system using BSC approach	Linking performance measurement system to green/lean practices can benefit firms for better positioning to succeed in their supply chain initiatives
Dües et al. (2011)	To explore and evaluate previous work focusing on the relationship and links between Lean and Green supply chain management practices.	Lean is beneficial for Green practices and the implementation of Green practices in turn also has a positive influence on existing Lean business practices
Gangele and Verma (2011)	To survey current green practices in Indian pharmaceutical manufacturers and GSCM evaluation.	Environmental management systems are given top attention than Green Purchasing, Customer Cooperation, Investment Recovery and Eco-Design is given low priority. Influence of GSCM on Performance factors such as Environmental and positive economic are relatively significant. Top two GCSM drivers are Pressure from environmental regulations and export pressure.
Luthra et al. (2011)	To develop a structural model using Interpretive Structural Modeling of the barriers to GCSM implementation in Indian automobile industry	Eleven variables were identified from literature review and expert opinions **Top Level Barriers:** Market competition and uncertainty; Lack of implementing green practices; Cost implications; Unawareness of customers **Bottom Level Barriers:** Lack of government support systems

continued on following page

Table 1. Continued

Author(s), Year	Research Objective	Research findings
Singh et al. (2011)	To assess the role of logistics & transportation in GCSM in the context of Indian retail industry	Technological integration with primary suppliers and with major customers was positively linked to environmental monitoring and environmental collaboration. However logistical integration only has an impact on GCSM with primary suppliers but not with the major customers.
Yung et al. (2011)	To examine the impact of environmental regulations on green supply chain management	1. EU directives foster green partnerships among manufacturing firms of all sizes located at different positions in international supply chains 2. Firms environmental management strategy affects its regulation compliance practices
Soler et al. (2010)	To describe the use of environmental information at different stages of the Swedish food supply chain	Consumer must be perceived as close to supply chain actors, enabling a correct transaction of consumer preferences into relevant green supply chain practice to avoid distortion of information.
Shukla et al. (2009)	To identify implementation level, major drivers, various practices and performance of environmentally and socially conscious SCM in the context of Indian automobile industry	Environmentally and socially responsive supply chains are in the early adoption stages in India. Actual implementation lacks a holistic approach
Darnall et al. (2008)	To empirically evaluate the relationship between EMS and GCSM practices	Environmental management systems are more likely to adopt GCSM practices.
Simpson and Samson (2008)	To develop strategies for GSCM	Described four GCSM strategies: A. Risk-based Strategies B. Efficiency-based Strategies C. Innovation-based Strategies D. Closed-loop Strategies
Seuring and Muller (2008)	To develop a conceptual framework for sustainable SCM	Identified dimensions are 1. Suppliers management for risks and performance 2. SCM for sustainable products
Zhu et al. (2008)	To evaluate perceived GCSM practices in four different Chinese manufacturing firms and relate them to closing the supply chain loop	Adoption of GCSM varies in different industry context. GCSM can be used as an environmental tool to improve the environmental image and gain competitiveness within the international business arena.
Field and Sroufe (2007)	To examine the implications of using recycled materials on operations strategy with a focus on the corrugated cardboard industry	The benefits of the changes in the supply chain and supplier relationships accrue primarily to nonintegrated firms and managers should expect the use of recycled material inputs to be dominated by nonintegrated firms with decreasing capital costs over time.
Vachon (2007)	To determine if there is a link between green supply chain practices (environmental collaboration and environmental monitoring) and environmental technologies selection	The results suggest that environmental collaboration with suppliers is positively associated with greater investment in pollution prevention technologies while such collaborations with customers has no impact on the adoption and the implementation of pollution prevention technologies
Zhu et al. (2007)	To examine the relationships between GCSM practices, environmental and economic performance in Chinese manufacturing firms, incorporating three moderating factors i.e. market, regulatory and competitive institutional pressures	1. Chinese firms have experienced increasing environmental pressure to implement GCSM practices. 2. The existence of market and regulatory pressures influences organizations to have improved environmental pressures especially in case of eco design and green purchasing 3. Manufacturers facing higher regulatory pressure tend to implement green purchasing and investment recovery 4. Competitive pressure significantly improves the economic benefits from adoption of different GCSM practices 5. None of the institutional pressures contribute to or lessen possible "win-win" situations for organizations.

continued on following page

Table 1. Continued

Author(s), Year	Research Objective	Research findings
Srivastava (2007)	To present a comprehensive integrated view of the published literature on all dimensions of GSCM, primarily taking a reverse logistics angle for facilitating future research directions	GCSM can reduce the ecological impact of industrial activity without sacrificing quality, cost, reliability, performance or energy utilization efficiency.
Sarkis (2003)	To present a strategic decision framework that will aid managerial decision making	Strategic & Operational elements were structured for evaluating green supply chain alternatives

(Source: Dubey and Bag, 2013)

Figure 1. Green manufacturing

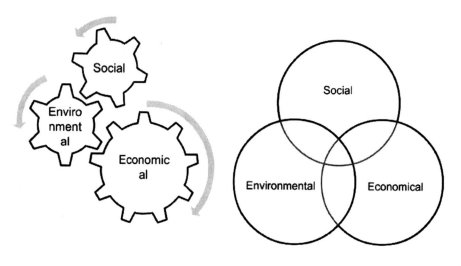

Drivers of GCSM identified in Malaysian context are Regulations, Customer requirements, expected business gains and social responsibility, (Seman et al. 2012). GCSM awareness is high and is perceived as a competitive advantage for companies in India. Adoption of GSCM practices is highest in areas where there is a correlation to efficiency and cost savings and vice versa (Bhateja et al., 2012).

The dimensions of GSCM are identified through literature review which has been discussed critically in literature review section as shown in Table 2.

By discussing with industry experts and literature review the following variables are used to define the construct and capture the drivers of green procurement: (1) Customer pressure (2) Social responsibility (3) Regulatory pressure (4) quality management;

The above dimensions will be used in developing a structured questionnaire for conducting survey among firms which are practicing Green Procurement practice.

3.2 Assumptions of Study

To test hypothesis the researcher has made certain assumptions are;

Antecedents of Green Manufacturing Practices

Figure 2. A conceptual model showing the relationship between green supply chain practices drivers and firm performance

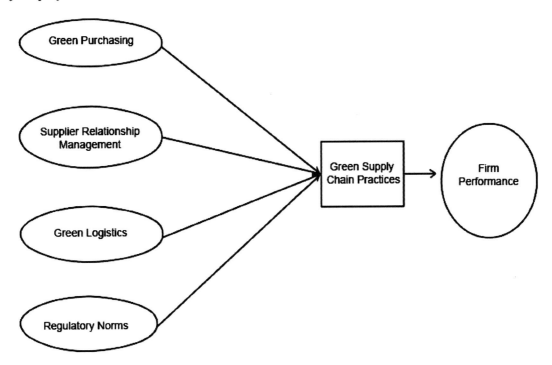

Table 2. Dimensions of green supply chain practices

Green Procurement drivers/Variables	Authors(Year)	Dimensions/Sub-Variables
Green Purchasing	Yang and Zhang(2012); Wahid et al.(2011); Bjorklund (2011);Zhu et al.(2007)	• Supplier pressure; • Consumer pressure; • CSR activity; • Brand image; • Green management cost;
Supplier Relationship Management	Bierma and Waterstraat,1999;Vachon and Klassen,2006;Hsu and Hu,2009;Bai and Sarkis,2010;Ku et al.,2010;Testa and Iraldo,2010;Hoof and Lyon,2013	• Bringing Suppliers together; • Guiding Suppliers to to implement their environmental programs; • Holding awareness seminars for suppliers/contractor; • Sending in-house companys auditor to appraise environmental performance of the suppliers
Green Logistics	Yang and Zhang(2012); Wahid et al.(2011); Bjorklund (2011);Zhu et al.(2007)	• Recovery of the company's end of life products; • Taking back of material; • Green fuels; • Eco product lifecycle approach for distribution; • Change for more environment friendly transportation;
Regulatory Norms	Yang and Zhang(2012); Wahid et al.(2011); Bjorklund (2011);Zhu et al.(2007); EPP (2011); ACC sustainability report(2010)Green Purchasing Australia report(2009);European Communities(2004);	• Green Labeling; • Disposal norms; • Waste water treatment and recycling; • Green mining; • Carbon emission norms; • Sox, NOx emission norms;
Firm performance	Raghavendran et al.(2012); Dess and Robinson (1984);	• customer satisfaction ; • profitability; • market-share;

1. The variables are assumed to be orthogonal in nature;
2. Macro variables are assumed to constant;
3. To measure firm performance researcher has assumed perception of senior manager or plant head of company;
4. There will be response and non-response error. However researcher has taken utmost care to minimize these sampling errors by designing questionnaire scientifically and adopting random sampling;

3.3 Hypotheses of Study

A hypothesis is a tentative statement, about the relationship between two or more variables. A hypothesis is a specific, testable prediction about what one expects to happen in the study. While the hypothesis predicts what the researchers expect to see, the goal of research is to determine whether this guess is right or wrong. When conducting an experiment, researchers might explore a number of different factors to determine the contribution of each to the ultimate outcome. While formulating the research hypotheses following questions are considered:

- Are these hypotheses based on the identified research study?
- Can these hypotheses be tested?
- Do the hypotheses include independent and dependent variables?

Based on the literature review conducted, research gaps and research objectives identified, following hypotheses are proposed for the present study:

Hypothesis 1: *There is a positive relationship between Green Purchasing and firm performance.*

Hypothesis 2: *There is a positive relationship between Supplier Relationship Management and firm performance.*

Hypothesis 3: *There is a positive relationship between Green Logistics and firm performance.*

Hypothesis 4: *There is a positive relationship between Regulatory norms and firm performance.*

4. RESEARCH DESIGN

4.1 Measurement Instrument Development

To carry out survey, an instrument for this study was developed scientifically using two approaches namely:

- Literature review;
- Pretest of questionnaire consisted of expert opinion and pilot survey;

To begin with an exhaustive literature survey has been conducted to identify key issues. The various critical factors are identified from various sources which are important for successful GSCM implementation. The various CSFs are identified. A pragmatic and grouping of dimensions is done. The various questionnaire instruments in the area are also identified. These included (e.g., Sarkis, 2003; Zhu et al., 2005; Vachon, 2007; Bhateja et al., 2012; Luthra et al., 2012). All of these researchers have developed their questionnaires differed from each other. However, the questionnaires developed by these researchers gave some useful insights into developing the questionnaire required for the identified research objectives.

4.1.1 Questionnaire Design

Resulting from the intensive literature review, the questionnaire is consolidated into two sections:

Section 1: Demographic profiles of the target firms;

Section 2: Includes questions on dimensions of GSCM in terms of 9 items of Green Purchasing, 8 items of Green Logistics, 7 items of Supplier Relationship Management, 8 items of Regulatory Frameworks and 12 items of Organizational performance.

4.1.2 Pretest

The pretest deals with how much measurement instrument is accurate in measuring the variables which is supposed to measure and how precise it is doing so. It is carried out in two phases:

- Expert opinion;
- Pilot Study;

Expert interviews, is performed to develop valid survey instrument for research. In order to test reliability and validity of the measurement instrument, an expert opinion was conducted.

Total five experts were invited to refine and validate measures for each concept. Two experts from academics and three from industry. The experts were asked to provide their opinion on initial 32 items of GSCM constructs and 12 items of organizational performance. The major comments were related to adjusting the details of wording (reworded or shortened) in some questionnaire items, to make it clearer and more concise. Suggestions were received to eliminate some overlap items.

Expert opinion resulted in 23 items for GSCM constructs and 8 items for organizational performance.

4.2 Sampling Design

The study is focused on manufacturing sector. In this case researchers have conducted study in Nasik and Pune which is one of the growing manufacturing hubs in India located in Maharashtra. There are over 2500 manufacturing companies however researchers initially aimed at 230 companies who are either practicing Green Procurement based on initial information provided. The response we received was 54 which are 32% of the total questionnaires mailed for survey. However 54 firms out of 230 firms is quite interesting number but from statistical perspective it may not be sufficient; however in such case PLSR (Partial Least Square Regression Analysis) will be highly suitable for hypothesis testing and developing model further, which we will discuss in our next section.

4.3 Sample Firm Identification

In order to study researcher has identified manufacturing companies who have implemented GSCM from CII database.

4.3.1 Data Collection

The structured questionnaire was e-mailed to over 230 companies listed in CII database. Out of 230 questionnaires only 54 questionnaires were returned after several follow up with respective companies. This represents 23.478% of total targeted sample. Researcher has taken utmost care to eliminate non-response error.

The respondents represent senior manager of the company who is responsible behind GSCM implementation or have implemented in their companies and possess professional qualification like MBA, CSCP, CPIM and CPSM.

4.3.2 Research Tool for Data Analysis

To begin with analysis, researcher used exploratory factor analysis to validate scale and constructs used in questionnaire. The exploratory factor analysis output was used as an input for regression analysis. Here the sample size is only 54 so in such case researcher prefer to use PLSR which is discussed in detail in next chapter.

5. SOLUTIONS AND RECOMMENDATIONS

5.1. Regression Modeling

Based upon EFA output (Refer to Table 9 and Table 10), there are two regression models to be tested are shown in Figures 3 and Figure 4.

In order to test the above to Regression Models, author has preferred to use PLSR (Partial Least Square Regression Analysis) using EFA output has input over Multiple Linear Regression Model (MLRA), due to two reasons;

Reason 1: The data distribution was not normal.
Reason 2: The data distribution is not normal as the value of Durbin-Watson statistics was found to be less than 1.5.

In such case when data distribution is not normal, autocorrelation effect is visible and certain degree of multicollinearity is said to have existed, PLSR is said to be much better technique which is variance based structural equation modeling (SEM) technique which is used for testing regression modeling.

Partial Least Square Regression (PLSR) Modeling

PLSR analysis is done using Minitab 15. It is statistical software that positions itself for the quality assurance market. It is specialized for many quality functions and is used by many industries for that purpose. Researcher has used Minitab 15 for PLSR analysis, on the collected data. Relationship is established between the factors selected from EFA and firm performance variable. The PLSR output is discussed below to understand how the output is different from above multiple linear regression output. In this case researcher has carried out PLSR analysis using selected four factors instead of performing on raw data to draw a comparison between result of both analysis and to eliminate the multicollinearity effect on variables so that conclusive remark can be drawn. In this cross-validation, is commonly used to determine the optimal number of components to take into account, is controlled by the validation argument in the modeling functions (mvr,plsr).

PLS Regression: Business Performance vs. Green Purchasing, Supplier Relationship Management, Green Logistics and Regulatory Norms

Table 3 indicates that explanatory variables identified as parsimonious and orthogonal factors (i.e.Green purchasing, Supplier relationship management, Green logistics and Regulatory norms) explain,22.6% of total variance of business performance.

Table 4 indicates that proposed regression model is statistically significant at 0.012 as the F-statistics value is greater than the $F_{cr}(4,49)$.

Table 5 shows standardized Beta coefficients of explanatory variables. The positive sign shows that each explanatory variable positively influence the business performance. It can be also seen that manufacturing firms focusing more on "supplier relationship management" and "Green purchasing" can expect to achieve superior environmental performance.

The PLS response plot (Figure 6) shows that there the calculated response plot vs actual response. The plot shows that the distance measured between two point increases towards right hand side as we move along the horizontal axis (Actual response axis).Thus we can see that the PLSR output is optimum up to 4 components. Thus here researcher has selected only four components output.

Business Performance = 0.89+0.11* Green Purchasing + 0.32* Supplier Relationship Management + 0.081 * Green Logistics +).065* Regulatory Norms + Error;

Figure 3. Business performance-GSCM model

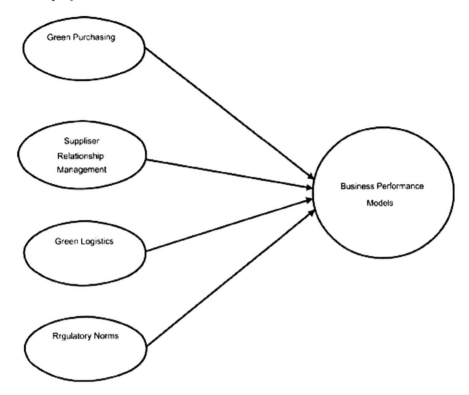

Figure 4. Environmental performance-GSCM model

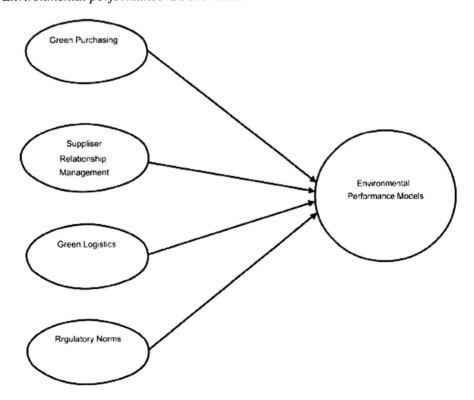

Table 3. Model selection and validation

Components	Variance	R-Sq
1	0.57864	0.221690
2	0.75135	0.224116
3	0.87368	0.225855
4	1.00000	0.225990

Table 4. ANOVA analysis

Source	DF	SS	MS	F	P
Regression	4	11.0584	2.7646	3.58	0.012
Residual Error	49	37.8749	0.7730		
Total	53	48.9333			

Table 5. Regression coefficient

Constant	0.893	Standardized coefficients
Green Purchasing		0.107
Supplier Relationship Management		0.321
Green Logistics		0.082
Regulatory Norms		0.0645

Figure 5. PLS Coefficient Plot

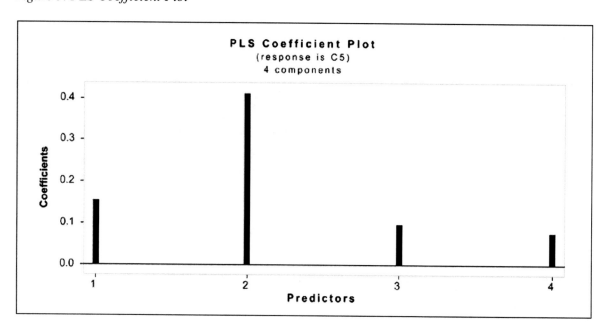

Figure 6. PLS Response Plot

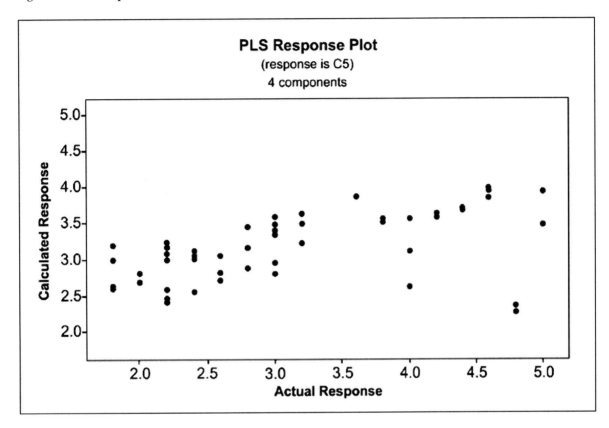

PLS Regression: Environmental Performance versus Green Purchasing, Supplier Relationship Management, Green Logistics and Regulatory Norms

Table 6 indicates that explanatory variables identified as parsimonious and orthogonal factors (i.e. Green purchasing, Supplier relationship management, Green logistics and Regulatory norms) explain, 33.2% (approximately) of total variance of environmental performance.

Table 7 indicates that proposed regression model is statistically significant at 0.000 as the F-statistics value is greater than the Fcr(4,49).

Table 8 shows that except supplier relationship management, other explanatory variables (i.e. green purchasing, green logistics and regulatory norms) are not supporting environmental performance. However the findings do not conform to the findings of other literatures. The present findings suggest that Indian manufacturing sector need to adopt green purchasing, green logistics or respect regulatory norms to achieve superior environmental performance. However it should be further explored.

Environmental Performance = 3.394 − 0.48*Green Purchasing + 0.81*Supplier Relationship Management − 0.061*Green Logistics − 0.208*Regulatory Norms + Error;

5.2. Recommendations

We can derive recommendations based on our literature review and solutions. The manufacturing firms in India focusing on green purchasing, supplier relationship management, green logistics and regulatory frameworks have experienced better business performance. It helps firm to improve

Table 6. Model selection and validation

Components	Variance	R-Sq
1	0.50521	0.191607
2	0.79837	0.263297
3	0.90112	0.318223
4	1.00000	0.331687

Table 7. ANOVA analysis

Source	DF	SS	MS	F	P
Regression	4	8.9601	2.24003	6.08	0.000
Residual Error	49	18.0537	0.36844		
Total	53	27.0138			

Table 8. Regression coefficient

Constant	3.39367	Standardized coefficients
Green Purchasing		-0.482951
Supplier Relationship Management		0.813714
Green Logistics		-0.060707
Regulatory Norms		-0.207767

their brand and enhance their reputation in market. It further helps to improve market share.

However, there are myths among manufacturers. According to some "green practices" require huge investment and in return it further reduces productivity. However, it is just reverse. It helps to cut down wastages resulting during each stage of supply chain, which further impact bottom line of the firm.

On the other hand supplier relationship has major impact on environmental performance which is based on perceptions of the managers. Based on their perceptions we can further conclude regarding Indian manufacturing firms based on response that we have received is quite interesting. Indian firms are yet to score in terms of environmental performance. Though there is growing awareness towards environmental performance but still a long way to go. Our key recommendations to Indian manufacturing firms are:

- Reduce carbon footprints. Indian firms need to give equal weightage to environmental parameters along with cost, quality and reliability while selecting suppliers;
- Road transportation which constitutes one of the dominant modes of transportation in India is one of the major sources of carbon footprints. A proper initiative need to be adopted by the concerned manufacturers to reduce carbon emissions by truck through proper training of truck drivers and helpers, prevent overloading, proper maintenance of engine and use of alternative fuels;
- Indian judicial system need to be further strengthened in enforcing laws and strict action to be initiated against defaulters will further help to prevent damages.

Figure 7. PLS coefficient plot

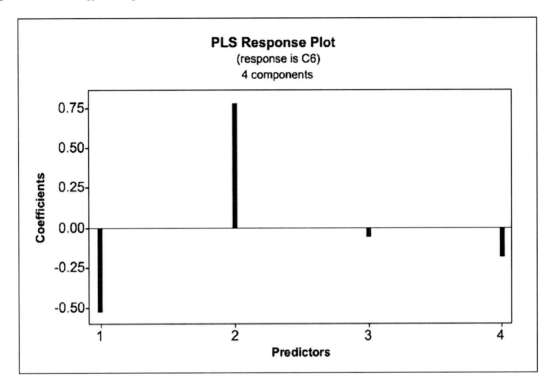

Figure 8. PLS response plot

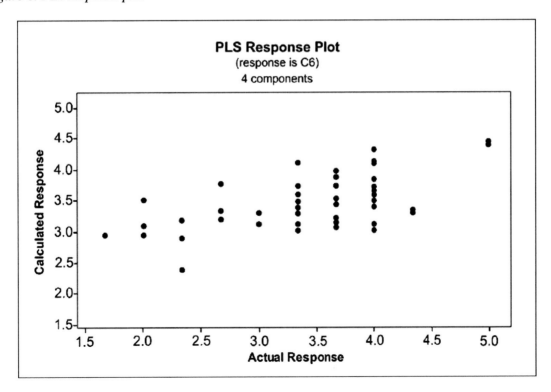

Table 9. Parsimonious orthogonal factors

	F1	F2	F3	F4	
GP1		.695			
GP2		.843			
GP3			.599		
GP4	.521				
GP5			.727		
GL1			.814		
GL3	.579				
GL4	.823				
GL5	.859				
SRM1	.900				
SRM2	.780				
SRM3		.567			
SRM4		.888			
SRM5		.800			
REG1		.725			
REG2				.653	
REG3				.693	
REG4				.716	
REG5				.783	
IF6	.656				
Eigen Value	3.869442299	3.470511	1.549396	2.032484	10.92183
% of Variance	19.34721149	17.35255	7.74698	10.16242	54.60916

Table 10. Parsimonious performance variables

	F1	F2	
P1		.792	
P2		.823	
P3		.705	
P4	.576		
P5	.837		
P6	.900		
P7	.916		
P8	.897		
Eigen Value	3.485348215	1.800812463	5.286161
Variance%	43.56685269	22.51015579	66.07701

5.3. Future Research Directions

There is an immense opportunity to extend present work. In order to validate empirical findings a case study approach can be adopted as there are few players who have successfully implemented GSCM practices in India. Hence, survey methodology will have its own limitation. At present random sampling technique will be serious challenge as GSCM practices in India is in infancy stage.

6. CONCLUSION

We can now conclude that the chapter is based on our empirical research using a structured questionnaire followed by survey using random sampling technique. In our study we have performed non-response bias test to understand the impact of non-response bias on the final regression analysis. The findings of our study have supported extant literatures and have provided interesting insights into Indian green manufacturing practices. Like any other research, the present study has its own limitations which will become the future research directions. The present study outcome is based upon survey conducted among 54 manufacturing companies. However in order to take care of the small sample size researcher has used PLSR to test hypotheses.

Secondly, R-Sq value of the conclusive model is 22.6% and 33.2% which indicates that there are other variables which are not explored in present study. However, the study can be further explored with large sample size.

REFERENCES

Bai, C., & Sarkis, J. (2010). Greener supplier development: Analytical evaluation using rough set theory. *Journal of Cleaner Production, 17*(2), 255–264.

Bhateja, A. K., Babbar, R., Singh, S., & Sachdeva, A. (2011). Study of green supply chain management in Indian manufacturing industries: A literature review cum an analytical approach for the measurement of performance. *International Journal of Computational Engineering & Management, 13*, 84–99.

Bhateja, A. K., Babbar, R., Singh, S., & Sachdeva, A. (2012). Study of the critical factor finding's regarding evaluation of green supply chain performance of Indian scenario for manufacturing sector. *International Journal of Computational Engineering & Management, 15*(1), 74–80.

Bierma, T. J., & Wasterstraat, F. L. (1999). Cleaner production from chemical suppliers: Understanding shared savings contracts. *Journal of Cleaner Production, 7*(2), 145–158. doi:10.1016/S0959-6526(98)00073-0

Bjorklund, M., & Martinsen, U., & Abrahamson. (2012). Performance measurements in the greening of supply chains, *Supply Chain Management. International Journal (Toronto, Ont.), 17*(1), 29–39.

Darnall, N., Jolley, G. J., & Handfield, R. (2008). Environmental management systems and green supply chain management: Complements for sustainability? *Business Strategy and the Environment, 17*(1), 30–45. doi:10.1002/bse.557

Dess, G. G., & Robinson, R. B. (1984). Measuring organizational performance in the absence of objective measures: the case of the privately-held firm and conglomerate business unit. *Strategic Management Journal, 5*(3), 265–273. doi:10.1002/smj.4250050306

Duarte, S., Cabrita, R., & Machado, V. C. (2011). Exploring lean and green supply chain performance using balanced score card perspective. In *Proceedings of the 2011 International Conference on Industrial Engineering and Operations Management,* Kuala Lumpur, Malaysia.

Dubey, R., & Bag, S. (2013). Exploring the dimensions of sustainable practices: An empirical study on Indian manufacturing firms. *International Journal of Operations and Quantitative Management, 19*(2), 123–146.

Dües, C. M., Tan, K. H., & Lim, M. (2013). Green as the new lean: How to use lean practices as a catalyst to greening your supply chain. *Journal of Cleaner Production, 40,* 93–100. doi:10.1016/j.jclepro.2011.12.023

European Commission. (2006, March). *Environmental fact sheet: Industrial development.* Retrieved April 8, 2013, from http://ec.europa.eu.pdf

Field, J., & Sroufe, R. (2007). The use of recycled materials in manufacturing: Implications for supply chain management. *International Journal of Production Research, 45*(18/19), 4439–4463. doi:10.1080/00207540701440287

Gangele, A., & Verma, A. (2011). The investigation of green supply chain management practices in pharmaceutical manufacturing industry through waste minimization. *International Journal of Industrial Engineering and Technology, 3*(4), 403–415.

Gavronski, I. (2011). A resource-based view of green supply management. *Transportation Research Part E, Logistics and Transportation Review, 47*(6), 872–885. doi:10.1016/j.tre.2011.05.018

Giovanni, P. D., & Vinzi, V. E. (2012). Covariance versus component-based estimates of performance in green supply chain management. *International Journal of Production Economics, 135*(2), 907–916. doi:10.1016/j.ijpe.2011.11.001

Hoof, B. V., & Lyon, T. P. (2013). Cleaner production in smaller firms taking part in Mexico's sustainable supplier program. *Journal of Cleaner Production, 41,* 270–282. doi:10.1016/j.jclepro.2012.09.023

Hsu, C. W., & Hu, A. H. (2008). Green supply chain management in the electronic industry. *International Journal of Science and Technology, 5*(2), 205–216. doi:10.1007/BF03326014

Hsu, C. W., & Hu, A. H. (2009). Applying hazardous substance management to supplier selection using analytic network process. *Journal of Cleaner Production, 17*(2), 255–264. doi:10.1016/j.jclepro.2008.05.004

Kim, J. H., Youn, S., & Roh, J. J. (2011). Green supply chain management orientation and firm performance: An evidence from South Korea. *International Journal of Services and Operations Management, 8*(3), 283–304. doi:10.1504/IJSOM.2011.038973

Kumar, R., & Chandrakar, R. (2012). Overview of green supply chain management: Operation and environmental impact at different stages of the supply chain. *International Journal of Engineering and Advanced Technology, 1*(3), 1–6.

Li, S., Nathan, B., Ragu-Nathan, T. S., & Rao, S. S. (2006). The impact of supply chain management practices on competitive advantage and organizational performance. *Omega, 34*(2), 107–124. doi:10.1016/j.omega.2004.08.002

Luthra, S. et al. (2012). Barriers to implement green supply chain management in automobile industry using interpretative structural modeling technique-An Indian perspective. *Journal of Industrial Engineering and Management, 4*(2), 231–257.

Nelson, D. M., Marsillac, E., & Rao, S. S. (2012). Antecedents and evolution of the green supply chain. *Journal of Operations and Supply Chain Management,* Special Issue, 29–43.

Raghavendran, P. S., Xavier, M. J., & Israel, D. (2012). Green purchasing practices: A study of eprocurement in B2B buying in Indian small and medium enterprises. *Journal of Supply Chain and Operations Management, 10*(1), 3–23.

Sarkis, J. (2003). A strategic framework for green supply chain management. *Journal of Cleaner Production, 11*(4), 397–409. doi:10.1016/S0959-6526(02)00062-8

Seman. (2012). Green supply chain management: A review and research direction. *International Journal of Managing Value and Supply Chains, 3*(1), 1-18.

Seuring, S., & Muller, M. (2008). From a literature review to a conceptual framework for supply chain management. *Journal of Cleaner Production, 16*(15), 1699–1710. doi:10.1016/j.jclepro.2008.04.020

Shukla, A. C., Deshmukh, S. G., & Kanda, A. (2009). Environmentally responsive supply chains: Learning from the Indian auto sector. *Journal of Advances in Management Research, 6*(2), 154–171. doi:10.1108/09727980911007181

Simpson, D., & Sampson, D. (2008). Developing strategies for green supply chain management. *Decision Sciences, 39*(4), 12–15.

Singh, A., Singh, B., & Dhingra, A. K. (2012). Drivers and barriers of green manufacturing practices: A survey of Indian industries. *International Journal of Engineering Science, 1*(1), 5–19.

Singh, L. P. (2011). Role of logistics and transportation in green supply chain management: An exploratory study of courier service industry in India. *International Journal of Advanced Engineering Technology, 2*(1), 260–269.

Soler, C., Bergstrom, K., & Shanahan, H. (2010). Green supply chains and the missing link between environmental information and practice. *Business Strategy and the Environment, 19*(1), 14–25.

Srivastava, S. K. (2007). Green supply chain management: A state of the art literature review. *International Journal of Management Reviews, 9*(1), 53–80. doi:10.1111/j.1468-2370.2007.00202.x

Testa, F., & Iraldo, F. (2010). Shadows and lights of GSCM (green supply chain management) determinants and effects of these practices based on a multinational study. *Journal of Cleaner Production, 18*(10/11), 953–962. doi:10.1016/j.jclepro.2010.03.005

UNEP. (2011). *Towards a green economy: Pathways to sustainable development and poverty eradication*. Retrieved November 11, 2012, from www.unep.org/greeneconomy

Vachon, S. (2007). Green supply chain practices and the selection of environmental technologies. *International Journal of Production Research, 45*(18-19), 4357–4379. doi:10.1080/00207540701440303

Vachon, S., & Klassen, R. D. (2006). Green project partnership in the supply chain: The case of the package printing industry. *Journal of Cleaner Production, 14*(6/7), 661–671. doi:10.1016/j.jclepro.2005.07.014

Wahid, N. A., Rahbar, E., & Shyan, T. S. (2011). Factors influencing the green purchase behavior of Penang environmental volunteers. *Journal of International Business Management, 5*(1), 38–49. doi:10.3923/ibm.2011.38.49

Whitelock, V. G. (2012). Alignment between green supply chain management strategy and business strategy. *International Journal of Procurement Management, 5*(4), 430–451. doi:10.1504/IJPM.2012.047198

World Commission for Environment and Development WCED. (1987). *Our common future*. Oxford: Oxford University Press.

Yang, C. L., & Sheu, C. (2011). The effects of environmental regulations on green supply chains. *African Journal of Business Management, 5*(26), 10601–10614.

Yang, W., & Zhang, Y. (2012). Research on factors on green purchasing practices of Chinese. *Journal of Business Management and Economics*, *3*(5), 222–231.

Yen, Y., & Yen, S. (2011). Top-management's role in green purchasing standards in high-tech industrial firms. *Journal of Business Research*, *65*(7), 951–959. doi:10.1016/j.jbusres.2011.05.002

Zhu, Q., & Sarkis, J. (2007). The moderating effects of institutional pressures on emergent green supply chain practices and performance. *International Journal of Production Research*, *45*(18-19), 4333–4355. doi:10.1080/00207540701440345

Zhu, Q., Sarkis, J., & Geng, Y. (2005). Green supply chain management in China: Pressure, practices and performance. *International Journal of Operations & Production Management*, *25*(5), 449–468. doi:10.1108/01443570510593148

Zhu, Q., Sarkis, J., & Lai, K. (2008). Green supply chain management implications for closing the loop. *Transportation Research Part E, Logistics and Transportation Review*, *44*(1), 1–8. doi:10.1016/j.tre.2006.06.003

Zhu, Q., Sarkis, J., Lai, K., & Geng, Y. (2008). The role of organizational size in the adoption of green supply chain management practices in China. *Corporate Social Responsibility and Environmental Management*, *15*(6), 322–337. doi:10.1002/csr.173

Zhu, Q., Sarkis, J., & Lai, K. H. (2012). Examining the effects of green supply chain management practices and their mediations on performance improvements. *International Journal of Production Research*, *50*(5), 1377–1394. doi:10.1080/00207543.2011.571937

KEY TERMS AND DEFINITIONS

Exploratory Factor Analysis: A multivariate statistics method used to uncover the underlying structure of a relatively large set of variables. It is a technique within factor analysis whose overarching goal is to identify the underlying relationships between measured variables.

Green Supply Chain: A set of activities right from sourcing, purchasing raw materials, inbound movement of materials and storage from source to manufacturing units, conversion of these raw materials into desired finished goods, distribution of finished goods from manufacturing units to respective distribution centers or directly to the end users, and finally disposing the waste generated after consumption in an environmental friendly manner.

PLSR: A structural equation modeling (SEM) technique that is used to model a response variable when there are large numbers of predictor variables and those predictors are highly correlated or even collinear.

Compilation of References

Aarts, E., & Korst, J. (1989). *Simulated annealing and boltzmann machines: A stochastic approach to combinatorial optimization and neural computing*. New York, NY: John Wiley & Sons Inc.

Aase, G., John, R. O., & Schniederjans, M. J. (2004). U-shaped assembly line layouts and their impact on labor productivity: An experimental study. *European Journal of Operational Research*, *156*(3), 698–711. doi:10.1016/S0377-2217(03)00148-6

Abbott, J., Stone, M., & Buttle, F. (2001). Customer relationship management in practice – A qualitative study. *Journal of Database Marketing*, *9*(1), 24–34. doi:10.1057/palgrave.jdm.3240055

ACC Sustainability Report. (2010). Retrieved July 11, 2012, from http://www.acclimited.com/newsite/pdf/Sustainable_Report-2010.pdf

Acevedo, O., Ambrose, Z., Flaherty, P. T., Aamer, H., Jain, P., & Sambasivarao, S. V. (2012). Identification of HIV inhibitors guided by free energy perturbation calculations. *Current Pharmaceutical Design*, *18*(9), 199–216. doi:10.2174/138161212799436421 PMID:22316150

Acharya, C., Coop, A., Polli, J. E., & Mackerell, A. D. (2011). Recent advances in ligand-based drug design: relevance and utility of the conformationally sampled pharmacophore approach. *Curr Comp-Aid Drug Design*, *7*(1), 10–22. doi:10.2174/157340911793743547 PMID:20807187

Aerens, R., Eyckens, P., Van Bael, A., & Duflou, J. R. (2009). Force prediction for single point incremental forming deduced from experimental and FEM observations. *International Journal of Advanced Manufacturing Technology*. doi: doi:10.1007/s00170-009-2160-2

Agarwal, A., Harding, D. P., & Schumacher, J. R. (2004). Organizing for CRM. *The McKinsey Quarterly*, *3*, 80–91.

Akshay, R. R., & Monroe, K. B. (1988). The moderating effect of prior knowledge on cue utilization in product evaluations. *The Journal of Consumer Research*, *15*(2), 253–264. doi:10.1086/209162

Al Geddawy, T., & El Maraghy, H. (2010). Design of single assembly line for the delayed differentiation of product variants. *Flexible Services and Manufacturing Journal*, *22*(3-4), 163–182. doi:10.1007/s10696-011-9074-7

Allen, T., & Yu, L. (2007). Paintshop production line optimization using response surface methodology. In *Proceedings of the 2007 Winter Simulation Conference* (pp. 1667-1672). Washington, DC: IEEE Press Piscataway.

Allwood, J. M., Houghton, N. E., & Jackson, K. P. (2005). The design of an incremental forming machine. In *Proceedings of the 11th Conference on Sheet Metal*, Erlangen, (pp. 471-478).

Allwood, J. M., King, G., & Duflou, J. R. (2004). Structured search for applications of the incremental sheet forming process by product segmentation. *IMECH E Proceedings Part B. Journal of Engineering and Manufacture*, *219*(B2), 239–244. doi:10.1243/095440505X8145

Al-Zuheri, A. (2013). *Modelling and optimisation of walking worker assembly line for productivity and ergonomics improvement* (PhD thesis). University of South Australia, Australia.

Ambrogio, G., Filice, L., & Micari, F. De Napol,i L., & Muzzupappa, M. (2006) Some considerations on the precision of incrementally formed double-curvature sheet components. In *Proceedings of the ESAFORM Conference*, Glasgow, UK.

Ambrogio, G., Filice, L., De Napoli, L., & Muzzupappa, M. (2003). Analysis of the influence of some process parameters on the dimensional accuracy in incremental forming by using a reverse engineering technique. In *Proceedings of the AED 2003 Conference*, Prague.

Ambrogio, G., Costantino, I., De Napoli, L., Filice, L., & Muzzupappa, M. (2004). Influence of some relevant process parameters on the dimensional accuracy in incremental forming: A numerical and experimental investigation. *Int. J. Mater. Process. Technol.*, *153–154*, 501–507. doi:10.1016/j.jmatprotec.2004.04.139

Ambrogio, G., De Napoli, L., Filice, L., Gagliardi, F., & Muzzupappa, M. (2005). Application of incremental forming process for high customized medical product manufacturing. *Journal of Materials Processing Technology*, *162*, 156–162. doi:10.1016/j.jmatprotec.2005.02.148

Ambrogio, G., Filice, L., & Micari, F. (2006). A force measuring based strategy for failure prevention in incremental forming. *Journal of Materials Processing Technology*, *177*, 413–416. doi:10.1016/j.jmatprotec.2006.04.076

Amino, H., Lu, Y., Maki, T., Osawa, S., & Fukuda, K. (2002). Dieless NC forming, prototype of automotive service parts. In *Proceedings of the 2nd International Conference on Rapid Prototyping and Manufacturing*, Beijing 2002.

Anderson, A. C. (2003). The process of structure-based drug design. *Chemistry & Biology*, *10*, 787–797. doi:10.1016/j.chembiol.2003.09.002 PMID:14522049

Anderson, E. J., & Ferris, M. C. (1994). Genetic algorithms for combinatorial optimisation: The assembly line balancing problem. *Journal on Computing*, *6*(2), 161–173.

Anderson, E. W., Fornell, C., & Lehman, D. R. (1994). Customer satisfaction, market share, and profitability: Findings from Sweden. *Journal of Marketing*, *58*(3), 53–66. doi:10.2307/1252310

Anderson, E., & Sullivan, M. (1993). The antecedents and consequences of customer satisfaction for firms. *Marketing Science*, *12*(2), 125–143. doi:10.1287/mksc.12.2.125

Anderson, R. E. (1973). Consumer dissatisfaction: The effect of disconfirmed expectancy on perceived product performance. *JMR, Journal of Marketing Research*, *10*(1), 38–44. doi:10.2307/3149407

Andrade-Campos, A. (2011). Development of an optimization framework for parameter identification and shape optimization problems in engineering. *International Journal of Manufacturing, Materials, and Mechanical Engineering*, *1*(1), 57–79. doi:10.4018/ijmmme.2011010105

Andreassen, T. W. (2000). Antecedents to satisfaction with service recovery. *European Journal of Marketing*, *34*(1-2), 156–175. doi:10.1108/03090560010306269

Ang, A. T. H., & Sivakumar, A. I. (2007). Online multiobjective single machine dynamic scheduling with sequence-dependent setups using simulation-based genetic algorithm with desirability function. In *Proceedings of the 2007 Winter Simulation Conference* (pp.1828-1834). Washington, DC: IEEE Press Piscataway.

Antari, J., Samira, C., & Zeroual, A. (2011). Modeling non linear real processes with ANN techniques. In *Proceedings of the 2011 International Conference on Multimedia Computing and Systems* (pp.1-5). Ouarzazate, Morocco: IEEE Press.

Antweiler, W. (1996). The pollution terms of trade. *Economic Systems Research*, *8*, 361–365. doi:10.1080/09535319600000027

Aruldoss, A. V., & Ebenezer, J. A. (2005). A modified hybrid EP-SQP approach for dynamic dispatch with valve-point effect. *International Journal of Electrical Power & Energy Systems*, *27*(8), 594–601. doi:10.1016/j.ijepes.2005.06.006

Ashburn, T. T., & Thor, K. B. (2004). Drug repositioning: Identifying and developing new uses for existing drugs. *Nature Reviews. Drug Discovery*, *3*(8), 673–683. doi:10.1038/nrd1468 PMID:15286734

Assael, H. (1992). *Consumer behavior and marketing action*. Boston, MA: PWS-Kent.

Ata, U. Z., & Toker, A. (2012). The effect of customer relationship management adoption in business-to-business markets. *Journal of Business and Industrial Marketing*, *27*(6), 497–507. doi:10.1108/08858621211251497

Athanassipoulos, A. D. (2000). Customer satisfaction: Cues to support market segmentation and explain switching behavior. *Journal of Business Research*, *47*(3), 191–207. doi:10.1016/S0148-2963(98)00060-5

Attaviriyanupap, K. H., Tanaka, E., & Hasegawa, J. (2002). A hybrid EP – SQP for dynamic economic dispatch with nonsmooth incremental fuel cost function. *IEEE Transactions on Power Systems*, *17*(2), 411–416. doi:10.1109/TPWRS.2002.1007911

Azaouzi, M., & Lebaal, N. (2012). Tool path optimization for single point incremental sheet forming using response surface method. *Simulation Modelling Practice and Theory*, *24*, 49–58. doi:10.1016/j.simpat.2012.01.008

Bacharoudis, E. C., Filios, A. E., Mentzos, M. D., & Margaris, D. P. (2008). Parametric study of a centrifugal pump impeller by varying the outlet blade angle. *The Open Mechanical Engineering Journal*, *2*, 75–83. doi:10.2174/1874155X00802010075

Back, T., Fogel, D. B., & Michalewicz, Z. (1997). *Handbook of evolutionary computation*. New York: Oxford University Press and Institute of Physics. doi:10.1887/0750308958

Bai, C., & Sarkis, J. (2010). Greener supplier development: Analytical evaluation using rough set theory. *Journal of Cleaner Production*, *17*(2), 255–264.

Bailetti, A. J., & Litva, P. F. (1995). Integrating customer requirements into product designs. *Journal of Product Innovation Management*, *12*(1), 3–15. doi:10.1016/0737-6782(94)00021-7

Baines, T. S., & Kay, J. M. (2002). Human performance modelling as an aid in the process of manufacturing system design: A pilot study. *International Journal of Production Research*, *40*(10), 2321–2334. doi:10.1080/00207540210128198

Bakanauskas, A., & Jakutis, A. (2010). Customer value: Determination in undefined environment. *Management of Organizations: Systematic Research*, *53*, 7–18.

Baldwin, C. Y., & Clark, K. B. (2000). *Design rules: The power of modularity*. Cambridge, MA: MIT Press.

Bambach, M., Hirt, G., & Ames, J. (2005). Quantitative validation of FEM simulations for incremental sheet forming using optical deformation measurement. *Advanced Materials Research*, *6-8*, 509–516. doi:10.4028/www.scientific.net/AMR.6-8.509

Band, W. A. (1991). *Creating value for customers*. New York, NY: John Wiley & Sons.

Banerjee, P., Zhou, Y., & Montreuil, B. (1997). Genetically assisted optimization of cell layout and material flow path skeleton. *IIE Transactions*, *29*(4), 277–291. doi:10.1080/07408179708966334

Banker, R. D., & Kauffman, R. J. (1991). *Quantifying the business value of information technology: An illustration of the business value linkage framework*. New York, NY: Stern School of Business, New York University.

Barsky, J. (1995). *World-class customer satisfaction*. Burr Ridge, IL: Irwin Professional.

Bastianoni, S., Pulselli, F. M., & Tiezzi, E. (2004). The problem of assigning responsibility for greenhouse gas emissions. *Ecological Economics*, *49*, 253–257. doi:10.1016/j.ecolecon.2004.01.018

Battini, D., Faccio, M., Ferrari, E., Persona, A., & Sgarbossa, F. (2007). Design configuration for a mixed-model assembly system in case of low product demand. *International Journal of Advanced Manufacturing Technology*, *34*(1-2), 188–200. doi:10.1007/s00170-006-0576-5

Bayazit, N. (2004). Investigating design: A review of forty years of design research. *Design Issues*, *20*(1), 16–29. doi:10.1162/074793604772933739

Beamon, B. M. (1998). Supply chain design and analysis: Models and methods. *International Journal of Production Economics*, *55*(3), 281–294. doi:10.1016/S0925-5273(98)00079-6

Behera, A., Vanhove, H., Lauwers, B., & Duflou, J. (2011). Accuracy improvement in single point incremental forming through systematic study of feature interactions. *Key Engineering Materials*, *473*, 881–888. doi:10.4028/www.scientific.net/KEM.473.881

Bellman, R. E., & Zadeh, L. A. (1970). Decision making in a fuzzy environment. *Management Science*, *17*, 141–164. doi:10.1287/mnsc.17.4.B141

Ben-Gal, I., & Bukchin, J. (2002). The ergonomic design of workstations using virtual manufacturing and response surface methodology. *IE Transactions*, *34*(4), 375–391. doi:10.1080/07408170208928877

Berning, G., Brandenburg, M., Gürsoy, K., Mehta, V., & Tölle, F. J. (2002). An integrated system solution for supply chain optimization in the chemical process industry. *OR-Spektrum*, *24*(4), 371–401. doi:10.1007/s00291-002-0104-4

Besseris, G. J. (2008). Multi-response optimisation using Taguchi method and super ranking concept. *Journal of Manufacturing Technology Management*, *19*(8), 1015–1029. doi:10.1108/17410380810911763

Besterfield, D. H. (1994). *Quality control*. Englewood Cliffs, NJ: Prentice-Hall.

Bettencourt, L. A. (1997). Customer voluntary performance: Customers as partners in service delivery. *Journal of Retailing*, *73*(3), 383–406. doi:10.1016/S0022-4359(97)90024-5

Beume, N., Naujoks, B., & Emmerich, M. (2007). SMS-EMOA: Multiobjective selection based on dominated hypervolume. *European Journal of Operational Research*, *181*(3), 1653–1669. doi:10.1016/j.ejor.2006.08.008

Bezdek, J. C. (1994). What is computational intelligence. In J. M. Zurada, R. J. Marks, & C. J. Robinson (Eds.), *Computational intelligence: Imitating life* (pp. 1–12). New York: IEEE Press.

Bhateja, A. K., Babbar, R., Singh, S., & Sachdeva, A. (2011). Study of green supply chain management in Indian manufacturing industries: A literature review cum an analytical approach for the measurement of performance. *International Journal of Computational Engineering & Management*, *13*, 84–99.

Bhateja, A. K., Babbar, R., Singh, S., & Sachdeva, A. (2012). Study of the critical factor finding's regarding evaluation of green supply chain performance of Indian scenario for manufacturing sector. *International Journal of Computational Engineering & Management*, *15*(1), 74–80.

Bhattacharya, A., Abraham, A., Vasant, P., & Grosan, C. (2007). Meta-learning evolutionary artificial neural network for selecting FMS under disparate level-of-satisfaction of decision maker. *International Journal of Innovative Computing, Information, & Control*, *3*(1), 131–140.

Bhattacharya, A., & Vasant, P. (2007). Soft-sensing of level of satisfaction TOC product-mix decision heuristic using robust fuzzy-LP. *European Journal of Operational Research*, *177*(1), 55–70. doi:10.1016/j.ejor.2005.11.017

Bhattacharya, A., Vasant, P., Sarkar, B., & Mukherjee, S. K. (2006). A fully fuzzified, intelligent theory-of-constraints product-mix decision. *International Journal of Production Research*, *46*(3), 789–815. doi:10.1080/00207540600823187

Bhattacharya, A., Vasant, P., & Susanto, S. (2007). Simulating theory of constraint problem with a novel fuzzy compromise linear programming model. In A. Elsheikh, A. T. Al Ajeeli, & E. M. Abu-Taieh (Eds.), *Simulation and modeling: Current technologies and applications* (pp. 307–336). Hershey, PA: IGI Publishing. doi:10.4018/978-1-59904-198-8.ch011

Bick, G. (2009). Increasing shareholder value through building customer and brand equity. *Journal of Marketing Management*, *25*(1), 117–141. doi:10.1362/026725709X410061

Bierma, T. J., & Wasterstraat, F. L. (1999). Cleaner production from chemical suppliers: Understanding shared savings contracts. *Journal of Cleaner Production*, *7*(2), 145–158. doi:10.1016/S0959-6526(98)00073-0

Bitran, G. R., & Mondschein, S. V. (1997). A comparative analysis of decision making procedures in the catalog sales industry. *European Management Journal*, *15*(2), 105–116. doi:10.1016/S0263-2373(96)00080-1

Bjorklund, M., & Martinsen, U., & Abrahamson. (2012). Performance measurements in the greening of supply chains, Supply Chain Management. *International Journal (Toronto, Ont.)*, *17*(1), 29–39.

Blattberg, R. C., & Deighton, J. (1996). Manage marketing by the customer equity test. *Harvard Business Review*, *74*(4), 136–144. PMID:10158473

Blattberg, R. C., Getz, G., & Thomas, J. S. (2001). *Customer equity: Building and managing relationships as valuable assets*. Boston, MA: Harvard Business School Press.

Bloch, P. H. (1995). Seeking the ideal form: Product design and consumer response. *Journal of Marketing*, *59*(3), 16–29. doi:10.2307/1252116

Bloch, P. H. (2011). Product design and marketing: Reflections after fifteen years. *Journal of Product Innovation Management*, *28*(3), 378–380. doi:10.1111/j.1540-5885.2011.00805.x

Bloemer, J. M. M., & Kasper, H. D. P. (1994). The impact of satisfaction on brand loyalty: Urging on classifying satisfaction and brand loyalty. *Journal of Consumer Satisfaction. Dissatisfaction and Complaining Behavior*, *7*, 152–160.

Blum, C., & Rolli, A. (2003). Metaheuristics in combinatorial optimization: Overview and conceptual comparison. *ACM Computing Surveys*, *35*(3), 268–308. doi:10.1145/937503.937505

Boër, C. R., El-Chaar, J., Imperio, E., & Avai, A. (1991). Criteria for optimum layout design of assembly systems. *CIRP Annals - Manufacturing Technology*, *40*(1), 415-418.

Bohling, T., Bowman, D., LaValle, S., Mittal, V., Narayandas, D., Ramani, G., & Varadarajan, R. (2006). CRM implementation: Effectiveness, issues and insights. *Journal of Service Research*, *9*(2), 184–194. doi:10.1177/1094670506293573

Böhringer, C., Carbone, J. C., & Rutherford, T. F. (2011). *Embodied carbon tariffs* (National Bureau of Economic Research Working Paper No. 17376).

Bojanic, D. C. (1996). Consumer perceptions of price, value and satisfaction in the hotel industry: An exploratory study. *Journal of Hospitality & Leisure Marketing*, *4*(1), 5–22. doi:10.1300/J150v04n01_02

Bok, J. K., Grossmann, I. E., & Park, S. (2000). Supply chain optimization in continuous flexible process networks. *Industrial & Engineering Chemistry Research*, *39*(5), 1279–1290. doi:10.1021/ie990526w

Bolton, R., & Drew, J. H. (1991). A multistage model of customers' assessment of service quality and value. *The Journal of Consumer Research*, *17*(4), 375–384. doi:10.1086/208564

Borenstein, D. (2000). Implementation of an object-oriented tool for the simulation of manufacturing systems and its application to study the effects of flexibility. *International Journal of Production Research*, *38*(9), 2125–2152. doi:10.1080/002075400188537

Borenstein, D., Becker, J. L., & Santos, E. R. (1999). A systemic and integrated approach to flexible manufacturing systems design. *Integrated Manufacturing Systems*, *10*(1), 6–14. doi:10.1108/09576069910370639

Borja de Mozota, B. (2003). *Design management*. New York: Allworth Press.

Bottani, E., & Bertolini, M. (2009). Technical and economic aspect of RFID implementation for asset tracking. *International Journal of RF Technologies: Research and Applications*, *1*(3), 169–193. doi:10.1080/17545730903159034

Bottegoni, G., Kufareva, I., Totrov, M., & Abagyan, R. (2009). Four-dimensional docking: A fast and accurate account of discrete receptor flexibility in ligand docking. *Journal of Medicinal Chemistry*, *52*(2), 397–406. doi:10.1021/jm8009958 PMID:19090659

Bouffioux, C., Eyckens, P., Henrard, C., Aerens, R., Van Bael, A., Sol, H., et al. (2007). Identification of material parameters to predict single point incremental forming forces. In *Proceedings of IDDRG Conference*, Gyor, 2007.

Boulding, W., Kalra, A., Staelin, R., & Zeithaml, V. A. (1993). A dynamic process model of service quality: From expectation. *JMR, Journal of Marketing Research*, *30*(1), 7–27. doi:10.2307/3172510

Boulding, W., Staelin, R., Ehret, M., & Johnston, W. J. (2005). A customer relationship management roadmap: What is known, potential pitfalls, and where to go. *Journal of Marketing*, *69*(4), 155–166. doi:10.1509/jmkg.2005.69.4.155

Box, G. E. P., & Hunter, J. S. (1987). *Empirical model-building and response surface*. New York: John Wiley.

Box, G. E. P., & Wilson, K. B. (1951). On the experimental attainment of optimum conditions. *Journal of the Royal Statistical Society. Series A (General)*, *13*(1), 1–45.

Boyle, M. J. (2004). Using CRM software effectively. *The CPA Journal*, *74*(7), 17.

Bradshaw, D., & Brash, C. (2001). Managing customer relationships in the e-business world: How to personalize computer relationships for increased profitability. *International Journal of Retail and Distribution Management*, *29*(12), 520–529. doi:10.1108/09590550110696969

Bredström, D., Lundgren, J. T., Rönnqvist, M., Carlsson, D., & Mason, A. (2004). Supply chain optimization in the pulp mill industry—IP models, column generation and novel constraint branches. *European Journal of Operational Research*, *156*(1), 2–22. doi:10.1016/j.ejor.2003.08.001

Brenton, P., Edwards-Jones, G., & Jensen, M. F. (2009). Carbon labelling and low-income country exports: A review of the development issues. *Development Policy Review*, *27*, 243–267. doi:10.1111/j.1467-7679.2009.00445.x

Brewton, J., & Schiemann, W. (2003). Measurement: The missing ingredient in today's CRM strategy. *Coastal Management*, *17*(1), 5–14.

British Standard Institution. (2011). [—*Specification for the assessment of the life cycle greenhouse gas emissions of goods and services*. London: British Standards Institution.]. *PAS*, *2050*, 2011.

Brooks, B. R., Bruccoleri, R. E., Olafson, B. D., States, D. J., Swaminathan, S., & Karplus, M. (1983). CHARMM: A program for macromolecular energy, minimization, and dynamics calculations. *Journal of Computational Chemistry*, *4*, 187–217. doi:10.1002/jcc.540040211

Brown, J., Johnson, M., & Reingen, P. H. (1987). Social ties and word-of-mouth referral behavior. *The Journal of Consumer Research*, *14*(3), 350–362. doi:10.1086/209118

Bueren, A., Schierholz, R., Kolbe, L. M., & Brenner, W. (2005). Improving performance of customer-processes with knowledge management. *Business Process Management Journal*, *11*(5), 573–588. doi:10.1108/14637150510619894

Bukchin, J. (1998). A comparative study of performance measures for throughput of a mixed model assembly line in a JIT environment. *International Journal of Production Research*, *36*(10), 2669–2685. doi:10.1080/002075498192427

Bulgak, A. A., & Sanders, J. L. (1990). An analytical assembly systems stations performance model for assembly systems with automatic inspection and repair loops. *Computers & Industrial Engineering*, *18*(3), 373–380. doi:10.1016/0360-8352(90)90059-U

Bulgak, A. A., Tarakc, Y., & Verter, V. (1999). Robust design of asynchronous flexible assembly systems. *International Journal of Production Research*, *4*(3), 3169–3184. doi:10.1080/002075499190220

Butter, R., & Krippendorff, K. (1984). Product semantics—Exploring the symbolic qualities of form. *Journal of the Industrial Designers Society of America*, *3*, 4–9.

Butz, H. E., & Goodstein, L. D. (1996). Measuring customer value: Gaining the strategic advantage. *Organizational Dynamics*, *24*(3), 63–77. doi:10.1016/S0090-2616(96)90006-6

Buzzel, R. D., & Gale, B. T. (1987). *The PIMS principles: Linking strategy to performance*. New York, NY: Free Press.

Byrne, M. D., & Bakir, M. A. (1999). Production planning using a hybrid simulation-analytical approach. *International Journal of Production Economics*, *59*(1-3), 305–311. doi:10.1016/S0925-5273(98)00104-2

Caccetta, L., & Kusumah, Y. S. (2001). *Graph theoretic based heuristics for the facility layout design problems*. Retrieved from orsnz.org.nz

Cadarso, M.-A., López, L.-A., Gómez, N., & Tobarra, M.-A. (2012). International trade and shared environmental responsibility by sector. An application to the Spanish economy. *Ecological Economics*, *83*, 221–235. doi:10.1016/j.ecolecon.2012.05.009

Cailleux, H., Mignot, C., & Kapferer, J. (2009). Is CRM for luxury brands? *Journal of Brand Management*, *16*(5-6), 406–412. doi:10.1057/bm.2008.50

Cai, S. (2009). The importance of customer focus for organizational performance: A study of Chinese companies. *International Journal of Quality & Reliability Management*, *26*(4), 369–379. doi:10.1108/02656710910950351

Callegari, M., Amodio, D., Ceretti, E., & Giardini, C. (2006). *Industrial robotics: Programming, simulation and applications*. Germany: Pro Literatur Verlag.

Carlile, P. R. (2002). A pragmatic view of knowledge and boundaries: Boundary objects in new product development. *Organization Science*, *13*(4), 442–455. doi:10.1287/orsc.13.4.442.2953

Carlyle, W. M., Montgomery, D. C., & Runger, G. C. (2000). Optimization problem and method in quality control and improvement. *Journal of Quality Technology*, *32*(1), 1–17.

Carothers, G. H., & Adams, M. (1991). Competitive advantage through customer value: The role of value-based strategies. In M. Stahl, & G. M. Bounds (Eds.), *Competing globally through customer value* (pp. 32–66). Westport, CT: Quorum Books.

Carson, Y., & Maria, A. (1997). Simulation optimization: Methods and applications. In *Proceedings of the 29th conference on Winter Simulation* (pp. 118-126). Washington, DC, USA: IEEE Computer Society.

Caruana, A., Money, A. H., & Berthon, P. R. (2000). Service quality and satisfaction: The moderating role of value. *European Journal of Marketing*, *34*(11-12), 1338–1352. doi:10.1108/03090560010764432

Caserta, M., & Voß, S. (2010). Metaheuristics: Intelligent problem solving. In V. Maniezzo (Ed.), *Matheuristics* (pp. 1–38). New York: Springer US.

Castillo, E. D., & Montgomery, D. C. (1996). Modified desirability functions for multiple response optimization. *Journal of Quality Technology*, *28*(3), 337–345.

Centeno, G., Silva, M. B., Cristino, V. A. M., Vallellano, C., & Martins, P. A. F. (2012). Hole-flanging by incremental sheet forming. *International Journal of Machine Tools & Manufacture*, *59*, 46–54. doi:10.1016/j.ijmachtools.2012.03.007

Ceretti, E., Giardini, C., & Attanasio, A. (2004). Experimental and simulative results in sheet incremental forming on CNC machines. *Journal of Materials Processing Technology*, *152*, 176–184. doi:10.1016/j.jmatprotec.2004.03.024

Chalmeta, R. (2006). Methodology for customer relationship management. *Journal of Systems and Software*, *79*(7), 1015–1024. doi:10.1016/j.jss.2005.10.018

Chams, M., Hertz, A., & Werra, D. (1987). Some experiments with simulated annealing for coloring graphs. *European Journal of Operational Research*, *32*, 260–266. doi:10.1016/S0377-2217(87)80148-0

Chan, F. T. S., & Smith, A. M. (1993). Simulation approach to assembly line modification: A case study. *Journal of Manufacturing Systems*, *12*(3), 239–245. doi:10.1016/0278-6125(93)90334-P

Chang, T. Z., & Chen, S. J. (1988). Market orientation, service quality and business profitability: A conceptual model and empirical evidence. *Journal of Services Marketing*, *12*(4), 246–264. doi:10.1108/08876049810226937

Chan, J. O. (2005). Toward a unified view of customer relationship management. *Journal of American Academy of Business*, *6*(1), 32–38.

Chao, C. C., Yang, J. M., & Jen, W. Y. (2007). Determining technology trends and forecasts of RFID by a historical review and bibliometric analysis from 1991 to 2005. *Technovation*, *27*(5), 268–279. doi:10.1016/j.technovation.2006.09.003

Chattopadhyay, S. P. (2001). Relationship marketing in an enterprise resource planning environment. *Marketing Intelligence & Planning*, *19*(2), 136–139. doi:10.1108/02634500110385444

Chaudhuri, A., & Holbrook, M. B. (2001). The chain of effects from brand trust and brand affect to brand performance: The role of brand loyalty. *Journal of Marketing*, *65*(2), 81–93. doi:10.1509/jmkg.65.2.81.18255

Chen, S. L., Lin, K. H., & Mittra, R. (2009). A low profile RFID tag designed for metallic objects. In *Proceedings of Microwave Conference, Singapore*, 226-228.

Chen, C. L., & Lee, W. C. (2004). Multi-objective optimization of multi-echelon supply chain networks with uncertain product demands and prices. *Computers & Chemical Engineering*, *28*(6), 1131–1144. doi:10.1016/j.compchemeng.2003.09.014

Chen, G. (1996). Conventional and fuzzy PID controller: An overview. *International Journal of Intelligent Control and Systems*, *1*, 235–246. doi:10.1142/S0218796596000155

Cheng, J. H., Chen, F. Y., & Chang, Y. H. (2008). Airline relationship quality: An examination of Taiwanese passengers. *Tourism Management*, *29*(3), 487–499. doi:10.1016/j.tourman.2007.05.015

Cheng, T., Li, Q., Zhou, Z., Wang, Y., & Bryant, S. H. (2012). Structure-based virtual screening for drug discovery: A problem-centric review. *The AAPS Journal*, *14*(1), 133–141. doi:10.1208/s12248-012-9322-0 PMID:22281989

Chen, H., & Tsao, Y. (2010). Broadband capacitively coupled patch antenna for RFID tag mountable on metallic objects. *Antennas and Wireless Propagation Magazine*, *9*, 489–492. doi:10.1109/LAWP.2010.2050854

Chen, I. J., & Popovich, K. (2003). Understanding customer relationship management (CRM): People, process and technology. *Business Process Management Journal*, *9*(5), 672–688. doi:10.1108/14637150310496758

Chen, J. J., Li, M. Z., Liu, W., & Wang, C. T. (2005). Sectional multipoint forming technology for large-size sheet metal. *J. Adv. Manuf. Technol.*, *25*, 935–939. doi:10.1007/s00170-003-1924-3

Chen, S. L., & Lin, K. H. (2008). A slim RFID tag antenna design for metallic object applications. *IEEE Antennas and Wireless Propagation Letters*, *7*, 729–732. doi:10.1109/LAWP.2008.2009473

Chen, X., Cheng, L. K., Fung, K. C., Lau, L. J., Sung, Y.-W., & Zhu, K. et al. (2012). Domestic value added and employment generated by Chinese exports: A quantitative estimation. *China Economic Review*, *23*, 850–864. doi:10.1016/j.chieco.2012.04.003

Chesborough, H. (2003). *Open innovation: The new imperative for creating and profiting from technological.* Boston, MA: Harvard Business School Press.

China 863. (2013). Retrieved October 31, 2013, from http://www.863.gov.cn/

China 973. (2013). Retrieved October 31, 2013, from http://www.973.gov.cn

China Briefing. (2012). *China releases 12th five-year plan for energy saving and emission reduction.* Retrieved May 18, 2013, from http://www.china-briefing.com/news/2012/05/04/china-releases-12th-five-year-plan-for-foreign-trade-development.html

China NSFC. (2013). Retrieved October 31, 2013, from http://www.nsfc.gov.cn

Choo, H. J., Moon, H., Kim, H., & Yoon, N. (2012). Luxury customer value. *Journal of Fashion Marketing and Management*, *16*(1), 81–101. doi:10.1108/13612021211203041

Chorin, A. J., & Marsden, J. E. (1979). *A mathematical introduction to fluid mechanics.* New York: Springer-Verlag. doi:10.1007/978-1-4684-0082-3

Christen, M., Hünenberger, P. H., Bakowies, D., Baron, R., Bürgi, R., & Geerke, D. P. et al. (2005). The GROMOS software for biomolecular simulation: GROMOS05. *Journal of Computational Chemistry*, *26*(16), 1719–1751. doi:10.1002/jcc.20303 PMID:16211540

Christensen, C. M. (2003). *The innovator's dilemma: When new technologies cause great firms to fail.* Boston, MA: Harvard Business School Press.

Ciborra, C., & Failla, A. (2000). Infrastructure as a process: The case of CRM in IBM. In C. Ciborra (Ed.), *From control to drift: The dynamics of corporate information infrastructures* (pp. 105–124). Oxford, UK: Oxford University Press.

Clark, K., & Fujimoto, T. (1990). The power of product integrity. *Harvard Business Review*, (November-December): 107–118. PMID:10107956

Clean Energy Regulator. (2013). *About the carbon pricing mechanism.* Australian Government, Australia. Retrieved May 18, 2013, from http://www.cleanenergyregulator.gov.au/Carbon-Pricing-Mechanism/About-the-Mechanism/Pages/default.aspx

Cohen, M. A., & Vandenbergh, M. P. (2012). The potential role of carbon labelling in a green economy. *Energy Economics*, *34*, 553–563. doi:10.1016/j.eneco.2012.08.032

Coit, D., Jackson, B. T., & Smith, A. E. (1998). Static neural network process models: Considerations and case studies. *International Journal of Production Economics*, *36*(11), 2953–2967. doi:10.1080/002075498192229

Cole, M. A. (2004). Trade, the pollution haven hypothesis and the environmental Kuznets curve: Examining the linkages. *Ecological Economics*, *48*, 71–81. doi:10.1016/j.ecolecon.2003.09.007

Colorni, A., Dorigo, M., & Maniezzo, V. (1991). distributed optimization by ant colonies. In *Proceedings of the first European Conference of Artificial Intelligence.* (pp. 134-142). Paris: Elsevier Publishing.

Commission of the European Communities. (2008). *Communication from the Commission to the European Parliament, the Council, the European Economic and Social Committee and the Committee of the Regions on the Sustainable Consumption and Production and Sustainable Industrial Policy action plan.* Brussels, Belgium: Author.

Cooper, R., Prendiville, A., & Jones, T. (1995). High technological NPD. *CoDesign, 3*, 14–22.

Cox, D., & William, L. (1987). Product novelty: Does it moderate the relationship between ad attitudes and brand attitudes? *Journal of Advertising, 16*, 39–44. doi:10.1080/00913367.1987.10673084

Coyne, K. (1989). Beyond service fads – Meaningful strategies for the real world. *Sloan Management Review, 30*(4), 69–76.

CPLEX. (2009). *Using the CPLEX callable library.* Incline Village, NV: CPLEX Optimization, Inc.

Creusen, H., & Schoormans, L. (2005). The different role of product appearance in consumer choice. *Journal of Product Innovation Management, 22*(1), 63–82. doi:10.1111/j.0737-6782.2005.00103.x

Creusen, M. E. H. (2011). Research opportunities related to consumer response to product design. *Journal of Product Innovation Management, 28*(3), 405–408. doi:10.1111/j.1540-5885.2011.00812.x

Cronin, J. J., Brady, M. K., & Hult, G. T. M. (2000). Assessing the effects of quality, value, and customer satisfaction on consumer behavioral intentions in service environments. *Journal of Retailing, 76*(2), 193–218. doi:10.1016/S0022-4359(00)00028-2

Crosby, L. A., Evans, K. R., & Cowles, D. (1990). Relationship quality in services selling: An interpersonal influence perspective. *Journal of Marketing, 54*(3), 68–81. doi:10.2307/1251817

Csikszentmihalyi, M., & Rochberg-Halton, E. (1981). *The meaning of things: domestic symbols and the self.* Cambridge, UK: Cambridge University Press. doi:10.1017/CBO9781139167611

Ćuković, S., Devedžić, G., & Ghionea, I. (2010). Automatic determination of grinding tool profile for helical surfaces machining using CATIA/VB Interface. *U.P.B. Scientific Bulletin, Series D, 72*(2), 85–96.

Curtis, C. (2013). *A plan to revitalize American manufacturing.* The White House Blog. Retrieved Oct 31, 2013, from http://www.whitehouse.gov/blog/2013/02/13/plan-revitalize-american-manufacturing

D'Angelo, A., Gastaldi, M., & Levialdi, N. (1998). Performance analysis of a flexible manufacturing system: A statistical approach. *International Journal of Production Economics, 56-57*, 47–59. doi:10.1016/S0925-5273(96)00115-6

Dahl, D. W. (2011). Clarity in defining product design: inspiring research opportunities for the design process. *Journal of Product Innovation Management, 28*(3), 425–427. doi:10.1111/j.1540-5885.2011.00816.x

Dahl, D. W., Chattopdhyay, A., & Gorn, G. (1999). The use of visual mental imagery in new product design. *JMR, Journal of Marketing Research, 36*(1), 18–28. doi:10.2307/3151912

Dai, Q. Y., Zhong, R. Y., Huang, G. Q., Qu, T., Zhang, T., & Luo, T. Y. (2012). Radio frequency identification-enabled real-time manufacturing execution system: A case study in an automotive part manufacturer. *International Journal of Computer Integrated Manufacturing, 25*(1), 51–56. doi:10.1080/0951192X.2011.562546

Dan Zhang, P. H., & Kotabe, M. (2011). Marketing–industrial design integration in new product development: The case of China. *Journal of Product Innovation Management, 28*(3), 360–373. doi:10.1111/j.1540-5885.2011.00803.x

Dan, W. (2010). Carbon emissions embodied in China's international trade based on input-output method. In *Proceedings of the 8th International Conference on Innovation & Management,* (pp. 81-85).

Darnall, N., Jolley, G. J., & Handfield, R. (2008). Environmental management systems and green supply chain management: Complements for sustainability? *Business Strategy and the Environment, 17*(1), 30–45. doi:10.1002/bse.557

Dashchenko, A. I., & Loladze, T. N. (1991). Choice of optimal configurations for flexible (readjustible) assembly lines by purposeful search. *CIRP Annals - Manufacturing Technology, 40*(1), 13-16.

Das, I., & Dennis, J. E. (1998). Normal-boundary intersection: A new method for generating the Pareto surface in nonlinear multicriteria optimization problems. *SIAM Journal on Optimization, 8*(3), 631–657. doi:10.1137/S1052623496307510

Dawkins, R. (1976). *The selfish gene*. Oxford: Oxford Press.

Day, G. S. (1990). *Market driven strategy: Processes for creating value*. New York, NY: Free Press.

Day, G. S., & Wensley, R. (1988). Assessing advantage: A framework for diagnosing competitive advantage. *Journal of Marketing, 52*(2), 1–20. doi:10.2307/1251261

De Ruyter, K., Bloemer, J., & Pascal, P. (1997). Merging service quality and service satisfaction: An empirical test of an integrative model. *Journal of Economic Psychology, 18*(4), 187–206. doi:10.1016/S0167-4870(97)00014-7

Dearnley, R. W., & Barel, A. F. (1989). A broad-band transmission line model for a rectangular microstrip antenna. *IEEE Transactions on Antennas and Propagation, 37*(1), 6–15. doi:10.1109/8.192158

Deb, K., & Jain, S. (2002). *Running performance metrics for evolutionary multiobjective optimization* (KanGAL Report No. 2002004). Kanpur, India: Indian Institute of Technology.

Deb, K., Pratap, A., Agarwal, S., & Meyarivan, T. (2002). A fast and elitist multiobjective genetic algorithm: NSGA-II. *IEEE Transactions on Evolutionary Computation, 6*(2), 182–197. doi:10.1109/4235.996017

Deb, S., Parra-Castillo, J. R., & Ghosh, K. (2011). An integrated and intelligent computer-aided process planning methodology for machined rotationally symmetrical parts. *International Journal of Advanced Manufacturing Systems, 13*(1), 1–26.

Decultot, N., Velay, V., Robert, L., Bernhart, G., & Massoni, E. (2008). *Behaviour modeling of aluminium alloy sheet for single point incremental forming*. Paper presented at the 11th Esaform Conference on Material Forming, Lyon.

Dejardin, S., Thibaud, S., & Gelin, J. C. (2009). Experimental investigations and numerical analysis for improving knowledge of incremental sheet forming process for sheet metal parts. *Journal of Materials Processing Technology*. doi: doi:10.1016/j.jmatprotec.2009.09.025

Dell'Era, C., & Verganti, R. (2007). Strategies of innovation and imitation of product languages. *Journal of Product Innovation Management, 24*, 580–599. doi:10.1111/j.1540-5885.2007.00273.x

Dennett, D. C. (1995). *Darwin's dangerous idea*. New York: Touchstone.

Derringer, G. C., & Suich, R. (1980). Simultaneous optimization of several response variables. *Journal of Quality Technology, 12*(4), 214–219.

Dess, G. G., & Robinson, R. B. (1984). Measuring organizational performance in the absence of objective measures: the case of the privately-held firm and conglomerate business unit. *Strategic Management Journal, 5*(3), 265–273. doi:10.1002/smj.4250050306

DiGiampaolo, E., & Martinelli, F. (2012). A passive UHF-rfid system for the localization of an indoor autonomous vehicle. *IEEE Transactions on Industrial Electronics, 9*(10), 3961–3970. doi:10.1109/TIE.2011.2173091

Dikos, A., Nelson, P. C., Tirpak, T. M., & Wang, W. (1997). Optimization of high-mix printed circuit card assembly using genetic algorithms. *Annals of Operations Research, 75*(0), 303–324. doi:10.1023/A:1018919815515

Dimitriadis, S. (2010). Testing perceived relational benefits as satisfaction and behavioral outcomes drivers. *International Journal of Bank Marketing, 28*(4), 207–213. doi:10.1108/02652321011054981

Ding, Y., Ceglarek, D., & Shi, J. (2002). Design evaluation of multi-station assembly processes by using state space approach. *Journal of Mechanical Design, 124*(3), 408–418. doi:10.1115/1.1485744

Dissou, Y., & Eyland, T. (2011). Carbon control policies, competitiveness, and border tax adjustment. *Energy Economics, 33*, 556–564. doi:10.1016/j.eneco.2011.01.003

Dobkin, D. M., & Weigand, S. M. (2005). Environmental effects on RFID tag antennas. *IEEE MTT-S International Microwave Symposium Digest, 135-138.*

Dodds, W. B., Monroe, K. B., & Grewal, D. J. (1991). Effects of price, brand, and store information on buyers' product evaluations. *JMR, Journal of Marketing Research, 28*(3), 307–319. doi:10.2307/3172866

Donaldson, B., & O' Toole, T. (2002). *Strategic marketing relationship.* Chichester, UK: John Wiley & Sons.

Dorsch, M. J., Swanson, S. R., & Kelley, S. W. (1998). The role of relationship quality in the satisfaction of vendors as perceived by customers. *Journal of the Academy of Marketing Science, 26*(2), 128–142. doi:10.1177/0092070398262004

Dror, R. O., Pan, A. C., Arlow, D. H., Borhani, D. W., Maragakis, P., & Shan, Y. et al. (2011). Pathway and mechanism of drug binding to G-protein-coupled receptors. *Proceedings of the National Academy of Sciences of the United States of America, 108*(32), 13118–13123. doi:10.1073/pnas.1104614108 PMID:21778406

Duarte, S., Cabrita, R., & Machado, V. C. (2011). Exploring lean and green supply chain performance using balanced score card perspective. In *Proceedings of the 2011 International Conference on Industrial Engineering and Operations Management,* Kuala Lumpur, Malaysia.

Dubey, R., & Bag, S. (2013). Exploring the dimensions of sustainable practices: An empirical study on Indian manufacturing firms. *International Journal of Operations and Quantitative Management, 19*(2), 123–146.

Dües, C. M., Tan, K. H., & Lim, M. (2013). Green as the new lean: How to use lean practices as a catalyst to greening your supply chain. *Journal of Cleaner Production, 40*, 93–100. doi:10.1016/j.jclepro.2011.12.023

Duflou, J. R., Lauwers, B., & Verbert, J. (2005b). Medical application of single point incremental forming: Cranial plate manufacturing. In *Proceedings of the 2005 VRAP Conference,* Leiria, Portugal. (pp. 161-164).

Duflou, J. R., Sol, H., Van Bael, A., & Habraken, A. M. (2003). *Description of the SeMPeR projet (Sheet Metal oriented Prototyping and Rapid manufacturing).* SBO-project financed by the IWT institute, 2003.

Duflou, J., Verbert, B., Belkassem, J., Gu, Sol, H., Henrard, C., & Habraken, A. M. (2008). Process window enhancement for single point incremental forming through multi-step toolpaths. *CIRP Annals Manufacturing Technology, 57,* 253-256.

Duflou, J. R., Szekeres, A., & VanHerck, A. (2005a). force measurements for single point incremental forming and experimental study. *Journal of Advanced Materials Research, 6-8,* 441–448. doi:10.4028/www.scientific.net/AMR.6-8.441

Duflou, J., Kellens, K., & Dewulf, W. (2011). Unit process impact assessment for discrete part manufacturing: A state of the art. *CIRP Journal of Manufacturing Science and Technology, 4*(2), 129–135. doi:10.1016/j.cirpj.2011.01.008

Duflou, J., Tuncol, Y., Szekeres, A., & Vanherck, P. (2007). Experimental study on force measurements for single point incremental forming. *Journal of Materials Processing Technology, 189,* 65–72. doi:10.1016/j.jmatprotec.2007.01.005

Durante, M., Formisano, A., Langella, A., & Minutolo, F. (2008). The influence of tool rotation on an incremental forming process. *Journal of Materials Processing Technology,* 2008.

Durrant, J. D., & McCammon, J. A. (2011). Molecular dynamics simulations and drug discovery. *BMC Biology, 9,* 71. doi:10.1186/1741-7007-9-71 PMID:22035460

Eglese, R. W. (1990). Simulated annealing: A tool for operational research. *European Journal of Operational Research, 46,* 271–281. doi:10.1016/0377-2217(90)90001-R

Eguia, I., Lozano, S. R. J., & Guerrero, F. (2011). A methodological approach for designing and sequencing product families in reconfigurable disassembly systems. *Journal of Industrial Engineering and Management, 4*(3), 418–435. doi:10.3926/jiem.2011.v4n3.p418-435

Emmens, W. C., & van dn Boogaard, A. H. (2008). Tensile tests with bending: A mechanism for incremental forming. In *Proceedings of the 11th ESAFORM conference on material forming*, Lyon, France.

Engel, J. F., Blackwell, R. D., & Miniard, P. W. (1990). *Consumer behavior*. Chicago, IL: Dryden Press.

Eschenauer, H., Koski, J., & Osyczka, A. (1990). *Multicriteria design optimization*. Berlin: Springer-Verlag. doi:10.1007/978-3-642-48697-5

Eskafi, M., Hosseini, S., & Yazd, A. M. (2013). The value of telecom subscribers and customer relationship management. *Business Process Management Journal, 19*(4), 737–748. doi:10.1108/BPMJ-Feb-2012-0016

European Commission. (2006, March). *Environmental fact sheet: Industrial development*. Retrieved April 8, 2013, from http://ec.europa.eu.pdf

Faheem, W., Castano, J. F., Hayes, C. C., & Gaines, D. M. (1998). What is a manufacturing interaction? In *Proceedings of 1998 ASME Design Engineering Technical Conferences* (pp. 1-6), Atlanta, GA.

Farhan, U. H., Tolourei-Rad, M., & O'Brien, S. (2012). An automated approach for assembling modular fixtures using SolidWorks. *World Academy of Science. Engineering and Technology, 72*, 394–397.

Farias, D. P. D. (2002). *The linear programming approach to approximate dynamic programming: Theory and application* (Doctoral dissertation). Stanford University, Palo Alto, California.

Fecikova, I. (2004). An index method for measurement of customer satisfaction. *The TQM Magazine, 16*(1), 57–66. doi:10.1108/09544780410511498

Feng, C.-X. J., & Wang, X.-F. (2004). Data mining techniques applied to predictive modeling of the knurling process. *IIE Transactions, 36*(3), 253–263. doi:10.1080/07408170490274214

Ferng, J.-J. (2003). Allocating the responsibility of CO_2 over-emissions from the perspectives of benefit principle and ecological deficit. *Ecological Economics, 46*, 121–141. doi:10.1016/S0921-8009(03)00104-6

Feyzbakhsh, S. A., & Matsui, M. (1999). Adam—Eve-like genetic algorithm: A methodology for optimal design of a simple flexible assembly system. *Computers & Industrial Engineering, 36*(2), 233–258. doi:10.1016/S0360-8352(99)00131-X

Field, J., & Sroufe, R. (2007). The use of recycled materials in manufacturing: Implications for supply chain management. *International Journal of Production Research, 45*(18/19), 4439–4463. doi:10.1080/00207540701440287

Filice, L., Ambrogio, G., & Micari, F. (2006). On-line control of single point incremental forming operations through punch force monitoring. *CIRP Annals - Manufacturing Technology, 55*(1), 245–248.

Filice, L., Fratini, L., & Micari, F. (2002). Analysis of material formability in incremental forming. *Annals of the CIRP, 51*, 199–202. doi:10.1016/S0007-8506(07)61499-1

Fishburn, P. C. (1967). *Additive utilities with incomplete product set: Applications to priorities and assignments*. Baltimore, MD: Operations Research Society of America.

Fitzell, P. (1998). *The explosive growth of private labels in North America*. New York, NY: Global Books.

Flake, G. W. (1998). *The computational beauty of nature: Computer explorations of fractals, chaos, complex systems, and adaptation*. Cambridge, MA: MIT Press.

Flint, D. J., Woodruff, R. B., & Gardial, S. F. (1997). Customer value change in industrial marketing relationships. *Industrial Marketing Management, 26*(2), 163–175. doi:10.1016/S0019-8501(96)00112-5

Fogel, D. B. (1995). *Evolutionary computation: Toward a new philosophy of machine intelligence*. Piscataway, NJ: IEEE Press. doi:10.1109/ICEC.1995.489143

Fornell, C. (1992). A national customer satisfaction barometer: The Swedish experience. *Journal of Marketing, 56*(1), 6–21. doi:10.2307/1252129

Fornell, C., Johnson, M. D., Anderson, E. W., Cha, J., & Bryant, B. E. (1996). The American customer satisfaction index: Nature, purpose, and findings. *Journal of Marketing, 60*(4), 7–18. doi:10.2307/1251898

Fornell, C., & Wernerfelt, B. (1987). Defensive marketing strategy by customer complaint management: A theoretical analysis. *JMR, Journal of Marketing Research, 24*(4), 337–346. doi:10.2307/3151381

Forza, C., & Filippini, R. (1998). TQM impact on quality conformance and customer satisfaction: A causal model. *International Journal of Production Economics, 55*(1), 1–20. doi:10.1016/S0925-5273(98)00007-3

Fountas, N. A. (2008). *Advanced operations supporting process planning for aircraft structural parts in aerospace manufacturing* (MSc Thesis). Faculty of Science Engineering and Computing, Kingston University, London, UK.

Fountas, N. A., Krimpenis, A. A., Vaxevanidis, N. M., & Davim, J. P. (2012). Single and multi-objective optimization methodologies in CNC machining. In J. P. Davim (Ed.), *Statistical and computational techniques in manufacturing* (pp. 187–218). New York: Springer. doi:10.1007/978-3-642-25859-6_5

Franzen, V., Kwiatkowski, L., Martins, P. A. F., & Tekkaya, A. E. (2009). Single point incremental forming of PVC. *Journal of Materials Processing Technology, 209*, 462–469. doi:10.1016/j.jmatprotec.2008.02.013

Fratini, L., Ambrogio, G., Di Lorenzo, R., Filice, L., & Micari, F. (2004). Influence of mechanical properties of the sheet material on formability in single point incremental forming. *Journal of Materials Processing Technology, 153–154*, 501–507.

Fu, L. (1994). *Neural networks in computer intelligence*. New York: McGraw Hill.

Gadidov, R., & Wilhelm, W. (2000). A cutting plane approach for the single-product assembly system design problem. *International Journal of Production Research, 38*(8), 1731–1754. doi:10.1080/002075400188564

Galbreath, J., & Rogers, T. (1999). Customer relationship leadership. *The TQM Magazine, 11*(3), 161–171. doi:10.1108/09544789910262734

Gale, B. T. (1994). *Managing customer value*. New York, NY: Free Press.

GAMS. (2009). *GAMS- A user' s guide*. Washington, DC: GAMS Development Corporation.

Ganesan, T., Vasant, P., & Elamvazuthi, I. (2011). Optimization of nonlinear geological structure mapping using hybrid neuro-genetic techniques. *Mathematical and Computer Modelling, 54*(11-12), 2913–2922. doi:10.1016/j.mcm.2011.07.012

Ganesan, T., Vasant, P., & Elamvazuthi, I. (2012). Hybrid neuro-swarm optimization approach for design of distributed generation power systems. *Neural Computing & Applications, 23*(1), 105–117. doi:10.1007/s00521-012-0976-4

Gangele, A., & Verma, A. (2011). The investigation of green supply chain management practices in pharmaceutical manufacturing industry through waste minimization. *International Journal of Industrial Engineering and Technology, 3*(4), 403–415.

Gaukler, G. M. (2011). Item-level RFID in a retail supply chain with stock-out-based substitution. *IEEE Trans. Industrial Informatics, 7*(2), 362–370. doi:10.1109/TII.2010.2068305

Gavrilas, M. (2010). Heuristic and metaheuristic optimization techniques with application to power systems. In the *Proceedings of 12th WSEAS International Conference on Mathematical Methods and Computational Techniques in Electrical Engineering* (pp. 95-103). Timisoara, Romania: WSEAS Press.

Gavronski, I. (2011). A resource-based view of green supply management. *Transportation Research Part E, Logistics and Transportation Review, 47*(6), 872–885. doi:10.1016/j.tre.2011.05.018

Geels, F. W. (2004). From sectoral systems of innovation to socio- technical systems. Insights about dynamics and change from sociology and institutional theory. *Research Policy, 33*, 897–920. doi:10.1016/j.respol.2004.01.015

Gee, R., Coates, G., & Nicholson, M. (2008). Understanding and profitably managing customer loyalty. *Marketing Intelligence & Planning, 25*(4), 359–374. doi:10.1108/02634500810879278

Gemser, G., & Leenders, M. A. A. M. (2001). How integrating industrial design in the product development process impacts on company performance. *Journal of Product Innovation Management, 18*(1), 28–38. doi:10.1016/S0737-6782(00)00069-2

Ghosh, M., Luo, D., Siddiqui, M. S., & Zhu, Y. (2012). Border tax adjustments in the climate policy context: CO2 versus broad-based GHG emission targeting. *Energy Economics, 34*, 5154–5167. doi:10.1016/j.eneco.2012.09.005

Giovanni, P. D., & Vinzi, V. E. (2012). Covariance versus component-based estimates of performance in green supply chain management. *International Journal of Production Economics, 135*(2), 907–916. doi:10.1016/j.ijpe.2011.11.001

Glover, F. (1989). Tabu search: Part 1. *ORSA Journal on Computing, 1*(3), 190–206. doi:10.1287/ijoc.1.3.190

Godard, O. (2007). *Unilateral European post-Kyoto climate policy and economic adjustment at EU borders* (EDF — Ecole Polytechnique, Cahier no. DDX 07-15).

Goguen, J. A. (1969). The logic of inexact concepts. *Syntheses, 19*, 325–373. doi:10.1007/BF00485654

Goldberg, D. E. (1989). *Genetic algorithms in search, optimization and machine learning*. Boston, MA: Longman.

Goldberg, D. E. (1994). Genetic and evolutionary algorithms come of age. *Communications of the ACM, 37*(3), 113–119. doi:10.1145/175247.175259

Goodrich, L. L. (1995). The design of the decade: Quantifying design impact over ten years. *Design Management Journal, 5*(2), 47.

Gorb, P., & Dums, A. (1987). Silent design. *Design Studies, 8*, 150–156. doi:10.1016/0142-694X(87)90037-8

Gottwald, S. (1993). *Fuzzy sets and fuzzy logic: The foundations of application—From a mathematical point of view*. Wiesbaden, Germany: Vieweg & Sohn Verlagsgesellschaft mbH. doi:10.1007/978-3-322-86812-1

Goulder, L. H., & Parry, I. W. H. (2008). *Instrument choice in environmental policy* (Discussion Paper DP 08-07). Washington, DC: Resources for the Future.

Gounaris, S. P., & Stathakopoulos, V. (2004). Antecedents and consequences of brand loyalty: An empirical study. *Journal of Brand Management, 11*(4), 283–306. doi:10.1057/palgrave.bm.2540174

Government of UK. (2005). *Securing the future: Delivering UK sustainable development strategy*. United Kingdom: Government of UK.

Grant, A. W. H., & Schlesinger, L. A. (1995). Realize your customer's full profit potential. *Harvard Business Review, 73*(5), 59–62.

Graves, S. C., Kletter, D. B., & Hetzel, W. B. (1998). A dynamic model for requirements planning with application to supply chain optimization. *Operations Research, 46*(3-Supplement-3), S35-S49.

Greenhouse Gas Protocol. (2011). *Product life cycle accounting and reporting standard*. Washington, DC: World Resources Institute and World Business Council for Sustainable Development.

Greve, H. R. (2007). Exploration and exploitation in product innovation. *Industrial and Corporate Change, 16*(5), 945–975. doi:10.1093/icc/dtm013

Griffin, A., & Hauser, J. R. (1996). Integrating R&D and marketing: A review and analysis of the literature. *Journal of Product Innovation Management, 13*, 191–213. doi:10.1111/1540-5885.1330191

Gronroos, C. (1990). *Service management and marketing: Managing the moments of truth in service competition*. Lexington, MA: Lexington Books.

Gros, D., & Egenhofer, C. (2010). *Global welfare implications for carbon border taxes climate change and trade taxing carbon at the border?* Brussels, Belgium: Centre for European Policy Studies.

Grossbart, S., Mittelstaedt, R. A., Curtis, W. W., & Rogers, R. D. (1975). Environmental sensitivity and shopping behaivor. *Journal of Business Research, 3*(4), 281–294. doi:10.1016/0148-2963(75)90010-7

Gummesson, E. (2004). Return on relationships (ROR), The value of relationship marketing and CRM in business-to-business contexts. *Journal of Business and Industrial Marketing, 19*(2), 136–148. doi:10.1108/08858620410524016

Gunter, B. (1988). Signal-to-noise ratios, performance criteria, and transformations[Discussion]. *Technometrics*, *30*(1), 32–35. doi:10.2307/1270316

Guo, J., Zou, L.-L., & Wei, Y.-M. (2010). Impact of intersectoral trade on national and global CO2 emissions: An empirical analysis of China and US. *Energy Policy*, *38*, 1389–1397. doi:10.1016/j.enpol.2009.11.020

Guo, X., Hao, A. W., & Shang, X. (2011). Consumer perceptions of brand functions: An empirical study in China. *Journal of Consumer Marketing*, *28*(4), 169–279. doi:10.1108/07363761111143169

Guo, Z. X., Wong, W. K., Leung, S. Y. S., & Fan, J. T. (2009). Intelligent production control decision support system for flexible assembly lines. *Expert Systems with Applications*, *36*(3), 4268–4277. doi:10.1016/j.eswa.2008.03.023

Guzmán, C., Gu, J., Duflou, J., Vanhove, H., Flores, P. A., & Habraken, A. M. (2012). Study of the geometrical inaccuracy on a SPIF two-slope pyramid by finite element simulations. *International Journal of Solids and Structures*, *49*, 3594–3604. doi:10.1016/j.ijsolstr.2012.07.016

Hall, J. (2000). Environmental supply chain dynamics. *Journal of Cleaner Production*, *8*, 455–471. doi:10.1016/S0959-6526(00)00013-5

Hallowell, R. (1996). The relationships of customer satisfaction, customer loyalty, and profitability: An empirical study. *International Journal of Service Industry Management*, *7*(4), 27–42. doi:10.1108/09564239610129931

Ham, M., & Jeswiet, J. (2007). Forming limit curves in single point incremental forming. *Annals of the CIRP, 56*.

Hamel, G., & Prahalad, C. K. (1994). *Computing for the future*. Boston, MA: Harvard Business School Press.

Hansemark, O. C., & Albinsson, M. (2004). Customer satisfaction and retention: The experience of individual employees. *Managing Service Quality*, *14*(1), 40–57. doi:10.1108/09604520410513668

Hargadon, A., & Sutton, R. I. (1997). Technological brokering and innovation in a product development firm. *Administrative Science Quarterly*, *42*(4), 716–749. doi:10.2307/2393655

Harik, R. F., Derigent, W. J. E., & Ris, G. (2008). Computer aided process planning in aircraft manufacturing. *Computer-Aided Design & Applications*, *5*(6), 953–962. doi:10.3722/cadaps.2008.953-962

Harrington, E. C. J. (1965). The desirability function. *Industrial Quality Control*, *21*(10), 494–498.

Hart, W. E. (1998). On the application of evolutionary pattern search algorithms. In *Proceedings of the 7th International Conference on Evolutionary Programming* (pp. 301-312). Berlin: Springer.

Hartley, R. (1992). *Concurrent engineering*. Combridge, USA: Productivity Press.

Hartline, M. D., Maxham, J. G., & McKee, D. O. (2000). Corridors of influence in the dissemination of customer-oriented strategy to customer contact service employees. *Journal of Marketing*, *64*(2), 35–50. doi:10.1509/jmkg.64.2.35.18001

Hart, S., & Service, L. (1988). The effects of managerial attitude to design on company. *Journal of Marketing Management*, *4*(2), 217–230. doi:10.1080/0267257X.1988.9964070

Hart, T. N., & Read, R. J. (1992). A multiple-start Monte Carlo docking method. *Proteins*, *13*(3), 206–222. doi:10.1002/prot.340130304 PMID:1603810

Hayes, R. (1990). Design: Putting class into 'world class'. *Design Management Journal*, *1*(2), 8–14.

Henrard, C., Bouffioux, C., Eyckens, P., Soly, H., Duflou, J., & Van Houtte, P. et al. (2011). Forming forces in single point incremental forming: Prediction by finite element simulations, validation and sensitivity. *Computational Mechanics*, *47*(5), 573–590. doi:10.1007/s00466-010-0563-4

Hererra, F., & Lozano, M. (1996). Adaptation of genetic algorithm parameters based on fuzzy logic controllers. In F. Hererra, & J. L. Verdegay (Eds.), *Genetic algorithms and soft computing* (pp. 95–125).

Herstatt, C., Verworn, B., & Nagahira, A. (2004). Reducing project related uncertainty in the "fuzzy front end" of innovation: A comparison of German and Japanese product innovation projects. *International Journal of Product Development*, *1*(1), 43–65. doi:10.1504/IJPD.2004.004890

Hertenstein, J. H., & Platt, M. B. (1997). Developing a strategic design culture. *Design Management Journal*, *2*(2), 10–19.

Hertenstein, J. H., Platt, M. B., & Veryzer, R. W. (2005). The impact of industrial design effectiveness on corporate financial performance. *Journal of Product Innovation Management*, *22*(1), 3–21. doi:10.1111/j.0737-6782.2005.00100.x

Heskett, J. L., Sasser, W. E., & Schlesinger, L. A. (1997). *The service profit chain: How leading companies link profit and growth to loyalty, satisfaction and value*. New York, NY: Free Press.

Higgins, K. T. (1998). The value of customer value analysis. *Marketing Research*, *10*(4), 39–44.

Hiller, F. S., & Liebermann, G. J. (1999). *Operations research*. New Delhi, India: CBS Publications and Distributions.

Hillier, F. S., & So, K. C. (1996). On the simultaneous optimization of server and work allocations in production line systems with variable processing times. *Operations Research*, *44*(3), 435–443. doi:10.1287/opre.44.3.435

Hinkelmann, K., & Kempthorne, O. (2008). *Design and analysis of experiments*. Hoboken, NJ: John Wiley and Sons.

Hirschman, E. C. (1982). Symbolizm and technology as sources for the generation of innovations. *Advances in Consumer Research. Association for Consumer Research (U. S.)*, *9*(1), 537–541.

Hirvonen, M., & Jaakkola, K. (2006). Dual-band platform tolerant antennas for radio-frequency identification. *IEEE Transactions on Antennas and Propagation*, *54*(9), 2632–2637. doi:10.1109/TAP.2006.880726

Hoegg, J., & Alba, J. W. (2011). Seeing is believing (too much): The influence of product form on perceptions of functional performance. *Journal of Product Innovation Management*, *28*(3), 346–359. doi:10.1111/j.1540-5885.2011.00802.x

Hoffman, T., & Kashmeri, S. (2000). Coddling the customer. *Computerworld*, *34*(50), 58–60.

Holbrook, M. B. (1996). Customer value – A framework for analysis and research. *Advances in Consumer Research. Association for Consumer Research (U. S.)*, *23*(1), 138–142.

Holland, J. H. (1975). *Adaptation in natural and artificial systems*. Ann Arber, MI: University of Michigan Press.

Homburg, C., Hoyer, W. D., & Fassnacht, M. (2002). Service orientation of a retailer's business strategy: Dimensions, antecedents, and performance outcomes. *Journal of Marketing*, *66*(4), 86–101. doi:10.1509/jmkg.66.4.86.18511

Hoof, B. V., & Lyon, T. P. (2013). Cleaner production in smaller firms taking part in Mexico's sustainable supplier program. *Journal of Cleaner Production*, *41*, 270–282. doi:10.1016/j.jclepro.2012.09.023

Horne, S. (2003). Needed: A cultural change. *Target Marketing*, *26*(8), 53–56.

Hou, T., Wang, J., Li, Y., & Wang, W. (2011). Assessing the performance of the MM/PBSA and MM/GBSA methods. 1. The accuracy of binding free energy calculations based on molecular dynamics simulations. *Journal of Chemical Information and Modeling*, *51*(1), 69–82. doi:10.1021/ci100275a PMID:21117705

Ho, W., & Ji, P. (2005). PCB assembly line assignment: A genetic algorithm approach. *Journal of Manufacturing Technology Management*, *16*(6), 682–692. doi:10.1108/17410380510609519

Howard, J. A., & Sheth, J. N. (1969). *The theory of buyer behavior*. New York, NY: John Wiley & Sons.

Hsieh, S.-J. (2002). Hybrid analytic and simulation models for assembly line design and production planning. *Simulation Modelling Practice and Theory*, *10*(1-2), 87–108. doi:10.1016/S1569-190X(02)00063-1

Hsu, C. W., & Hu, A. H. (2008). Green supply chain management in the electronic industry. *International Journal of Science and Technology*, *5*(2), 205–216. doi:10.1007/BF03326014

Hsu, C. W., & Hu, A. H. (2009). Applying hazardous substance management to supplier selection using analytic network process. *Journal of Cleaner Production*, *17*(2), 255–264. doi:10.1016/j.jclepro.2008.05.004

Huang, J. Z., Yang, P. H., Chew, W. C., & Ye, T. T. (2009). A compact broadband patch antenna for UHF RFID tags. In *Proceedings of 2009 Asia Pacific Microwave Conference*, 1044–1047.

Hubbard, B. (2012). Is there a future for carbon footprint labeling in the UK? *Ecologist*. Retrieved May 19, 2013, from http://www.theecologist.org/News/news_analysis/1231410/is_there_a_future_for_carbon_footprint_labelling_in_the_uk.html

Hu, C., Dou, X., Wu, Y., Zhang, L., & Hu, Y. (2012). Design, synthesis and CoMFA studies of N1-amino acid substituted 2,4,5-triphenyl imidazoline derivatives as p53-MDM2 binding inhibitors. *Bioorganic & Medicinal Chemistry*, *20*(4), 1417–1424. doi:10.1016/j.bmc.2012.01.003 PMID:22273545

Hui, P. C.-L., Chan, K. C. C., Yeung, K. W., & Ng, F. S.-F. (2002). Fuzzy operator allocation for balance control of assembly lines in apparel manufacturing. *IEEE Transactions on Engineering Management*, *49*(2), 173–180. doi:10.1109/TEM.2002.1010885

Hunt, H. K. (1977). CS/D - overview and future research directions. In H. K. Hunt (Ed.), *Conceptualization and measurement of CS/D* (pp. 455–488). Cambridge, MA: Marketing Science Institute.

Hussain, G., & Gao, L. (2007). A novel method to test the thinning limits of sheet metals in negative incremental forming. *International Journal of Machine Tools & Manufacture*, *47*, 419–435. doi:10.1016/j.ijmachtools.2006.06.015

Hussain, G., Gao, L., Hayat, N., Cui, Z., Pang, Y. C., & Dar, N. U. (2008). Tool and lubrication for negative incremental forming of a commercially pure titanium sheet. *Journal of Materials Processing Technology*, *203*, 193–201. doi:10.1016/j.jmatprotec.2007.10.043

Imai, K., Nonaka, I., & Takeuchi, H. (1985). *Managing new product development process: How Japanese learn and unlearn*. Boston, MA: Harvard Business School Press.

International Organization for Standardization (ISO). (2012). *ISO/TS 14067 Greenhouse gases -- Carbon footprint of products -- Requirements and Guidelines for Quantification and Communication*. ISO.

Iseki, H. (2001). Flexible and incremental bulging of sheet metal using high-speed water jet. *JSME International Journal, Series C*, *44*(2), 486–493. doi:10.1299/jsmec.44.486

Iseki, H., & Kumon, H. (1994). Forming limit of incremental sheet metal stretch forming using spherical rollers. *J. JSTP*, *35*, 1336.

Ismail, A. R., Haniff, M. H. M., Deros, B. M., Rani, M. R. A., Makhbul, Z. K. M., & Makhtar, N. K. (2010). The optimization of environmental factors at manual assembly workstation by using Taguchi method. *Journal of Applied Sciences*, *10*(13), 1293–1299. doi:10.3923/jas.2010.1293.1299

Ispas, C., Anania, F. D., Mohora, C., & Ivan, I. (2007). New methods for compensating the machining errors by CAD modeling of the machining surfaces on milling machines in coordinates. *Annals of the University of Petroşani. Mechanical Engineering (New York, N.Y.)*, *9*, 169–184.

Ivetac, A., & McCammon, J. A. (2010). Mapping the druggable allosteric space of G-protein coupled receptors: a fragment-based molecular dynamics approach. *Chemical Biology & Drug Design*, *76*(3), 201–217. PMID:20626410

Iwasaki, Y., & Havitz, M. E. (1998). A path-analytic model of the relationship between involvement, psychological commitment, and loyalty. *Journal of Leisure Research*, *30*(2), 337–347.

Izamshah, R. A., Mo, J. P. T., & Ding, S. (2011). task automation for modeling deflection prediction on machining thin-wall part with Catia V5. *Advances in Mechanical Engineering*, *1*(1), 8–14.

Jackson, K. P., Allwood, J. M., & Landert, M. (2008). Incremental forming of sandwich panels. *Journal of Materials Processing Technology*, *204*, 290–303. doi:10.1016/j.jmatprotec.2007.11.117

Jackson, K., & Allwood, J. (2009). The mechanics of incremental sheet forming. *Journal of Materials Processing Technology*, *209*, 1158–1174. doi:10.1016/j.jmatprotec.2008.03.025

Jacoby, J., & Kyner, D. B. (1973). Brand loyalty vs. repeat purchasing behavior. *JMR, Journal of Marketing Research*, *10*(1), 1–9. doi:10.2307/3149402

Jain, R., Jain, S., & Dhar, U. (2003). Measuring customer relationship management. *Journal of Service Research*, *2*(2), 97–109.

Jakobson, M. (1981). Absolutely continuous invariant measures for one-parameter families of one-dimensional maps. *Communications in Mathematical Physics*, *81*, 39–38. doi:10.1007/BF01941800

Jayaraman, A., & Gunal, A. K. (1997). Applications of discrete event simulation in the design of automotive powertrain manufacturing systems. In the *Proceedings of 29th Conference on Winter Simulation* (pp. 758–764). Atlanta, GA: IEEE Computer Society Washington.

Jayaram, S., Jayaram, U., Kim, Y. J., DeChenne, C., Lyons, K. W., Palmer, C., & Mitsui, T. (2007). Industry case studies in the use of immersive virtual assembly. *Virtual Reality (Waltham Cross)*, *11*(4), 217–228. doi:10.1007/s10055-007-0070-x

Jeba Singh, K. D., & Jebaraj, C. (2005). Feature-based design for process planning of machining processes with optimization using genetic algorithms. *International Journal of Production Research*, *43*(18), 3855–3887. doi:10.1080/00207540500032160

Jensen, H. R. (2001). Antecedents and consequences of consumer value assessments: Implications for marketing strategy and future research. *Journal of Retailing and Consumer Services*, *8*(6), 299–310. doi:10.1016/S0969-6989(00)00036-9

Jeswiet, J., Micari, F., Hirt, G., Bramley, A., Duflou, J., & Allwood, J. (2005). Asymmetric single point incremental forming of sheet metal. *CIRP Annals Manufacturing Technology*, *54*(2), 88–114. doi:10.1016/S0007-8506(07)60021-3

Jimenez, F., Sanchez, G., Vasant, P., & Verdegay, J. (2006). A multi-objective evolutionary approach for fuzzy optimization in production planning. In *Proceedings of IEEE International Conference on Systems, Man, and Cybernetics* (pp. 3120-3125). USA: IEEE Press.

Jiménez, F., Gómez-Skarmeta, A. F., & Sánchez, G. (2004). Nonlinear optimization with fuzzy constraints by multi-objective evolutionary algorithms. *Computational Intelligence. Theory and Applications: Advances in Soft Computing*, *33*, 713–722.

Jin, F., Matsushita, O., Katayama, S., Jin, S., Matsushita, C., Minami, J., & Okabe, A. (1996). Purification, characterization, and primary structure of Clostridium perfringens lambda-toxin, a thermolysin-like metalloprotease. *Infection and Immunity*, *64*(1), 230–237. PMID:8557345

Ji, Y. H., & Park, J. J. (2008). Formability of magnesium AZ31 sheet in the incremental forming at warm temperature. *Journal of Materials Processing Technology*, *201*, 254–358. doi:10.1016/j.jmatprotec.2007.11.206

Johne, A. F., & Snelson, P. A. (1988). Success factors in product innovation: A selective review of the literature. *Journal of Product Innovation Management*, *5*(2), 114–128. doi:10.1016/0737-6782(88)90003-3

Johnson, D. S., Aragon, C. R., McGeoch, L. A., & Schevon, C. (1987). Optimization by simulated annealing: An experimental evaluation, Part 1. *Operations Research*, *37*, 865–892. doi:10.1287/opre.37.6.865

Johnson, M. D., Anderson, E. W., & Fornell, C. (1995). Rational and adaptive performance expectations in a customer satisfaction framework. *The Journal of Consumer Research*, *21*(4), 128–140. doi:10.1086/209428

Johnson, M. D., & Fornell, C. (1991). A framework for comparing customer satisfaction across individuals and product categories. *Journal of Economic Psychology*, *12*(2), 267–286. doi:10.1016/0167-4870(91)90016-M

Jones, T. J., Reidsema, C., & Smith, A. (2006). Automated feature recognition system for supporting conceptual engineering design. *International Journal of Knowledge-Based and Intelligent Engineering Systems*, *10*(6), 477–492.

Joreskog, K. G., & Sorbom, D. (1993). *LISREL 8: User's reference guide*. Chicago, IL: Scientific Software International.

Kalagnanam, J. R., & Diwekar, U. M. (1997). An efficient sampling technique for off-line quality control. *Technometrics*, *39*(3), 308–319. doi:10.1080/00401706.1997.10485122

Kamath, M., Suri, R., & Sanders, J. L. (1988). Analytical performance models for closed-loop flexible assembly systems. *International Journal of Flexible Manufacturing Systems*, *1*(1), 51–84. doi:10.1007/BF00713159

Kanji, G., & Moura, P. (2002). Kanji's business scorecard. *Total Quality Management, 13*(1), 13–27. doi:10.1080/09544120120098537

Kaplan, R. S., & Norton, D. P. (1992). The balanced scorecard – Measures that drive performance. *Harvard Business Review, 70*(1), 71–79. PMID:10119714

Karr, C. L. (1991). Genetic algorithms for fuzzy controllers. *Artificial Intelligence Expert, 6*(2), 27–33.

Kasper, H. (1988). On problem perception, dissatisfaction and brand loyalty. *Journal of Economic Psychology, 9*(3), 387–397. doi:10.1016/0167-4870(88)90042-6

Katayama, H., & Bennett, D. (1996). Lean production in a changing competitive world: A Japanese perspective. *International Journal of Operations & Production Management, 16*(2), 8–23. doi:10.1108/01443579610109811

Katz, A. (1993). Measuring technology's business value. *Information Systems Management, 10*(1), 33–39. doi:10.1080/10580539308906910

Kaukonen, M., Söderhjelm, P., Heimdal, J., & Ryde, U. (2008). QM/MM-PBSA method to estimate free energies for reactions in proteins. *The Journal of Physical Chemistry B, 112*(39), 12537–12548. doi:10.1021/jp802648k PMID:18781715

Kaynak, H. (2003). The relationship between total quality management practices and their effects on firm performance. *Journal of Operations Management, 21*(4), 405–435. doi:10.1016/S0272-6963(03)00004-4

Kellen, V. (2002). *CRM measurement frameworks* (Working paper). DePaul University, Chicago, IL.

Keller, K. L. (1993). Conceptualizing, measuring and managing customer-based brand equity. *Journal of Marketing, 57*(1), 1–22. doi:10.2307/1252054

Kelley, T. (2001). *The Art of innovation*. New York: Curreny.

Kennedy, J., & Eberhart, R. (1995). Particle swarm optimization. In *IEEE Proceedings of the International Conference on Neural Networks: Perth, Australia*, (pp. 1942-1948).

Khalifa, A. S. (2004). Customer value: A review of recent literature and an integrative configuration. *Management Decision, 42*(5), 645–666. doi:10.1108/00251740410538497

Khan Mahmud, T. H., & Ingebrigt, S. (2009). Multivariate linear regression models based on ADME descriptors and predictions of ADMET profile for structurally diverse thermolysin inhibitors. *Letters in Drug Design & Discovery, 6*(6), 428–436. doi:10.2174/157018009789057607

Khurana, A., & Rosenthal, S. R. (1998). Towards holistic "front ends" in new product development. *Journal of Product Innovation Management, 15*(1), 57–74. doi:10.1016/S0737-6782(97)00066-0

Khuri, A. I., & Conlon, M. (1981). Simultaneous optimization of multiple responses represented by polynomial regression functions. *Technometrics, 23*(4), 363–375. doi:10.1080/00401706.1981.10487681

Khuri, A. I., & Mukhopadhyay, S. (2010). Part I. The foundational years: 1951-1975. *WIREs Computational Statistics, 2*, 128–149. doi:10.1002/wics.73

Kim, C. K., Han, D., & Park, S. B. (2001). The effect of brand personality and brand identification on brand loyalty: Applying the theory of social identification. *The Japanese Psychological Research, 43*(4), 195–206. doi:10.1111/1468-5884.00177

Kim, C., Nam, S. Y., Park, D. J., Park, I., & Hyun, T. Y. (2007). Product control system using RFID tag information and data mining. *Lecture Notes in Computer Science, 4412*, 100–109. doi:10.1007/978-3-540-71789-8_11

Kim, D., & Yeo, J. (2008). Low-profile RFID tag antenna using compact AMC substrate for metallic objects. *IEEE Antennas and Wireless Propagation Letters, 7*, 718–720. doi:10.1109/LAWP.2008.2000813

Kim, D., & Yeo, J. (2012). Dual-band long range passive RFID tag antenna using an AMC ground plane. *IEEE Transactions on Antennas and Propagation, 60*(6), 2620–2626. doi:10.1109/TAP.2012.2194638

Kim, H. S., & Kim, Y. G. (2009). A CRM performance measurement framework: Its development process and application. *Industrial Marketing Management, 38*(4), 477–489. doi:10.1016/j.indmarman.2008.04.008

Kim, J. H., Youn, S., & Roh, J. J. (2011). Green supply chain management orientation and firm performance: An evidence from South Korea. *International Journal of Services and Operations Management, 8*(3), 283–304. doi:10.1504/IJSOM.2011.038973

Kim, J., Suh, E., & Hwang, H. (2003). A model for evaluating the effectiveness of CRM using balanced scorecard. *Journal of Interactive Marketing, 17*(2), 5–19. doi:10.1002/dir.10051

Kim, J.-Y., & Kim, Y.-D. (1995). Graph theoretic heuristics for unequal-sized facility layout problems. *Omega, 23*(4), 391–401. doi:10.1016/0305-0483(95)00016-H

Kim, T. J., & Yang, D. Y. (2000). Improvement of formability for the incremental sheet metal forming process. *International Journal of Mechanical Sciences, 42*(7), 1271–1286. doi:10.1016/S0020-7403(99)00047-8

Kim, Y. H., & Park, J. J. (2002). Effect of process parameters on formability in incremental forming of sheet metal. *Journal of Materials Processing Technology*, 130–131.

Kim, Y. K., Kim, Y. J., & Kim, Y. (1996). Genetic algorithms for assembly line balancing with various objectives. *Computers & Industrial Engineering, 30*(3), 397–409. doi:10.1016/0360-8352(96)00009-5

Kirkpatrick, S., Gelatt, C. D., & Vecchi, M. P. (1983). Optimization by simulated annealing. *Science, 220*(4598), 671–680. doi:10.1126/science.220.4598.671 PMID:17813860

Kirschning, M., Jansen, R. H., & Koster, N. H. L. (1981). Accurate model for open end effect of microstrip lines. *Electronics Letters, 17*(3), 123–124. doi:10.1049/el:19810088

Klampfl, E., Gusikhin, O., & Rossi, G. (2006). Optimization of workcell layouts in a mixed-model assembly line environment. *International Journal of Flexible Manufacturing Systems, 17*(4), 277–299.

Klir, G. J., Clair, U. H. S., & Yuan, B. (1997). *Fuzzy set theory: Foundations and applications.* Upper Saddle River, NJ: Prentice Hall.

Kohli, R., Piontek, F., Ellington, T., VanOsdol, T., Shepard, M., & Brazel, G. (2001). Managing customer relationships through e-business decision support applications: A case of hospital-physician collaboration. *Decision Support Systems, 32*(2), 171–187. doi:10.1016/S0167-9236(01)00109-9

Komunda, M., & Osarenkhoe, A. (2012). Remedy or cure for service failure? Effects of service recovery on customer satisfaction and loyalty. *Business Process Management Journal, 18*(1), 82–103. doi:10.1108/14637151211215028

Konak, A., Coit, D. W., & Smith, A. E. (2006). Multi-objective optimization using genetic algorithms: A tutorial. *Reliability Engineering & System Safety, 91*(9), 992–1007. doi:10.1016/j.ress.2005.11.018

Kondo, Y., Moriguchi, Y., & Shimizu, H. (1998). CO_2 emissions in Japan: Influences of imports and exports. *Applied Energy, 59*, 163–174. doi:10.1016/S0306-2619(98)00011-7

Koopmans, T. C., & Beckmann, M. (1957). Assignment problems and the location of economic activities. *Econometrica, 25*(1), 53–76. doi:10.2307/1907742

Kopac, J., & Kampus, Z. (2005). Incremental sheet metal forming on CNC milling machine-tool. *Journal of Materials Processing Technology*, 622–628. doi:10.1016/j.jmatprotec.2005.02.160

Kordupleski, R. T., & Laitamaki, J. (1997). Building and deploying profitable growth strategies based on the waterfall of customer value added. *European Management Journal, 15*(2), 158–166. doi:10.1016/S0263-2373(96)00085-0

Kotler, P. (1997). *Marketing management: Analysis, planning, implementation, and control.* Upper Saddle River, NJ: Prentice-Hall.

Kotler, P. (2000). *Marketing management: The millennium edition.* Upper Saddle River, NJ: Prentice-Hall.

Kotler, P., & Levy, S. J. (1969). Broadening the concept of marketing. *Journal of Marketing, 33*(1), 10–15. doi:10.2307/1248740 PMID:12309673

Kouvelis, P., Chiang, W.-C., & Kiran, A. S. (1992). A survey of layout issues in flexible manufacturing systems. *Omega*, *20*(3), 375–390. doi:10.1016/0305-0483(92)90042-6

Krakauer, J. (1987). *Smart manufacturing with AI*. Dearborn, MI: Society of Manufacturing Engineers.

Krasnikov, A., Jayachandran, S., & Kumar, V. (2009). The impact of customer relationship management implementation on cost and profit efficiencies: Evidence from the US commercial banking industry. *Journal of Marketing*, *73*(6), 61–76. doi:10.1509/jmkg.73.6.61

Krauss, M. (2002). At many firms, technology obscures CRM. *Marketing News*, *36*(6), 5.

Kreuzbauer, R., & Malter, A. J. (2005). Embodied cognition and new product design: changing product form to influence brand categorization. *Journal of Product Innovation Management*, *22*(2), 165–176. doi:10.1111/j.0737-6782.2005.00112.x

Krimpenis, A. (2008) *CNC rough milling optimization of complex sculptured surface parts using artificial intelligence algorithms* (PhD Thesis). National Technical University of Athens, Athens, Greece.

Krimpenis, A. A., Fountas, N. A., Skolias, J., Tzivelekis, C., & Vaxevanidis, N. M. (2011). Intelligent post-processor creation for sculptured surfaces in CAM software. In *Proceedings of the 4th International Conference on Manufacturing Engineering -ICMEN* (pp. 287-294), Thessaloniki, Greece.

Krippendorff, K. (1989). On the essential contexts of artifacts or on the proposition that "design is making sense (of things)". *Design Issues*, *5*(2), 9–38. doi:10.2307/1511512

Kroemer, R. T. (2007). Structure-based drug design: Docking and scoring. *Current Protein & Peptide Science*, *8*(4), 312–328. doi:10.2174/138920307781369382 PMID:17696866

Kubinyi, H. (2008). Comparative molecular field analysis (CoMFA). In J. Gasteiger (Ed.), *Handbook of Chemoinformatics: From Data to Knowledge in 4 Volumes*. Germany: Wiley-VCH.

Kuehn, A. (1962). Consumer brand choice as a learning process. *Journal of Advertising Research*, *2*(4), 10–17.

Kumar, N., Scheer, L. K., & Steenkamp, J. E. M. (1995). The effects of supplier fairness on vulnerable resellers. *JMR, Journal of Marketing Research*, *32*(1), 54–65. doi:10.2307/3152110

Kumar, R., & Chandrakar, R. (2012). Overview of green supply chain management: Operation and environmental impact at different stages of the supply chain. *International Journal of Engineering and Advanced Technology*, *1*(3), 1–6.

Kumar, V., & Whitney, P. (2003). Faster, deeper user research. *Design Management Journal*, *14*(2), 50–55.

Kung, H.-K., & Changchit, C. (1991). Just-in-time simulation model of a PCB assembly line. *Computers & Industrial Engineering*, *20*(1), 17–26. doi:10.1016/0360-8352(91)90036-6

Kuo, S. K., Chen, S. L., & Lin, C. T. (2010). Design and development of RFID label for steel coil. *IEEE Transactions on Industrial Electronics*, *57*(6), 2180–2186. doi:10.1109/TIE.2009.2034174

Kuo, S. K., & Liao, L. G. (2010). an analytic model for impedance calculation of an RFID metal tag. *IEEE Antennas and Wireless Propagation Letters*, *9*, 603–607. doi:10.1109/LAWP.2010.2053511

Kurtoglu, A. (2004). Flexibility analysis of two assembly lines. *Robotics and Computer-integrated Manufacturing*, *20*(3), 247–253. doi:10.1016/j.rcim.2003.10.011

Kurttila, M., Pesonen, M., & Kangas, J. (2000). Utilizing the analytic hierarchy process (AHP) in SWOT analysis – A hybrid method and its application to a forest-certification case. *Forest Policy and Economics*, *1*, 41–52. doi:10.1016/S1389-9341(99)00004-0

Lagace, R. R., Dahlstrom, R., & Gassenheimer, J. B. (1991). The relevance of ethical salesperson behavior on relationship quality: The pharmaceutical industry. *Journal of Personal Selling & Sales Management*, *11*(4), 39–47.

Lagaros, N. D., Papadrakakis, M., & Kokossalakis, G. (2002). Structural optimization using evolutionary algorithms. *Computers & Structures*, *80*, 571–589. doi:10.1016/S0045-7949(02)00027-5

Lai, M., & Li, R. (2010). A low-profile broadband RFID tag antenna for metallic objects. In *Proceedings of 2010 International Conference on Microwave and Millimeter Wave Technology,* 1891–1893.

Lam, A., Wilton, S. J. E., Leong, P., & Luk, W. (2008). An analytical model describing the relationships between logic architecture and FPGA density. In *the Proceedings of the International Conference on Field Programmable Logic and Applications* (pp. 221-226). Heidelberg, Germany: IEEE.

Lamarche, B., & Rivest, L. (2007). Dynamic product modeling with inter-features associations: comparing customization and automation. *Computer-Aided Design & Applications, 4*(6), 877–886. doi:10.1080/16864360.2007.10738519

Lambert, D. M. (2004). *Supply chain management: Processes, partnerships, performance.* Sarasota, FL: Supply Chain Management Institute.

Lambert, D. M. (2010). Customer relationship management as a business process. *Journal of Business and Industrial Marketing, 25*(1), 4–17. doi:10.1108/08858621011009119

Lamminen, L., Tuominen, T., & Kivivuori, S. (2005). Incremental sheet forming with and industrial robot – Forming limits and their effects on component design. In *Proceedings of 3rd International Conference on Advanced Materials Processing,* Finland, (pp. 331).

Lan, C.-H., & Kang, C.-J. (2006). Constrained spatial layout and simultaneous production evaluation for a production system. *Concurrent Engineering, 14*(2), 111–120. doi:10.1177/1063293X06065529

Lapierre, J. (2000). Customer-perceived value in industrial contexts. *Journal of Business and Industrial Marketing, 15*(2-3), 122–140. doi:10.1108/08858620010316831

Latour, B. (1987). *Science in action: How to follow scientists and engineers through society.* Cambridge, MA: Harvard University Press.

Le, V. S., Ghiotti, A., & Lucchetta, G. (2008). *Preliminary studies on single point incremental forming for thermoplastic materials.* Paper presented at the 11th ESAFORM 2008 Conference on Material Forming, Lyon, France.

Lee, H. F., & Stecke, K. E. (1995). *An integrated design support method for flexible assembly systems* (Working paper no. 681-d). University of Michigan, Ann Arbor, MI.

Lee, M. A., & Takagi, H. (1993). Dynamic control of genetic algorithms using fuzzy logic techniques. In S. Forrest (Ed.), *Proceedings of the 5th International Conference on Genetic Algorithms* (pp. 76-83). San Mateo, CA: Morgan Kaufmmann.

Lee, J. S., & Back, K. J. (2010). Reexamination of attendee-based brand equity. *Tourism Management, 31*(3), 395–401. doi:10.1016/j.tourman.2009.04.006

Lee, K., Jeong, K. W., Lee, Y., Song, J. Y., Kim, M. S., Lee, G. S., & Kim, Y. (2010). Pharmacophore modeling and virtual screening studies for new VEGFR-2 kinase inhibitors. *European Journal of Medicinal Chemistry, 45*(11), 5420–5427. doi:10.1016/j.ejmech.2010.09.002 PMID:20869793

Lee, S. G., Khoo, L. P., & Yin, X. F. (2000). Optimising an assembly line through simulation augmented by genetic algorithms. *International Journal of Advanced Manufacturing Technology, 16*(3), 220–228. doi:10.1007/s001700050031

Lee, Y. K., Back, K. J., & Kim, J. Y. (2009). Family restaurant brand personality and its impact on consumer's emotion, satisfaction, and band loyalty. *Journal of Hospitality & Tourism Research (Washington, D.C.), 33*(3), 305–328. doi:10.1177/1096348009338511

Leigh, T. W., & Tanner, J. F. (2004). Introduction: JPSSM special issue on customer relationship management. *Journal of Personal Selling & Sales Management, 24*(4), 259–262.

Lennard-Jones, J. E. (1924). On the determination of molecular fields: I: from the variation of the viscosity of a gas with temperature. *Proc Royal Soc A, 106*(738), 441–462. doi:10.1098/rspa.1924.0081

Lenzen, M. (2007). Aggregation (in-)variance of shared responsibility: A case study of Australia. *Ecological Economics, 64,* 19–24. doi:10.1016/j.ecolecon.2007.06.025

Lenzen, M., & Murray, J. (2010). Conceptualizing environmental responsibility. *Ecological Economics, 70,* 261–270. doi:10.1016/j.ecolecon.2010.04.005

Lenzen, M., Murray, J., Sack, F., & Wiedmann, T. (2007). Shared producer and consumer responsibility — Theory and practice. *Ecological Economics*, *61*, 27–43. doi:10.1016/j.ecolecon.2006.05.018

Lenzen, M., Pade, L. L., & Munksgaard, J. (2004). co2 mulitipliers in multi-region input-output models. *Economic Systems Research*, *16*, 391–412. doi:10.1080/0953531042000304272

Leszak, E. (1967). *Apparatus and process for incremental dieless* forming (Patent US3342051A1, published 1967-09-19).

Levine, S. (2000). The rise of CRM. *America's Network*, *104*(6), 34.

Levy, S. (2006). *The perfect thing: How the iPod shuffles commerce, culture, and coolness*. New York: Simon & Schuster.

Liang, T. F. (2008). Interactive multi-objective transportation planning decisions using fuzzy linear programming. *Asia Pacific Journal of Operational Research*, *25*(1), 11–31. doi:10.1142/S0217595908001602

Ligas, M. (2000). People, products, and pursuits: Exploring the relationship between consumer goals and product meanings. *Psychology and Marketing*, *17*(11), 983–1003. doi:10.1002/1520-6793(200011)17:11<983::AID-MAR4>3.0.CO;2-J

Lindgreen, A., Palmer, R., Vanhamme, J., & Wouters, J. (2006). A relationship management assessment tool: Questioning, identifying, and prioritizing critical aspects of customer relationships. *Industrial Marketing Management*, *35*(1), 57–71. doi:10.1016/j.indmarman.2005.08.008

Lindgreen, A., & Wynstra, F. (2005). Value in business markets: What do we know? Where are we going? *Industrial Marketing Management*, *34*(7), 732–748. doi:10.1016/j.indmarman.2005.01.001

Lin, F. T., & Yao, J. S. (2002). Applying genetic algorithms to solve the fuzzy optimal profit problem. *Journal of Information Science and Engineering*, *18*, 563–580.

Ling, R., & Yen, D. C. (2001). Customer relationship management: An analysis framework and implementation strategies. *Journal of Computer Information Systems*, *41*(3), 82–97.

Lin, L., & Cochran, D. K. (1987). Optimization of a complex flow line for printed circuit board fabrication by computer simulation. *Journal of Manufacturing Systems*, *6*(1), 47–57. doi:10.1016/0278-6125(87)90049-5

Linzmayer, O. W. (2004). *Apple confidential 2.0: The definitive history of the world's most colorful company*. San Francisco: No Starch Press.

Li, S., Nathan, B., Ragu-Nathan, T. S., & Rao, S. S. (2006). The impact of supply chain management practices on competitive advantage and organizational performance. *Omega*, *34*(2), 107–124. doi:10.1016/j.omega.2004.08.002

Liu, W. N., Zheng, L. J., Sun, D. H., Liao, X. Y., Zhao, M., & Su, J. M. et al. (2012). RFID-enabled real-time production management system for Loncin motorcycle assembly line. *International Journal of Computer Integrated Manufacturing*, *25*(1), 86–99. doi:10.1080/0951192X.2010.523846

Li, W. D., & Qiu, Z. (2006). State-of-the-art technologies and methodologies for collaborative product development systems. *Computer Aided Design*, *37*(9), 931–940. doi:10.1016/j.cad.2004.09.020

Li, Y., & Hewitt, C. N. (2008). The effect of trade between China and the UK on national and global carbon dioxide emissions. *Energy Policy*, *36*, 1907–1914. doi:10.1016/j.enpol.2008.02.005

Lo, A. S., Stalcup, L. D., & Lee, A. (2010). Customer relationship management for hotels in Hong Kong. *International Journal of Contemporary Hospitality Management*, *22*(2), 139–159. doi:10.1108/09596111011018151

Lojacono, G., & Zaccai, G. (2004). The evolution of the design-inspired enterprise. *Sloan Management Review*, *45*, 75–79.

Longo, F. G. M., & Papoff, E. (2006). Effective design of an assembly line using modeling & simulation. In the *Proceedings of the 38th Conference on Winter Simulation* (pp. 1893-1898). Monterey, California: Winter Simulation Conference.

Lorenz, E. N. (1963). Deterministic non-periodic flow. *Journal of the Atmospheric Sciences*, *20*(2), 130–141. doi:10.1175/1520-0469(1963)020<0130:DNF>2.0.CO;2

Lozano, M., Herrera, F., Krasnogor, N., & Molina, D. (2004). Real coded memetic algorithms with crossover hill climbing. *Evolutionary Computational Journal*, *12*(3), 273–302. doi:10.1162/1063656041774983 PMID:15355602

Luchs, M., & Scott Swan, K. (2011). The emergence of product design as a field of marketing inquiry. *Journal of Product Innovation Management*, *28*(3), 327–345. doi:10.1111/j.1540-5885.2011.00801.x

Lukas, B. A., & Ferrell, O.bC. (2000). The effect of market orientation on product innovation. *Journal of the Academy of Marketing Science*, *28*(2), 239–247. doi:10.1177/0092070300282005

Luo, X., Li, W., Tu, Y., Xue, D., & Tang, J. (2011). Operator allocation planning for reconfigurable production line in one-of-a-kind production. *International Journal of Production Research*, *49*(3), 689–705. doi:10.1080/00207540903555486

Luo, Z. (2013). Introduction to mechanism design for sustainability. In Z. Luo (Ed.), *Mechanism design for sustainability: Techniques and cases*. Dordrecht, The Netherlands: Springer. doi:10.1007/978-94-007-5995-4_1

Luthra, S. et al. (2012). Barriers to implement green supply chain management in automobile industry using interpretative structural modeling technique-An Indian perspective. *Journal of Industrial Engineering and Management*, *4*(2), 231–257.

Luyben, M. L., & Floudas, C. A. (1994). Analyzing the interaction of design and control. 1. A multiobjective framework and application to binary distillation synthesis. *Computers & Chemical Engineering*, *18*(10), 933–969. doi:10.1016/0098-1354(94)E0013-D

Maa, C., & Shanblatt, M. (1992). A two phase optimization neural network. *IEEE Transactions on Neural Networks*, *3*(6), 1003–1009. doi:10.1109/72.165602 PMID:18276497

Ma, D. L., Chan, D. S., & Leung, C. H. (2013). Drug repositioning by structure-based virtual screening.[Epub ahead of print]. *Chemical Society Reviews*. doi:10.1039/c2cs35357a PMID:23288298

Madden, T. J., Fehle, F., & Fournier, S. (2006). Brands matter: An empirical demonstration of the creation of shareholder value through branding. *Journal of the Academy of Marketing Science*, *34*(2), 224–235. doi:10.1177/0092070305283356

Mager, P. P. (1982). Theoretical approaches to drug design and biological activity: critical comments to the use of mathematical methods applied to univariate and multivariate quantitative structure-activity relationships (QSAR). *Medicinal Research Reviews*, *2*(1), 93–121. doi:10.1002/med.2610020106 PMID:7109783

Maidique, M. A., & Zigger, B. J. (1985). The new product learning cycle. *Research Policy*, *14*(6), 299–313. doi:10.1016/0048-7333(85)90001-0

Malai, V., & Speece, M. (2005). Cultural impact on the relationship among perceived service quality, brand name value, and customer loyalty. *Journal of International Consumer Marketing*, *17*(4), 7–39. doi:10.1300/J046v17n04_02

Manco, L., Filice, L, & Ambrogio, G. (2009). Analysis of the thickness distribution varying tool trajectory in single-point incremental forming. *Proc. IMechE Vol. 224 Part B: J. Engineering Manufacture*. DOI: 10.1177/09544054JEM1958

Mani, M., Govindarajalu, C., Kee, H. L., Kishore, S., Tatsui, E., & Seki, C. et al. (2008). Climate change policies and international trade: Challenges and opportunities. In *International trade and climate change: Economic, legal, and institutional perspective*. Washington, DC: The World Bank.

Manziniy, R., Gamberiy, M., Regattieriy, A., & Personaz, A. (2004). Framework for designing a flexible cellular assembly system. *International Journal of Production Research*, *42*(17), 3505–3528. doi:10.1080/00207540410001696023

Marabuto, S. R., Afonso, D., Ferreira, J. A. F., Melo, R. Q., Martins, M., & Alves de Sousa, R. J. (2011). Finding the best machine for SPIF operations. A brief discussion. *Key Engineering Materials*, *473*, 861–868. doi:10.4028/www.scientific.net/KEM.473.861

March, J. G. (1991). Exploration and exploitation in organizational learning. *Organization Science*, *2*(1), 71–87. doi:10.1287/orsc.2.1.71

Markham, C. E., Barber, D. G. B., Hise, J. H., Ihde, S. A., Lindsay, J. D., Nygaard, K. S., & Yosten, R. D. (2011). *U.S. patent no. 7,882,438*. Washington, DC: U.S. Patent and Trademark Office.

Markwick, R. L., Cervantes, C. F., Abel, B. L., Komives, E. A., Blackledge, M., & McCammon, J. A. (2010). Enhanced conformational space sampling improves the prediction of chemical shifts in proteins. *Journal of the American Chemical Society*, *132*(4), 1220–1221. doi:10.1021/ja9093692 PMID:20063881

Martín-del-Campo, C., François, J. L., & Morales, L. B. (2002). BWR fuel assembly axial design optimization using Tabu search. *Nuclear Science and Engineering*, *142*(1), 107–115.

Martin, G. E. (1994). Optimal design of production lines. *International Journal of Production Research*, *32*(5), 989–1000. doi:10.1080/00207549408956983

Massey, A. P., Montoya-Weiss, M., & Holcom, K. (2001). Re-engineering the customer relationship: Leveraging knowledge assets at IBM. *Decision Support Systems*, *32*(2), 155–170. doi:10.1016/S0167-9236(01)00108-7

Mazumdar, T. (1993). A value-based orientation to new product planning. *Journal of Consumer Marketing*, *10*(1), 28–41. doi:10.1108/07363769310026557

McCarthy, E. J., & Perreault, W. D. (1987). *Basic marketing: A managerial approach*. Homewood, IL: Irwin.

McCollough, M. A., Berry, L. L., & Yadav, M. S. (2000). An empirical investigation of customer satisfaction after service failure and recovery. *Journal of Service Research*, *3*(2), 121–137. doi:10.1177/109467050032002

McDougall, H. G., & Levesque, T. (2000). Customer satisfaction with services: Putting perceived value into equation. *Journal of Services Marketing*, *14*(5), 392–410. doi:10.1108/08876040010340937

McGovern, T., & Panaro, J. (2004). The human side of customer relationship management. *Benefits Quarterly*, *20*(3), 26–33.

Meier, H., Buff, B., Laurischkat, R., & Smukala, V. (2009). Increasing the part accuracy in dieless robot-based incremental sheet metal forming. *CIRP Annals – Manufacturing Technology*, *58*. Doi:10.1016/j.cirp.2009.03.056

Meier, H., Dewald, O., & Zhang, J. (2005). Development of a robot-based sheet metal forming process. *Steel Research International*, *76*(2-3), 167–170.

Mendoza, L. E., Marius, A., Perez, M., & Griman, A. C. (2007). Critical success factors for a customer relationship management strategy. *Information and Software Technology*, *49*(8), 913–945. doi:10.1016/j.infsof.2006.10.003

Mentzer, J. T., Rutner, S. M., & Matsuno, K. (1997). Application of the means-end value hierarchy model to understanding logistics service value. *International Journal of Physical Distribution & Logistics Management*, *27*(9-10), 630–643. doi:10.1108/09600039710188693

Meredith, J. R., McCutcheon, D. M., & Hartley, J. (1994). Enhancing competitiveness through the new market value equation. *International Journal of Operations & Production Management*, *14*(11), 7–22. doi:10.1108/01443579410068611

Mertes, J. (1965). Visual design and marketing manager. *California Management Review*, *8*(2), 29–39. doi:10.2307/41165669

Micari, F., Ambrogio, G., & Filipe, L. (2007). Shape and dimensional accuracy in single point incremental forming: State of the art and future trends. *Journal of Materials Processing Technology*, *191*, 390–395. doi:10.1016/j.jmatprotec.2007.03.066

Michalewicz, Z. (1993). A hierarchy of evolution programs: An experimental study. *Evolutionary Computation*, *1*(1), 51–76. doi:10.1162/evco.1993.1.1.51

Mihelis, G., Grigoroudis, E., Siskos, Y., Politis, Y., & Malandrakis, Y. (2001). Customer satisfaction measurement in the private bank sector. *European Journal of Operational Research*, *130*(2), 347–360. doi:10.1016/S0377-2217(00)00036-9

Milgrom, P., & Roberts, J. (1995). Complemenarties and fit: Strategy, structure and organizational change in manufacturing. *Journal of Accounting and Economics*, *19*(2-3), 179–208. doi:10.1016/0165-4101(94)00382-F

Miller, B. R., McGee, T. D., Swails, J. M., Homeyer, N., Gohlke, H., & Roitberg, A. E. (2012). MMPBSA.py: an efficient program for end-state free energy calculations. *Journal of Chemical Theory and Computation*, *8*(9), 3314–3321. doi:10.1021/ct300418h

Miller, J., & Muir, D. (2004). *The business of brands*. Chichester, UK: John Wiley & Sons.

Minami, C., & Dawson, J. (2008). The CRM process in retail and service sector firms in Japan: loyalty development and financial return. *Journal of Retailing and Consumer Services*, *15*(5), 375–385. doi:10.1016/j.jretconser.2007.09.001

Mitchell, M. (1996). *An introduction to genetic algorithms*. Cambridge, UK: The MIT Press.

Mithas, S., Krishnan, M. S., & Fornell, C. (2005). Why do customer relationship management applications affect customer satisfaction? *Journal of Marketing*, *69*(4), 201–209. doi:10.1509/jmkg.2005.69.4.201

Mittal, B., & Lassar, W. M. (1998). Why do consumers switch? The dynamics of satisfaction versus loyalty. *Journal of Services Marketing*, *12*(3), 177–194. doi:10.1108/08876049810219502

Mittal, V., & Kamakura, W. A. (2001). Satisfaction, repurchase intent, and repurchase behavior: Investigating the moderating effect of customer characteristics. *JMR, Journal of Marketing Research*, *38*(1), 131–142. doi:10.1509/jmkr.38.1.131.18832

Moenaert, R. K., De Meyer, A., Souder, W. E., & Deschoolmeester, D. (1995). R&D/marketing, communications during the fuzzy front-end. *IEEE Transactions on Engineering Management*, *42*(3), 243–258. doi:10.1109/17.403743

Mohebbi, N., Hoseini, M. A., & Esfidani, M. R. (2012). Identification & prioritization of the affecting factors on CRM implementation in edible oil industry. *Journal of Basic and Applied Scientific Research*, *2*(5), 4993–5001.

Mo, L., Zhang, H., & Zhou, H. (2008). Broadband UHF RFID tag antenna with a pair of U slots mountable on metallic objects. *Electronics Letters*, *44*(20), 5–6. doi:10.1049/el:20089813

Monroe, K. (2005). *Pricing: Making profitable decisions*. New York, NY: McGraw-Hill.

Montgomery, D. C., Peck, E. A., & Vining, G. G. (2001). *Introduction to linear regression analysis*. New York, NY: Wiley.

Morgan, R. M., & Hunt, S. D. (1994). The commitment - trust theory of relationship marketing. *Journal of Marketing*, *58*(3), 20–38. doi:10.2307/1252308

Morris, G. M., Goodsell, D. S., Halliday, R. S., Huey, R., Hart, W. E., Belew, R. K., & Olson, A. J. (1998). Automated docking using a Lamarckian genetic algorithm and an empirical binding free energy function. *Journal of Computational Chemistry*, *19*, 1639–1662. doi:10.1002/(SICI)1096-987X(19981115)19:14<1639::AID-JCC10>3.0.CO;2-B

Morris, G. M., Huey, R., Lindstrom, W., Sanner, M. F., Belew, R. K., Goodsell, D. S., & Olson, A. J. (2009). AutoDock4 and AutoDockTools4: Automated docking with selective receptor flexibility. *Journal of Computational Chemistry*, *30*(16), 2785–279. doi:10.1002/jcc.21256 PMID:19399780

Morrisson, O., & Huppertz, J. W. (2010). External equity, loyalty program membership, and service recovery. *Journal of Services Marketing*, *24*(3), 244–254. doi:10.1108/08876041011040640

Moscato, P. (1999). Memetic algorithms: A short algorithms. In D. Corne (Ed.), *New ideas in optimization* (pp. 219–234).

Mostaghim, S., & Teich, J. (2003). Strategies for finding good local guides in multiobjective particle swarm optimization. In *IEEE Swarm Intelligence Symposium*, Indianapolis, USA, (pp. 26-33).

Mostaghim, S., & Teich, J. (2005). A new approach on many objective diversity measurement. In *Dagstuhl Seminar Proceedings 04461, Practical Approaches to Multiobjective Optimization*, (pp. 1-15).

Motameni, R., & Shahrokhi, M. (1998). Brand equity valuation: A global perspective. *Journal of Product and Brand Management*, *7*(4), 275–290. doi:10.1108/10610429810229799

Mózner, Z. F. V. (2013). A consumption-based approach to carbon emission accounting - Sectoral differences and environmental benefits. *Journal of Cleaner Production*, *42*, 83–95. doi:10.1016/j.jclepro.2012.10.014

Mukherjee, I., & Ray, P. K. (2006). A review of optimization techniques in metal cutting processes. *Computers & Industrial Engineering*, *50*(1-2), 15–34. doi:10.1016/j.cie.2005.10.001

Munksgaard, J., Wier, M., Lenzen, M., & Dey, C. (2005). using input-output analysis to measure the environmental pressure of consumption at different spatial level. *Journal of Industrial Ecology*, *9*, 169–185. doi:10.1162/1088198054084699

Munnukka, F., & Farvi, P. (2012). The price-category effect and the formation of customer value of hightech products. *Journal of Consumer Marketing*, *29*(4), 293–301. doi:10.1108/07363761211237362

Muñoz, P., & Steininger, K. W. (2010). Austria's CO2 responsibility and the carbon content of its international trade. *Ecological Economics*, *69*, 2003–2019. doi:10.1016/j.ecolecon.2010.05.017

Muradian, R., O'Connor, M., & Martinez-Alier, J. (2002). Embodied pollution in trade: Estimating the environmental load displacement of industrialised countries. *Ecological Economics*, *41*, 51–67. doi:10.1016/S0921-8009(01)00281-6

Myers, R. H., Walter, H., & Carter, J. R. (1973). Response surface techniques for dual response systems. *Technometrics*, *15*(2), 301–317. doi:10.1080/00401706.1973.10489044

Myung, H., Kim, J. H., & Fogel, D. (1995). Preliminary investigation into a two-stage method of evolutionary optimization on constrained problems. In J. R. MacDonnell, R. G. Reynolds, & D. B. Fogel (Eds.), *Proceedings of the 4th Annual Conference on Evolutionary Programming* (pp. 449-463). Cambridge, MA: MIT Press.

Nagano, H., Ishida, S., & Ikeda, J. (2011). A study about boundaries of firm in FPD industry. *Journal of Japan Association for Management Systems*, *28*(1), 1–8.

Nair, V. N., Abraham, B., MacKay, J., & Nelder, J. A., Box, Ge., Phadke, M. S., ... Wu, C. F. J. (1992). Taguchi's parameter design: A panel discussion. *Technometrics*, *34*(2), 127–161. doi:10.1080/00401706.1992.10484904

Narver, J. C., & Slater, S. F. (1990). The effect of a market orientation on business profitability. *Journal of Marketing*, *54*(4), 20–35. doi:10.2307/1251757

Naumann, E. (1995). *Creating customer value*. Cincinnati, OH: Thompson Executive Press.

Neddermeijer, H. G., van Oortmarssen, G. J., Piersma, N., & Dekker, R. (2000). A framework for response surface methodology for simulation optimization. In the *Proceedings 32nd Conference on Winter Simulation* (pp. 129–136). Orlando, FL: IEEE.

Nelson, D. M., Marsillac, E., & Rao, S. S. (2012). Antecedents and evolution of the green supply chain. *Journal of Operations and Supply Chain Management*, Special Issue, 29 – 43.

Newell, F. (2000). *Loyalty.com: Customer relationship management in the new era of Internet marketing*. New York, NY: McGraw-Hill.

Newman, J. W., & Werbel, R. A. (1973). Multivariate analysis of brand loyalty for major household appliances. *JMR, Journal of Marketing Research*, *10*(4), 404–409. doi:10.2307/3149388

NFC Forum. (2013). RetrievedOctober31, 2013, fromhttp://www.nfc-forum.org

Ngai, E. W. T., Moon, K. K. L., Riggins, F. J., & Yi, C. Y. (2008). RFID research: An academic literature review (1995-2005) and future research directions. *International Journal of Production Economics*, *112*(2), 510–520. doi:10.1016/j.ijpe.2007.05.004

Niedbala, H., Polanski, J., Gieleciak, R., Musiol, R., Tabak, D., & Podeszwa, B. et al. (2006). Comparative molecular surface analysis (CoMSA) for virtual combinatorial library screening of styrylquinoline HIV-1 blocking agents. *Combinatorial Chemistry & High Throughput Screening*, *9*(10), 753–770. doi:10.2174/138620706779026042 PMID:17168681

NIKKEI BIZTECH. (2004, August 15). Study finds free care used more, pp. 168-175.

NIKKEI BIZTECH. (2004, December 20). Study finds free care used more, pp. 20-25.

NIKKEI BUSINESS. (2003, October 6). Study finds free care used more, pp. 28-33.

NIKKEI BUSINESS. (2006, January 16). Study finds free care used more, pp. 13.

NIKKEI BUSINESS. (2007, December 24/31). Study finds free care used more, pp. 46-49.

NIKKEI BUSINESS. (2008, January 28). Study finds free care used more, pp. 53-55.

NIKKEI BUSINESS. (2009, March 31). Study finds free care used more, pp. 7-9.

NIKKEI BUSINESS. (2010, July 5). Study finds free care used more, pp. 34-40.

NIKKEI BUSINESS. (2011, January 31). Study finds free care used more, pp. 22-24.

NIKKEI BUSINESS. (2011, October 31). Study finds free care used more, pp. 16.

NIKKEI DESIGN. (2004, October). Study finds free care used more, pp. 50-61.

NIKKEI ELECTRONICS. (2005, September 26). Study finds free care used more, pp. 112-119.

NIKKEI ELECTRONICS. (2006, October 9). Study finds free care used more, pp. 112-121.

NIKKEI ELECTRONICS. (2007, August 27). Study finds free care used more, pp. 168-170.

NIKKEI ELECTRONICS. (2007, November 19). Study finds free care used more, pp. 54-61.

NIKKEI MONOZUKURI. (2006, November). Study finds free care used more, pp. 54-55.

Noble, C. H. (2011). On elevating strategic design research. *Journal of Product Innovation Management*, *28*, 389–393. doi:10.1111/j.1540-5885.2011.00808.x

Nomura, J., & Takakuwa, S. (2006). Optimization of a number of containers for assembly lines: The fixed-course pick-up system. *International Journal of Simulation Modelling*, *5*(4), 155–166. doi:10.2507/IJSIMM05(4)3.066

Noorossana, R., Tajbakhsh, S. D., & Saghaei, A. (2009). An artificial neural network approach to multiple-response optimization. *International Journal of Advanced Manufacturing Technology*, *40*(11), 1227–1238. doi:10.1007/s00170-008-1423-7

Normann, R., & Ramirez, R. (1993). From value chain to value constellation: Designing interactive strategy. *Harvard Business Review*, *71*(4), 65–77. PMID:10127040

Nowakowska, N. (1977). Methodological problems of measurement of fuzzy concepts in the social sciences. *Behavioral Science*, *22*, 107–115. doi:10.1002/bs.3830220205

Nykamp, M. (2001). *The customer differential: The complete guide to implementing customer relationship management*. New York, NY: AMACOM.

Oakley, M. (1986). *Managing design. Product design and technological innovation*. University Press.

Oikonomou, D., Moulianitis, V., Lekkas, D., & Koutsabasis, P. (2011). DSS for health emergency response: A contextual, user-centred approach. *International Journal of User-Driven Healthcare*, *1*(2), 39–56. doi:10.4018/IJUDH.2011040120110401

Okimoto, N., Futatsugi, N., Fuji, H., Suenaga, A., Morimoto, G., & Yanai, R. et al. (2009). High-performance drug discovery: computational screening by combining docking and molecular dynamics simulations. *PLoS Computational Biology*, *5*(10), e1000528. doi:10.1371/journal.pcbi.1000528 PMID:19816553

Olhoff, A., Simmons, B., Abaza, H., Tamiotti, L., Teh, R., & Kulaçoğlu, V. (2009). *Trade and climate change: A report by the United Nations Environment Programme and the World Trade Organization*. Geneva, Switzerland: WTO and UNEP.

Oliva, T., Oliver, R. L., & McMillan, I. (1992). A catastrophe model for developing service satisfaction strategies. *Journal of Marketing*, *56*(3), 83–95. doi:10.2307/1252298

Oliver, R. L. (1977). Effects of expectations and disconfirmation on post exposure product evaluation. *The Journal of Applied Psychology*, *62*(4), 480–486. doi:10.1037/0021-9010.62.4.480

Oliver, R. L. (1980). A cognitive model of the antecedents and consequences of satisfaction decisions. *JMR, Journal of Marketing Research*, *17*(4), 460–469. doi:10.2307/3150499

Oliver, R. L. (1981). Measurement and evaluation of satisfaction process in retail settings. *Journal of Retailing*, *57*(1), 25–48.

Oliver, R. L. (1993). A conceptual model of service quality and service satisfaction: Compatible goals, different concepts. In T. A. Swartz, D. E. Bowen, & S. W. Brown (Eds.), *Advances in marketing and management* (pp. 65–85). Greenwich, CT: JAI Press.

Oliver, R. L. (1997). *Satisfaction – A behavioral perspective on the consumer*. New York, NY: McGraw-Hill.

Oliver, R. L. (1999). Whence customer loyalty? *Journal of Marketing, 63*(4), 33–44. doi:10.2307/1252099

Ooi, K. T. (2005). Design optimization of a rolling piston compressor for refrigerators. *Applied Thermal Engineering, 25*(5), 813–829. doi:10.1016/j.applthermaleng.2004.07.017

Organization of Economic Cooperation and Development (OECD). (2012a). *Measuring trade in value added: An OECD-WTO joint initiative*. Industry and globalisation. OECD. Retrieved February 14, 2013, from http://www.oecd.org/industry/industryandglobalisation/measuring-tradeinvalue-addedanoecd-wtojointinitiative.htm

Organization of Economic Cooperation and Development (OECD). (2013). *China and OECD - Trade*. Retrieved February 23, 2013, from http://www.oecd.org/tad/chinaandoecd-trade.htm

Ortiz, F. Jr, & Simpson, J. R. (2002). *A genetic algorithm with a modified desirability function approach to multiple response optimization*. Florida: College of Engineering.

Ozgener, S., & Iraz, R. (2006). Customer relationship management in small-medium enterprises: The case of Turkish tourism industry. *Tourism Management, 27*(6), 1356–1363. doi:10.1016/j.tourman.2005.06.011

Padmavathy, C., Balaji, M. S., & Sivakumar, V. J. (2012). Measuring effectiveness of customer relationship management in Indian retail banks. *International Journal of Bank Marketing, 30*(4), 246–266. doi:10.1108/02652321211236888

Pamsari, M. B., Dehban, M., & Lulemani, H. K. (2013). Assessment of the key success factors of customer relationship management. *Universal Journal of Management and Social Sciences, 3*(4), 23–29.

Pan, S. L., Tan, C. W., & Lim, E. T. K. (2006). Customer relationship management (CRM) in e-government: A relational perspective. *Decision Support Systems, 42*(1), 237–250. doi:10.1016/j.dss.2004.12.001

Papadrakakis, M., Tsompanakis, Y., & Lagaros, N. D. (1999). Structural shape optimization using evolution strategies. *Engineering Optimization, 31*, 515–540. doi:10.1080/03052159908941385

Papageorgiou, L. G., Rotstein, G. E., & Shah, N. (2001). Strategic supply chain optimization for the pharmaceutical industries. *Industrial & Engineering Chemistry Research, 40*(1), 275–286. doi:10.1021/ie990870t

Parappagoudar, M. B., Pratihar, D. K., & Datta, G. L. (2007). Non-linear modeling using central composite design to predict green sand mould properties. *Proceedings of the Institution of Mechanical Engineers. Part B, Journal of Engineering Manufacture, 221*, 881–894. doi:10.1243/09544054JEM696

Parasuraman, A. (1997). Reflections on gaining competitive advantage through customer value. *Journal of the Academy of Marketing Science, 25*(2), 154–161. doi:10.1007/BF02894351

Parasuraman, A., Berry, L. L., & Zeithaml, V. A. (1991). Perceived service quality as a customer-focused performance measure: An empirical examination of organizational barriers using and extended service quality model. *Human Resource Management, 30*(3), 335–364. doi:10.1002/hrm.3930300304

Parasuraman, A., & Grewal, D. (2000). The impact of technology on the quality-value-loyalty chain: A research agenda. *Journal of the Academy of Marketing Science, 28*(1), 168–175. doi:10.1177/0092070300281015

Parasuraman, A., Zeithaml, V. A., & Berry, L. L. (1988). SERVQUAL: A multiple-item scale for measuring consumer perceptions of service. *Journal of Retailing, 64*(1), 12–40.

Parvatiyar, A., & Sheth, J. N. (2001). Customer relationship management: Emerging practice, process, and discipline. *Journal of Economic & Social Research, 3*(2), 1–34.

Payne, A., Christopher, M., Clark, M., & Peck, H. (1999). *Relationship marketing for competitive advantage*. Oxford, UK: Butterworth Heinemann.

Payne, A., & Frow, P. (2005). A strategic framework for customer relationship management. *Journal of Marketing, 69*(4), 652–671. doi:10.1509/jmkg.2005.69.4.167

Payne, A., & Frow, P. (2006). Customer relationship management: From strategy to implementation. *Journal of Marketing Management, 22*(1-2), 135–168. doi:10.1362/026725706776022272

Peng, M. W., & Health, P. S. (1996). The growth of firm in planned economies in transition: Institutions, organizations, and strategic choice. *Academy of Management Review, 21*(2), 492–528.

Perea-Lopez, E., Ydstie, B. E., & Grossmann, I. E. (2003). A model predictive control strategy for supply chain optimization. *Computers & Chemical Engineering, 27*(8), 1201–1218. doi:10.1016/S0098-1354(03)00047-4

Peters, G. P., & Hertwich, E. G. (2008). CO2 embodied in international trade with implications for global climate policy. *Environmental Science & Technology, 42*, 1401–1407. doi:10.1021/es072023k PMID:18441780

Phadke, M. S. (1989). *Quality engineering using robust design*. Englewood Cliffs, NJ: Prentice Hall.

Phillips, L. W., Chang, D. R., & Buzzel, R. D. (1983). Product quality, cost position and business performance: A test of some key hypotheses. *Journal of Marketing, 47*(2), 26–43. doi:10.2307/1251491

Pierreval, H., C. C., Paris, J. L., & Viguier, F. (2003). Evolutionary approaches to the design and organization of manufacturing systems. *Computers & Industrial Engineering, 44*(3), 339–364. doi:10.1016/S0360-8352(02)00195-X

Pinto, J. K., & Slevin, D. P. (1987). Critical factors in successful project implementation. *IEEE Transactions on Engineering Management, 34*(1), 22–27. doi:10.1109/TEM.1987.6498856

Pisano, G. P., & Verganti, R. (2008). Which kind of collaboration is right for you? The new leaders in innovation will be those who figure out the best way to leverage a network of outsiders. *Harvard Business Review, 84*(12), 78–86.

Plakoyiannaki, E., & Saren, M. (2006). Time and the customer relationship management process: Conceptual and methodological insights. *Journal of Business and Industrial Marketing, 21*(4), 218–230. doi:10.1108/08858620610672588

Porter, M. E. (1985). *Competitive advantage*. New York, NY: Free Press.

Porter, M. E. (1996). What is strategy? *Harvard Business Review, 74*(6), 61–78. PMID:10158474

Pradyumn, S. K. (2007). On the normal boundary intersection method for generation of efficient front. In Y. Shi et al. (Eds.), *ICCS 2007, Part I, LNCS 4487* (pp. 310–317). Berlin, Heidelberg: Springer-Verlag.

Prakhov, N. D., Chernorudskiy, A. L., & Gainullin, M. R. (2010). VSDocker: A tool for parallel high-throughput virtual screening using AutoDock on Windows-based computer clusters. *Bioinformatics (Oxford, England), 26*(10), 1374–1375. doi:10.1093/bioinformatics/btq149 PMID:20378556

Prothro, J. T., Durgin, G. D., & Griffin, J. D. (2006). The effects of a metal ground plane on RFID tag antennas. *IEEE Antennas and Propagation Society International Symposium, Albuquerque*, 3241-3244.

Pues, H., & Van, A. (1984). Accurate transmission-line model for the rectangular microstrip antenna. *IEE Proceedings H: Microwaves Optics and Antennas, 131*(6), 334-340.

Qi, T., Winchester, N., Karplus, V. J., & Zhang, X. (2012). *CO2 emissions embodied in China's trade and reduction policy assessment global trade analysis project (GTAP), United States*. Retrieved February 14, 2013, from https://www.gtap.agecon.purdue.edu/resources/download/5991.pdf

Raghavendran, P. S., Xavier, M. J., & Israel, D. (2012). Green purchasing practices: A study of eprocurement in B2B buying in Indian small and medium enterprises. *Journal of Supply Chain and Operations Management, 10*(1), 3–23.

Raidl, G. R. (2006). A unified view on hybrid metaheuristics. In F. Almeida (Ed.), *Hybrid metaheuristics* (pp. 1–12). Heidelberg, Germany: Springer Berlin Heidelberg. doi:10.1007/11890584_1

Ramik, J., & Vlach, M. (2002). Fuzzy mathematical programming: A unified approach based on fuzzy relations. *Fuzzy Optimization and Decision Making, 1,* 335–346. doi:10.1023/A:1020978428453

Ramnath, B. V., Elanchezhian, C., & Kesavan, R. (2010). Suitability assessment of lean kitting assembly through fuzzy based simulation model. *International Journal of Computers and Applications, 4*(1), 25–31. doi:10.5120/795-1129

Rao, K., Nikitin, P. V., & Lam, S. F. (2005). Antenna design for UHF RFID tags: A review and a practical application. *IEEE Transactions on Antennas and Propagation, 53*(12), 3870–3876. doi:10.1109/TAP.2005.859919

Rao, R. V., & Patel, B. K. (2011). Material selection using a novel multiple attribute decision making method. *International Journal of Manufacturing, Materials, and Mechanical Engineering, 1*(1), 43–56. doi:10.4018/ijmmme.2011010104

Rashedi, E., Nezamabadi-pour, H., & Saryazdi, S. (2009). GSA: A gravitational search algorithm. *Information Sciences, 179,* 2232–2248. doi:10.1016/j.ins.2009.03.004

Rauch, M., Hascoet, J. Y., Hamann, J. C., & Plennel, Y. (2009). Tool path programming optimization for incremental sheet forming applications. *Computer Aided Design.* doi:10.1016/j.cad.2009.06.006

Ravald, A., & Gronroos, C. (1996). The value concept and relationship marketing. *European Journal of Marketing, 30*(2), 19–30. doi:10.1108/03090569610106626

Rave, J. I. P., & Álvarez, G. P. J. (2011). Application of mixed-integer linear programming in a car seats assembling process. *Pesquisa Operacional, 31*(3), 593–610. doi:10.1590/S0101-74382011000300011

Reddy, B., & Brioso, R. G. (2011). *Automated and generic finite element analysis for industrial robot design* (MSc thesis). Linköping University, Department of Management and Engineering, Sweden.

Reddy, A. S., Pati, S. P., Kumar, P. P., Pradeep, H. N., & Sastry, G. N. (2007). Virtual screening in drug discovery- a computational perspective. *Current Protein & Peptide Science, 8*(4), 329–351. doi:10.2174/138920307781369427 PMID:17696867

Redstrom, J. (2006). Towards user design? On the shift from object to user as the subject of design. *Design Studies, 27*(2), 123–139. doi:10.1016/j.destud.2005.06.001

Reed, R., Lemak, D. J., & Montgomery, J. C. (1996). Beyond process: TQM content and firm performance. *Academy of Management Review, 21*(1), 173–202.

Reichheld, F. F. (1996). *The loyalty effect.* Boston, MA: Harvard Business School Press.

Reichheld, F. F., & Sasser, W. E. (1990). Zero defections: Quality comes to services. *Harvard Business Review, 68*(5), 105–111. PMID:10107082

Reichheld, F. F., & Teal, T. (1996). *The loyalty effect: The hidden force behind growth, profits, and lasting value.* Boston, MA: Harvard Business School Press.

Reid, S. E., & De Brentani, U. (2004). The fuzzy front end of new product development for discontinuous innovation: a theoretical model. *Journal of Product Innovation Management, 21*(3), 170–184. doi:10.1111/j.0737-6782.2004.00068.x

Rekiek, B., De Lit, P., & Delchambre, A. (2000). Designing mixed-product assembly lines. *IEEE Transactions on Robotics and Automation, 16*(3), 268–280. doi:10.1109/70.850645

Rekiek, B., De Lit, P., & Delchambre, A. (2002). Hybrid assembly line design and user's preferences. *International Journal of Production Research, 40*(5), 1095–1111. doi:10.1080/00207540110116264

Reynolds, K., & Arnold, M. (2000). Customer loyalty to the salesperson and the store: Examining relationship customers in an upscale retail context. *Journal of Personal Selling & Sales Management, 20*(2), 89–97.

Reynolds, K., & Beatty, S. (1999). Customer benefits and company consequences of customer-salesperson relationships in retailing. *Journal of Retailing, 75*(1), 11–32. doi:10.1016/S0022-4359(99)80002-5

RF-SIM. (2013). Retrieved October 31, 2013, from http://www.directel.hk/rfsim.php

Richards, K. A., & Jones, E. (2008). Customer relationship management: Finding value drivers. *Industrial Marketing Management, 37*(2), 120–130. doi:10.1016/j.indmarman.2006.08.005

Richardson, M. L., & Gartner, W. H. (1999). Contemporary organizational strategies for enhancing value in healthcare. *International Journal of Health Care Quality Assurance, 12*(5), 183–189. doi:10.1108/09526869910280339

Ritchie, M., Dewar, R., & Simmons, J. (1999). The generation and practical use of plans for manual assembly using immersive virtual reality. *Proceedings of the Institution of Mechanical Engineers. Part B, Journal of Engineering Manufacture, 213*(5), 461–474. doi:10.1243/0954405991516930

Roberts, K., Varki, S., & Brodie, R. (2003). Measuring the quality of relationships in customer services: An empirical study. *European Journal of Marketing, 37*(1-2), 169–196. doi:10.1108/03090560310454037

Rosenberg, R. S. (1967). *Simulation of genetic populations with biochemical properties* (Ph.D. thesis). University of Michigan, MI.

Rosendahl, K. E., & Strand, J. (2009). *Simple model frameworks for explaining inefficiency of the clean development mechanism*. Washington, DC: The World Bank. doi:10.1596/1813-9450-4931

Rosenthal, S. R., & Capper, M. (2006). Ethnographies in the front end: designing for enhanced customer experiences. *Journal of Product Innovation Management, 23*(3), 215–237. doi:10.1111/j.1540-5885.2006.00195.x

Ross, P. J. (1995). *Taguchi techniques for quality engineering*. New York, NY: McGraw-Hill Professional.

Ross, T. J. (2010). *Fuzzy logic with engineering applications*. Chicester, UK: Wiley-Blackwell. doi:10.1002/9781119994374

Rowley, J. (2005). The four Cs of customer loyalty. *Marketing Intelligence & Planning, 23*(6), 574–581. doi:10.1108/02634500510624138

Rust, R. T., Danaher, P., & Varki, S. (2000). Using service quality data for competitive marketing decisions. *International Journal of Service Industry Management, 11*(5), 438–469. doi:10.1108/09564230010360173

Rust, R. T., & Oliver, R. L. (1994). Service quality: Insights and managerial implications from the frontier. In R. T. Rust, & R. L. Oliver (Eds.), *Service quality: New directions in theory and practice* (pp. 1–19). London, UK: Sage. doi:10.4135/9781452229102.n1

Rust, R. T., & Zahorik, A. J. (1993). Customer satisfaction, customer retention and market share. *Journal of Retailing, 69*(2), 145–156. doi:10.1016/0022-4359(93)90003-2

Rust, R. T., Zahorik, A. J., & Keiningham, T. L. (1995). Return on quality (ROQ), Making service quality financially accountable. *Journal of Marketing, 59*(2), 58–70. doi:10.2307/1252073

Rust, R. T., Zeithaml, V. A., & Lemon, K. N. (2000). *Driving customer equity: How customer lifetime value is reshaping corporate strategy*. New York, NY: Free Press.

Rusut, R. T., Thompson, D. V., & Hamilton, R. W. (2006). Defeating feature fatigue. *Harvard Business Review, 84*(2), 98–107. PMID:16485808

Ryals, L., & Knox, S. (2001). Cross-functional issues in the implementation of relationship marketing through customer relationship management. *European Management Journal, 19*(5), 534–542. doi:10.1016/S0263-2373(01)00067-6

Saad, S. S., & Nakad, Z. S. (2011). A standalone RFID indoor positioning system using passive tags. *IEEE Transactions on Industrial Electronics, 58*(5), 1961–1970. doi:10.1109/TIE.2010.2055774

Saarijarvi, H., Karjaluoto, H., & Kuusela, H. (2013). Extending customer relationship management: From empowering firms to empowering customers. *Journal of Systems and Information Technology, 15*(2), 140–158. doi:10.1108/13287261311328877

Saaty, T. (1980). *The analytic hierarchy process: Planning, priority setting, resource allocation*. New York: McGraw-Hill.

Saaty, T. L. (1990). How to make a decision: The analytic hierarchy process. *European Journal of Operational Research, 48*(1), 9–26. doi:10.1016/0377-2217(90)90057-I

Sakkiah, S., Thangapandian, S., & Lee, K. W. (2012). Ligand-based virtual screening and molecular docking studies to identify the critical chemical features of potent cathepsin D inhibitors. *Chemical Biology & Drug Design*, *80*(1), 64–79. doi:10.1111/j.1747-0285.2012.01339.x PMID:22269155

Samson, D., & Terziovski, M. (1999). The relationship between total quality management practices and operational performance. *Journal of Operations Management*, *17*(4), 393–409. doi:10.1016/S0272-6963(98)00046-1

Sanchez, G., Jimenez, F., & Vasant, P. (2007). Fuzzy optimization with multi-objective evolutionary algorithms: A case study. In *Proceedings of the 2007 IEEE Symposium on Computational Intelligence in Multi-criteria Decision Making* (pp. 58-64). Honolulu, Hawaii.

Sarkis, J. (2003). A strategic decision framework for green supply chain management. *Journal of Cleaner Production*, *11*(4), 397–409. doi:10.1016/S0959-6526(02)00062-8

Sato, M. (2012). *Embodied carbon in trade: A survey of the empirical literature* (Centre for Climate Change Economics and Policy Working Paper No. 89, Grantham Research Institute on Climate Change and the Environment Working Paper No. 77). United Kingdom: Centre for Climate Change Economics and Policy and Grantham Research Institute on Climate Change and the Environment.

Schafer, T., & Schraft, R. D. (2004). *Incremental sheet forming by industrial robots using a hammering tool.* Paper presented at the 10th European Forum on Rapid Prototyping, Association Française de Prototypage Rapid.

Schwefel, H. P. (1979). Direct search for optimal parameters within simulation models. In *Proceedings of the 12th Annual Simulation Symposium* (pp.91-102), Tampa, Florida.

Seijo-Vidal, R. L., & Bartolomei-Suarez, S. M. (2010). Testing line optimization based on mathematical modeling from the metamodels obtained from a simulation. In the *Proceedings of the 2010 Winter Simulation Conference* (pp. 1739 – 1749). Baltimore: IEEE.

Seman. (2012). Green supply chain management: A review and research direction. *International Journal of Managing Value and Supply Chains*, *3*(1), 1-18.

Sena, J. I. V., Alves de Sousa, R. J., & Valente, R. A. F. (2011). On the use of EAS solid-shell formulations in the numerical simulation of incremental forming processes. *Engineering Computations*, *28*, 287–313. doi:10.1108/02644401111118150

Seuring, S., & Muller, M. (2008). From a literature review to a conceptual framework for supply chain management. *Journal of Cleaner Production*, *16*(15), 1699–1710. doi:10.1016/j.jclepro.2008.04.020

Sewall, M. A. (1978). Market segmentation based on consumer ratings of proposed product designs. *JMR, Journal of Marketing Research*, *15*(4), 557–564. doi:10.2307/3150625

Shafer, S. M., & Smunt, T. L. (2004). Empirical simulation studies in operations management: Context, trends, and research opportunities. *Journal of Operations Management*, *22*(4), 345–354. doi:10.1016/j.jom.2004.05.002

Sheth, J. N. (2002). The future of relationship marketing. *Journal of Services Marketing*, *16*(7), 590–592. doi:10.1108/08876040210447324

Sheth, J. N., Newman, B. I., & Gross, B. L. (1991). *Consumption values and market choice*. Cincinnati, OH: South Western.

Sheth, J. N., & Sisodia, R. S. (2002). Marketing productivity: Issues and analysis. *Journal of Business Research*, *55*(5), 349–362. doi:10.1016/S0148-2963(00)00164-8

Sheth, J. N., Sisodia, R. S., & Sharma, A. (2000). The antecedents and consequences of customer-centric marketing. *Journal of the Academy of Marketing Science*, *28*(1), 55–66. doi:10.1177/0092070300281006

Shim, M., & Park, J. (2001). The formability of aluminum sheet in incremental forming. *Journal of Materials Processing Technology*, *113*, 654–658. doi:10.1016/S0924-0136(01)00679-3

Shirts, M., Mobley, D., & Chodera, J. (2007). Alchemical free energy calculations: Ready for prime time? *Annual Reports in Comput Chem*, *3*, 41–59. doi:10.1016/S1574-1400(07)03004-6

Showalter, S. A., & Brüschweiler, R. (2007). Validation of molecular dynamics simulations of biomolecules using NMR spin relaxation as benchmarks: Application to the AMBER99SB force field. *Journal of Chemical Theory and Computation, 3*(3), 961–975. doi:10.1021/ct7000045

Shtub, A., & Zimerman, Y. (1993). A neural-network-based approach for estimating the cost of assembly systems. *International Journal of Production Economics, 93*(3), 189–207. doi:10.1016/0925-5273(93)90068-V

Shudong, C. S., Wenju, Z., Fanyuan, M., Jianhua, S., & Minglu, L. (2004). The design of a grid computing system for drug discovery and design. *Lecture Notes in Computer Science, 3251*, 799–802. doi:10.1007/978-3-540-30208-7_108

Shukla, A. C., Deshmukh, S. G., & Kanda, A. (2009). Environmentally responsive supply chains: Learning from the Indian auto sector. *Journal of Advances in Management Research, 6*(2), 154–171. doi:10.1108/09727980911007181

Siebers, P.-O. (2004). *The impact of human performance variation on the accuracy of manufacturing system simulation models* (PhD thesis). Cranfield University, UK.

Sievenpiper, D. F. (1999). *High-impedance electromagnetic surfaces* (Ph.D. dissertation). University of California, Los Angeles, CA.

SIMpass. (2013). Retrieved October 31, 2013, from http://www.watchdata.com/telecom/10022.html

Simpson, D., & Sampson, D. (2008). Developing strategies for green supply chain management. *Decision Sciences, 39*(4), 12–15.

Singh, A., Singh, B., & Dhingra, A. K. (2012). Drivers and barriers of green manufacturing practices: A survey of Indian industries. *International Journal of Engineering Science, 1*(1), 5–19.

Singh, L. P. (2011). Role of logistics and transportation in green supply chain management: An exploratory study of courier service industry in India. *International Journal of Advanced Engineering Technology, 2*(1), 260–269.

Singholi, A., Chhabra, D., & Bagai, S. (2010). Performance evaluation and design of flexible manufacturing system: A case study. *Global Journal of Enterprise Information System, 2*(1), 24–34.

Singh, S. P., & Sharma, R. R. K. (2006). A review of different approaches to the facility layout problems. *International Journal of Advanced Manufacturing Technology, 30*(5), 425–433. doi:10.1007/s00170-005-0087-9

Sin, L. Y. M., Tse, A. C. B., & Yim, F. H. K. (2005). CRM: Conceptualization and scale development. *European Journal of Marketing, 39*(11-12), 1264–1290. doi:10.1108/03090560510623253

Sirovetnukul, R., & Chutima, P. (2010). The impact of walking time on U-shaped assembly line worker allocation problems. *English Journal, 14*(2), 53–78.

Sivadas, E., & Baker-Prewitt, J. (2000). An examination of the relationship between service quality, customer satisfaction, and store loyalty. *International Journal of Retail & Distribution Management, 28*(2), 73–82. doi:10.1108/09590550010315223

Skjoedt, M., Silva, M. B., Martins, P. A. F., & Bay, N. (2010). Strategies and limits in multi-stage single-point incremental forming. *The Journal of Strain Analysis for Engineering Design, 45*(1), 33–44. doi:10.1243/03093247JSA574

Slater, S. F. (1997). Developing a customer value-based theory of the firm. *Journal of the Academy of Marketing Science, 25*(2), 162–167. doi:10.1007/BF02894352

Slater, S. F., & Narver, J. C. (1995). Market orientation and the learning organization. *Journal of Marketing, 59*(3), 63–74. doi:10.2307/1252120

Slater, S. F., & Narver, J. C. (1998). Customer-led and market-oriented: Let's not confuse the two. *Strategic Management Journal, 19*(10), 1001–1006. doi:10.1002/(SICI)1097-0266(199810)19:10<1001::AID-SMJ996>3.0.CO;2-4

Smed, J., Johnsson, M., Johtela, T., & Nevalainen, O. (1999). *Techniques and applications of production planning in electronics manufacturing systems* (Technical Report). Turku Centre for Computer Science.

Smith, J. B., & Colgate, M. (2007). Customer value creation: A practical framework. *Journal of Marketing Theory and Practice, 15*(1), 7–23. doi:10.2753/MTP1069-6679150101

Smith, M., & Chang, C. (2010). Improving customer outcomes through the implementation of customer relationship management: Evidence from Taiwan. *Asian Review of Accounting, 18*(3), 260–285. doi:10.1108/13217341011089658

Soler, C., Bergstrom, K., & Shanahan, H. (2010). Green supply chains and the missing link between environmental information and practice. *Business Strategy and the Environment, 19*(1), 14–25.

Solomon, M. R. (1992). *Consumer behavior: Buying, having and being*. Needham Heights, MA: Allyn and Bacon.

Solot, P., & van Vliet, M. (1994). Analytical models for FMS design optimization: A survey. *International Journal of Flexible Manufacturing Systems, 6*(3), 209–233. doi:10.1007/BF01328812

Song, X. M., & Parry, M. E. (1996). What separates Japanese new product winners from losers. *Journal of Product Innovation Management, 13*(5), 422–439. doi:10.1016/0737-6782(96)00055-0

Song, X. M., & Parry, M. E. (1997). A cross-national comparative study of new product development processes: Japan and the United States. *Journal of Marketing, 61*(2), 1–18. doi:10.2307/1251827

Song, X. M., & Parry, M. E. (1997). The determinants of Japanese new product successes. *JMR, Journal of Marketing Research, 34*(1), 64–76. doi:10.2307/3152065

Son, H., Choi, G., & Pyo, C. (2006). Design of wideband RFID tag antenna for metallic surfaces. *Electronics Letters, 42*(5), 2–3. doi:10.1049/el:20064323

Son, H., & Jeong, S. (2011). Wideband RFID Tag Antenna for Metallic Surfaces Using Proximity-Coupled Feed. *IEEE Antennas and Wireless Propagation Letters, 10*, 377–380. doi:10.1109/LAWP.2011.2148151

Sousa, R., & Voss, C. A. (2009). The effects of service failures and recovery on customer loyalty in e-services: An empirical investigation. *International Journal of Operations & Production Management, 29*(8), 834–864. doi:10.1108/01443570910977715

Spedding, T. A., De Souza, R., Lee, S. S. G., & Lee, W. L. (1998). Optimizing the configuration of a keyboard assembly cell. *International Journal of Production Research, 36*(8), 2131–2144. doi:10.1080/002075498192814

Spieckermann, S., Gutenschwager, K., Heinzel, H., & Voß, S. (2000). Simulation-based optimization in the automotive industry - A case study on body shop design. *Simulation, 75*(5), 276–286.

Srivastava, S. K. (2007). Green supply chain management: A state of the art literature review. *International Journal of Management Reviews, 9*(1), 53–80. doi:10.1111/j.1468-2370.2007.00202.x

Stampfer, M. (2004). Integrated setup and fixture planning system for gearbox casings. *International Journal of Advanced Manufacturing Technology, 26*(4), 310–318. doi:10.1007/s00170-003-1997-z

Starkey, M. W., Williams, D., & Stone, M. (2002). The state of customer management performance in Malaysia. *Marketing Intelligence & Planning, 20*(6), 378–385. doi:10.1108/02634500210445437

Statnikov, R. B., & Matusov, J. B. (1995). *Multicriteria optimization and engineering*. New York: Chapman and Hall. doi:10.1007/978-1-4615-2089-4

Steiner, C. J. (1995). A Philosophy for innovation: The role of unconventional individuals in innovations success. *Journal of Product Innovation Management, 12*(5), 431–440. doi:10.1016/0737-6782(95)00058-5

Stone, M., Woodcock, N., & Wilson, M. (1996). Managing the change from marketing planning to customer relationship management. *Long Range Planning, 29*(5), 675–683. doi:10.1016/0024-6301(96)00061-1

Storbacka, K., Strandvik, T., & Gronroos, C. (1994). Managing customer relationships for profit: The dynamics of relationship quality. *International Journal of Service Industry Management, 5*(8), 21–38. doi:10.1108/09564239410074358

Storn, R., & Price, K. V. (1995). *Differential evolution – A simple and efficient adaptive scheme for global optimization over continuous spaces* (ICSI, Technical Report TR-95-012), pp. 1-12.

Strauss, B., & Neuhaus, P. (1997). The qualitative satisfaction model. *International Journal of Service Industries Management, 8*(3), 236–249. doi:10.1108/09564239710185424

Suenaga, A., Okimoto, N., Hirano, Y., & Fukui, K. (2012). An efficient computational method for calculating ligand binding affinities. *PLoS ONE, 7*(8), e42846. doi:10.1371/journal.pone.0042846 PMID:22916168

Sukthomya, W., & Tannock, J. D. T. (2005). Taguchi experimental design for manufacturing process optimisation using historical data and a neural network process model. *International Journal of Quality & Reliability Management, 22*(5), 485–502. doi:10.1108/02656710510598393

Sung, W. T., & Chen, J. H. (2012). Enhancing Molecular Docking Efficiency for Computer-Aided Drug Design via Systems Theorem. *Int J Comp Consumer Control, 1*(1), 54–61.

Surekha, B., Kaushik, L. K., Panduy, A. K., Vundavilli, A. P. R., & Parappagoudar, M. B. (2011). Multiobjective optimization of green sand mould system using evolutionary algorithms. *International Journal of Advanced Manufacturing Technology, 58*, 1–9.

Sushil, K., Satsangi, P. S., & Prajapati, D. R. (2010). Optimization of green sand casting process parameters of a foundry by using Taguchi method. *International Journal of Advanced Manufacturing Technology, 55*, 23–34.

Swain, A. K., & Morris, A. S. (2000). A novel hybrid evolutionary programming method for function optimization. In *Proceedings of the Congress on Evolutionary Computation* (pp. 1369-1376).

Swan, J. E., & Oliver, R. L. (1989). Postpurchase communications by consumers. *Journal of Retailing, 65*(4), 516–533.

Sweeney, J. C., & Soutar, G. N. (2001). Consumer-perceived value: The development of a multiple-item scale. *Journal of Retailing, 77*(2), 203–220. doi:10.1016/S0022-4359(01)00041-0

Swift, P. W. (1997). Science drives creativity: A methodology for quantifying perceptions. *Design Management Journal, 8*(2), 51–57.

Szymanski, D. M., & Henard, D. H. (2001). Customer satisfaction: A meta-analysis of the empirical evidence. *Journal of the Academy of Marketing Science, 29*(1), 16–35. doi:10.1177/0092070301291002

Tabucanon, T. T. (1996). Multi objective programming for industrial engineers. In M. Avriel, & B. Golany (Eds.), *Mathematical programming for industrial engineers* (pp. 487–542). New York: Marcel Dekker, Inc.

Tanaka, S., Nakamura, T., & Hayakawa, K. (1999). Incremental sheet metal forming using elastic tools. In *Proceeding of the Sixth International Conference of Technology of Plasticity*, Nuremberg, (pp. 1477–1482).

Tanaka, S., Nakamura, T., Hayakawa, K., Nakamura, H., & Motomura, K. (2005). Incremental sheet metal forming process for pure titanium denture plate. In *Proceedings of the 8th International Conference on Technology of Plasticity* (pp. 135-136).

Tang, K., & Li, T. (2003). Comparison of different partial least-squares methods in quantitative structure–activity relationships. *Analytica Chimica Acta, 476*(1), 85–92. doi:10.1016/S0003-2670(02)01257-6

Tang, L. C., Goh, T. N., Yam, H. S., & Yoap, T. (2006). A unified approach for dual response surface optimization. *Journal of Quality Technology, 34*(4), 37–52.

Tan, M. Q. B., Tan, R. B. H., & Khoo, H. H. (2012). Prospects of carbon labelling - A life cycle point of view. *Journal of Cleaner Production*, 1–13.

Taylor, S., & Baker, T. (1994). An assessment of the relationship between service quality and customer satisfaction in the formation of consumers' purchase intentions. *Journal of Retailing, 70*(2), 163–178. doi:10.1016/0022-4359(94)90013-2

Teas, K. R., & Agarwal, S. (2000). The effects of extrinsic product cues on consumers' perceptions of quality, sacrifice, and value. *Journal of the Academy of Marketing Science, 28*(2), 278–290. doi:10.1177/0092070300282008

Tentzeris, M. M. (2010). Low-profile broadband RFID tag antennas mountable on metallic objects. *IEEE Antennas and Propagation Society International Symposium*, 1–4.

Testa, F., & Iraldo, F. (2010). Shadows and lights of GSCM (green supply chain management) determinants and effects of these practices based on a multinational study. *Journal of Cleaner Production, 18*(10/11), 953–962. doi:10.1016/j.jclepro.2010.03.005

The Gallup Organisation. (2009). *Europeans' attitudes towards the issue of sustainable consumption and production: Analytical report*. The Gallup Organisation.

The White House. (2012). *President Obama speaks on manufacturing*. Retrieved from http://www.whitehouse.gov/photos-and-video/video/2012/03/09/president-obama-speaks-manufacturing#transcript

Tirpak, T. M. (2008). Developing and deploying electronics assembly line optimization tools: A Motorola case study. *Decision Making in Manufacturing and Services, 2*(1-2), 63–78.

Tjahjono, B., Ball, P., Ladbrook, J., & Kay, J. (2009). Assembly line design principles using six sigma and simulation. In *the Proceedings of the 2009 Winter Simulation Conference* (pp. 3066- 3076). Austin, TX: IEEE.

Torczon, V. (1997). On the convergence of pattern search methods. *SIAM Journal on Optimization, 7*, 1–25. doi:10.1137/S1052623493250780

Torres, E. N., & Kline, S. (2006). From satisfaction to delight: A model for the hotel industry. *International Journal of Contemporary Hospitality Management, 18*(4), 290–301. doi:10.1108/09596110610665302

Townsend, J. D., Montoya, M. M., & Calantone, R. J. (2011). Form and function: A matter of perspective. *Journal of Product Innovation Management, 28*(3), 327–345. doi:10.1111/j.1540-5885.2011.00804.x

Triantaphyllou, E. (2000). *Multi-criteria decision making: A comparative study*. Dordrecht, The Netherlands: Kluwer Academic Publishers. doi:10.1007/978-1-4757-3157-6

Triantaphyllou, E., & Mann, S. H. (1995). Using the analytic hierarchy process for decision making in engineering applications: Some challenges. *International Journal of Industrial Engineering: Applications and Practice, 2*(1), 35–44.

Tronvoll, B. (2011). Negative emotions and their effect on customer complaint behavior. *Journal of Service Management, 22*(1), 111–134. doi:10.1108/09564231111106947

Tsao, H. Y., & Chen, L. W. (2005). Exploring brand loyalty from the perspective of brand switching costs. *International Journal of Management, 22*(3), 436–441.

Turabieh, H., Sheta, A., & Vasant, P. (2007). Hybrid optimization genetic algorithm (HOGA) with interactive evolution to solve constraint optimization problems for production systems. *International Journal of Computational Science, 1*(4), 395–406.

Ukkonen, L., Engels, D., Sydänheimo, L., & Kivikoski, M. (2004). Planar wire-type inverted-F RFID tag antenna mountable on metallic objects. *IEEE Antennas and Propagation Society International Symposium, 1*, 101-104

Ukkonen, L., Sydanheirno, L., & Kivikoski, M. (2004). A novel tag design using inverted-F antenna for radio frequency identification of metallic objects. *IEEE/Sarnoff Symposium on Advances in Wired and Wireless Communication*, 91-94.

Unal, R., & Dean, E. B. (1991). Taguchi approach to design optimization for quality and cost: An overview. In the *Proceedings of 13th Annual Conference of the International Society of Parametric Estimators* (pp. 1-9). New Orleans, LA: NASA Technical Documents.

UNEP. (2011). *Towards a green economy: Pathways to sustainable development and poverty eradication*. Retrieved November 11, 2012, from www.unep.org/greeneconomy

United Nations. (1992). *Agenda 21*. United Nations Conference on Environment & Development, Rio de Janerio, Brazil.

Upham, P., Dendler, L., & Bleda, M. (2011). Carbon labelling of grocery products: Public perceptions and potential emissions reductions. *Journal of Cleaner Production, 19*, 348–355. doi:10.1016/j.jclepro.2010.05.014

Utterback, J. M., Vedin Bengt-Arne, A. E., Ekman, S., Sanderson, S., Tether, B., & Verganti, R. (2006). *Design-Inspired Innovation*. New York: World Scientific.

Vachon, S. (2007). Green supply chain practices and the selection of environmental technologies. *International Journal of Production Research*, *45*(18-19), 4357–4379. doi:10.1080/00207540701440303

Vachon, S., & Klassen, R. D. (2006). Green project partnership in the supply chain: The case of the package printing industry. *Journal of Cleaner Production*, *14*(6/7), 661–671. doi:10.1016/j.jclepro.2005.07.014

Vasant, P. (2012). A novel hybrid genetic algorithms and pattern search techniques for industrial production planning. *International Journal of Modeling, Simulation, and Scientific Computing*, *3*(4). DOI No: 10.1142/S1793962312500201

Vasant, P. (2006). Fuzzy production planning and its application to decision making. *Journal of Intelligent Manufacturing*, *17*(1), 5–12. doi:10.1007/s10845-005-5509-x

Vasant, P. (2013). Hybrid optimization techniques for industrial production planning. In Z. Li, & A. Al-Ahmari (Eds.), *Formal methods in manufacturing systems: Recent advances* (pp. 84–111). Hershey, PA: Engineering Science Reference. doi:10.4018/978-1-4666-4034-4.ch005

Vasant, P., & Barsoum, N. N. (2006). Fuzzy optimization of units products in mix- product selection problem using FLP approach. *Soft Computing. A Fusion of Foundations. Methodologies and Applications*, *10*, 144–151.

Vasant, P., Barsoum, N., Kahraman, C., & Dimirovski, G. (2007). Application of fuzzy optimization in forecasting and planning of construction industry. In D. Vrakas, & I. Vlahavas (Eds.), *Artificial intelligence for advanced problem solving techniques* (pp. 254–265). Hershey, PA: IGI Publishing.

Vasant, P., Bhattacharya, A., Sarkar, B., & Mukherjee, S. K. (2007). Detection of level of satisfaction and fuzziness patterns for MCDM model with modified flexible S-curve MF. *Applied Soft Computing*, *7*, 1044–1054. doi:10.1016/j.asoc.2006.10.005

Vasant, P., & Kale, H. (2007). Introduction to fuzzy logic and fuzzy linear programming. In A. Frederick, & P. Humphreys (Eds.), *Encyclopedia of decision making and decision support technologies* (pp. 1–15). Hershey, PA: IGI Publishing.

Vasquez-Parraga, A. Z., & Alonso, S. (2000). Antecedents of customer loyalty for strategic intent. In J. P. Workman, & W. D. Perreault (Eds.), *Marketing theory and applications* (pp. 82–83). Chicago, IL: American Marketing Association.

Verbert, J., Belkassem, B., Henrard, C., Habraken, A. M., Gu, J., & Sol, H. et al. (2008). Multi-Step toolpath approach to overcome forming limitations in single point incremental forming. *International Journal of Material Forming*, *1*, 1203–1206. doi:10.1007/s12289-008-0157-2

Verganti, R. (2006). Innovating through design. *Harvard Business Review*, *84*(12), 114–122.

Verganti, R. (2008). Design, meanings, and radical innovation: A metamodel and a research agenda. *Journal of Product Innovation Management*, *25*(3), 436–456. doi:10.1111/j.1540-5885.2008.00313.x

Verganti, R. (2009). *Design-driven innovation: Changing the rules of competition by radically innovating what things mean*. Boston, MA: Harvard Business Press.

Verganti, R. (2011). Designing breakthrough products. *Harvard Business Review*, *89*(10), 115–120.

Verganti, R. (2011). Radical design and technological epiphanies: A new focus for research on design management. *Journal of Product Innovation Management*, *28*(3), 384–388. doi:10.1111/j.1540-5885.2011.00807.x

Verhoef, P. C., & Donkers, B. (2001). Predicting customer potential value: An application in the insurance industry. *Decision Support Systems*, *32*(2), 189–199. doi:10.1016/S0167-9236(01)00110-5

Verworn, B., Herstatt, C., & Nagahira, A. (2008). The fuzzy front end of Japanese new product development projects: impact on success and differences between incremental and radical projects. *R & D Management*, *38*(1), 1–19. doi:10.1111/j.1467-9310.2007.00492.x

Veryzer, R. W. (1993). Aesthetic response and the influence of design principles on product preferences. *Advances in Consumer Research. Association for Consumer Research (U. S.)*, *20*(1), 224–228.

Veryzer, R. W. (1995). The place of product design and aesthetics. *Advances in Consumer Research. Association for Consumer Research (U. S.)*, *22*(1), 641–645.

Veryzer, R. W. (1998). Discontinuous innovation and the new product development process. *Journal of Product Innovation Management*, *15*(4), 304–321. doi:10.1016/S0737-6782(97)00105-7

Veryzer, R. W. (2005). The roles of marketing and industrial design in discontinuous new product development. *Journal of Product Innovation Management*, *22*(1), 22–41. doi:10.1111/j.0737-6782.2005.00101.x

Veryzer, R. W., & Borja de Mozota, B. (2005). The impact of user-oriented design on new product development. *Journal of Product Innovation Management*, *22*(1), 128–143. doi:10.1111/j.0737-6782.2005.00110.x

Veryzer, R. W., & Hutchinson, W. (1998). The influence of utility and prototypicality on aesthetic responses to new product designs. *The Journal of Consumer Research*, *24*(4), 224–228. doi:10.1086/209516

Vredenburg, K., Isensee, S., & Righi, C. (2002). *User-centered design: An integrated approach*. Upper Saddle River, NJ: Prentice Hall.

Waagen, D., Diercks, P., & McDonnell, J. (1992). The stochastic direction set algorithm: A hybrid techniques for finding function extreme. In D. B. Fogel and W. Atmar (Eds.), *Proceedings of the 1st Annual Conference on Evolutionary Programming* (pp. 35-42). Evolutionary Programming Society.

Wah, B. W., & Chen, Y. (2003). Hybrid evolutionary and annealing algorithms for nonlinear discrete constrained optimization. *International Journal of Computational Intelligence and Applications*, *3*(4), 331–355. doi:10.1142/S1469026803001063

Wahid, N. A., Rahbar, E., & Shyan, T. S. (2011). Factors influencing the green purchase behavior of Penang environmental volunteers. *Journal of International Business Management*, *5*(1), 38–49. doi:10.3923/ibm.2011.38.49

Wang, X., Tang, D., & Loua, P. (2009). An ergonomic assembly workstation design using axiomatic design theory. In the *Proceedings of the 16th ISPE International Conference on Concurrent Engineering* (pp. 403-412). London, UK: Springer London.

Wang, A., Koc, B., & Nagi, R. (2005). Complex assembly variant design in agile manufacturing. Part I: System architecture and assembly modeling methodology. *IIE Transactions*, *37*(1), 1–15. doi:10.1080/07408170590516764

Wang, J. (2005). *A review of operations research applications in workforce planning and potential modeling of military training. (Land operations division: Systems sciences laboratory report)*. Australian Government Department of Defense.

Wang, J., Wolf, R. M., Caldwell, J. W., Kollman, P. A., & Case, D. A. (2004). Development and testing of a general amber force field. *Journal of Computational Chemistry*, *25*(9), 1157–1174. doi:10.1002/jcc.20035 PMID:15116359

Wang, Q., & Chatwin, C. R. (2005). Key issues and developments in modelling and simulation-based methodologies for manufacturing systems analysis, design and performance evaluation. *International Journal of Advanced Manufacturing Technology*, *25*(11-12), 1254–1265. doi:10.1007/s00170-003-1957-7

Wang, Q., Lassalle, S., Mileham, A. R., & Owen, G. W. (2009). Analysis of a linear walking worker line using a combination of computer simulation and mathematical modeling approaches. *Journal of Manufacturing Systems*, *28*(2-3), 64–70. doi:10.1016/j.jmsy.2009.12.001

Wang, T. Y., Wu, K. B., & Liu, Y. W. (2001). A simulated annealing algorithm for facility layout problems under variable demand in cellular manufacturing systems. *Computers in Industry*, *46*(2), 181–188. doi:10.1016/S0166-3615(01)00107-5

Wang, Y. G., & Lo, H. P. (2002). Service quality, customer satisfaction, customer value and behavior intentions: Evidence from China's telecommunication industry. *Info – The Journal of Policy. Regulation and Strategy for Telecommunications*, *4*(6), 50–60. doi:10.1108/14636690210453406

Wang, Y., Lo, H. P., Chi, R., & Yang, Y. (2004). An integrated framework for customer value and customer-relationship-management performance: A customer-based perspective from China. *Managing Service Quality*, *14*(2-3), 169–182. doi:10.1108/09604520410528590

Wanner, E. F., Guimarae, F. G., Saldanha, R. R., Takahashi, R. H., & Fleming, P. J. (2005). Constraint quadratic approximation operator for treating equality constraints with genetic algorithms. In *Proceedings of the 2005 IEEE Congress on Evolutionary Computation* (pp. 2255-2262).

Wayzode, N. D., & Tupkar, A. B. (2013). Customization of Catia V5 for design of shaft coupling. In *Proceedings of the International Conference on Emerging Frontiers in Technology for Rural Area 2012*, (pp. 30-33).

Wayzode, N. D., & Wankhade, N. (2013). Design of flange coipling using CATSCript. *Indian Streams Research Journal*, 2(12), 1–7.

Webster, F. E. (1988). Determining the characteristics of the socially conscious consumer. *Business Horizons*, 31(3), 29–39. doi:10.1016/0007-6813(88)90006-7

Whalsh, V. (1995). The evaluation of design. *International Journal of Technological Management*, 10(4/5/6), 489-509.

Whalsh, V. (1996). Design, innovation and the boundaries of the firm. *Research Policy*, 25(4), 509–529. doi:10.1016/0048-7333(95)00847-0

Whalsh, V., Roy, R., & Bruce, M. (1998). Competitive by design. *Journal of Marketing Management*, 4(2), 201–216. doi:10.1080/0267257X.1988.9964069

Whitelock, V. G. (2012). Alignment between green supply chain management strategy and business strategy. *International Journal of Procurement Management*, 5(4), 430–451. doi:10.1504/IJPM.2012.047198

Wiedmann, T., Wood, R., Lenzen, M., Minx, J., Guan, D., & Barrett, J. (2008). *Development of an embedded carbon emissions indicator – Producing a time series of input-output tables and embedded carbon dioxide emissions for the UK by using a MRIO data optimisation system*. London: Stockholm Environment Institute at the University of York and Centre for Integrated Sustainability Analysis at the University of Sydney.

Wikström, J. (2011). *3D model of fuel tank for system simulation - A methodology for combining CAD models with simulation tools* (MSc thesis). Linköping University, Department of Management and Engineering, Sweden.

Wikstrom, S., & Normann, R. (1994). *Knowledge and value: A new perspective on corporate transformation*. London, UK: Routledge.

Willemain, T. R., Smart, C. N., Shockor, J. H., & DeSautels, P. A. (1994). Forecasting intermittent demand in manufacturing: A comparative evaluation of Croston's method. *International Journal of Forecasting*, 10(4), 529–538. doi:10.1016/0169-2070(94)90021-3

Wilson, G. M., & Muftuoglu, Y. (2012). Computational strategies in cancer drug discovery. In R. Mohan (Ed.), *Advances in cancer management* (pp. 237–254).

Winfield, L. L., Inniss, T. R., & Smith, D. M. (2009). Structure activity relationship of antiproliferative agents using multiple linear regression. *Chemical Biology & Drug Design*, 74(3), 309–316. doi:10.1111/j.1747-0285.2009.00863.x PMID:19703034

Wold, S. (1987). Principal component analysis. *Chemometrics and Intelligent Laboratory Systems*, 2, 37–52. doi:10.1016/0169-7439(87)80084-9

Wong, C. K. K., Mok, P. Y., Ip, W. H., & Chan, C. K. (2005). Optimization of manual fabric-cutting process in apparel manufacture using genetic algorithms. *International Journal of Advanced Manufacturing Technology*, 27(1-2), 152–158. doi:10.1007/s00170-004-2161-0

Wong, P. Y. M., & Leung, S. Y. S. (2006). Developing a genetic optimisation approach to balance an apparel assembly line. *International Journal of Advanced Manufacturing Technology*, 28(3-4), 387–394. doi:10.1007/s00170-004-2350-x

Woodcock, N., Stone, M., & Foss, B. (2003). *The customer management scorecard: Managing CRM for profit*. London, UK: Kogan Page.

Woodruff, R. B. (1997). Customer value: The next source for competitive advantage. *Journal of the Academy of Marketing Science*, 25(2), 139–153. doi:10.1007/BF02894350

Woodruff, R. B., & Gardial, S. (1996). *Know your customer: New approaches to understanding customer value and satisfaction*. Oxford, UK: Blackwell.

World Commission for Environment and Development WCED. (1987). *Our common future*. Oxford: Oxford University Press.

Wu, J. Y., Li, J. X., Cui, X. S., & Mao, L. H. (2011). miniaturized dual-band patch antenna mounted on metallic plates for RFID passive tag. In *Proceedings of International Conference on Control, Automation and Systems Engineering (CASE)*, Singapore, 1-4.

Wu, F.-C. (2009). Robust design of nonlinear multiple dynamic quality characteristics. *Computers & Industrial Engineering*, *56*(4), 1328–1332. doi:10.1016/j.cie.2008.08.001

Wyner, G. A. (1996). Customer valuation: Linking behavior and economics. *Marketing Research*, *8*(2), 36–38.

Xu, M., & Walton, J. (2005). Gaining customer knowledge through analytical CRM. *Industrial Management & Data Systems*, *105*(7), 955–971. doi:10.1108/02635570510616139

Xu, X. (2012). From cloud computing to cloud manufacturing. *Robotics and Computer-integrated Manufacturing*, *28*(1), 75–86. doi:10.1016/j.rcim.2011.07.002

Xu, Y., Yen, D. C., Lin, B., & Chou, D. C. (2002). Adopting customer relationship management technology. *Industrial Management & Data Systems*, *102*(8-9), 442–452. doi:10.1108/02635570210445871

Xu, Z., & Liang, M. (2006). Integrated planning for product module selection and assembly line design / reconfiguration. *International Journal of Production Research*, *44*(11), 2091–2117. doi:10.1080/00207540500357146

Yamane, T. (1970). *Statistics – An introductory analysis*. Tokyo, Japan: John Weatherhill.

Yang, C. L., & Sheu, C. (2011). The effects of environmental regulations on green supply chains. *African Journal of Business Management*, *5*(26), 10601–10614.

Yang, P. H., Li, Y., Jiang, L. J., Chew, W. C., & Ye, T. T. (2011). Compact metallic RFID tag antennas with a loop-fed method. *IEEE Transactions on Antennas and Propagation*, *59*(12), 4454–4462. doi:10.1109/TAP.2011.2165484

Yang, W., & Zhang, Y. (2012). Research on factors on green purchasing practices of Chinese. *Journal of Business Management and Economics*, *3*(5), 222–231.

Yan, Y., & Yang, L. (2010). China's foreign trade and climate change: A case study of CO_2 emissions. *Energy Policy*, *38*, 350–356. doi:10.1016/j.enpol.2009.09.025

Yen, Y., & Yen, S. (2011). Top-management's role in green purchasing standards in high-tech industrial firms. *Journal of Business Research*, *65*(7), 951–959. doi:10.1016/j.jbusres.2011.05.002

Yin, R. (1994). *Case study research: Design and METHODS* (2nd ed.). Thousand Oaks, CA: Sage Publications Inc.

Yoshimura, M., Yoshida, S., Konish, I., Izui, Y., Nishiwaki, K., & Inamor, S. et al. (2006). A rapid analysis method for production line design. *International Journal of Production Research*, *44*(6), 1171–1192. doi:10.1080/00207540500336355

Yu, B., Kim, S. J., Jung, B., Harackiewicz, F. J., & Lee, B. (2006). RFID tag antenna using two-shorted microstrip patches mountable on metallic objects. *Microwave and Optical Technology Letters*, *49*(2), 414–416. doi:10.1002/mop.22159

Zablah, A. R., Bellenger, D. N., & Johnston, W. J. (2004). An evaluation of divergent perspectives on customer relationship management: Towards a common understanding of an emerging phenomenon. *Industrial Marketing Management*, *33*(6), 475–489. doi:10.1016/j.indmarman.2004.01.006

Zadeh, L. A. (1965). Fuzzy sets. *Information and Control*, *8*(3), 338–353. doi:10.1016/S0019-9958(65)90241-X

Zadeh, L. A. (1971). Similarity relations and fuzzy orderings. *Information Sciences*, *3*, 177–206. doi:10.1016/S0020-0255(71)80005-1

Zamora, J., Vasquez-Parraga, A. Z., Morales, F., & Cisternas, C. (2004). Formation process of guest loyalty: Theory and empirical test. *Estudios y Perspectivas en Turismo*, *13*(3-4), 197–221.

Zeithaml, V. A. (1988). Consumer perceptions of price, quality and value: A means-end model and synthesis of evidence. *Journal of Marketing*, *52*(3), 2–22. doi:10.2307/1251446

Zeithaml, V. A., Berry, L., & Parasuraman, A. (1996). The behavioral consequences of service quality. *Journal of Marketing*, *60*(2), 31–46. doi:10.2307/1251929

Zeithaml, V. A., Parasuraman, A., & Berry, L. (1990). *Delivering quality service: Balancing customer perceptions and expectations*. New York, NY: Free Press.

Zelinka, I. (2002). Analytic programming by means of SOMA Algorithm. In *Proc. 8th International Conference on Soft Computing Mendel'02, Brno, Czech Republic*, (pp. 93-101).

Zhang, C. H., & Huang, S. H. (1995). Application of neural network in manufacturing—A state of art survey. *International Journal of Production Research, 33*(3), 705–728. doi:10.1080/00207549508930175

Zhang, Q., & Doll, W. J. (2001). The fuzzy front end and success of new product development: A causal model. *European Journal of Innovation Management, 4*(2), 95–112. doi:10.1108/14601060110390602

Zha, X. F., & Lim, S. Y. E. (2003). Intelligent design and planning of manual assembly workstations: A neuro-fuzzy approach. *Computers & Industrial Engineering, 44*(4), 611–632. doi:10.1016/S0360-8352(02)00238-3

Zhou, S. Q., Ling, W. Q., & Peng, Z. X. (2007). An RFID based remote monitoring system for enterprise internal production management. *International Journal of Advanced Manufacturing Technology, 33*(7-8), 837–844. doi:10.1007/s00170-006-0506-6

Zhou, X., & Kojima, S. (2010). *Carbon emissions embodied in international trade: An assessment from the Asian perspective*. Japan: Institute for Global Environmental Strategies.

Zhou, Z., Cheng, S., & Hua, B. (2000). Supply chain optimization of continuous process industries with sustainability considerations. *Computers & Chemical Engineering, 24*(2), 1151–1158. doi:10.1016/S0098-1354(00)00496-8

Zhu, Q., & Sarkis, J. (2007). The moderating effects of institutional pressures on emergent green supply chain practices and performance. *International Journal of Production Research, 45*(18-19), 4333–4355. doi:10.1080/00207540701440345

Zhu, Q., Sarkis, J., & Geng, Y. (2005). Green supply chain management in China: Pressure, practices and performance. *International Journal of Operations & Production Management, 25*(5), 449–468. doi:10.1108/01443570510593148

Zhu, Q., Sarkis, J., & Lai, K. (2008). Green supply chain management implications for closing the loop. *Transportation Research Part E, Logistics and Transportation Review, 44*(1), 1–8. doi:10.1016/j.tre.2006.06.003

Zhu, Q., Sarkis, J., & Lai, K. H. (2012). Examining the effects of green supply chain management practices and their mediations on performance improvements. *International Journal of Production Research, 50*(5), 1377–1394. doi:10.1080/00207543.2011.571937

Zhu, Q., Sarkis, J., Lai, K., & Geng, Y. (2008). The role of organizational size in the adoption of green supply chain management practices in China. *Corporate Social Responsibility and Environmental Management, 15*(6), 322–337. doi:10.1002/csr.173

Zimmermann, H. J. (1996). *Fuzzy set theory and its applications*. Boston, MA: Kluwer. doi:10.1007/978-94-015-8702-0

Zineldin, M. (2006). The royalty of loyalty: CRM, quality and retention. *Journal of Consumer Marketing, 23*(7), 430–437. doi:10.1108/07363760610712975

Zitzler, E., & Thiele, L. (1998). Multiobjective optimization using evolutionary algorithms - A comparative case study. In *Conference on Parallel Problem Solving from Nature (PPSN V)*, (pp. 292–301).

Zitzler, E., Knowles, J., & Thiele, L. (2008). Quality assessment of Pareto set approximations. In J. Branke et al. (Eds.), *Multiobjective optimization, LNCS 5252* (pp. 373–404). Berlin, Heidelberg: Springer-Verlag. doi:10.1007/978-3-540-88908-3_14

About the Contributors

Zongwei Luo is a senior researcher at the E-business Technology Institute, The University of Hong Kong (China). Before that, he was working at the IBM TJ Watson Research Center in Yorktown Height (NY, USA). He also served as the Affiliate Senior Consultant to ETI Consulting Limited. His research has been supported by various funding sources, including China NSF, HKU seed funding, HK RGC, and HK ITF. His research results have appeared in major international journals and leading conferences. He is the founding Editor-in-Chief of the *International Journal of Applied Logistics* and serves as an associate editor and editorial advisory board member in many international journals. Dr. Luo's recent interests include applied research and development in the area of robotics and automation, service science and computing, innovation management and sustainable development, technology adoption and risk management, and e-business model and practices, especially for logistics and supply chain management.

***

Atiya Al-Zuheri received his PhD in Manufacturing Engineering at the University of South Australia. He also holds a Master degree in industrial engineering from University of Technology, Iraq. He has over 18 years' experience in design, manufacturing, and operations management in various firms in Iraq. His main research interests in modelling and simulation of flexible assembly systems and ergonomics. Atiya has 17 publications to his credit in international conferences and journals.

A. Andrade-Campos received his PhD in Mechanical Engineering from the University of Aveiro, Portugal in 2005. He is an Assistant Professor of Mechanical Engineering at the University of Aveiro, research collaborator of CEMUC (Mechanical Engineering Center from University of Coimbra, Portugal) and LiMATb (Laboratory of Materials Engineering of Brittany, France). His research interests include inverse methods, identification and determination of constitutive model parameters, optimization methods, the use of optimization methods in mechanical systems and shape optimization in metal forming problems.

Saikat Kumar Basu recieved his two Masters in Botany (Specialization: Microbiology) from the University of Calcutta (India) and Agricultural Studies from the University of Lethbridge (Canada). Currently a doctoral candidate of Biomolecular Sciences in the Department of Biological Sciences (University of Lethbridge), he has several peer-reviewed articles and book chapters published in reputed national and international journals. He has written, edited and co-edited over 20 books till date. He has participated in a number of national and international seminars and conferences presenting papers and posters and is an active member of a number of organizations.

About the Contributors

Surajit Bag is presently working as Procurement manager at Tega Industries India Limited and at the same time pursuing his executive PhD in the field of SCM from University of Petroleum & Energy Studies, Dehradun. He has published two papers in Scopus indexed journals and presented more than 3 papers at reputed national and international conferences.

Tianxin Cai is currently researching dynamic & steady state modeling/large scale global optimization algorithm and application on industrial operation/meteorology/environment impact/ scientific legislation/economic analysis/operation safety. This research has covered three major areas: optimization, simulation and complex systems. The application field has focused on optimization including nonlinear, mixed-integer, disjunctive programming, and global optimization; optimization of differential algebraic systems; synthesis of energy and water systems, and metabolic networks; planning, and scheduling; enterprise-wide optimization; optimization methods for data-handling problems. Simulation involves the modeling and optimization with Aspen, HYSYS, ProII,Matlab and gPROMS involving design and verification of process operating systems; and fault tree analysis for safety. Finally, topics in complex systems include microscale chemical synthesis and sensing; agent systems in engineering design and optimization; synthesis and design of chemical processes.

Xiuhua Chen completed her master and doctoral degrees in Botany (Specialization: plant molecular biology and bioengineering) from Northeast Agricultural University, Harbin, China in Jan. 2010 and Jun. 2012, respectively. She did a joint PhD fellowship in Agricultural and Agri-Food Canada for two years between Jan. 2010 and Jan. 2012. Currently she is working as a research assistant on maize molecular breeding in Institute of Food Crops, Yunnan Academy of Agricultural Sciences since Aug. 2012. She has organized and participated in several national and international trainings, conferences and seminars. She is interested on molecular biology, bioengineering, breeding and bioinformatics.

Sumukh Deshpande is a Bioinformatics Scientist at Premier Biosoft (India). He received his Masters and a Bachelors degree in Bioinformatics from Glasgow (UK) in 2009 and from Vellore (India) in 2008, respectively. Before working at Premier Biosoft (India), he has also worked as Bioinformatics Research Associate at Jaivik Data Consulting (India) from July 2012 to November 2013 and has worked as Research Assistant at University of Calgary (Canada) from May 2010 to April 2012. He is involved in the development of cloud-based NGS software using novel bioinformatics algorithms and scripting languages (Unix and R) for the development of pipeline for NGS data processing and analysis. He is also involved in providing whole-genome and exome sequencing data analysis support along with development of data analysis pipeline using various bioinformatics tools and scripting languages which includes linux shell, perl and R programming.

Rameshwar Dubey is actively involved in research and full time teaching at Symbiosis Institute of Operations Management, Nasik & as an adjunct faculty at University of Petroleum & Energy Studies. He hold a PhD in Operations Management and DBA (Manufacturing Management), with over 15 papers (published & finally accepted for publications in Scopus & SCI listed Journals), 5 papers in other reputed refereed Journals and 5 book chapters published by Springer, IGI and McGraw Hill. He is an editor-in-chief of Journal of Supply Chain Management Systems, editorial board member of International Journal of Innovation Science (Scopus indexed) and AIMS International Journal. He is one of the founding members

About the Contributors

of International Association of Innovation Professionals (USA) and an elected Fellow member of Indian Institution of Production Engineers & a proud recipient of AIMS International-IMT youngest research award for exemplary work towards promoting research in the year 2011 and best faculty award for the year 2008 by University of Petroleum & Energy Studies, Dehradun. He is presently an active reviewer of The TQM Journal, Journal of Humanitarian Logistics and Supply Chain Management, Benchmarking: An International Journal, Supply Chain Management, International Journal of Production Research, International Journal of Innovation Science, AIMS International Journal, International Journal of Indian Culture and Business Management and Oxford University Press.

Irraivan Elamvazuthi obtained his PhD from the Department of Automatic Control & Systems Engineering, University of Sheffield, UK in 2002. He is currently an Associate Professor at the Department of Electrical and Electronic Engineering, Universiti Teknologi PETRONAS (UTP), Malaysia. His research interests include Control, Robotics, Mechatronics, Power Systems and Bio-medical Applications.

João Batista Sá de Farias graduated in Mechanic Engineering from the University of Vale do Rio dos Sinos, Brazil, in 2003. He obtained the M.Sc. degree in Mechanic Engineering in 2008 at the University Federal do Rio Grande do Sul. He is now a PhD student in the Mechanical Engineering Department at the University of Aveiro. His research interests are CAD/CAM strategies for single point incremental forming and incremental forming of high strength steels.

Jorge A.F. Ferreira graduated from the Electronics and Telecommunications Engineering program at the University of Aveiro, Portugal, in 1990. He obtained the M.Sc. degree in Electronics and Telecommunications Engineering in 1994 and the Ph.D. degree in Mechanical Engineering in 2003, both from the University of Aveiro. He is now an Assistant Professor in the Mechanical Engineering Department at the University of Aveiro. His research interests transverse the industrial automation field in areas such as modeling and simulation of physical systems, fluid power systems, hardware-in-the-loop simulation or instrumentation and control systems.

Nikolaos A. Fountas received his mechanical engineering diploma from the School of Pedagogical & Technological Education (ASPETE) of Athens, Greece in 2004 and his MSc degree from Kingston University in 2008. Since 2004, he has worked as a design and manufacturing engineer for several industries mainly in the fields of CNC programming, machine tools / manufacturing and product design. Currently he is a PhD student at Kingston University-UK (engineering faculty) and he works as laboratory assistant under contract, teaching CNC machine tools - CNC programming, materials processing technology and CAD/CAM/CAE systems in the Department of Mechanical Engineering Technology Educators, School of Pedagogical & Technological Education (ASPETE), Athens, Greece. His main research interests fall into the areas of CNC programming, CAD/CAM/CAE systems and software automation, sculptured surface machining and optimization through artificial intelligence / soft computing techniques.

T. Ganesan is currently a doctoral candidate with the Department of Chemical Engineering Universiti Teknologi PETRONAS (UTP), Tronoh, Malaysia. He holds a bachelor's degree in Chemical Engineering (Hons.) and a Master of Science in Computational Fluid Dynamics from UTP. His research interests include multi-objective optimization and computational intelligence.

Kiminori Gemba is a Professor at the Graduate School of Technology Management, Ritsumeikan University. His research areas are innovation theories and strategic management. In 1992, he was awarded a masters degree in engineering at the University of Tokyo. Subsequently, he joined Sanwa Research Institute as a researcher. During seven years, in the research division, his work has focused on environmental management and new industry creation. He received a Ph.D. degree from University of Tokyo in 1999. His Doctoral thesis was "The dynamics of diversification of Japanese industries". Afterward, he became a research associate and an associate professor at the University of Tokyo.

Satoru Goto is a Ph.D candidate in Department of Technology Management at Ritsumeikan University, Japan. He has an undergraduate degree and master degree in Science & Engineering at Ritsumeikan University. His research for his master degree was in robotics technology. He has a business experience as a mechanical engineer at HORIBA, Ltd. and engaged in the development of X-Ray fluorescence analyzer. His current research interests are the interaction between design and technology, and the role of industrial designers in New Product Development and Design Driven Innovation. He has had articles published and submitted in Japan Association for Management System (JAMS). He translated Design driven innovation, which Robert Verganti published in 2009, into Japanese.

Shuichi Ishida is currently a Professor of Technology Management at Ristumeikan University, Japan. He has been a visiting researcher in Institute for Manufacturing and visiting fellow in St Edmund's College at University of Cambridge, UK. His main research interests concern innovation dynamics, technological entrepreneurships and management systems. He received his Ph.D. in the field of Business Administration from Hokkaido University and Doctor of Engineering from Kyoto University respectively. He has a business experience as a Li-ion-battery-engineer at SONY. He is also a co-founder and Chairman of the society for diffusion of low-carbon-emission-cars in Osaka supported by Ministry of Land, Infrastructure, Transport and Tourism (MLIT), and a director of several Japanese academic societies. He has had articles published and submitted in *Meso-Organizations and the creation of knowledge (PRAEGER)*, *RADMA* and *International Society for Professional Innovation Management (ISPIM)*.

Kijpokin Kasemsap received his BEng degree in Mechanical Engineering from King Mongkut's University of Technology Thonburi, his MBA degree from Ramkhamhaeng University, and his DBA degree in Human Resource Management from Suan Sunandha Rajabhat University. Now he is a Special Lecturer at Faculty of Management Sciences, Suan Sunandha Rajabhat University based in Bangkok, Thailand. He is a Member of International Association of Engineers (IAENG), International Association of Engineers and Scientists (IAEST), International Economics Development and Research Center (IEDRC), International Association of Computer Science and Information Technology (IACSIT), International Foundation for Research and Development (IFRD), and International Innovative Scientific and Research Organization (IISRO). He also serves on an Editorial Board of International Journal of Management Sciences. He has numerous original research articles in top international journals, conference proceedings, and book chapters on business management, human resource management, and knowledge management published internationally.

About the Contributors

Agathocles A. Krimpenis received his PhD (2008) from National Technical University of Athens (Mechanical Engineering department). He is currently a lecturer under contract, teaching CNC Machine Tools - CNC programming, robotics and CAD/CAM/CAE systems in the Department of Mechanical Engineering Technology Educators, School of Pedagogical & Technological Education (ASPETE), Athens, Greece. His main research interests fall into the areas of CNC programming, CAD/CAM/CAE systems and software automation, sculptured surface machining and optimization through artificial intelligence / soft computing techniques.

Xianping Li obtained his Master in Botany from Yunnan University in 2002 (China). He did his doctor thesis in Lethbridge research center, AAFC and got his PhD degree in Botany from the Northeast Agricultural University in 2012 (China). He is director of potato research and development center, Industrial Crop Research Institute, Yunnan Academy of Agricultural Sciences. He has taken on several national and provincial projects and has released 18 potato new varieties till date. He has several articles published in national journals. Xianping enjoys reading, traveling, and nature photography during his leisure.

Lee Luong is a Professor in the School of Engineering, Division of Information Technology, Engineering and the Environment, University of South Australia. Lee has had extensive experience in teaching, research and academic development within the tertiary sector as well as with industry. Prior to joining the University of South Australia, he worked for more than ten years outside the academic environment, including the mining industry, Commonwealth Scientific and Industrial Research Organisation (CSIRO), and Defence Science and Technology Organisation (DSTO). Lee is one of a few pioneers in the applications of Artificial Intelligence techniques in engineering with research and PhD programs in the areas of Logistics and Supply Chain, Cellular and Flexible Manufacturing Systems, and Sustainable Product Design and Manufacture. Lee has an extensive list of publications in international journals and conferences with ten book chapters, forty-five journal articles and more than one hundred refereed papers in international conferences. He was the Asia-Pacific Editor of the International Journal of Robotics and Computer Integrated Manufacturing for five years and currently on its Editorial Board.

Sónia Marabuto received her Masters of Science in Mechanical Engineering in 2010 from the University of Aveiro. After obtaining her degree, she remained at the University for three years working as a researcher in the field of Incremental Forming and Automation.

Miguel António Martins received his Masters of Science in Mechanical Engineering in 2011 from the University of Aveiro. After obtaining his degree, he remained at the University for two years working as a researcher in the field of Incremental Forming and Automation.

Ricardo Alves de Sousa (São Paulo, 7th October, 1977) has obtained BSc and MSc degrees in Mechanical Engineering from the University of Porto in 2000 and 2003, respectively, being awarded as the top ten student of the 2000 class in Mechanical Engineering. In 2006, he obtained the PhD degree in Mechanical Engineering from the University of Aveiro, Portugal. In 2011, he received the international scientific ESAFORM prize for outstanding contributions in the field of material forming. In 2013, he received the innovation prize from APCOR, Portugal.

Hu Sun received the B.S. degree from Huazhong University of Science and Technology, Hubei, China, in 2007. He is currently a Ph.D. student in mechanical engineering at Huazhong University of Science and Technology (HUST), and is an exchange Ph.D. student at the electrical and computer engineering department of the University of Waterloo. His current research interests include RFID design and applications, electromagnetic metamaterials.

Bo Tao received his B.S., and Ph.D. degrees in mechanical engineering from Huazhong University of Science and Technology (HUST), in 1999 and 2007, respectively. From 2007 to 2009, he was a post-doctor in the Department of Electronics Science and Technology, HUST. After that, he has been an Associate Professor in School of Mechanical Science and Engineering and State Key Laboratory of Digital Manufacturing Equipment and Technology, HUST. He has published more than 20 papers in international journals such as ASME Transaction; Sensors and Actuators. His research interests include electronic manufacturing equipment and technology, RFID technology and applications, intelligent manufacturing and applications.

Pandian Vasant was born in Sungai Petani, Malaysia in 1961. Currently, he is a Senior Lecturer of Mathematics and Optimization at Fundamental & Applied Sciences Department at University Technology Petronas in Tronoh, Perak, Malaysia and a member of American Mathematical Society since 1997. He has graduated in 1986 from University of Malaya (MY) in Kuala Lumpur, obtaining his BSc Degree with Honors (II Class Upper) in Mathematics, and in 1988 also obtained a Diploma in English for Business from Cambridge Tutorial College, Cambridge, England. In the year 2002 he obtained his MSc (By Research) in Engineering Mathematics from the School of Engineering & Information Technology of University of Malaysia Sabah, Malaysia, and he also has a Doctoral Degree (Ph.D, 2004-2008) from University Putra Malaysia in Malaysia. After graduation, during 1987-88 he was Tutor in Operational Research at University Science Malaysia in Alor Setar, Malaysia and during 1989-95 he was Tutor of Engineering Mathematics at the same university but with Engineering Campus at Tronoh, Malaysia. Thereafter during 1996-2003 he became a lecturer in Advanced Calculus and Engineering Mathematics at Mara University of Technology, in Kota Kinabalu, Malaysia. He became Senior Lecturer of Engineering Mathematics in American Degree Program at Nilai International College (Malaysia), during 2003-2004 before taking his present position at Universiti Teknologi Petronas in Malaysia. His main research interests are in the areas of Optimization Methods and Applications to Decision Making and Industrial Engineering, Fuzzy Optimization, Computational Intelligence, and Hybrid Soft Computing. Vasant has co-authored 200 research papers and articles in national journals, international journals, conference proceedings, conference paper presentation, and special issue guest editor, lead guest editor for book chapters' project, conference abstract, edited book and book chapters. In the year 2009, Vasant was awarded top reviewer for the journal Applied Soft Computing (Elsevier). He has been Co-editor for AIP Conference Proceedings of PCO (Power Control and Optimization) conferences since 2008 and editorial board member of international journals in the area of Soft Computing, Optimization and Computer Applications. Currently he's a lead managing editor for GJTO (Global Journal Technology & Optimization), Editor-in-Chief of IJEOE (IGI Global) and organizing committee member (PCO Global) for PCO Global conferences.

Nikolaos M. Vaxevanidis is a Professor of manufacturing technology in the Department of Mechanical Engineering, School of Pedagogical & Technological Education (ASPETE), Athens, Greece and Director of the Laboratory of Manufacturing Processes & Machine Tools (LMProMaT) in the aforementioned Department. He received his diploma (1985) and Ph.D. (1996) in mechanical engineering from National Technical University of Athens, Greece. He served as Lecturer in the Hellenic Air Force Academy (2000-2006) and he was also assoc. Professor under contract (2003-2010) in the Department of Mechanical Engineering/University of Thessaly, Greece. He is member of ASME, ASM and ASQ. His scientific interests include manufacturing technology, surface engineering, tribology and quality management. He has published more than 100 papers on these topics in International Journals and International Conference Proceedings. He is member of the Editorial Board of the Int. J. of Machining and Machinability of Materials and the Int. J. of Manufacturing, Materials, and Mechanical Engineering.

Yu Mei Wong graduated with a bachelor degree in environmental life science from the University of Hong Kong in 2009 and has involved in corporate social responsibility field for several years. Her professional experience includes social assessment, environmental assessment, carbon auditing, capacity building, key performance indicators, etc. She has worked very closely with different internal and external stakeholders such as sourcing, suppliers/vendors, industry groups, NGOs, etc. Her current pursuit in the supply chain responsibility department in an international apparel retailer has equipped her with solid experience in incorporating sustainability into business model. In 2013, she completed a Master of Science in Environmental Management with Distinction and acquired both practical skills and environmental theories. Yu Mei continues to increase her passion for environment and her enthusiasm for supply chain sustainability.

Ke Xing received his PhD in Manufacturing Engineering at the University of South Australia. He is a program director in the School of Engineering, Division of Information Technology, Engineering and the Environment, University of South Australia. His research interests include issues related to sustainability in product and service development, design for reliability and maintainability, lean and green supply chain modelling, applications of artificial intelligence in system modelling and design optimisation. He has published research papers at national and international journals, conference proceedings as well as chapters of books.

Kazar Yaegashi is an associate professor of the college of business administration at Ritsumeikan University. He has undergraduate degree in the college of art and design at Musashino Art University and master degree in Graduate school of Interdisciplinary Information Studies at the University of Tokyo. Prior to being an associate professor at Ritsumeikan University, he worked as a designer in media industry, a research associate at Musashino Art University, and a lecture at Fukuyama University. His current research interests are Design driven innovation, design in business administration, design education and Project Based Learning. He has published in the Journal of Educational Technology Research and the Journal of Japanese Society for The Science of Design. He translated Design driven innovation, which Robert Verganti published in 2009, into Japanese.

Zhouping Yin received his B.S., M.S. and Ph.D. degrees in mechanical engineering from Huazhong University of Science and Technology (HUST), in 1994, 1996 and 2000, respectively. He is a PROFESSOR in School of Mechanical Science and Engineering, HUST. Since 2005, he has been Vice Head of the State Key Laboratory of Digital Manufacturing Equipment and Technology at HUST. He was awarded the China National Funds for Distinguished Young Scientists in 2006. He is a "Cheung Kong" Chair Professor of HUST since 2009. He has been principal investigator for projects sponsored by National Science Foundation of China (NSFC), National Basic Research Project (973) of China and others. He has published 2 monographs, 3 chapters in English books and more than 80 papers in international journals such as IEEE Transactions, ASME Transaction, and Computer-Aided Design. His research interests include electronic manufacturing equipment and technology, RFID technology and applications, digital manufacturing and applications.

KuZilati KuShaari is an Associate Professor at the Chemical Engineering Department, Faculty of Engineering, Universiti Teknologi PETRONAS (UTP). She holds a degree in Chemical Engineering (Hons.) (University of Detroit Mercy, MI, USA) and a Masters in Chemical Engineering (West Virginia University, WV, USA). She completed her PhD in 2007 at West Virginia University, WV, USA. Her areas of expertise are powder technology, particle coating and computational fluid dynamics.

Jixuan Zhu graduated in 2010 from the School of Mechanical Science and Engineering at Huazhong University of Science and Technology (HUST) in specialty of machine design & manufacturing and automation. He has been a PhD student in School of Mechanical Science and Engineering at HUST from 2010. His main research interests include embedded intelligent sensing technology, RFID design and application.

Index

A

Ant Colony Optimization 60, 88-89
Anti-Metallic RFID Tag 127, 129, 136, 151, 153, 155
Artificial Intelligence (AI) 192, 223
Artificial Neural Networks (ANNs) 191, 223
Assembly Operation 125
Assembly Workstations 102, 110, 115, 125

B

Brand 14, 16-17, 229, 232, 245, 247, 250-261, 263-265, 267-276, 280-284, 348
Brand Loyalty 252-254, 257-260, 263-265, 267-270, 272-273, 276, 280, 284
Brand Management 250, 270, 272, 276, 282, 284

C

Carbon Emissions 303-311, 313-315, 317-329, 332, 334, 348
Carbon Leakage 304, 310-313, 321, 323, 328, 332
Chaotic Particle Swarm Optimization (Ch-PSO) 38, 44-45, 47, 51-52, 54, 58
Chemical Industry 21-24, 35
Ch-PSO 38, 44-45, 47, 51-52, 54, 58
Computational Drug Design 285-287, 299
Computer Aided Design (CAD) 8, 159, 161-164, 174, 184-185, 189-194, 197-201, 206, 220, 223-224
Computer Aided Manufacturing (CAM) 8, 159, 161, 164, 184-185, 189-194, 207, 210-217, 222-224
Computer Aided Process Planning (CAPP) 8, 190, 194, 199, 202, 206, 224
Computer Numerical Control (CNC) 161, 170, 173, 175, 178, 184-186, 188-189, 191, 194, 204, 210, 212-213, 217, 222-224
Consumer Responsibility 303-309, 313, 323-324, 328, 330, 332
Crossover Operation 95
Customer 5-6, 9, 11, 13, 15, 125, 160, 245, 248, 252-284, 337, 340
Customer Relationship Management (CRM) 252-254, 256-284
Customer Satisfaction 252-260, 262-268, 270-273, 275, 278-280, 282-284, 337
Customer Value 252-259, 264-273, 276-277, 279-281, 283-284
Cutter Location Data (CL-data) 224

D

Design Research 226-229, 231-233, 236, 239-245, 248, 250-251
Differential Evolution (DE) 40-41, 56-58
Digital Manufacturing 2, 4-5, 7, 19
Drug Design 285-287, 289-301

E

Economy by Export 317-322, 332
Embodied Carbon Emissions 303, 305, 311, 313-314, 316-328, 332
Emergency Response Plan 21
ergonomics 96-97, 105-106, 110-117
Evolutionary Algorithms (EAs) 63, 95, 224
Evolutionary Pattern Search Algorithms (EPSAs) 63, 95
Exploitation 61, 95, 226, 229, 232, 243, 246-247, 250-251
Exploration 11, 16, 61, 95, 226, 229, 231-232, 238-239, 243, 246-247, 250-251
Exploratory Factor Analysis 333, 343, 354

F

Firm Performance 257, 264, 273, 278, 333, 336-337, 341-342, 344, 352

Flat Panel Display (FPD) Industry 226, 228, 234-243, 247
Fuzzy Front End 233, 241, 244, 246, 248-251
Fuzzy Logic 57-58, 60-61, 64-72, 89-95, 100, 102, 120-121, 123, 125, 233-234, 241, 244-251

G

Genetic Algorithm (GA) 39-40, 55, 57-58, 60-65, 73, 90, 92, 94-95, 103, 108-109, 113-114, 117, 119-120, 123, 192, 297, 300
Gravitational Search Algorithms (GSA) 58
Green Manufacturing 8, 333-336, 340, 351, 353
Green Mould Sand System 58
Green Supply Chain 8, 333-334, 337-338, 341, 351-354
Green Supply Chain Managment (GSCM) 333, 336-337, 340, 342-343, 351, 353

H

High Impendence Surface (HIS) 127, 158
Hybrid Evolutionary Methods 95

I

Index 36, 43, 45-47, 177, 271, 331, 337
Innovation 1-3, 5-7, 9, 17-19, 23, 93, 226-227, 229-231, 234, 242, 245-251, 283, 308, 312, 329, 334
Intelligence 1, 6, 19, 23, 55, 57-59, 66-67, 89-94, 101, 119, 192, 220, 223, 259, 270-271, 278-279
Intelligent Manufacturing 2-8, 18-19, 23, 92
Interconnection 1, 8, 19, 287
Internet of Things (IoT) 1, 3-4, 6-8, 18-19

L

Lean Assembly Line 125
Ligands 285-287, 289-292, 294-295, 299
Low Carbon Manufacturing 303-305, 320, 322, 324, 327-328

M

Manual Assembly Line 96, 101, 125
Manufacturing Execution System (MES) 128
Manufacturing Feature Recognition (MFR) 206, 224
Meaning 227-229, 231-233, 238-239, 243, 245, 250-251, 255
Measurement Metrics 39, 47, 54, 58
Metaheuristic 38-39, 58, 60, 120
Micro Electromechanical Systems (MEMS) 7

Mixed-Integer Linear Programming (MILP) 21, 24-26, 28, 35
Modelling 55, 94-102, 105-106, 109-112, 114, 117-118, 120, 122, 124-125, 186
modern manufacturing 128, 190, 219
Molecular Docking 285-286, 288-289, 294-295, 298-299
Molecular Dynamics (MD) 55, 285-286, 289-292, 294-296, 298-300
Multiobjective Optimization 38-39, 55-58
multiple objectives 96, 105
Multiregional Input Output Model 332
Multivariate Linear Regression Analysis (MLA) 286, 300

N

New Product Development (NPD) 226-229, 233-234, 237-241, 243, 245, 251

O

Optimization 5, 8, 21-26, 28, 34-36, 38-39, 42-44, 46-47, 54-66, 68-69, 72-73, 77, 82, 84-85, 88-98, 100, 102-115, 117-125, 148, 151, 160, 169-170, 185-186, 189, 192, 213, 222-224, 244, 332

P

Partial Least Square Regression 333, 343-344
Particle Swarm 38, 42-44, 55, 58, 60, 88-89
Particle Swarm Optimization (PSO) 42-44, 58, 60
passive RFID Tags 127
Pattern Search Method 64, 95
Pharmacophore 285-288, 292-293, 295-297, 299
PLSR 333, 343-344, 351, 354
Producer Responsibility 304-307, 309, 328, 332
Productivity 2, 23, 96-97, 104-106, 111-117, 212, 215, 246, 265, 279, 348
Programming Application Interfaces 190

Q

Quantitative Structure Activity Relationship (QSAR) 286, 292-293, 295, 297, 301

R

Radio Frequency Identification (RFID) 1, 4, 6-7, 9-12, 15, 17, 19, 127-129, 131, 134, 136, 140, 142-144, 148, 151, 153-158
RF-SIM 9-12, 16-19
Robust Optimization 21-24, 35

S

Service Manufacturing 2, 4-6, 8-9, 19
SIMpass 9-12, 15-19
Single Point Incremental Forming (SPIF) Process 159-166, 168-170, 173-175, 178-179, 181-182, 184-185, 187-189
Smart Card Manufacturing 2, 10, 17
Smart Manufacturing 1-4, 7, 17, 19, 21-24, 35-36, 127, 159-160, 169-170, 185, 252, 254, 259, 262, 265, 267-268, 285, 287
Software automation 190-191
Strategy Management 19
Supply Chain 8-9, 21-26, 28-29, 35-36, 94, 156, 274, 303-310, 313, 318-324, 326-329, 333-334, 337-338, 341, 348, 351-354
Supply Chain Design 21-24, 35

T

Technological Research 226, 228-229, 231-235, 238-244, 250-251
Trade in Value Added 315, 317, 330, 332
Transformation 1-4, 12, 17, 19, 22-25, 105, 280, 291
Transmission Line Model 129, 131-134, 136, 138, 144, 146-148, 151, 155-156, 158

W

Walking Worker Assembly Line (WWAL) 96-98, 111-112, 114-117
Weighted Sum Approach 46, 58
Worker 96-98, 101-102, 104, 107, 111, 115-117, 124-125

CPSIA information can be obtained at www.ICGtesting.com
Printed in the USA
BVOW06*1442020314
346245BV00008B/109/P

	DATE DUE		